ANALYSIS OF
LINEAR NETWORKS AND SYSTEMS

A Matrix-Oriented Approach with Computer Applications

ANALYSIS OF
LINEAR NETWORKS AND SYSTEMS

A Matrix-Oriented Approach with Computer Applications

SHU-PARK CHAN, Ph.D.
University of Santa Clara

SHU-YUN CHAN, Ph.D.
San Diego State College

SHU-GAR CHAN, Ph.D.
Naval Postgraduate School

ADDISON-WESLEY PUBLISHING COMPANY

Reading, Massachusetts · Menlo Park, California · London · Don Mills, Ontario

This book is in the
ADDISON-WESLEY SERIES IN ELECTRICAL ENGINEERING

Consulting Editors
David K. Cheng, Leonard A. Gould, Fred K. Manasse

To the memory of our father

The late General of the Army, Chi-Tong Chan 陳濟棠
—a soldier, statesman, and educator—
who taught us the four principles of goodness:

Set a *good* goal in mind; ─────────────── 立好志

Acquire a *good* wealth of knowledge; ──────── 讀好書

Exercise *good* self-discipline; ──────── 做好人

Perform only *good* deeds. ──────── 行好事

for which we are forever grateful.

PREFACE

In the past two decades, the undergraduate curriculum in Electrical Engineering has undergone many changes—partly due to the availability of high-speed computers and new techniques for system analysis and design. As a consequence, the E.E. departments in various colleges and universities are shifting their emphasis on subject matter and changing their methods of approach in basic courses such as the analysis of linear networks and systems.

This book is written primarily as a text at the advanced undergraduate or first-year graduate level for courses on the analysis of linear networks and systems using a modern matrix approach. Because it is both comprehensive and self-contained, it can be used either for self-study or for reference. We have tried to maintain mathematical rigor so that the text will not unduly burden the reader with mathematical abstractness, and we have used many examples throughout the text as aids in explaining various techniques and algorithms.

A primary objective of this book is to give a unified treatment of matrix formulation of network equations and illustrate the interrelationships of the loop (mesh) equations, cutset (node) equations, and the state equations. While the modern concept of state-variable analysis is treated in considerable detail, the conventional frequency-domain analysis techniques—namely the loop and the node analysis using Laplace transform methods—are also discussed in depth with the introduction of the cutset analysis of which the node analysis is a special case. Also, linear system theory is introduced as a generalization of linear network theory. In treating it so, we have attempted to provide a smooth transition from the analysis of networks to that of systems by using the state-variable method as a bridge joining together the two theories—namely, the circuit concept and the system concept.

The fundamentals of network and system analysis are briefly discussed in Chapter 1 as a review—as most of the topics presented in this chapter are covered in a first course in basic electrical engineering. Chapter 2 is devoted to the discussion of elementary network topology, which provides the theoretical basis for the matrix formulation of network equations.

In Chapter 3, three analysis methods—namely the loop method, the node method, and the state-variable method—are presented in the form of matrix formulation of network equations. By using elementary network topology it is shown that the less familiar cutset method is actually a natural generalization of the node method. It is also pointed out that all three analysis methods are developed from the same set of network postulates—Kirchhoff's voltage-law and current-law equations and voltage-current-relationship equations.

In Chapter 4 several useful network theorems are presented and their applications are illustrated by means of examples. Also the concept of using transform networks in determining network responses is discussed and initial conditions, especially for degenerate networks, are studied in considerable detail.

Chapter 5 presents classical methods for the solution of network equations—namely the solutions of loop and node equations by Laplace transformation. Also, network functions, network responses, sinusoidal steady-state analysis, frequency response plots, and Fourier-series analysis are discussed in this chapter with emphasis on the study of sinusoidal steady-state responses of linear networks.

Chapter 6 introduces the theory of two-ports, in which two-port parameters, equivalent networks, interconnections of two-ports, and two-port devices are studied at considerable length. Then n-ports and n-terminal networks are briefly discussed as extensions of the two-port theory. Also included is a section on scattering parameters, the understanding of which is essential for network analysis and design at high frequencies.

Chapter 7 is devoted to the study of state equations for linear systems and methods of solution. Several basic techniques are presented for deriving and solving the state equations for different systems. A brief discussion of time-varying systems is also included. Chapter 8 presents the fundamentals of discrete-time systems which are necessary in developing an insight into the understanding of many systems used in daily life. Some of the timely topics such as digital filters are also introduced in this chapter.

Chapters 9 and 10 are concerned with computer-aided analysis which should provide a working knowledge of the subject. Chapter 9 deals with the application of some of the basic numerical techniques to network analysis. By using the Newton-Raphson iterative formula for root finding and the partial-fraction expansion technique, the method of Laplace transform is applied. The convolution of two functions and the solution of state equations using the Runge-Kutta Adams-Moulton technique, among others, demonstrate the power of a digital computer as an aid to network (and system) analysis. Chapter 10 introduces four general-purpose network analysis programs, each of which represents one of the following analysis techniques: (a) the node-analysis method, (b) the (k-tree) topological technique, (c) the state-variable approach, and (d) the flowgraph technique. Applications of each of these programs are illustrated with examples.

Selected topics in physical systems, such as controlability, observability, stability, and sensitivity are briefly discussed in Chapter 11. These topics serve as an introduction to some of the more advanced courses in system theory.

Although we assume the reader has been exposed to matrix analysis, Laplace transformation, Fourier series, and numerical technique, a profound knowledge on these topics is not required. For review purposes brief discussions on these topics are included in Appendixes A1, A2, and A3, respectively.

Most of the chapters of this book have been classroom tested since 1968 in network or system analysis courses at the University of Santa Clara, San Diego State College, and Naval Postgraduate School. The following table illustrates some possible ways of making different combinations of chapters that may service different types of courses:

Course title	No. of units (hours/week)	Selection of chapters in logical order	Remarks
Linear Network (System) Analysis	3 quarter units	1, 2, 3, 4, 5, 6 (6–1, 6–2, 6–3, 6–6), 7 (7–1, 7–2, 7–3), and selected coverage of A1 and A2	For a conventional analysis course omitting coverage of computer applications
Same as above	4 quarter units or 3 semester units	1, 2, 3, 4, 5, 6 (6–1, 6–2, 6–3, 6–6), 7 (7–1, 7–2, 7–3), 9, and selected coverage of A3 and 10	
Analysis of Linear Systems	4 quarter units or 3 semester units	1, 2, 3, 4, 5, 6 (6–1, 6–2, 6–3, 6–6), 7, and selected coverage of 8 and 11	For a course with emphasis on the systems approach and omitting coverage of computer applications
Linear Network (System) Analysis	4 semester units	1, 2, 3, 4, 5, 6 (6–1, 6–2, 6–3, 6–6), 7, 9, selected coverage of A3 and 10, and selected coverage of 8 and 11	
Analysis of Linear Networks and Systems	Two 3 quarter-unit course sequence	Entire text with the following divisions: *First course*: 1, 2, 3, 4, 5, 6 (6–1, 6–2, 6–3, 6–6), 7 (7–1, 7–2, 7–3) *Second course*: 6 (6–4, 6–5, 6–7, 6–8, 6–9, 6–10), 7 (7–4 through 7–11), 8, 9, 10, 11	
Same as above	5 semester units	Entire text	

We are indebted to many of our friends, colleagues, and students for their assistance during the writing of this book. We are grateful to Professor David K. Cheng of Syracuse University who did the initial review of our manuscript, to Professor Sanjit K. Mitra of University of California at Davis who did the detailed final review and made many valuable suggestions and critical comments, and to Professor Roy M. Johnson of Montana State University for his contribution of subroutines RKAM and VPLOT.

In addition, Shu-Park Chan would like to express his personal thanks to the following people of the University of Santa Clara: Dr. Robert J. Parden, Dean of the School of Engineering, for his continuous encouragement; Dr. C. K. Sun and Mr. Philip T. F. Wong, for carefully proofreading the manuscript and working out the problem solutions; Mr. George R. Banner, for his contribution in the section of scattering parameters; Mr. Fouad Y. Matouk, for making all the fine drawings, Mrs. Donald R. Schlotterbeck, for her excellent work in typing the manuscript, Miss Diana Sanchez, who spent many long hours in producing copies of the chapters to be used as classnotes; and his secretary Miss Jackie Wagner, for her efficient handling of all the correspondence. Shu-Yun Chan would like to thank Professor Martin P. Capp, Dean of the Engineering School of San Diego State College, for his encouragement and support, and Shu-Gar Chan would like to express thanks to Professor Sydney R. Parker of the Naval Postgraduate School for his encouragement and advice.

We would also like to acknowledge our appreciation for the assistance given us by the staff of Addison-Wesley Publishing Company, Inc., in the production of this book.

Finally, our special thanks go to our wives Stella, Lillian, and Shui-Ming for their continual encouragement, constant care and assistance, and infinite patience in making the writing of this book possible.

November, 1971   S.-P.C.
                                                                           S.-Y.C.
                                                                           S.-G.C.

CONTENTS

---

* Various FORTRAN programs will be included in this chapter. Complete card decks will be supplied by the authors to the instructor upon adoption of this text.

---

† An introduction to various methods of approach used in several large, general-purpose analysis programs, together with their basic input requirements, capabilities and limitations.

CHAPTER 1

FUNDAMENTALS OF NETWORK AND SYSTEM ANALYSIS

1.1 INTRODUCTION

This chapter is intended to provide a brief review of the definitions and fundamental concepts used in the study of linear networks and systems. Most of the
topics discussed in this chapter are covered in a first course in basic electrical
engineering. They are included only to serve as a reinforcement of the ideas of
these important concepts of network and system theory.

The general concept of a network or system will first be discussed. Then,
various ways of classifying (electrical) networks will be described, and the two
terms, network theory and system theory, will also be discussed briefly.

Next the concepts of analysis and synthesis will be introduced and examples
given to illustrate the difference between them. Then a brief review of the most
commonly used models for electrical components and their voltage-current (v-i)
relations will be presented. This will be followed by discussions on basic network
concepts including idealized sources, terminal constraints, reduction of series-
parallel connections of sources, singularity functions, the concepts of linearity and
time invariance, Kirchhoff's voltage and current postulates, among others.

In the final section, the scope of the text will be outlined to indicate the types
of networks and systems that will be studied and the extent to which such topics as
computer-aided network analysis and modern system theory will be covered.

1.2 NETWORKS AND SYSTEMS

A *network* or *system* may be described, in a broad sense, as a collection of objects
called *elements* (*components*, *parts*, or *subsystems*) which form an entity governed
by certain *laws* or *constraints*. Thus a *physical system* is an entity made up of
physical objects as its elements or components. A familiar example of a physical
system is an automobile, the components of which are the motor, the transmission
system, the exhaust system, etc. It should be noted that a subsystem of a given
system can be considered as a system itself. For example, the transmission system
of an automobile is a system with gears and other mechanical parts as its components, although it is a subsystem of the automobile.

A *mathematical model* describes the behavior of a physical system or device in
terms of a set of equations, together with a schematic diagram of the device con-

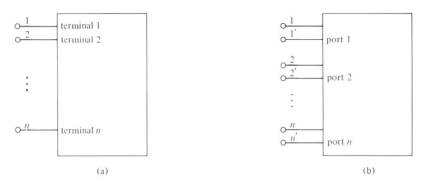

Fig. 1.2.1 (a) An n-terminal network, and (b) an n-port.

taining the symbols of its elements, their connections, and numerical values. As an example, a physical electrical system can be represented graphically by a network which consists of resistors, inductors, and capacitors, etc. as its components. Such an illustration, together with a set of linear differential equations, is referred to as the model of the system.

Electrical networks may be classified into various categories. Four of the more familiar classifications are: (a) linear and nonlinear networks, (b) time-invariant and time-varying networks, (c) passive and active networks, and (d) lumped and distributed networks.

A *linear* network is one which can be described by a set of linear (differential) equations; otherwise it is a nonlinear network. A *time-invariant* network implies that none of the network components has parameters that vary with time; otherwise it is a *time-varying* system. If the total energy delivered into a given network is nonnegative at any time instant, the network is said to be *passive*; otherwise it is *active*. Finally, if the dimensions of the network components are small compared to the wavelength of the highest of the signal frequencies applied to the network, it is called a *lumped* network; otherwise it is referred to as a *distributed* network.

There are, of course, other ways of classifying networks. For example, one might wish to classify networks according to the number of accessible terminals or terminal-pairs (ports). Thus, terms such as *n-terminal networks* and *n-ports* are commonly used terms in network theory (Fig. 1.2.1). Another method of classification is one based on network configurations (topology)* which gives rise to such terms as *ladders, lattices, bridged-T networks*, etc. (Fig. 1.2.2).

As indicated earlier, although the words *networks* and *systems* are synonyms that will be used interchangeably throughout the text, the terms *network theory* and *system theory* sometimes denote different points of view in the study of networks or systems. Roughly speaking, *network theory* is mainly concerned with

---

* Network topology deals with the way in which the network elements are interconnected. A detailed discussion on elementary network topology will be given in Chapter 2.

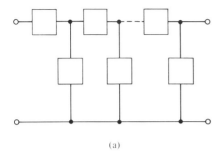

(a)

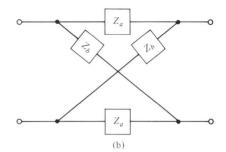

(b)

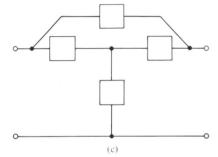

(c)

Fig. 1.2.2 Three different network configurations: (a) a ladder, (b) a balanced lattice, and (c) a bridged-T network.

interconnections of components (network topology) within a given system, whereas *system theory* attempts to attain generality by means of abstraction through a generalized (input-output-state) model.

One of the goals of this text is to present a unified treatment in the study of linear networks and systems. That is, while the study of linear networks with regard to their topological properties is treated as an important phase of the entire development of the theory, efforts are made to attain generality from such a study.

1.3 ANALYSIS AND SYNTHESIS

The subject of network theory can be divided into two main parts, namely, analysis and synthesis. In a broad sense, the *analysis* process may be defined as "the separating of a given material or abstract entity (system) into its constituent elements." On the other hand, the *synthesis* process is "the combining of the constituent elements of separate material or abstract entities into a single or unified entity (system)."* However, in terms of (electrical) systems, these two terms may be described using the excitation–response system theoretic approach as follows.

---

* The definitions of analysis and synthesis are quoted directly from *The Random House Dictionary of the English Language*, Random House, New York, 1967.

Fig. 1.3.1 Three key words: excitation, system, response in describing analysis and synthesis.

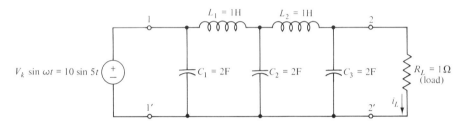

Fig. 1.3.2 The LC filter with resistive load as illustration for an analysis problem.

Fig. 1.3.3 Illustration for a typical network synthesis problem.

Consider the diagrammatic sketch shown in Fig. 1.3.1 in which a driving force or *excitation* (*input*) has been applied to a given *system* (or *network*), causing certain behavior in the form of a *response* (*output*) of the system to the specific excitation.

In analysis, the system is given, and the problem is to determine the response of the system to a specified excitation. In synthesis, both the excitation and response (or the ratio, response/excitation, as a function of frequency) are specified and *one* system is to be found which satisfies the excitation–response specifications.

For example, a network analysis problem may be stated as follows: Given the lossless (LC) ladder filter terminated at a resistive load R_L (the whole unit as a system) which is depicted in Fig. 1.3.2, find the load current i_L (response) for a sinusoidal voltage $10 \sin 5t$ (specified excitation) applied across the input terminals 1–$1'$.

On the other hand, a typical synthesis problem may be stated as follows: Find *a* network realization of the lossless filter terminated at $R_L = 1\Omega$ (Fig. 1.3.3).

if the transfer impedance is to be

$$Z_{21} \triangleq \frac{V_2}{I_1} = \frac{1}{s^5 + a_4 s^4 + a_3 s^3 + a_2 s^2 + a_1 s + 1},$$

where $a_i (i = 1, 2, 3, 4)$ are known numerical constants, and s is the complex frequency variable.

It is worth noting that in an analysis problem, the solution is always *unique* no matter how difficult it may be, whereas in a synthesis problem there might exist *infinitely many* solutions or, sometimes, *none at all*!

Finally, it should be pointed out that in some network theory texts the words "synthesis" and "design" might be used interchangeably throughout the entire discussion of the subject. However, the term *synthesis* is generally used to describe *analytical* procedures that can usually be carried out step by step, whereas the term *design* includes *practical* (design) procedures (such as trial-and-error techniques which are based, to a great extent, on the experience of the designer) as well as analytical methods.

1.4 REVIEW OF NETWORK MODELS FOR DEVICES AND THEIR *v-i* RELATIONS

To analyze the behavior of a given physical system, the first step is to establish a mathematical model. This model is usually in the form of a set of either differential or difference equations (or a combination of them), the solution of which accurately describes the motion of the physical system.

There is, of course, no exception in the field of electrical engineering. A physical electrical system such as an amplifier circuit, for example, is first represented by a network drawn on paper. This network is composed of resistors, capacitors, inductors, voltage and/or current sources,* and each of these network elements is given a symbol together with a mathematical expression (i.e., the voltage-current or simply *v-i* relation) relating its terminal voltage and current at every instant of time. Once the network and the *v-i* relation for each of its elements are specified, we can apply Kirchhoff's voltage and current laws, possibly together with the physical principles to be introduced in Chapter 4, Section 4.3, to establish the mathematical model in the form of differential equations.

In the present section, we shall review the characteristics of the two-terminal elements—resistor, capacitor, and inductor—as well as the magnetically coupled coils. (Other two-terminal-pair models—the ideal transformer, gyrator, and negative impedance inverter—will be introduced in Chapter 6.) In the following section, a discussion of both independent and dependent sources will be given.

---

* Here, of course, active elements such as transistors are represented by their equivalent circuits as combinations of resistors and dependent sources. See examples in Section 6.8, for instance.

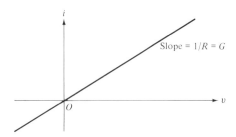

Fig. 1.4.1 Symbolic representation
of a resistor.

Fig. 1.4.2 The voltage-current characteristic
of a linear time-invariant resistor.

a) Resistor The first type of network elements to be discussed is the resistor
represented symbolically in Fig. 1.4.1. If the resistor is linear, the relation between
its terminal voltage $v(t)$ and current $i(t)$ with the directions as shown at any instant
of time t is governed by *Ohm's law*

$$v(t) = Ri(t) \qquad \text{or} \qquad i(t) = Gv(t), \tag{1.4.1}$$

where the quantities R and G are called the *resistance* and the *conductance* of the
resistor, respectively, and are related by the expression

$$R = \frac{1}{G}. \tag{1.4.2}$$

The units for R and G are ohms (Ω) and mhos ($\mho$), respectively, if the voltage $v(t)$ is
in volts (V) and the current $i(t)$, in amperes (A).

   If the resistance R is *time invariant* (i.e., a constant), then the relation between
the voltage $v(t)$ and current $i(t)$ at any time t is characterized by a *straight line* of a
constant slope in the v-i plane as depicted in Fig. 1.4.2, and the corresponding
resistor is defined as a *linear time-invariant resistor*. If, however, the resistance R
varies as a function of time t, but satisfies Eq. (1.4.1), then the resistor is called a
linear time-varying resistor.

   As shown in Fig. 1.4.2, the slope of the straight line is equal to the value of the
conductance G. Two special cases can be derived by an inspection of this figure:

1. A horizontal line is obtained when G is zero ($R = \infty$). Any device having this
   v-i characteristic is referred to as an *open circuit*.

2. On the other hand, a vertical line results when the value of G is infinite ($R = 0$),
   which corresponds to the v-i characteristic of a device referred to as a *short
   circuit*.

   Any device whose v-i characteristic can be described by a curve (other than a
straight line) in the v-i plane is defined as a *nonlinear resistor*.

   As an example, consider the v-i characteristic shown in Fig. 1.4.3, which is a
typical characteristic of a tunnel diode. Thus, a tunnel diode can be regarded as a

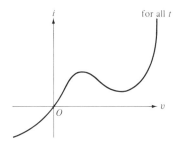

Fig. 1.4.3 The *v-i* characteristic of a tunnel diode.

Fig. 1.4.4 Symbolic representation of a capacitor.

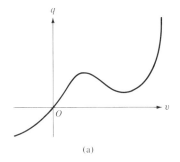

(a)

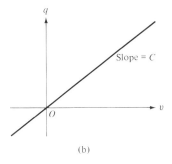

(b)

Fig. 1.4.5 Characteristic of (a) a nonlinear and (b) a linear capacitor.

nonlinear time-invariant resistor since its *v-i* characteristic is nonlinear and is not a function of time. Moreover, it is evident from the figure that for every value of the voltage there is only one value of current corresponding to it. Thus, the voltage v uniquely defines the current i^* and a device having such a characteristic is called a *voltage-controlled resistor*. On the other hand, if in a *v-i* characteristic, i uniquely defines v, then the device is defined as a *current-controlled resistor*.

b) Capacitor Similar to the definition of a nonlinear resistor, a capacitor, represented symbolically in Fig. 1.4.4, is defined as a device whose terminal characteristic can be described by a curve in the charge-voltage or q-v plane, where q and v represent, respectively, the electric charge stored in, and the voltage across, the device as shown. One such characteristic is depicted in Fig. 1.4.5 (a). As in the case of a resistor, a capacitor may be a linear or nonlinear, time-invariant or time-varying capacitor depending upon how q and v are related. Thus, if the charge q and voltage v are governed by a straight line as depicted in Fig. 1.4.5 (b), then we may write

$$v = Sq \quad \text{or} \quad q = Cv, \tag{1.4.3}$$

_____

* Whereas the reverse is not true; that is, i does not uniquely define v.

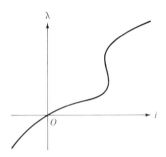

Fig. 1.4.6 Symbolic representation of an inductor.

Fig. 1.4.7 Characteristic of a nonlinear inductor.

where S and C, being reciprocals to one another, are called the *elastance* and the *capacitance* of the capacitor, and are in darafs and farads (F), respectively, provided that q is in coulombs and v is in volts. If, in addition to satisfying Eq. (1.4.3), the capacitance is independent of time (i.e., C = constant), then the capacitor is said to be a linear time-invariant capacitor.

As an example, consider a capacitor with the q-v characteristic shown in Fig. 1.4.5 (a). If the q-v curve does not vary with time, the capacitor is a nonlinear, time-invariant, voltage-controlled capacitor.

For a linear time-varying capacitor, one can obtain the voltage-current characteristic by differentiating (1.4.3) with respect to t. Thus

$$i(t) = \frac{dq(t)}{dt} = \frac{d}{dt}[C(t)v(t)]$$
$$= C(t)\frac{dv(t)}{dt} + v(t)\frac{dC(t)}{dt},$$

(1.4.4)

which, in the case of a linear time-invariant capacitor, reduces to

$$i(t) = C\frac{dv(t)}{dt}$$

or

$$v(t) = \frac{1}{C}\int_{-\infty}^{t} i(\tau)\,d\tau = v(0) + \frac{1}{C}\int_{0}^{t} i(\tau)\,d\tau$$

(1.4.5)

since C is a constant and hence $dC/dt = 0$. Equation (1.4.5) is usually taken as the mathematical definition of a linear time-invariant capacitor.

c) Inductor The third network element, the inductor, has the symbolic representation shown in Fig. 1.4.6. In general, an inductor is characterized by a curve in the magnetic flux linkage-current or simply λ-i plane such as the one depicted in Fig. 1.4.7. The voltage across the inductor is governed by Faraday's law of

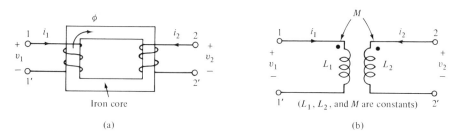

Fig. 1.4.8 (a) A system of magnetically coupled coils, and (b) its linear time-invariant model.

induction; that is,

$$v(t) = \frac{d\lambda(t)}{dt}. \tag{1.4.6}$$

If the inductor is linear but time varying, its characteristic is a straight line in the λ-i plane with the slope of the line varying as a function of time. Mathematically, λ and i are related by

$$\lambda(t) = L(t)i(t), \tag{1.4.7}$$

where $L(t)$, the slope of the line at time t, is referred to as the *inductance*. Inductance is in henrys (H) when λ is in volt-seconds and i is in amperes. Substituting (1.4.7) into (1.4.6), we obtain

$$v(t) = L(t)\frac{di(t)}{dt} + i(t)\frac{dL(t)}{dt}, \tag{1.4.8}$$

which, for a time-invariant inductor, reduces to

$$v(t) = L\frac{di(t)}{dt} \tag{1.4.9}$$

or

$$i(t) = \frac{1}{L}\int_{-\infty}^{t} v(\tau)\,d\tau = i(0) + \frac{1}{L}\int_{0}^{t} v(\tau)\,d\tau$$

since L is a constant in this case. Equation (1.4.9) is usually taken to be the mathematical definition of a linear time-invariant inductor.

d) Magnetically Coupled Coils Figures 1.4.8 (a) and (b) depict, respectively, a system of two magnetically coupled coils and its linear time-invariant model. At the time of manufacture, reference dots are placed on the coupled coils to indicate the relative winding sense between them. The rule for determining the dot location

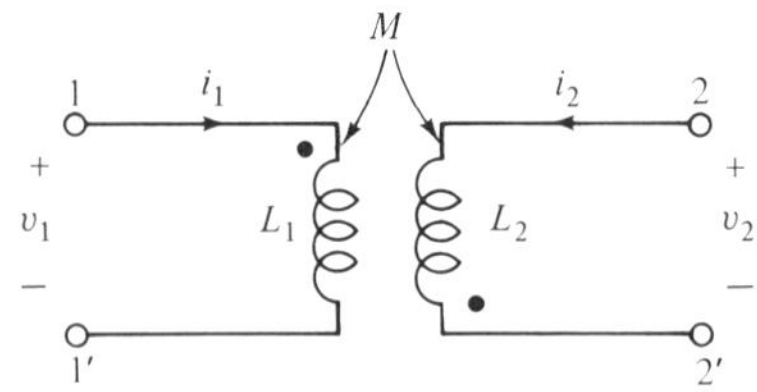

Fig. 1.4.9 Illustration of the use of dots.

is called the *dot convention* which can be described as follows: A dot is arbitrarily placed at one end of one winding and a current is allowed to enter at this end, producing a flux ϕ in the iron core with the direction governed by the right-hand rule (i.e., if the thumb of the right hand indicates the direction of the flux ϕ, the fingers wrapping around the winding indicate the direction of current). Next determine at which end of the other winding to place a dot so that when a current enters at the dotted end a flux will be produced in the same direction as that of ϕ. Thus, if a dot is placed at node 1 of Fig. 1.4.8 (a), the other dot must be placed at node 2 as shown in Fig. 1.4.8 (b). Equivalently, the same relative winding sense will be indicated by placing the dots at terminals $1'$ and $2'$ instead.

Once the dots are assigned, the iron core in the system can be removed for convenience without losing the relative winding sense. For the linear time-invariant model shown in (b) of Fig. 1.4.8, the terminal voltages and currents are described mathematically by the equations

$$v_1(t) = L_1 \frac{di_1}{dt} + M \frac{di_2}{dt}, \tag{1.4.10}$$

$$v_2(t) = M \frac{di_1}{dt} + L_2 \frac{di_2}{dt}, \tag{1.4.11}$$

where M is the *mutual inductance* (also in henrys) associated with the flux linking the two inductances L_1 and L_2, and is related to them by a constant k called the *coefficient of coupling* in the expression

$$M = k \sqrt{L_1 L_2} \tag{1.4.12}$$

with $0 \le k \le 1$. If one dot is placed at the other end of one winding and the second dot is left unchanged such as the one shown in Fig. 1.4.9, then the equations (1.4.10) and (1.4.11) must be modified to yield the correct *v-i* relations; viz.,

$$v_1(t) = L_1 \frac{di_1}{dt} - M \frac{di_2}{dt}, \tag{1.4.13}$$

$$v_2(t) = - M \frac{di_1}{dt} + L_2 \frac{di_2}{dt}. \tag{1.4.14}$$

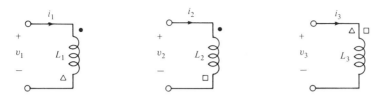

Fig. 1.4.10 Linear time invariant model for three coupled coils with L's and M's being constants.

Whether or not a negative sign should be placed before the mutual inductance M in the equations depends upon both the dot locations and the current directions. The general rule for the sign of a mutual inductance can be stated as follows: *If both currents enter (or leave) the windings at (from) the dotted terminals, a plus sign is placed before the corresponding mutual inductance M; otherwise a negative sign is used.* The same procedure can be used for a system of a finite number of coils as illustrated in the following example.

Example 1.4.1 Consider the linear time-invariant model for a set of three coils depicted in Fig. 1.4.10 with a set of dots to indicate the relationship between each pair of windings. That is, the dots with the shape ● identify the winding sense between windings 1 and 2, △ for windings 1 and 3, and □ for windings 2 and 3. Let M_{12}, M_{13}, and M_{23} denote the mutual inductances associated with L_1 and L_2, L_1 and L_3, and L_2 and L_3, respectively. Then, using the rule stated above, the voltage v_1 can be expressed as

$$v_1(t) = L_1 \frac{di_1}{dt} + M_{12} \frac{di_2}{dt} - M_{13} \frac{di_3}{dt}. \tag{1.4.15}$$

The reader is invited to complete the expressions for v_2 and v_3.

Since we are mainly concerned with linear time-invariant networks and systems, it would be advantageous at this time to complete the description of these network elements in the s-domain (that is, the Laplace transforms of the v-i relations) so that we can establish a table of reference on the v-i relations of these elements in both the t- and s-domains. For the sake of preserving clarity when we introduce the impedance definition for single passive elements, we shall assume for the time being that both energy storage elements L and C are initially de-energized; the equivalent s-domain representations of these elements with initial conditions will be treated in great detail when we study the transform networks in Chapter 4 (Section 4.4).

For a linear time-invariant resistor, the terminal characteristic is described by Ohm's law with the parameter R being a constant; the Laplace transform is

$$V(s) = RI(s). \tag{1.4.16}$$

Table 1.4.1 Description of Linear Time-Invariant Network Elements.*

Element	Symbolic representation in t-domain and s-domain		Terminal characteristics t-domain	s-domain
Resistor (Ω)	$i(t)$ $v(t)$ R	$I(s)$ $V(s)$ $Z_R = R$	$v(t) = Ri(t)$	$V(s) = RI(s)$
Capacitor (F)	$i(t)$ $v(t)$ C	$I(s)$ $V(s)$ $Z_C = \dfrac{1}{sC}$	$v(t) = \dfrac{1}{C}\displaystyle\int_0^t i(\tau)\,d\tau$	$V(s) = \dfrac{1}{sC}I(s)$
Inductor (H)	$i(t)$ $v(t)$ L	$I(s)$ $V(s)$ $Z_L = sL$	$v(t) = L\dfrac{di(t)}{dt}$	$V(s) = sLI(s)$
Coupled Coils (H)	$i_1(t)$ M $i_2(t)$ $v_1(t)$ L_1 L_2 $v_2(t)$	$I_1(s)$ sM $I_2(s)$ $V_1(s)$ sL_1 sL_2 $V_2(s)$	$v_1(t) = L_1\dfrac{di_1}{dt} + M\dfrac{di_2}{dt}$ $v_2(t) = M\dfrac{di_1}{dt} + L_2\dfrac{di_2}{dt}$	$V_1(s) = sL_1I_1(s) + sMI_2$ $V_2(s) = sMI_1(s) + sL_2I_2$

* The equivalent representations of the energy storage elements as well as the coupled coils with nonzero initial conditions will be treated in Chapter 4 (Section 4.4).

In like manner, the Laplace transform of (1.4.5) with the initial voltage $v(0)$ set to zero is

$$V(s) = \frac{1}{sC}I(s) \qquad (1.4.17)$$

which is the description of an initially uncharged capacitor in the s-domain.

Likewise, taking the Laplace transform of (1.4.9) with the initial current $i(0)$ set to zero yields the description for an initially de-energized inductor in the s-domain; that is,

$$V(s) = sLI(s). \qquad (1.4.18)$$

The *transform impedance* $Z(s)$ of a passive (linear time-invariant) element is defined as the ratio of the transform voltage across the element to the transform current through that element with the initial condition set to zero. Thus, in light of (1.4.16) through (1.4.18), the transform impedances for the resistor, the capacitor, and the inductor are, respectively, given by

$$Z_R(s) = R, \qquad Z_c(s) = \frac{1}{sC}, \qquad \text{and} \qquad Z_L(s) = sL. \qquad (1.4.19)$$

For the magnetically coupled coils of Fig. 1.4.8, its s-domain representation in the s-domain can be obtained by taking the Laplace transforms of Eqs. (1.4.10) and (1.4.11) with the initial conditions again set to zero, yielding, respectively,

$$V_1(s) = sL_1 I_1(s) + sMI_2(s), \qquad (1.4.20)$$

$$V_2(s) = sMI_1(s) + sL_2 I_2(s). \qquad (1.4.21)$$

A summary of the description of the linear time-invariant network elements is given in Table 1.4.1 for reference.

In the next section we shall present a brief discussion on both independent and dependent sources.

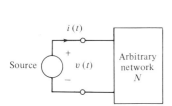

Fig. 1.5.1 Classification of a source.

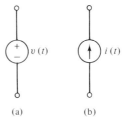

(a)   (b)

Fig. 1.5.2 Schematic representations of independent sources: (a) voltage source, and (b) current source.

1.5 INDEPENDENT AND DEPENDENT SOURCES

Consider a source of electrical energy connected to an arbitrary network resulting in a terminal voltage $v(t)$ and current $i(t)$ as depicted in Fig. 1.5.1. In general, both $v(t)$ and $i(t)$ will change in value if the network N is replaced by a different one of an arbitrary nature. However, if the terminal voltage $v(t)$ remains unchanged for any arbitrary network connected across it, then this source can be represented by a model called an *independent voltage source* as shown in (a) of Fig. 1.5.2. On the other hand, if the terminal current $i(t)$ is independent of what is connected across it, then the source can be represented by a model referred to as the *independent current source* as shown in Fig. 1.5.2 (b).

Let us study these two idealized situations more closely. Consider first an independent voltage source connected to a variable linear time-invariant resistor as depicted in Fig. 1.5.3, where the current $i(t)$ and the power $p(t)$ delivered by the

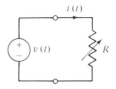

Fig. 1.5.3 Illustration of a generalized short circuit.

source are, respectively, given by

$$i(t) = \frac{v(t)}{R} \tag{1.5.1}$$

and

$$p(t) = v(t) \cdot i(t) = \frac{v^2(t)}{R}. \tag{1.5.2}$$

Let us assume, for the sake of clarity, that the source voltage $v(t)$ is time invariant; that is, $v(t)$ is a constant, say, $v(t) = K$. Then the energy delivered by the source in a time interval T is given by

$$\varepsilon(T) \triangleq \int_0^T p(t)\, dt$$

$$= \frac{K^2 T}{R} \tag{1.5.3}$$

which is inversely proportional to the value of the resistance R. Equation (1.5.3) reveals a very interesting fact: An independent voltage source is capable of delivering an infinite amount of energy in a finite interval T, since, in (1.5.3), $\varepsilon(T)$ approaches infinity as R tends to zero. The above argument explains why an independent voltage source is an idealized situation. The same argument applies to an independent current source.* However, physical generators may be approximated to a certain degree by one or the other of these two sources.

Another interesting property of an independent voltage source can be derived by observing that, in (1.5.1), as R tends to zero, the current in the resistor approaches infinity, indicating the fact that there is no resistance in the entire network—a short-circuit condition. In other words, an independent voltage source differs from a short circuit in that, in general, the terminal voltage is not equal to zero, and, instead, is a prescribed function of time $v(t)$. In the special case when $v(t) = 0$, the independent voltage source indeed reduces to a short circuit. Hence an independent voltage source can be regarded as a *generalized short circuit.*

A parallel argument leads to the interesting observation on an independent current source; that is, it differs from open circuit in that its terminal current $i(t)$ is

---

* See Problem 1.1.

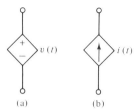

Fig. 1.5.4 (a) Symbolic representation of a dependent voltage source and (b) that of a dependent current source.

in general a prescribed function of time and is identical to an open circuit when $i(t) = 0$. Thus, an independent current source can be regarded as a *generalized open circuit*.

In general, a practical source may be approximated by a model consisting of interconnections of independent sources and passive elements. For example, a battery can be accurately represented by a model consisting of a constant independent voltage source in series with a linear time-invariant resistor. However, there are some electrical devices such as vacuum tubes and transistors that cannot be approximated by models consisting of only independent sources and passive elements. Thus, we need to define another type of source referred to as dependent sources.

A *dependent voltage (current) source* is a source whose terminal voltage (current) depends on another voltage or current. A dependent source is said to be a *voltage-controlled* (current-controlled) source if its terminal behavior is controlled by another voltage (current). Thus, for example, the terminal voltage $v(t)$ of a voltage-controlled source can be expressed as $v(t) = cv_1(t)$, where $v_1(t)$ is the voltage of another branch in the network and c is a constant of proportionality. The schematic representations of both dependent voltage and current sources are depicted as shown in Fig. 1.5.4.

In this text when we speak of a voltage source, an independent voltage source is implied. The same applies to a current source. The effects of inserting sources into a network of passive elements will be discussed in the following section.

1.6 NETWORK CONSTRAINTS

In any passive network the currents and voltages are free to assume any values subject only to certain laws and possibly some physical principles dictated by the geometry of that network. Without any excitation, all voltages and currents of the network remain zero. Suppose now a voltage source is connected to this network. As a result, one or more of the node potentials are clamped at certain voltage levels. Thus, the application of a voltage source to a network can be regarded as an applied constraint. For the same reasoning, the application of a current can likewise be regarded as an applied constraint.

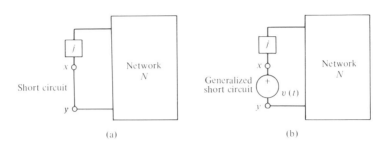

Fig. 1.6.1 Illustration of inserting a voltage source by the plier method.

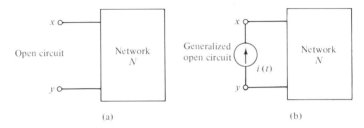

Fig. 1.6.2 Illustration of inserting a current source by the soldering-iron method.

Since a voltage source is a generalized short circuit, the application of a voltage source at any terminal-pair implies a short-circuit constraint at that terminal-pair. In like manner, an open-circuit constraint is implied at a terminal-pair to which a current source is applied.

The insertion of sources into a passive network may or may not result in a modification of the network geometry or topology depending upon how the sources are applied. The geometry of a given network will be invariant if the insertion of sources is done in either of two ways: (1) A voltage source is inserted into the gap formed by cutting a branch, and (2) a current source is connected to a selected terminal-pair. The former is referred to as *the plier method* and the latter is called *the soldering-iron method* as depicted in Figs. 1.6.1 and 1.6.2, respectively. A short-circuit condition is implied at *x-y* in Fig. 1.6.1 and an open circuit in Fig. 1.6.2. Further study on network constraints will be made when we discuss the reciprocity theorem in Chapter 4.

1.7 BASIC CONCEPTS

In this section some of the basic concepts that have not been discussed in the preceding sections will be briefly reviewed; namely, source reduction, linearity, and time invariance.

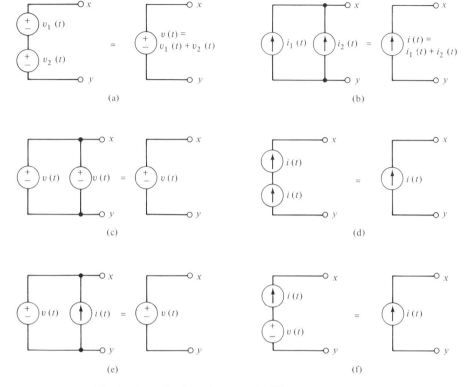

Fig. 1.7.1   Reduction of sources in different combinations.

I. Source Reduction   When a network consists of two or more sources, the solution of the network problem can often be obtained more easily by first reducing the number of sources whenever possible. To achieve this goal, we must consider the combination of independent sources both in series and in parallel since, among others, they are the basic tools for network reduction.

Figure 1.7.1 illustrates the following:

a) Two voltage sources connected in series can be added algebraically to yield a single equivalent voltage source.

b) A parallel combination of two current sources can be replaced by an equivalent single current source equal to the algebraic sum of the two.

c) A parallel combination of *identical* voltage sources is equivalent to a single voltage source.

d) A series combination of *identical* current sources is equivalent to only one current source.

Fig. 1.7.2 A network N considered as a single-input–single-output system.

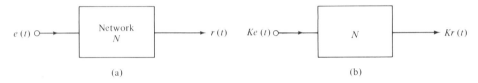

Fig. 1.7.3 Illustration of homogeneity condition.

e) A parallel combination of a voltage source and a current source is equivalent to a single voltage source.

f) A series combination of a voltage source and a current source is equivalent to a single current source.

The verification of each of these cases is quite straightforward and is hence left as an exercise.

The six cases described in Fig. 1.7.1, together with the source transformations to be studied in Chapter 3, often enable one to reduce a given network to a simpler one so that the solution can be simplified.

II. The Concept of Linearity In Section 1.4, the concept of a linear passive element was discussed in detail. To extend the concept of linearity to networks, consider a network N as a single-input–single-output system as depicted in Fig. 1.7.2, where $e(t)$ is the voltage or current excitation applied at some point of the network and $r(t)$ is the voltage or current response taken at the same or any other point of the network. The *zero-input* response of the system is defined as the response $r(t)$ when the input $e(t)$ is identically zero. The *zero-state response** is defined as the response $r(t)$ when the energy originally stored (initial conditions) in the network is set to zero.

A network is said to be *zero-state linear* if the following two conditions are satisfied for all possible inputs and all possible sets of initial conditions:

1. If the input is multiplied by an arbitrary constant K, the corresponding zero-state response is multiplied by the same constant K, as illustrated in (a) and (b) of Fig. 1.7.3. This condition is referred to as the *homogeneity condition.*

* The reason for using the term "state" will be obvious when the concept of state is introduced in Chapter 3.

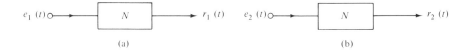

(a)   (b)

Fig. 1.7.4 Illustration of additivity condition.

(c)

2. The zero-state response to the sum of inputs is equal to the sum of the individual zero-state responses. The condition is referred to as the *additivity condition* and is depicted in Fig. 1.7.4.

Next we shall define the so-called zero-input linearity. Let $(x_1, x_2, \ldots, x_n)$ denote a set of initial conditions (initial state) of a network. A network is said to be *zero-input linear* if the following two conditions are satisfied.

1. If $r(t)$ is the zero-input response corresponding to any arbitrary set of initial conditions $(x_1, x_2, \ldots, x_n)$, then the zero-input response to the same set of initial conditions multiplied by any arbitrary constant K, i.e., $(Kx_1, Kx_2, \ldots, Kx_n)$, is given by $Kr(t)$—the *homogeneity condition.**

2. If $r_x(t)$ and $r_y(t)$ are the zero-input responses corresponding to any arbitrary two sets of initial conditions $(x_1, x_2, \ldots, x_n)$ and $(y_1, y_2, \ldots, y_n)$, respectively, then the zero-input response corresponding to the initial conditions $(x_1 + y_1, x_2 + y_2, \ldots, x_n + y_n)$ is $r_x(t) + r_y(t)$—the *additivity condition.**

Finally, a network is said to be *linear* if:

1. It is zero-state linear.

2. It is zero-input linear.

3. The complete response is equal to the sum of zero-state and zero-input responses.

III. The Concept of Time Invariance Consider again the network of Fig. 1.7.2. Let $r(t)$ be the complete response of the network to any arbitrary input $e(t)$ and any arbitrary set of initial conditions $(x_1, x_2, \ldots, x_n)$. The network is said to be *time invariant* if, for the same set of arbitrary initial conditions $(x_1, x_2, \ldots, x_n)$, the response $r(t)$ is independent of time of application of the input; that is, any arbitrary input $e(t - \tau)$ for any $\tau > 0$. A system with its input and output described by an

---

* If the initial conditions are regarded as (initial condition) sources, then each set of initial conditions is equivalent to a source vector, say, $X = (x_1, x_2, \ldots, x_n)$. With this notation of a vector, the homogeneity and additivity conditions are the same as their counterparts in the previous case.

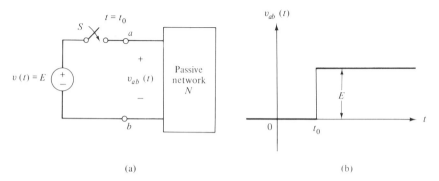

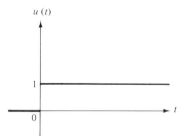

Fig. 1.8.1   (a) A network with switch closed at $t = t_0$ and (b) the waveform of $v_{ab}(t)$.

Fig. 1.8.2   Unit-step function.

ordinary differential equation with constant coefficients is a good example of a linear time-invariant system.

1.8 SINGULARITY FUNCTIONS

Consider the passive network N shown in (a) of Fig. 1.8.1. If the passive network is initially at rest (i.e., zero initial condition) and the closure of the switch occurs at $t = t_0$, the voltage waveform across the terminal-pair a-b is described graphically as shown in Fig. 1.8.1 (b), in which the switching action is assumed to take place at $t = t_0$ in zero time. If the values of E and t_0 are normalized to 1 and 0, respectively, the resulting waveform is given in Fig. 1.8.2 which is referred to as a unit-step function—one of the singularity functions to be discussed here.

The Unit-Step Function, denoted by $u(t)$, is defined by the equations

$$u(t) = \begin{cases} 0, & \text{for } t < 0; \\ 1, & \text{for } t \geq 0. \end{cases}$$

In terms of the unit-step function, the waveform in Fig. 1.8.1 (b) can be written as $v_{ab}(t) = Eu(t - t_0)$.

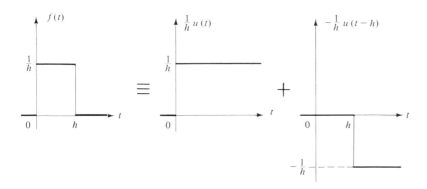

Fig. 1.8.3 A pulse function of unity area.

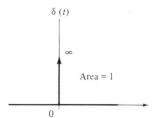

Fig. 1.8.4 The impulse function.

The Impulse Function Consider the pulse function of Fig. 1.8.3, which can be expressed as a linear combination of unit-step functions as

$$f(t) = \frac{1}{h}[u(t) - u(t - h)]. \tag{1.8.1}$$

Note that the area of this function is unity. As the value of h is reduced, the amplitude of the function increases over the interval $(0, h)$. However, the area of the function remains unchanged since it is equal to $(1/h) \times h = 1$. As h approaches zero, the limit of the waveform reduces to that shown in Fig. 1.8.4, which is referred to as the *unit-impulse (or delta) function* and is denoted by $\delta(t)$. Thus, in terms of $u(t)$, $\delta(t)$ can be written as

$$\begin{aligned} \delta(t) &= \lim_{h \to 0} f(t) \\ &= \lim_{h \to 0} \frac{[u(t) - u(t - h)]}{h}. \end{aligned} \tag{1.8.2}$$

The last expression in Eq. (1.8.2) is simply the definition of the derivative of the unit-step function. Hence, we have

$$\delta(t) = \frac{d}{dt} u(t). \tag{1.8.3}$$

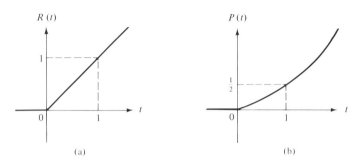

Fig. 1.8.5 (a) Unit ramp and (b) unit parabola.

Thus the unit-impulse function is zero everywhere except at $t = 0$; at this point the value of the function is infinite, and, in addition, the area of the unit impulse is unity, i.e.,

$$\int_{-\infty}^{\infty} \delta(t)\, dt = \int_{0^-}^{0^+} \delta(t)\, dt = 1, \tag{1.8.4}$$

where 0^- and 0^+ denote the time instants immediately to the left and right of the time instant $t = 0$, respectively.

Additional singularity functions can be derived by taking derivatives of the unit impulse of various orders. Thus, for example, the derivative of the unit-impulse function,

$$\delta_1(t) = \frac{d}{dt}\, \delta(t),$$

is called *unit doublet* which consists of a positive impulse immediately followed by a negative impulse. The unit doublet is similar in nature to a couple used in mechanics.

The Ramp Function The integral of a unit-step function yields the *unit-ramp function* and the integral of a unit ramp results in a *unit parabola* as depicted in (a) and (b) of Fig. 1.8.5, respectively. Denoting the unit ramp and unit parabola by $R(t)$ and $P(t)$, respectively, they can be expressed as

$$R(t) = \int_{-\infty}^{t} u(t) = tu(t) \tag{1.8.5}$$

and

$$P(t) = \int_{-\infty}^{t} R(t) = \frac{t^2}{2}\, u(t). \tag{1.8.6}$$

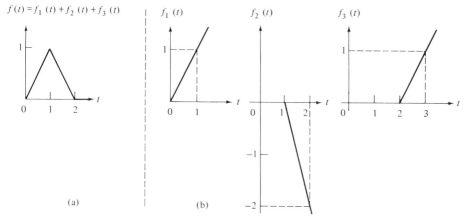

Fig. 1.8.6 (a) A triangular pulse and (b) its component functions.

The singularity functions such as the unit step, unit ramp, unit parabola, and unit impulse are often used as standard inputs in checking the response of a control system in the t-domain. Another important application of singularity functions is that they can be used as building blocks for the construction of arbitrary functions. Let us illustrate this idea by means of an example that follows.

Example 1.8.1 The triangular pulse of Fig. 1.8.6 (a) can be expressed as a linear combination of three ramp functions as depicted in Fig. 1.8.6 (b). In terms of the component functions $f_1(t), f_2(t),$ and $f_3(t),$ the triangular pulse $f(t)$ can be expressed as

$$f(t) = f_1(t) + f_2(t) + f_3(t)$$
$$= tu(t) - 2(t-1)u(t-1) + (t-2)u(t-2). \qquad (1.8.7)$$

More applications on singularity functions will be discussed when we study network responses in Chapter 5.

1.9 KIRCHHOFF'S VOLTAGE AND CURRENT LAWS

In the preceding sections, we have introduced some of the basic network elements (resistors, inductors, capacitors, sources, etc.) and discussed the voltage-current relationships (VCR) which characterize the network elements in terms of element voltages and currents. These element equations constitute one of the three postulates or laws from which the network equations are formulated. This postulate is referred to as the *VCR-postulate* and the equations as the *VCR-equations*.

    The other two basic network postulates are Kirchhoff's voltage and current laws which may be stated as follows.

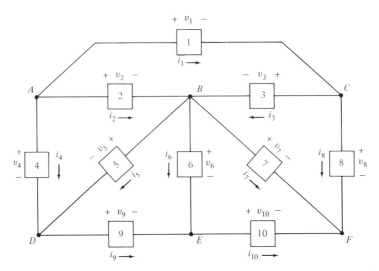

Fig. 1.9.1 Network N used to illustrate Kirchhoff's laws.

Kirchhoff's voltage law (KVL) states that the algebraic sum of all branch voltages around any circuit (i.e., closed loop) of a given network is zero at any given instant of time. Or, in mathematical terms, for any closed loop in the network,

$$\sum_k v_k = 0, \tag{1.9.1}$$

where v_k are the branch voltages around the loop.

Kirchhoff's current law (KCL) states that the algebraic sum of all branch currents leaving any node (i.e., junction point) of a given network is zero at any given instant of time. Mathematically, we have, for any node in the network

$$\sum_j i_j = 0, \tag{1.9.2}$$

where i_j are the branch currents connected to the node.

The following simple example will illustrate the application of the two Kirchhoff's laws.

Example 1.9.1 Consider the network N shown in Fig. 1.9.1. Let us assume that Kirchhoff's voltage and current laws are valid for N, and that each of the ten branches of N represents either a source, a passive element, or a combination of sources and/or passive elements. Then, application of Kirchhoff's voltage law around the outer loop in the clockwise direction through branches 1, 8, 10, 9, and 4 gives the following KVL-equation:

$$v_1 + v_8 - v_{10} - v_9 - v_4 = 0,$$

where a voltage drop has been considered positive. Next, by applying Kirchhoff's current law at node B, we have the following KCL-equation:

$$-i_2 + i_5 + i_6 + i_7 - i_3 = 0,$$

where a positive current has been taken to be one flowing away from the node.

It is interesting to note that the two laws of Kirchhoff are *duals* to each other. That is, if one replaces the words "voltages" by "currents," "around" by "leaving," and "circuit" by "node" in the KVL, one immediately obtains the statement of the KCL, and vice versa. Moreover, these two postulates are concerned only with the topology of the network (i.e., the way in which the network elements are interconnected). Thus they are completely independent from the electrical properties of the network elements, in contrast to the VCR-postulate which deals solely with such properties.

The three sets of equations associated with the three basic network postulates —the VCR-equations, the KVL-equations, and the KCL-equations—are known as the three *primary* sets of equations in the formulation of network equations. A detailed discussion of the matrix formulation of these primary equations, as well as the development of the network equations based on a particular analysis method (loop, node or cutset, or state-variable approach), will be represented in Chapter 4.

1.10 SUMMARY AND REMARKS

In this chapter we have presented a brief discussion on the fundamentals of network theory. We first reviewed the definitions of some of the basic terms that will be used throughout the entire discussion of this text. They are by no means exhaustive; other terms that have not been presented in this chapter will be introduced later in appropriate places during the development of the theory.

It should be pointed out that the types of networks or systems to be discussed in this text are mainly the lumped, linear, time-invariant networks which include active as well as passive elements.

The analysis methods to be presented in this book are the loop analysis, the cutset analysis (which includes the node analysis as a special case), and the state-space analysis, all three of which will be developed in the subsequent chapters using the matrix approach. Computer applications based on the three analysis methods together with the topological techniques will be discussed in the two chapters entitled Computer-Aided Analysis I and II. In these two chapters, several subroutines based on well-known numerical analysis algorithms will be developed which are useful in various phases of network analysis. Also four general-purpose computer programs will be presented and their applications to analyzing linear (active) networks discussed. These two chapters will serve *only* as an introduction to the subject of computer-aided analysis and design.

Finally, selected interesting topics in modern system theory such as the concepts of controllability and observability, and optimal control systems, among others, will be briefly discussed.

REFERENCES

1. C. A. Desoer and E. S. Kuh, *Basic Circuit Theory*, McGraw-Hill, 1969, Chapters 1 and 2.
2. M. E. VanValkenburg, *Network Analysis*, 2nd Ed., Prentice-Hall, 1964, Chapters 1–3.
3. E. A. Guillemin, *Introductory Circuit Theory*, Wiley, 1953, Chapter 1.
4. R. E. Scott, *Elements of Linear Circuits*, Addison-Wesley, 1965, Chapter 1.
5. C. W. Merriam III, *Analysis of Lumped Electrical Systems*, Wiley, 1969, Chapter 1.
6. J. B. Cruz, Jr. and M. E. VanValkenburg, *Introductory Signals and Circuits*, Blaisdell, 1967, Chapters 4, 5, and 7.
7. B. C. Kuo, *Linear Networks and Systems*, McGraw-Hill, 1967, Chapters 1 and 2.
8. B. J. Leon and P. A. Wintz, *Basic Linear Networks for Electrical and Electronics Engineers*, Holt, Rinehart & Winston, 1970, Chapters 1 and 2.
9. F. F. Kuo, *Network Analysis and Synthesis*, 2nd Ed., Wiley, 1966, Chapters 1 and 5.

PROBLEMS

1.1 For the three coupled coils of Fig. 1.4.10, determine the expressions for the voltages $v_2(t)$ and $v_3(t)$.

1.2 Show that an independent current source is capable of supplying an infinite amount of power.

1.3 Explain why an independent current source can be regarded as a generalized open circuit.

1.4 Verify each of the source reductions in Fig. 1.7.1.

1.5 Reduce the branches in Fig. P.1.5 to single sources.

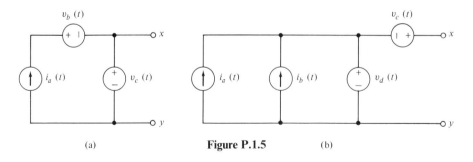

(a)   **Figure P.1.5**   (b)

1.6 The input and output of a certain system shown in Fig. P.1.6 are related by the equation

$$r(t) = \sqrt{\int_0^1 e^2(t)\,dt}.$$

Is it a linear system? Explain.

Figure P.1.6

1.7 Determine whether the network shown in Fig. P.1.7 is linear.

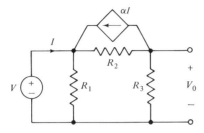

Figure P.1.7

1.8 The input and output of a certain system are related by

$$r(t) = \int_{t_0}^{t} e^{-(t-\lambda)} \lambda e(\lambda) \, d\lambda,$$

where t_0 is the initial time.

a) Is the system linear? Why?   b) Is it time invariant? Why?

1.9 Repeat Problem 1.8 for a system with an input-output relationship given by

$$\frac{dr(t)}{dt} + 4r(t) = te(t).$$

1.10 Express the functions with waveforms shown in Fig. P.1.10 in terms of unit-step functions.

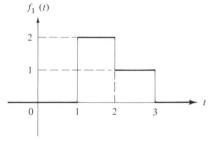

(a)

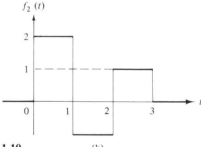

Figure P.1.10

(b)

1.11 Express the function having the waveform of Fig. P.1.11 in terms of ramp and step functions.

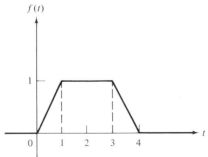

Figure P.1.11

CHAPTER 2

ELEMENTARY NETWORK TOPOLOGY

2.1 INTRODUCTION

Topology is a branch of mathematics; it may be described as "the study of those properties of geometric forms that remain invariant under certain transformations, such as bending, stretching, etc."* Network topology (or network graph theory) is a study of (electrical) networks in connection with their nonmetric geometrical (namely, topological) properties by investigating the interconnections between the branches and the nodes of the networks. As will be seen in later chapters, such a study will lead to important results in network theory such as the algorithms for formulating the loop equations, the node equations, the state equations, and the various theorems on planarity and duality, etc.

2.2 DEFINITIONS

Consider the passive network N shown in Fig. 2.2.1 (a). There are 6 branches $(R_1, R_2, L_3, C_4, L_5, R_6)$ and four nodes (A, B, C, D) in N. If each of the 6 branches is replaced by a line segment (or an arc) with the positions of the 4 nodes remaining unchanged, a topological representation, G, of N is obtained as shown in Fig. 2.2.1 (b). Such a representation G of a network N is referred to as the *graph* of the network. The line segments denoting the branches of N are called the *edges* of G and the endpoints of the edges (corresponding to the nodes of N) are the *vertices* of G.

If edge 2 is removed in graph G of Fig. 2.2.1 (b), we obtain a new graph G_1 [Fig. 2.2.1 (c)] which is a portion of G, and is therefore referred to as a *subgraph* of G. In general, if a subgraph does not contain the entire graph, it is called a *proper subgraph* of the graph (in the same sense as a proper subset of a given set in set theory).

Next, we observe that edges 1, 4, and 5 in G [Fig. 2.2.2 (d)] form a *path*, which is a sequence of edges connected in succession, between vertices A and C. Other paths between the same pair of vertices include the path containing edges 1 and 6, the one containing edges 3 and 5, the one containing edges 3, 4, and 6, and the one consisting of edge 2 alone. Edges 1, 3, and 4 together form a *closed path* or a *loop*

---

* This brief description of topology is quoted directly from *The Random House Dictionary of the English Language*, Random House, New York, 1967.

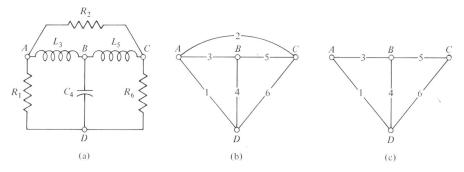

Fig. 2.2.1 (a) A passive network N, (b) the graph G of N; (c) a subgraph G_1 of G.

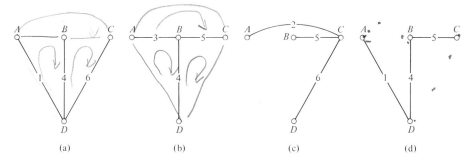

Fig. 2.2.2 Four trees of G: (a) $T^{(1)}$, (b) $T^{(2)}$, (c) $T^{(3)}$, and (d) $T^{(4)}$.

in G, and the corresponding branches form a *circuit* in N [Fig. 2.2.1 (a)]. Thus, a closed path in a graph is called a *circuit* (or a *loop*). Another circuit in G of Fig. 2.2.1 (b) is 2, 3, 5, namely the loop consisting of edges 2, 3, and 5. Two other circuits of G are 1, 3, 5, 6 and 4, 5, 6.*

The *degree* of a vertex in a graph is defined as the number of edges connected to (or incident with) that vertex. For example, the degree of each of the four vertices in G of Fig. 2.2.1 (b) is 3, since there are exactly 3 edges incident with each vertex of G. On the other hand, in G_1 of Fig. 2.2.1 (c), vertices A and C are of degree 2, whereas vertices B and D are of degree 3.

If edges 2, 3, and 5 are removed from G, a subgraph $T^{(1)}$ is formed by the remaining edges and vertices 1, 4, 6 as shown in Fig. 2.2.2 (a). This subgraph, which connects the 4 vertices of G in one piece with only 3 edges and without any circuits, is called a *tree* of G. The edges contained in the tree are referred to as the *tree-branches* (or simply *branches*) of that tree. Those edges *not* contained in a given tree, say $T^{(k)}$, form the complement set of edges, denoted by $T^{(k)'}$, which is called the *co-tree* (or *chord-set*) with respect to $T^{(k)}$. The edges in a co-tree are referred to

---

* Actually there are more than two, for example, 1, 2, 4, 5, etc.

as the *links* (or *chords*). Obviously, the edges in a tree together with those in the corresponding co-tree constitute the set of all edges in the given graph. For example, three other trees of G, as shown in parts (b), (c), and (d) of Fig. 2.2.2, are $T^{(2)} = 3, 4, 5$, $T^{(3)} = 2, 5, 6$, and $T^{(4)} = 1, 4, 5$, respectively. The co-trees corresponding to the 4 trees of G in Fig. 2.2.2 are $T^{(1)'} = 2, 3, 5$; $T^{(2)'} = 1, 2, 6$; $T^{(3)'} = 1, 3, 4$; and $T^{(4)'} = 2, 3, 6$.

From the four trees of G shown in Fig. 2.2.2, we observe the following common properties.

a) Each tree contains the same number of edges (3 for the graph G used in this example).

b) Each tree contains all the (4) vertices of the given graph G.

c) All of the edges contained in each tree are connected in one piece, that is, each tree is a one-piece subgraph of G.

d) None of the trees contains any circuits.

The four properties stated above are indeed the properties of any tree of a given graph. We should point out at this stage that when we speak about a tree of a given graph G, we assume that all the edges of G are connected together in one piece. Such a graph is called a *connected graph*. Thus, all the trees of G are connected subgraphs of G. In light of properties (a) and (b), the following theorem about trees and co-trees should be intuitively obvious.

Theorem 2.2.1* A tree $T^{(k)}$ of a connected graph G having e edges and v vertices contains exactly $v - 1$ branches, and the corresponding co-tree $T^{(k)'}$ of G contains exactly $e - v + 1$ links.

To see the validity of the above theorem, we only need to observe that the first edge connects two vertices and each additional edge will connect just one more vertex in forming a tree. For example, a path containing two edges will connect three vertices in one piece. Thus, by induction, it can be shown† that any tree $T^{(k)}$ of a connected graph G of e edges and v vertices contains $v - 1$ branches.

Finally, since co-tree $T^{(k)'}$ is the complement of $T^{(k)}$, it contains $e - (v - 1)$ or $e - v + 1$ links.

The concept of a tree is one of the most important ideas in the theory of network topology. Consider again the graph G of Fig. 2.2.1 (b). If we choose a tree, say $T^{(1)} = 1, 4, 6$ of Fig. 2.2.2 (a), we can obtain a set of three (the number of links) independent circuits each of which is formed by adding one link to the tree so that the circuit contains that link and the *unique* path consisting of some tree-branches in the tree. These independent circuits are referred to as the *fundamental circuits*

---

* See S. P. Chan [CH 1] for a detailed proof of this theorem.

† See S. P. Chan [CH 1] for a detailed proof of this statement.

(f-circuits) with respect to the chosen tree. For a connected graph G of e edges and v vertices, the number of fundamental circuits with respect to any given tree is obviously $e - v + 1$, which is the difference between e, the number of edges, and $v - 1$, the number of branches of a tree of G. Thus, for tree $T^{(1)}$, there are three fundamental circuits, namely, $C_{f_2} = 1, 2, 6$, $C_{f_3} = 1, 3, 4$, and $C_{f_5} = 4, 5, 6$, identifying links 2, 3, and 5, respectively.

An important result derived from the concept of fundamental circuits may now be stated in the following theorem.

Theorem 2.2.2   For a connected graph G of e edges and v vertices, the set of $(e - v + 1)$ fundamental circuits (f-circuits) with respect to a given tree T constitutes a complete set of independent circuits of G. That is, all other circuits are necessarily dependent on the $(e - v + 1)$ f-circuits.

As will be seen later, this theorem is very useful in determining a set of independent Kirchhoff voltage law (KVL) equations that completely describe the given network N. In other words, the KVL-equations may be obtained by first choosing a tree T from the graph G and then forming the set of $e - v + 1$ (independent) f-circuits from which the corresponding KVL-equations are written.

The fact that the f-circuits constitute an independent set is evident by noting that each f-circuit contains one distinct link which is not contained in any other f-circuit in the set and hence is guaranteed to be independent. That other circuits of G are necessarily dependent on the set of all f-circuits with respect to any given tree is by no means obvious. In fact, the proof is quite lengthy and is beyond the scope of this text. The interested reader is referred to literature elsewhere.*

Another important concept in network topology is that of a *cutset* which may be defined as a *minimal* set of edges, K, in a connected graph G such that the removal of all elements in K will result in two separate connected subgraphs. The word "minimal" used in the above definition implies that no proper subset of K will possess the same property as K. That is, if not all of the elements (edges) of K are removed from G, the resultant graph will remain a connected graph (not a disjoint two-piece graph). For example, consider the graph G of Fig. 2.2.1 (b). The set $K_1 = 2, 3, 4, 6$ which consists of edges 2, 3, 4, and 6 is a cutset, because the removal of these 4 edges from G results in a two-piece subgraph of G as shown in Fig. 2.2.3 (a).

It should be evident that no proper subset of K_1 will "cut" graph G into two pieces as described in the definition. In other words, the removal of any three members of K_1 (namely 234, 236, or 346) from G will still result in a connected (one-piece) subgraph of G, thus illustrating the minimality property of the cutset. Three other cutsets of G are $K_2 = 1, 2, 4, 5$, $K_3 = 1, 4, 6$, and $K_4 = 3, 4, 5$, as shown in parts (b), (c), and (d), respectively, of Fig. 2.2.3 in which the "cutline"—the line that cuts the graph into two pieces—is also indicated for each cutset. The

---

* See, for example, S. P. Chan [CH 1], or S. Seshu and M. B. Reed [SE 1].

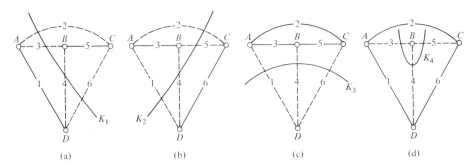

Fig. 2.2.3 Four cutsets of G: (a) K_1, (b) K_2, (c) K_3, and (d) K_4.

members of each cutset, which are those edges intersecting the cutline, are drawn in dotted lines for each cutset in the figure.

Note that in Fig. 2.2.3 (c), when edges 1, 4, and 6 of K_3 are deleted from G, a subgraph G_3 consisting of edges 2, 3, 5 (containing vertices A, B, and C) and the isolated vertex, D, results. Here, one of the two pieces of G_3 is the isolated vertex, D, and the other piece is the subgraph containing edges 2, 3, 5. Similar remarks apply to cutset K_4 in Fig. 2.2.3 (d) where vertex B is the isolated vertex. Thus, in the definition, one (or both, in a trivial case) of the two connected subgraphs which result from removing all the members of the cutset from a connected graph may be an isolated vertex.

As a parallel to the case of a circuit, a cutset of the graph G corresponds to a Kirchhoff current law (KCL) equation in the network N. For a tree T of G, the set of all *fundamental cutsets* with respect to T is defined to be the set of $v - 1$ cutsets each of which contains one distinct (tree-) branch of T along with possibly some links. For example, for tree $T^{(1)} = 1, 4, 6$ of Fig. 2.2.2 (a), the three fundamental cutsets of G with respect to $T^{(1)}$ are $K_{f_1} = 1, 2, 3$, $K_{f_4} = 3, 4, 5$, and $K_{f_6} = 2, 5, 6$ for branches 1, 4, and 6, respectively.

A statement about fundamental cutsets parallel to Theorem 2.2.2 concerning fundamental circuits is given in the following theorem.

Theorem 2.2.3 For a connected graph G of e edges and v vertices, the set of $(v - 1)$ fundamental cutsets (f-cutsets) with respect to a tree T constitutes a complete set of independent cutsets of G. That is, all other cutsets of G are necessarily dependent on the $(v - 1)$ f-cutsets.

Because of the striking parallellism between the circuits (KVL-equations) and the cutsets (KCL-equations), the remarks concerning the usefulness and the validity of Theorem 2.2.2 can be applied also to Theorem 2.2.3.

We should point out that the number $(v - 1)$ is sometimes referred to as the *rank* of the (connected) graph G and is denoted by $R(G)$, or simply r; whereas the number $(e - v + 1)$ is called the *nullity* of G and is denoted by $N(G)$, or simply n.

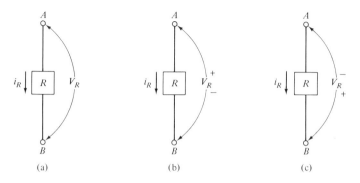

Fig. 2.3.1 Voltage polarity and current direction.

2.3 THE ORIENTED GRAPH

Consider a resistor R with terminals A and B. Let the voltage across R be denoted by v_R and the current through R be i_R as shown in Fig. 2.3.1 (a). Then, Ohm's law states that

$$v_R = \pm i_R R. \tag{2.3.1}$$

The $\pm$ sign preceding the product $i_R R$ in (2.3.1) is necessary because the polarity of v_R in Fig. 2.3.1 (a) has not been specified. However, if we assign the current direction and voltage polarity of i_R and v_R, respectively, as shown in Fig. 2.3.1 (b), then (2.3.1) becomes

$$v_R = i_R R. \tag{2.3.2}$$

In other words, if the voltage v_R across R is *assigned* to be positive at terminal A with respect to terminal B, and if the current i_R through R is considered positive in the direction from A to B, then v_R is equal to the product $i_R R$. On the other hand, if the current direction and the voltage polarity are assigned as shown in Fig. 2.3.1 (c), the correct expression of Ohm's law will be

$$v_R = -i_R R. \tag{2.3.2}$$

Thus, for simplicity, a convention can be made about the assignments of current arrows and voltage polarity marks so that only one type of symbols (either voltage polarity marks or current arrows) will be necessary for the branches in the network (or the edges in the graph).

Convention: If an arrow is assigned to a branch of a network to denote the positive direction of the current flow, the polarity of the voltages across that branch is also implicitly affixed such that the "$+$" mark is placed at the terminal identified by the tail of the arrow, and conversely.

For example, in Fig. 2.3.1 (b), if the current arrow is used, the voltage polarity marks are automatically implied and thus can be omitted as shown in Fig. 2.3.2 (a).

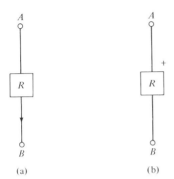

Fig. 2.3.2 Equivalent representations of Fig. 2.3.1 (b).

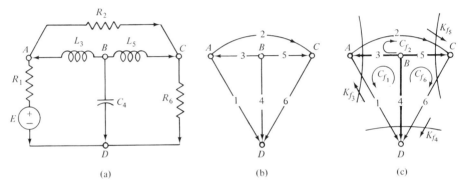

(a)   (b)   (c)

Fig. 2.3.3 (a) A network driven by a voltage source E_1, (b) the corresponding oriented graph, and (c) the f-circuits and f-cutsets of the graph.

Equivalently, to represent the same instance depicted in Fig. 2.3.1 (b), one can use the "$+$" mark alone and omit both the arrow and the "$-$" sign as shown in Fig. 2.3.2 (b).

Next, consider the network of Fig. 2.2.1 (a). Let it be driven by a voltage source E_1 connected in series with R_1 between nodes A and D, and arbitrarily assign a positive direction of the current flow in each branch as shown in Fig. 2.3.3 (a). The corresponding graph is shown in Fig. 2.3.3 (b).

Note that the graph of Fig. 2.3.3 (b) is exactly the same as that of Fig. 2.2.1 (b) except that each edge now has an *orientation* as indicated by the arrow. Such a graph is called an *oriented graph*. Thus, the same graph without the arrows is accordingly defined as a *nonoriented graph*.

Similarly, a circuit is said to be an *oriented circuit* if a positive direction (or orientation) is assigned to that circuit. For a fundamental circuit, it is convenient to assign the orientation in such a way that the circuit orientation is the same as that of the link defining the f-circuit. Likewise, for a cutset, an orientation (positive direction) can be assigned to it from one side to the other across the cutline, and the cutset becomes an *oriented cutset*. For convenience, the orientation of an

f-cutset is assigned such that the cutset orientation agrees with that of the (tree-) branch defining the f-cutset.

As an example, choose the tree $T = 3, 4, 5$ in the oriented graph of Fig. 2.3.3 (b). The three oriented f-circuits C_{f_1}, C_{f_2}, and C_{f_6}, defined by links 1, 2, and 6, respectively, and the three oriented $\tilde{f}$-cutsets K_{f_3}, K_{f_4}, and K_{f_5}, defined by branches 3, 4, and 5, respectively, are shown in $\widetilde{\text{Fig.}}$ 2.3.3 (c).

2.4 NETWORK MATRICES

Refer to the oriented graph of Fig. 2.3.3 (c). Following the convention established in the preceding section, we can write the three KVL-equations for the three f-circuits with respect to tree $T = 3, 4, 5$ as follows.

$$
\begin{array}{llll}
\text{For } C_{f_1}: & v_1 \quad + v_3 - v_4 \qquad\qquad = 0 & \text{(a)}; \\
\text{For } C_{f_2}: & \quad v_2 + v_3 \qquad - v_5 \qquad = 0 & \text{(b)}; \\
\text{For } C_{f_6}: & \qquad\qquad - v_4 + v_5 + v_6 = 0 & \text{(c)}.
\end{array}
\qquad (2.4.1)
$$

Here, the symbol v_j denotes the voltage across edge j. The sign is positive if the edge orientation is the same as that of the f-circuit, and is negative otherwise. When written in matrix form, (2.4.1) becomes

$$
\begin{array}{c}
\phantom{C_{f_1}}\begin{array}{cccccc} 1 & 2 & 3 & 4 & 5 & 6 \end{array} \\
\begin{array}{c} C_{f_1} \\ C_{f_2} \\ C_{f_6} \end{array}
\begin{bmatrix}
1 & 0 & 1 & -1 & 0 & 0 \\
0 & 1 & 1 & 0 & -1 & 0 \\
0 & 0 & 0 & -1 & 1 & 1
\end{bmatrix}
\end{array}
\begin{bmatrix} v_1 \\ v_2 \\ v_3 \\ v_4 \\ v_5 \\ v_6 \end{bmatrix} = \mathbf{0}.
\qquad (2.4.2)
$$

Equation (2.4.2) can be rewritten in symbolic matrix notation as

$$
\mathbf{B}_f \mathbf{v}_e = \mathbf{0}, \qquad (2.4.3)
$$

where $\mathbf{B}_f$ is called the *fundamental circuit matrix* and $\mathbf{v}_e$ is called the *element-voltage vector*. In general, the f-circuit matrix $\mathbf{B}_f = [b_{ij}]_{n \times e}$ of an oriented, connected graph G with e edges and v vertices is defined as a matrix of order $n \times e$ (where $n \equiv e - v + 1$ is the nullity of G). Each row of the matrix identifies an f-circuit (with respect to some tree T) with the circuit orientation identical to the link orientation,* and each column represents an edge so that

$b_{ij} = 1$,   if edge j is in f-circuit i and the edge orientation is identical to the circuit orientation;

$b_{ij} = -1$,   if edge j is in f-circuit i and the edge orientation opposes the circuit orientation;

$b_{ij} = 0$,   if edge j is not in f-circuit i.

---

* Such a circuit orientation will later lead to a convenient form of $\mathbf{B}_f$ as in (2.5.6) of Theorem 2.5.3.

Thus, (2.4.3) gives a set of n independent KVL-equations of the specified network and each row of $\mathbf{B}_f$ represents a statement of the Kirchhoff voltage law—namely, the algebraic sum of the element- (edge-) voltages around an f-circuit is equal to zero.

Next, refer again to the network of Fig. 2.3.3 (a) and consider node D as the reference node. Then we can write an equation for each of the remaining three nodes A, B, and C by applying the Kirchhoff current law which states that the algebraic sum of all the edge-currents at a node is equal to zero. Thus, considering currents flowing away from a node as positive, we obtain the following KCL-equations at nodes A, B, and C:

$$
\begin{array}{lll}
\text{For node } A: i_1 + i_2 - i_3 \hspace{3.5em} = 0 & \text{(a)}; \\
\text{For node } B: \hspace{3em} i_3 + i_4 + i_5 \hspace{1.5em} = 0 & \text{(b)}; \hspace{2em} (2.4.4) \\
\text{For node } C: \hspace{1em} -i_2 \hspace{3em} -i_5 + i_6 = 0 & \text{(c)}.
\end{array}
$$

Rewritten in matrix form, (2.4.4) becomes

$$
\begin{array}{c}
\hspace{2.5em} 1 \hspace{1em} 2 \hspace{1em} 3 \hspace{1em} 4 \hspace{1em} 5 \hspace{1em} 6 \\
\begin{array}{c} A \\ B \\ C \end{array}
\left[\begin{array}{cccccc}
1 & 1 & -1 & 0 & 0 & 0 \\
0 & 0 & 1 & 1 & 1 & 0 \\
0 & -1 & 0 & 0 & -1 & 1
\end{array}\right]
\begin{bmatrix} i_1 \\ i_2 \\ i_3 \\ i_4 \\ i_5 \\ i_6 \end{bmatrix} = \mathbf{0}.
\end{array}
\hspace{2em} (2.4.5)
$$

Equation (2.4.5), in symbolic matrix notation, is

$$
\mathbf{A} \mathbf{i}_e = \mathbf{0}, \hspace{8em} (2.4.6)
$$

where $\mathbf{A}$ is known as the *incidence* (or *vertex*) *matrix* and $\mathbf{i}_e$ is called the *element-current vector*. In general, the incidence matrix $\mathbf{A} = [a_{ij}]_{r \times e}$ of an oriented, connected graph G with e edges and v vertices is defined as a matrix of order $r \times e$ (where $r \equiv v - 1$ is the rank of G). Each row of such a matrix denotes an *incidence set* of edges—namely, the set of edges incident with (connected to) a vertex, and each column identifies an edge so that

$a_{ij} = 1$,   if edge j is incident with vertex i and the edge orientation is away from vertex i;

$a_{ij} = -1$,   if edge j is incident with vertex i and the edge orientation is toward vertex i;

$a_{ij} = 0$,   if edge j is not incident with vertex i.

Thus, (2.4.5) represents a set of r independent KCL-equations of the network and each row of $\mathbf{A}$ signifies a statement of the Kirchhoff current law at a node.

Equivalently, Eq. (2.4.5) may be generalized into a set of KCL-equations corresponding to the r f-cutsets for a chosen tree T. To illustrate this, let us consider the tree $T = 3, 4, 5$ of the graph in Fig. 2.3.3 (b), and write the KCL-equations for the f-cutsets [shown in Fig. 2.3.3 (c)] as follows.

$$
\begin{aligned}
\text{For } K_{f_3}: & \quad -i_1 - i_2 + i_3 & & & = 0; \\
\text{For } K_{f_4}: & \quad i_1 & + i_4 & + i_6 & = 0; \\
\text{For } K_{f_5}: & \quad i_2 & & + i_5 - i_6 & = 0.
\end{aligned}
\tag{2.4.7}
$$

In matrix form, (2.4.7) becomes

$$
\begin{array}{c}
\\
K_{f_3} \\
K_{f_4} \\
K_{f_5}
\end{array}
\begin{array}{cccccc}
1 & 2 & 3 & 4 & 5 & 6 \\
\end{array}
\begin{bmatrix}
-1 & -1 & 1 & 0 & 0 & 0 \\
1 & 0 & 0 & 1 & 0 & 1 \\
0 & 1 & 0 & 0 & 1 & -1
\end{bmatrix}
\begin{bmatrix}
i_1 \\ i_2 \\ i_3 \\ i_4 \\ i_5 \\ i_6
\end{bmatrix}
= \mathbf{0}.
\tag{2.4.8}
$$

Or, in symbolic matrix notation, we have

$$
\mathbf{Q}_f \mathbf{i}_e = \mathbf{0},
\tag{2.4.9}
$$

where $\mathbf{Q}_f$ is called the *fundamental cutset matrix* and $\mathbf{i}_e$ the element-current matrix as in (2.4.6). In general, the f-cutset $\mathbf{Q}_f = [q_{ij}]_{r \times e}$ of an oriented, connected graph of e edges and v vertices is defined as a matrix of order $r \times e$ in which each row identifies an f-cutset (with respect to some tree T) with the cutset orientation the same as the (tree-) branch orientation, and each column represents an edge so that

$q_{ij} = 1$,   if edge j is in f-cutset i and the edge orientation is the same as the cutset orientation;

$q_{ij} = -1$,   if edge j is in f-cutset i and the edge orientation opposes the cutset orientation;

$q_{ij} = 0$,   if edge j is not in f-cutset i.

From (2.4.8) and (2.4.9), we see that the f-cutset matrix $\mathbf{Q}_f$ can be interpreted as the incidence matrix $\mathbf{A}$ of the graph of Fig. 2.3.3 (b), with vertex B as reference since

$$
\mathbf{A} = \begin{array}{c} A \\ D \\ C \end{array}
\begin{array}{cccccc}
1 & 2 & 3 & 4 & 5 & 6 \\
\end{array}
\begin{bmatrix}
1 & 1 & -1 & 0 & 0 & 0 \\
-1 & 0 & 0 & -1 & 0 & -1 \\
0 & -1 & 0 & 0 & -1 & 1
\end{bmatrix}
\tag{2.4.10}
$$

which is the negative of $\mathbf{Q}_f$ in (2.4.8). It is worth noting that (2.4.6) is a special case of (2.4.9), which is a more general statement of the Kirchhoff current law about a linear network.

2.5 INTERRELATIONSHIPS OF NETWORK MATRICES

From Eqs. (2.4.2), (2.4.3), (2.4.5), and (2.4.6), we observe the very interesting fact that the product $\mathbf{A} \cdot \mathbf{B}_f^T$ is a zero (or null) matrix:

$$\mathbf{A} \cdot \mathbf{B}_f^T = \begin{bmatrix} 1 & 1 & -1 & 0 & 0 & 0 \\ 0 & 0 & 1 & 1 & 1 & 0 \\ 0 & -1 & 0 & 0 & -1 & 1 \end{bmatrix} \begin{bmatrix} 1 & 0 & 0 \\ 0 & 1 & 0 \\ 1 & 1 & 0 \\ -1 & 0 & -1 \\ 0 & -1 & 1 \\ 0 & 0 & 1 \end{bmatrix} = [0]_{3 \times 3}, \qquad (2.5.1)$$

where $\mathbf{B}_f^T$ is the transpose of $\mathbf{B}_f$.

That the product $\mathbf{A} \cdot \mathbf{B}_f^T$ is zero is by no means a coincidence even though it is observed from a specific example. In fact, this important property concerning matrices $\mathbf{A}$ and $\mathbf{B}_f$ can be formally stated in the following theorem.

Theorem 2.5.1 Given an oriented, connected graph G of e edges and v vertices, if the columns of the matrices $\mathbf{B}_f$, $\mathbf{A}$, and $\mathbf{Q}_f$ are arranged in the same edge order, the following relationships hold:

$$\mathbf{A} \cdot \mathbf{B}_f^T = \mathbf{0} \qquad \text{(or } \mathbf{B}_f \cdot \mathbf{A}^T = \mathbf{0}) \qquad\qquad (2.5.2)$$

and

$$\mathbf{Q}_f \cdot \mathbf{B}_f^T = \mathbf{0} \qquad \text{(or } \mathbf{B}_f \cdot \mathbf{Q}_f^T = \mathbf{0}). \qquad\qquad (2.5.3)$$

The validity of (2.5.2) can be seen by observing the fact that for any incidence set S_i (which consists of the set of all edges incident with some vertex i) and any f-circuit set S_j (which consists of the set of all edges contained in some f-circuit j), the number of edges common to these two sets is equal to either zero (meaning no edges in common) or two. If the columns of $\mathbf{A}$ and $\mathbf{B}_f$ are arranged in the same edge order, the product of the ith row-vector of $\mathbf{A}$ and the jth column-vector of $\mathbf{B}_f^T$ gives either the sum of two nonzero products if there are two common edges between S_i (corresponding to the ith row of $\mathbf{A}$) and S_j (corresponding to the jth column of $\mathbf{B}_f^T$), or a zero if there are no common edges between them. But the sum of two nonzero products is always zero because of our sign convention made about the typical elements a_{ij} of $\mathbf{A}$ and b_{ij} of $\mathbf{B}_f$ in their definitions. A detailed proof of Theorem 2.5.1 can be found in other literature.* Once $\mathbf{A} \cdot \mathbf{B}_f^T = \mathbf{0}$ is established, the result $\mathbf{B}_f \cdot \mathbf{A}^T = \mathbf{0}$ can be obtained by taking the transpose of $\mathbf{A} \cdot \mathbf{B}_f^T = \mathbf{0}$.

To establish $\mathbf{Q}_f \cdot \mathbf{B}_f^T = \mathbf{0}$, a similar development can be followed with S_i representing some f-cutset i (corresponding to the ith row of $\mathbf{Q}_f$) and S_j as before, but observing that the number of common edges between S_i and S_j is equal to either zero or a positive even integer.* Again, once $\mathbf{Q}_f \cdot \mathbf{B}_f^T = \mathbf{0}$ is established, $\mathbf{B}_f \cdot \mathbf{Q}_f^T = \mathbf{0}$ is seen to be true since one is the transpose of the other.

The next theorem states the rank of each of the three network matrices $\mathbf{B}_f$, $\mathbf{A}$, and $\mathbf{Q}_f$.

---

* See for example, S. P. Chan [CH 1], Chapter 2, or Seshu and Reed [SE 1], Chapter 4.

Theorem 2.5.2 For a connected graph G of e edges and v vertices, the ranks of matrices $\mathbf{B}_f$, $\mathbf{A}$, and $\mathbf{Q}_f$ are given by

$$R(\mathbf{B}_f) = e - v + 1 \equiv n, \tag{2.5.4}$$

$$R(\mathbf{A}) = R(\mathbf{Q}_f) = v - 1 \equiv r. \tag{2.5.5}$$

We can establish that $R(\mathbf{B}_f) = n$ by first observing that each row of $\mathbf{B}_f$ represents an f-circuit which contains a unique link that does not appear in any other f-circuits. Hence, all of the n rows of $\mathbf{B}_f$ are linearly independent, and by a theorem in matrix theory, the rank of $\mathbf{B}_f$ is equal to n.* Similarly, $R(\mathbf{Q}_f) = r$ can be established by observing that each row of $\mathbf{Q}_f$ denotes an f-cutset which contains a unique tree-branch that is not contained in any other f-cutset.

We can also establish that all of the r rows of $\mathbf{A}$ are linearly independent (so that $R(\mathbf{A}) = r$). If we add one additional row to $\mathbf{A}$ as the last row, which represents the incidence set of edges for the reference vertex, we form a new matrix $\mathbf{A}_a$.† Then each column of $\mathbf{A}_a$ contains exactly two nonzero elements, a "1" and a "-1". This is true because each edge (represented by a column of $\mathbf{A}_a$) is incident with exactly two vertices, say i and j (denoted by rows i and j of $\mathbf{A}_a$), such that if the edge orientation is away from vertex i (corresponding to the "1" in row i), it must be toward vertex j (with the "-1" in row j). Thus the sum of the v rows of $\mathbf{A}_a$ is always zero, indicating that at least one of the v rows is a dependent one. This is exactly why we can use $\mathbf{A}$ instead of $\mathbf{A}_a$ to describe the graph completely. The fact that any $r(= v - 1)$ rows of $\mathbf{A}_a$ constitute a linearly independent set of rows is intuitively evident by noting that the set of *any* of r rows of $\mathbf{A}_a$ completely describes the given graph G such that G can be constructed using the information about the incidence of the e edges with the $r = v - 1$ (and hence the v) vertices of G and that any $v - 2$ rows of $\mathbf{A}_a$ cannot completely specify G.‡

The interrelationships of matrices $\mathbf{B}_f$, $\mathbf{A}$, and $\mathbf{Q}_f$ are stated in the following theorem.

Theorem 2.5.3 For an oriented, connected graph G of e edges and v vertices, if matrices $\mathbf{B}_f$, $\mathbf{A}$, and $\mathbf{Q}_f$ are partitioned as

$$\mathbf{B}_f = [\mathbf{U}_n \quad \mathbf{B}_{f_{12}}], \tag{2.5.6}$$

$$\mathbf{A} = [\mathbf{A}_{11} \quad \mathbf{A}_{12}], \tag{2.5.7}$$

and

$$\mathbf{Q}_f = [\mathbf{Q}_{f_{11}} \quad \mathbf{U}_r], \tag{2.5.8}$$

where $\mathbf{U}_n$ and $\mathbf{U}_r$ denote unit matrices of orders n and r, respectively, such that the columns of the three matrices are arranged in the same edge order and such that

---

* Refer to, for example, F. E. Hohn [HO 1], Chapter 5.
† Matrix $\mathbf{A}_a$ is referred to as the (*complete*) *incidence matrix* in some texts.
‡ For a complete proof, refer to S. P. Chan [CH 1], Chapter 2.

the first $n(= e - v + 1)$ columns correspond to the n links with respect to a tree T and the other $r(= v - 1)$ columns correspond to the r branches of T, then

$$\mathbf{Q}_{f_{11}} = -\mathbf{B}_{f_{12}}^T = \mathbf{A}_{12}^{-1} \cdot \mathbf{A}_{11} \qquad (2.5.9)$$

and

$$\mathbf{Q}_f = [-\mathbf{B}_{f_{12}}^T \quad \mathbf{U}_r] = \mathbf{A}_{12}^{-1} \cdot \mathbf{A}, \qquad (2.5.10)$$

where $\mathbf{B}_{f_{12}}^T$ denotes the transpose of $\mathbf{B}_{f_{12}}$, and $\mathbf{A}_{12}^{-1}$ is the inverse of $\mathbf{A}_{12}$.

The proof of Theorem 2.5.3 is straightforward since (2.5.9) and (2.5.10) follow immediately from (2.5.2) and (2.5.3) with the matrix partitionings as specified by (2.5.6), (2.5.7), and (2.5.8). For example, to derive (2.5.9), we form the product $\mathbf{Q}_f \cdot \mathbf{B}_f^T$, according to (2.5.3), (2.5.6), and (2.5.8), as follows.

$$\mathbf{Q}_f \cdot \mathbf{B}_f^T = [\mathbf{Q}_{f_{11}} \quad \mathbf{U}_r]\begin{bmatrix} \mathbf{U}_n^T \\ \mathbf{B}_{f_{12}}^T \end{bmatrix} = \mathbf{Q}_{f_{11}} \cdot \mathbf{U}_n^T + \mathbf{U}_r \cdot \mathbf{B}_{f_{12}}^T = \mathbf{0}$$

or,

$$\mathbf{Q}_{f_{11}} + \mathbf{B}_{f_{12}}^T = \mathbf{0},$$

which yields

$$\mathbf{Q}_{f_{11}} = -\mathbf{B}_{f_{12}}^T. \qquad (2.5.11)$$

Next, using (2.5.2), (2.5.6), and (2.5.7), we can obtain, in a similar manner,

$$-\mathbf{B}_{f_{12}}^T = \mathbf{A}_{12}^{-1} \cdot \mathbf{A}_{11},$$

which, when combined with Eq. (2.5.11), gives Eq. (2.5.9). The fact that $\mathbf{A}_{12}^{-1}$ exists is evident by noting that $\mathbf{A}_{12}$ is the incidence matrix of the tree T which contains exactly those edges of G that are identified by the columns of $\mathbf{A}_{12}$. This implies that $\mathbf{A}_{12}$ is a square matrix of order r and that the rank of $\mathbf{A}_{12}$ is r, since it is the incidence matrix of a connected graph (T) of $r + 1(= v)$ vertices. Thus, $\mathbf{A}_{12}$ is nonsingular and $\mathbf{A}_{12}^{-1}$ exists.*

Theorem 2.5.3 shows not only that $\mathbf{Q}_f$ can be determined from $\mathbf{B}_f$, but also that the converse is true. That is, once $\mathbf{Q}_f$ is known, $\mathbf{B}_f$ can be found from the following relationship:

$$B_f = [\mathbf{U}_n \quad -\mathbf{Q}_{f_{11}}^T], \qquad (2.5.12)$$

which is an immediate consequence of (2.5.6) and (2.5.9).

Example 2.5.1   As an illustration of Theorem 2.5.3, consider once again the graph G of Fig. 2.3.3 (c) with tree $T = 3, 4, 5$. Matrices $\mathbf{B}_f$ and $\mathbf{Q}_f$ for this tree and

---

* A nonsingular matrix is a square matrix with a nonzero determinant. If a matrix is non-singular, it has an inverse. See F. E. Hohn [HO 1], for example, for the justification of these statements.

matrix $\mathbf{A}$ with vertex D as reference have been obtained as shown in Eqs. (2.4.2), (2.4.8), and (2.4.5), respectively. Rearranging the columns of these three matrices in the same order as described in the theorem, we have

$$
\mathbf{B}_f = [\mathbf{U}_n \quad \mathbf{B}_{f_{12}}] = \begin{matrix} \\ C_{f_1} \\ C_{f_2} \\ C_{f_6} \end{matrix}
\begin{array}{c}
\overbrace{\begin{matrix} 1 & 2 & 6 \end{matrix}}^{\text{Links}} \quad \overbrace{\begin{matrix} 3 & 4 & 5 \end{matrix}}^{\text{Branches}} \\
\left[\begin{array}{ccc|ccc}
1 & 0 & 0 & 1 & -1 & 0 \\
0 & 1 & 0 & 1 & 0 & -1 \\
0 & 0 & 1 & 0 & -1 & 1
\end{array}\right]
\end{array}, \tag{2.5.13}
$$

$$
\mathbf{Q}_f = [\mathbf{Q}_{f_{11}} \quad \mathbf{U}_r] = \begin{matrix} K_{f_3} \\ K_{f_4} \\ K_{f_5} \end{matrix}
\begin{array}{c}
\begin{matrix} 1 & 2 & 6 \end{matrix} \quad \begin{matrix} 3 & 4 & 5 \end{matrix} \\
\left[\begin{array}{ccc|ccc}
-1 & -1 & 0 & 1 & 0 & 0 \\
1 & 0 & 1 & 0 & 1 & 0 \\
0 & 1 & -1 & 0 & 0 & 1
\end{array}\right]
\end{array}, \tag{2.5.14}
$$

and

$$
\mathbf{A} = [\mathbf{A}_{11} \quad \mathbf{A}_{12}] = \begin{matrix} A \\ B \\ C \end{matrix}
\begin{array}{c}
\begin{matrix} 1 & 2 & 6 \end{matrix} \quad \begin{matrix} 3 & 4 & 5 \end{matrix} \\
\left[\begin{array}{ccc|ccc}
1 & 1 & 0 & -1 & 0 & 0 \\
0 & 0 & 0 & 1 & 1 & 1 \\
0 & -1 & 1 & 0 & 0 & -1
\end{array}\right]
\end{array}, \tag{2.5.15}
$$

where $n = e - v + 1 = 6 - 4 + 1 = 3$, $r = v - 1 = 4 - 1 = 3$. Also

$$
\mathbf{B}_{f_{12}} = \begin{bmatrix} 1 & -1 & 0 \\ 1 & 0 & -1 \\ 0 & -1 & 1 \end{bmatrix}; \tag{2.5.16}
$$

$$
\mathbf{Q}_{f_{11}} = \begin{bmatrix} -1 & -1 & 0 \\ 1 & 0 & 1 \\ 0 & 1 & -1 \end{bmatrix} = -\mathbf{B}_{f_{12}}^T; \tag{2.5.17}
$$

$$
\mathbf{A}_{11} = \begin{bmatrix} 1 & 1 & 0 \\ 0 & 0 & 0 \\ 0 & -1 & 1 \end{bmatrix}; \tag{2.5.18}
$$

and

$$
\mathbf{A}_{12} = \begin{bmatrix} -1 & 0 & 0 \\ 1 & 1 & 1 \\ 0 & 0 & -1 \end{bmatrix} \tag{2.5.19}
$$

which coincidentally is its own inverse (i.e., $\mathbf{A}_{12}^{-1} = \mathbf{A}_{12}$). Thus

$$
\mathbf{A}_{12}^{-1} \cdot \mathbf{A}_{11} = \begin{bmatrix} -1 & 0 & 0 \\ 1 & 1 & 1 \\ 0 & 0 & -1 \end{bmatrix} \begin{bmatrix} 1 & 1 & 0 \\ 0 & 0 & 0 \\ 0 & -1 & 1 \end{bmatrix} = \begin{bmatrix} -1 & -1 & 0 \\ 1 & 0 & 1 \\ 0 & 1 & -1 \end{bmatrix} = \mathbf{Q}_{f_{11}} \tag{2.5.20}
$$

which verifies (2.5.9).

Finally,

$$\mathbf{A}_{12}^{-1} \cdot \mathbf{A} = \begin{bmatrix} -1 & 0 & 0 \\ 1 & 1 & 1 \\ 0 & 0 & -1 \end{bmatrix} \begin{bmatrix} 1 & 1 & 0 & | & -1 & 0 & 0 \\ 0 & 0 & 0 & | & 1 & 1 & 1 \\ 0 & -1 & 1 & | & 0 & 0 & -1 \end{bmatrix}$$

$$= \begin{bmatrix} -1 & -1 & 0 & | & 1 & 0 & 0 \\ 1 & 0 & 1 & | & 0 & 1 & 0 \\ 0 & 1 & -1 & | & 0 & 0 & 1 \end{bmatrix} = \mathbf{Q}_f \qquad (2.5.21)$$

which verifies (2.5.10).

2.6 PLANARITY AND DUALITY

A network (graph) is said to be *planar* if it can be mapped onto a plane such that no two branches (edges) will intersect at a point which is not a node (vertex); otherwise it is *nonplanar*. Thus, a graph is planar if its corresponding network is planar.

Planar networks are of particular interest to the designers of printed circuits and integrated circuits. In the case of printed circuits, a planar design results in the minimum cost of production; and for integrated circuits, planarity of a network is almost a necessity for reasons of economy as well as technology of manufacturing.

The planarity property of a network plays an important role in network theory because of its close relationship with the duality property which may be described as follows. Two networks N_1 and N_2 are said to be *duals* of each other if any circuit (cutset) of N_1 corresponds uniquely to a cutset (circuit) of N_2, and vice versa, when appropriate interchanges of the types of network elements in the two networks are made—a process we shall discuss later in detail. Topologically speaking, two (planar) graphs G_1 and G_2 are duals of each other if the rank of G_1 is equal to the nullity of G_2, and vice versa; that is, if $R(G_1) = N(G_2)$ and $N(G_1) = R(G_2)$.*

The relationship between duality and planarity is now stated in the following theorem.

Theorem 2.6.1 (*Whitney*) A graph G has a dual if and only if G is planar.

This theorem implies that planarity is the necessary and sufficient condition for the existence of a dual of a given graph. That is, for any graph G, a dual of G exists if it is planar; otherwise it has no dual.

The proof of Theorem 2.6.1 is quite involved and is beyond the scope of this text. The interested reader is referred to literature elsewhere.†

Although sophisticated algorithms have been written for testing the planarity of complicated graphs, the mapping of a graph onto a plane usually can be done by inspection through the rearrangement of the edges and vertices for relatively simple graphs. The following examples serve to illustrate some of the points discussed thus far.

---

* For a more general definition of duality, see, for example, S. P. Chan [CH 1], Chapter 3.
† See, for example, Seshu and Reed [SE 1], Chapter 3.

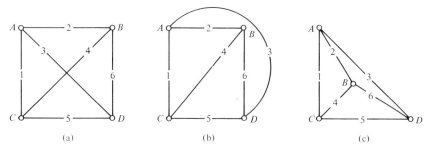

Fig. 2.6.1 (a) Planar graph G_1, (b) a redrawing G_2; and (c) another redrawing G_3 of G_1.

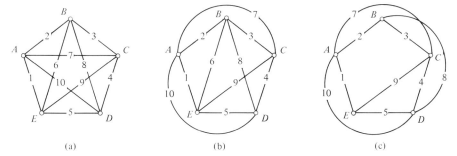

Fig. 2.6.2 The basic 5-vertex nonplanar graph G_{k_1} and its redrawings.

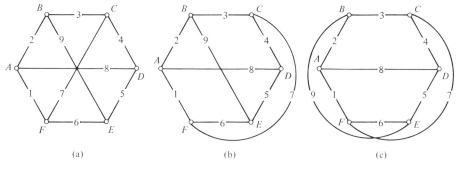

Fig. 2.6.3 The basic 6-vertex nonplanar graph G_{k_2} and its redrawings.

Example 2.6.1 Consider the graph, G_1, shown in Fig. 2.6.1 (a), which has six edges and four vertices. Obviously, there is a crossing in G_1 between edges 3 and 4. But this crossing can be avoided by rearranging (redrawing) either one of the two edges, 3 and 4, as shown in graph G_2 in Fig. 2.6.1 (b), or by rearranging vertex B as shown in graph G_3 in Fig. 2.6.1 (c).

Example 2.6.2 The graph, G_{k_1}, of Fig. 2.6.2 (a) is one of the two simplest nonplanar graphs. It has 10 edges and five vertices (the smallest number of vertices possible for a nonplanar graph); and evidently there are 5 crossings. If edges 7 and 10 are redrawn as shown in Fig. 2.6.2 (b), 4 of the crossings are eliminated with one remaining, a situation which cannot be avoided without creating new crossings. For example, if edge 8 in Fig. 2.6.2 (b) is redrawn, a new crossing between edges 7 and 8 results as indicated in Fig. 2.6.2 (c). It is interesting to note that no rearrangements of vertices are possible in order to eliminate all the 5 crossings. Thus, graph G_{k_1} is nonplanar.

Example 2.6.3   Consider the graph, G_{k_2}, of Fig. 2.6.3 (a) which is the other simplest nonplanar graph. Note that G_{k_2} has only 9 edges (the smallest number of edges possible for a nonplanar graph) and 6 vertices. Note that the ("triple") crossing which occurs at the center of G_{k_2} is also nonplanar.

In fact, graphs G_{k_1} and G_{k_2} are known as the two basic nonplanar graphs of Kuratowski,* who first stated and proved the following celebrated theorem about nonplanar graphs.

Theorem 2.6.2   (*Kuratowski*) A graph G is planar if and only if it does not contain either of the two basic nonplanar graphs G_{k_1} and G_{k_2} as its subgraphs, where each edge in G_{k_1} or G_{k_2} represents either a single edge or a series connection of edges in G.

The proof of Theorem 2.6.2 may be found either in Kuratowski's original work* or in a more advanced text.†

If a graph is planar, then, according to Theorem 2.6.1, it has a dual, and the dual can be found by the application of the so-called "dot method" which is now described.

The Dot Method for Constructing a Dual

When the given planar graph G_a of e edges and v vertices is mapped onto a plane with all crossings eliminated, the plane is divided into m "inside" regions (or "windows") and one outside region. The dual can be constructed through the following steps:

a) Place a dot in each region (including the outside region) of G_a.

b) Connect the $m + 1$ dots by line segments such that, for each edge j between two regions, there is a line segment j' which crosses edge j and is connected between the two dots corresponding to the two windows. Thus exactly e line segments are required to cross the e edges of G_a.

c) The resultant planar graph, G_b, consisting of the $m + 1$ dots as vertices and the e line segments as edges is the dual of G_a.

Example 2.6.4   Consider graph G_2 of Fig. 2.6.1 (b) which is evidently planar and divides the plane into 4 regions—3 inside regions and one outside region. Thus, following the three steps of the dot method, we construct the dual as shown in

---

* See C. Kuratowski [KU 7].
† Refer to C. Berge [BE 2], for example.

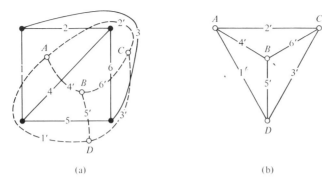

Fig. 2.6.4 Illustration of the application of the dot method for constructing the dual G'_2 of a planar graph G_2.

Fig. 2.6.4 (a). The dots for the inside regions are marked as vertices A, B, and C, and the dot for the outside region is identified as vertex D in the dual with the edges primed with numbers corresponding to the edge numbers in the original graph G_2. The dual is redrawn in Fig. 2.6.4 (b) for clarity.

From Example 2.6.4, it is evident that a circuit, say 1, 2, 5, 6, of G_2 corresponds to a cutset, namely 1', 2', 5', 6', of its dual G'_2, and a cutset, say 1, 3, 4, 6, of G_2 corresponds to a circuit, 1', 3', 4', 6', of G'_2, as expected.

Thus, for a planar network N, the dot method can be applied to obtain its dual network N' such that a circuit (cutset) of N corresponds to a cutset (circuit) of N'. Moreover, since a circuit represents a KVL-equation and a cutset identifies a KCL-equation in a network, it follows that the set of $e - v + 1$ independent KVL-equations of N corresponds to the set of $v' - 1$ independent KCL-equations of N' where e is the number of edges and v the number of vertices of N, and e' is the number of edges and v' the number of vertices of N'. Since v' must also be equal to the number of regions into which N divides, we conclude that the number of windows (inside regions) m bounded by the branches of N is equal to the nullity of N, which follows from the fact that

$$m = v' - 1 = e - v + 1. \tag{2.6.1}$$

This is now restated in the next theorem.

Theorem 2.6.3 For a planar network N, a complete set of independent KVL-equations can be obtained by writing the set of m KVL-equations around the m window frames (*meshes*), which are the boundaries of the inside regions (window-panes) into which the plane is divided by the branches of N.

Theorem 2.6.3 provides an easy means for picking a complete set of independent circuits for writing the KVL-equations, provided that the given network is planar. If, on the other hand, the network is nonplanar, we can always pick a tree first, and then write the KVL-equations for the f-circuits of that tree. This scheme is a general one, and can be applied to a nonplanar as well as a planar network.

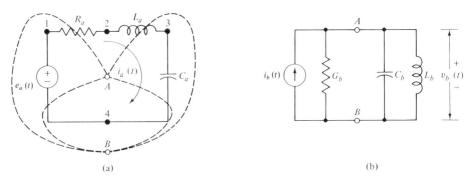

(a)   (b)

Fig. 2.6.5 (a) Planar RLC network N_a with a voltage source $e_a(t)$; (b) the dual N_b of N_a with a current source $i_b(t)$.

Since a KVL-equation in a planar network N corresponds to a KCL-equation in its dual N', it is evident that when the very same KVL-equation for N is used as a KCL-equation for N', the units for the terms in the equation must be appropriately changed to produce correct results. The following simple example will serve to illustrate the ideas involved.

Example 2.6.5 Refer to the planar series RLC network N_a with a voltage source $e_a(t)$ as shown in Fig. 2.6.5 (a). The KVL-equation around the only circuit is written, with the edge-voltages substituted by the corresponding voltage-current relationship (VCR-) equations, as

$$e_a(t) = R_a i_a(t) + L_a \frac{d}{dt} i_a(t) + \frac{1}{C_a} \int i_a(t)\, dt. \tag{2.6.2}$$

It is evident that if (2.6.2) is used to represent the KCL-equation for the corresponding node A in the dual N' of N [Fig. 2.6.5 (a)], then the symbols $e_a(t)$, $i_a(t)$, R_a, L_a, and C_a in (2.6.2) must be changed to $i_b(t)$, $v_b(t)$, G_b, C_b, and L_b, respectively, to give the correct expression

$$i_b(t) = G_b v_b(t) + C_b \frac{d}{dt} v_b(t) + \frac{1}{L_b} \int v_b(t)\, dt \tag{2.6.3}$$

for the dual network N_b of N_a shown in Fig. 2.6.5 (b).

Note that (2.6.2) and (2.6.3) are identical in form, but the symbols have been appropriately changed. This implies a very important fact that once the solution to (2.6.2) for network N_a is found, the solution to (2.6.3) of the dual N_b can be automatically written by merely transforming a set of symbols! This is indeed one of the important applications of duality.

We now generalize the above discussion on dual networks as follows. A (planar) network N_1 has a dual N_2 if the following conditions exist.

a) There exists a one-to-one correspondence between the branches of N_1 and those of N_2.

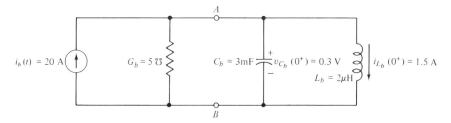

Fig. 2.6.6 The dual network N_b of N_a with initial conditions and numerical values.

b) The $v\text{-}i$ equations for the branches of N_1 become the $i\text{-}v$ equations for the corresponding branches of N_2 upon the interchange of symbols: for any branch j of N_1 and the corresponding branch j' of N_2, we have the following correspondence:

$$\begin{aligned}
v_j &\leftrightarrow i_{j'} & L_j &\leftrightarrow C_{j'} \\
i_j &\leftrightarrow v_{j'} & C_j &\leftrightarrow L_{j'} \\
R_j &\leftrightarrow G_{j'}
\end{aligned} \qquad (2.6.4)$$

where v stands for voltage, i for current, R for resistance, G for conductance, L for inductance, and C for capacitance.

c) The KVL- (KCL) equations of N_1 become the KCL- (KVL) equations of N_2 upon the interchange of symbols as described in (2.6.4).

It should be noted that since there is no such network element as mutual capacitance, a planar network has a dual only if it contains no mutual inductances, or if such mutual inductances are negligible when present so that they are ignored before the dual can be constructed.

As for associating voltage polarity marks and current directions for sources and/or initial conditions in the dual network, one can always use the following rules:

a) Treat a source exactly the same way as a passive element is treated using the convention discussed in Section 2.3 (Fig. 2.3.1).

b) Treat initial conditions as sources.

c) Numerical values for the corresponding elements should be the same for both the original network and its dual.

Example 2.6.6 Consider again the network of Fig. 2.6.5 (a). Let us assume the following numerical element values: $e_a(t) = 20$ V, $R_a = 5\Omega$, $L_a = 3$ mH, $C_a = 2\mu$F. In addition, assume that an initial current $i_{L_a}(0^+) = 0.3$A flows in the inductor L_a from node 2 to node 3, and an initial voltage $v_{C_a}(0^+) = 1.5$ V exists across the capacitor C_a with the positive polarity mark at node 3. The dual network N_b may be obtained following the three rules discussed above and is shown with initial conditions and numerical values in Fig. 2.6.6.

Note that in the original network N_a, the set of voltage polarity marks for $e_a(t)$ represents a positive current flowing from node 1 to node 4 through that branch, keeping in mind that the voltage source has been treated just like a passive element using the convention discussed in Section 2.3. This current is considered negative with respect to the loop current $i_a(t)$ since they are in opposite directions. Thus, as far as node A (which corresponds to loop a with i_a as loop current) of the dual network N_b is concerned, the current source (which corresponds to E_a in N_a) should be negative with respect to this node—namely, flowing toward node A. Similarly, a *positive* initial current $i_{L_a}(0^+) = 0.3$A with respect to loop a in N_a corresponds to a *positive* initial voltage $v_{C_b}(0^+) = 0.3$ V at node A in N_b (since the set of polarity marks for $v_{C_b}(0^+)$ represents a *positive* current flowing *away from* node A in N_b as shown in Fig. 2.6.6). Finally, the initial current $i_{L_b}(0^+) = 1.5$A flowing from node A to node B in N_b is obtained in a like manner. The numerical values of the corresponding elements and initial conditions are, of course, exactly the same for both the original network N_a and its dual network N_b.

REFERENCES

1. S. P. Chan, *Introductory Topological Analysis of Electrical Networks*, Holt, Rinehart & Winston, 1969, Chapters 1–3.

2. S. Seshu and M. B. Reed, *Linear Graphs and Electrical Networks*, Addison-Wesley, 1961, Chapters 1–4.

3. H. E. Koenig, Y. Tokad, and H. K. Kesavan, *Analysis of Discrete Physical Systems*, McGraw-Hill, 1967, Chapter 4.

4. P. H. O'N. Roe, *Networks and Systems*, Addison-Wesley, 1966, Chapter 1.

5. B. C. Kuo, *Linear Networks and Systems*, McGraw-Hill, 1967, Chapter 4.

6. W. A. Blackwell, *Mathematical Modeling of Physical Networks*, Macmillan, 1968, Chapters 4 and 5.

PROBLEMS

2.1 Prove that every tree T of a connected graph G of v vertices must contain $v - 1$ edges.

2.2 Obtain a graph from each of the networks shown in Fig. P.2.2. Label the vertices and the edges in the graph.

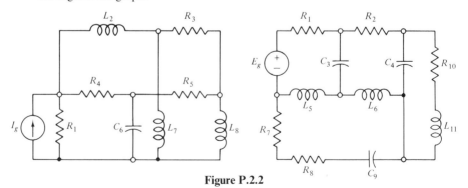

Figure P.2.2

2.3 List all the circuits and trees of each of the graphs shown in Fig. P.2.3.

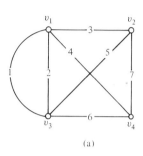

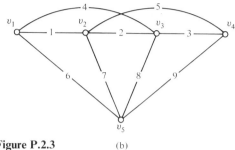

(a)   **Figure P.2.3**   (b)

2.4 Find all the paths between vertices v_a and v_b of each of the graphs shown in Fig. P.2.4.

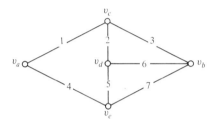

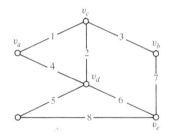

Figure P.2.4

2.5 For the graph shown in Fig. P.2.5, select the tree $T = 1, 3, 6, 7$ and show the fundamental cutsets and fundamental circuits with respect to tree T.

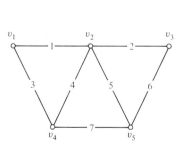

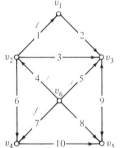

Figure P.2.5   **Figure P.2.6**

2.6 For the oriented graph G shown in Fig. P.2.6:

a) Find the rank and the nullity of G.

b) Write the independent Kirchhoff equations (KVL and KCL) for the meshes (window-frames) and the nodes of G with node v_6 as reference.

c) Repeat part (b) for the fundamental cutsets and fundamental circuits with respect to tree $T = 1, 3, 4, 6, 10$ of G.

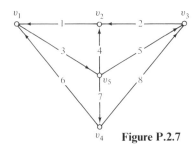

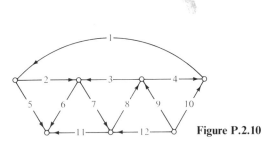

Figure P.2.7   **Figure P.2.10**

2.7. For the oriented graph G shown in Fig. P.2.7:

a) Write the KVL- and the KCL-equations for the meshes (window frames) and the nodes with node v_5 as reference.

b) Choose the tree $T = 3, 4, 5, 7$ and repeat part (a), using cutsets for KCL.

c) Compare the equations obtained in parts (a) and (b) and comment on the results.

2.8 Find the oriented graph from each of the following incidence matrices:

a) $A_1 = \begin{bmatrix} -1 & 1 & 1 & 0 & 0 \\ 0 & 0 & -1 & 1 & 0 \\ 0 & -1 & 0 & -1 & -1 \end{bmatrix}$

b) $A_2 = \begin{bmatrix} 1 & 1 & 1 & 0 & 0 & 0 \\ 0 & -1 & 0 & -1 & 0 & 0 \\ 0 & 0 & -1 & 1 & -1 & 0 \\ 0 & 0 & 0 & 0 & 1 & -1 \end{bmatrix}$

2.9 Refer to the oriented graph of Fig. P.2.7. Select the tree $T = 1, 2, 7, 8$ and verify the validity of the following matrix equations:

a) $B_f v_e = 0$ b) $Q_f i_e = 0$

c) $A i_e = 0$ d) $A B_f^T = 0$

e) $Q_f B_f^T = 0$

2.10 For the connected, oriented graph of Fig. P.2.10, select the tree $T = 2, 3, 4, 6, 7, 10$ and find the following matrices in the form specified:

a) $B_f = [U_n \quad B_{f_{12}}]$

b) $Q_f = [Q_{f_{11}} \quad U_r]$

c) $A = [A_{11} \quad A_{12}]$, where A_{12} is nonsingular.

2.11 A connected, oriented graph G has a fundamental circuit matrix as follows:

$$B_f = \begin{array}{c} \begin{matrix} 1 & 4 & 7 & 8 & 2 & 3 & 5 & 6 \end{matrix} \\ \begin{bmatrix} 1 & 0 & 0 & 0 & 1 & 1 & 0 & 0 \\ 0 & 1 & 0 & 0 & -1 & 0 & 1 & 0 \\ 0 & 0 & 1 & 0 & 0 & -1 & 0 & -1 \\ 0 & 0 & 0 & 1 & 0 & 0 & -1 & 1 \end{bmatrix} \end{array}$$

Find G.

2.12 In Fig. P.2.12 are given a battery E, a lamp R, and eight switches labeled 1 through 8. Determine the connection of these components such that the following four conditions

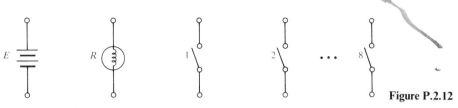

Figure P.2.12

are satisfied:

a) With switches 1, 3, and 5 open, the lamp will go on if and only if switches 2, 4, 6, 7, and 8 are closed.
b) With switches 1, 6, and 8 open, the lamp will go on if and only if switches 2, 3, 4, 5, and 7 are closed.
c) With switches 2, 4, and 6 open, the lamp will go on if and only if switches 1, 3, 5, 7, and 8 are closed.
d) With switches 3, 4, and 7 open, the lamp will go on if and only if switches 1, 2, 5, 6, and 8 are closed.

Show all your work in detail.

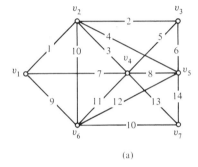

(a)

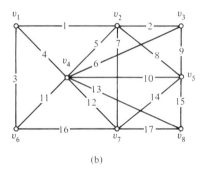

(b)

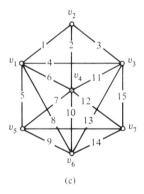

(c)

Figure P.2.13

2.13 Determine whether each of the graphs shown in Fig. P.2.13 is planar. If the graph is planar, redraw the graph to eliminate all crossings which are not vertices.

2.14 Using the dot method, construct a dual for each of the *planar* graphs found in Problem 2.13.

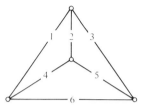

Figure P.2.15

2.15 Repeat Problem 2.14 for the graph shown in Fig. P.2.15, and then compare the two dual graphs. Comment on your observation.

2.16 If the dual graph G' of a planar graph G has the same topology (same incidence relationships between the edges and the vertices) as G, then G and G' are called *self-duals*. Show three different self-dual graphs.

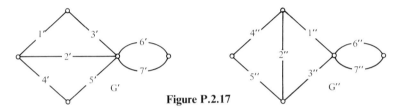

Figure P.2.17

2.17 Using the dot method, show that the two planar graphs G' and G'' shown in Fig. P.2.17 have the same dual G. This implies that a planar graph G may have more than one dual. (But the duals of a planar graph are *2-isomorphic* or *equivalent* graphs. For a detailed discussion of equivalent graphs, refer to S. P. Chan [CH 1]).

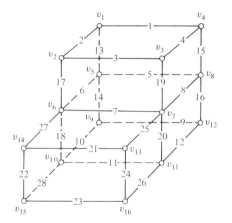

Figure P.2.18

2.18 In Fig. P.2.18 is shown the graph G of a network N such that each branch of N is represented by an edge of G and each node of N by a vertex of G.

a) How many independent node equations are necessary to completely describe N?
b) How many independent loop equations do we need to completely characterize N?
c) Is G (or N) planar or nonplanar?

Show your work in detail.

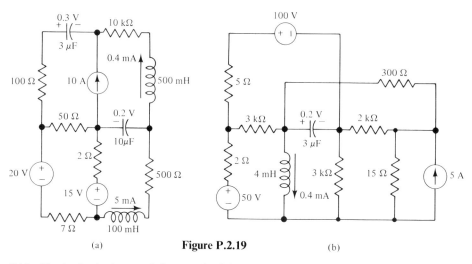

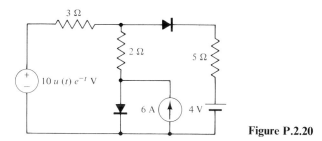

Figure P.2.19

(a)   (b)

2.19 Obtain the dual network from each of the *planar* networks shown in Fig. P.2.18.

Figure P.2.20

2.20 Repeat Problem 2.19 for the network of Fig. P.2.20, in which the two diodes are ideal. Show all your work in detail.

CHAPTER 3

MATRIX FORMULATION OF NETWORK EQUATIONS

3.1 INTRODUCTION

In Chapter 1 we introduced the three postulates, or primary systems of equations in network theory: the Kirchhoff voltage-law (KVL) equations, the Kirchhoff current-law (KCL) equations, and the voltage-current relationship (VCR) equations. In Chapter 2 we presented a brief discussion on elementary network topology. The main objective of this chapter is to develop three different systems of network equations—the loop equations, the node equations, and the state equations—in matrix form using the concepts of the network matrices $(\mathbf{A}, \mathbf{B}_f, \mathbf{Q}_f)$, and the relationships between them, as discussed in Chapter 2. For simplicity and clarity, the discussion presented in this chapter is limited to networks composed of resistances, capacitances, (self and mutual) inductances, and independent (voltage and current) sources.

One of the advantages of the matrix formulations of network equations is that matrix manipulations are concise in form and hence are especially convenient in the theoretical developments of the three systems of network equations. Perhaps more importantly, matrix representation of a given network in terms of either loop, node, or state equations is best suited for digital computation, especially when the network is a complex one.

3.2 THE LOOP EQUATIONS

In writing the loop equation for a particular loop of a given network, we first write the KVL-equation around that loop by summing algebraically the element (edge) voltages of the loop to zero. Then, the voltage across each element is expressed in terms of element current and the parameter value using the VCR-equation. Finally, the element current is expressed in terms of the loop current and, after simplification, the loop equation is obtained. This procedure is best illustrated by means of the examples which follow.

Example 3.2.1   Consider the network of Fig. 3.2.1. The KVL-equation for this network is simply

$$-e(t) + v_1(t) + v_2(t) + v_3(t) = 0. \tag{3.2.1}$$

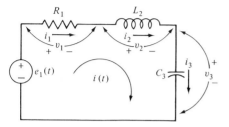

Fig. 3.2.1 The network for Example 3.2.1.

The VCR-equations for the three passive elements in the loop are

$$v_1(t) = R_1 i_1(t),$$

$$v_2(t) = L_2 \frac{d}{dt} i_2(t), \tag{3.2.2}$$

$$v_3(t) = \frac{1}{C_3} \int i_3(t)\, dt.$$

Upon substituting (3.2.2) into (3.2.1), we have

$$-e(t) + R_1 i_1(t) + L_2 \frac{d}{dt} i_2(t) + \frac{1}{C_3} \int i_3(t)\, dt = 0. \tag{3.2.3}$$

The loop current $i(t)$ is related to the element currents in the loop by

$$i(t) = i_1(t) = i_2(t) = i_3(t) \tag{3.2.4}$$

so that (3.2.3) may be rewritten as

$$e(t) = R_1 i(t) + L_2 \frac{d}{dt} i(t) + \frac{1}{C_3} \int i(t)\, dt, \tag{3.2.5}$$

which is the desired loop equation for the network.

When the given network contains two or more loops, the first question to be answered is "How many loop equations do we need to write?". This, in turn, amounts to determining the correct number of independent KVL-equations necessary (and sufficient) to represent a given network. From Section 2.4, we learned that for a (connected) network with a graph consisting of e edges and v vertices, the number of independent loops which give the same number of independent KVL-equations is exactly $e - v + 1 \equiv n$ (see Theorem 2.5.2). Thus, if the given network N is planar, we can simply choose the "window frames" or meshes as independent loops (see Theorem 2.6.3), and write the loop equation for each loop. If, however, the given network is nonplanar, we can first choose a tree T and then write the loop equation for each of the $e - v + 1$ independent fundamental circuits (Theorem 2.2.2) with respect to tree T.

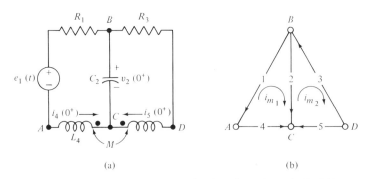

Fig. 3.2.2 (a) Network N, and (b) the oriented graph G of N.

Since the Laplace transform method is a very convenient means for the solution of a linear network, it is desirable that the loop equations for a given network be written in the complex frequency domain (the s-domain). The following example serves to illustrate the points discussed above.

Example 3.2.2 Let us obtain the loop equations for the network N of Fig. 3.2.2 (a), the oriented graph G of which is shown in Fig. 3.2.2 (b).

Note that the branch consisting of the voltage source $e_1(t)$ and the resistor R_1 between nodes A and B in N is represented as a single edge (edge 1) in G. The orientations of the edges and of the circuits in G are completely arbitrary. Since G is planar, the two $(e - v + 1 = 5 - 4 + 1 = 2)$ window frames are chosen as the independent loops for writing the loop equations. The VCR-equations for the five edges are

$$v_1(t) = e_1(t) + R_1 i_1(t),$$

$$v_2(t) = \frac{1}{C_2} \int_0^t i_2(t)\, dt + v_2(0^+),$$

$$v_3(t) = R_3 i_3(t),$$

$$v_4(t) = L_4 \frac{d}{dt} i_4(t) + M \frac{d}{dt} i_5(t),$$

$$v_5(t) = M \frac{d}{dt} i_4(t) + L_5 \frac{d}{dt} i_5(t),$$

$$(3.2.6)$$

where $v_k(t)$ and $i_k(t)$ are, respectively, the voltage across and current through edge k $(k = 1, 2, 3, 4, 5)$, $v_k(0^+)$ is the initial voltage for that edge, and $i_k(0^+)$ is the initial current. Taking the Laplace transform of each equation in (3.2.6), and putting the resulting equations in matrix form, we obtain

$$
\underbrace{\begin{bmatrix} V_1(s) \\ V_2(s) \\ V_3(s) \\ V_4(s) \\ V_5(s) \end{bmatrix}}_{\mathbf{V}_e(s)} = \underbrace{\begin{bmatrix} E_1(s) \\ 0 \\ 0 \\ 0 \\ 0 \end{bmatrix}}_{\mathbf{E}_e(s)} + \underbrace{\begin{bmatrix} R_1 & 0 & 0 & 0 & 0 \\ 0 & \dfrac{1}{sC_2} & 0 & 0 & 0 \\ 0 & 0 & R_3 & 0 & 0 \\ 0 & 0 & 0 & sL_4 & sM \\ 0 & 0 & 0 & sM & sL_5 \end{bmatrix}}_{\mathbf{Z}_e(s)} \underbrace{\begin{bmatrix} I_1(s) \\ I_2(s) \\ I_3(s) \\ I_4(s) \\ I_5(s) \end{bmatrix}}_{\mathbf{I}_e(s)}
$$

$$
+ \frac{1}{s} \underbrace{\begin{bmatrix} 0 \\ v_2(0^+) \\ 0 \\ 0 \\ 0 \end{bmatrix}}_{\mathbf{v}_e(0^+)} - \underbrace{\begin{bmatrix} 0 & 0 & 0 & 0 & 0 \\ 0 & 0 & 0 & 0 & 0 \\ 0 & 0 & 0 & 0 & 0 \\ 0 & 0 & 0 & L_4 & M \\ 0 & 0 & 0 & M & L_5 \end{bmatrix}}_{\mathbf{L}_e} \underbrace{\begin{bmatrix} 0 \\ 0 \\ 0 \\ i_4(0^+) \\ i_5(0^+) \end{bmatrix}}_{\mathbf{i}_e(0^+)}. \tag{3.2.7}
$$

In compact matrix notation, (3.2.7) becomes

$$
\mathbf{V}_e(s) = \mathbf{E}_e(s) + \mathbf{Z}_e(s)\mathbf{I}_e(s) + \frac{1}{s}\mathbf{v}_e(0^+) - \mathbf{L}_e\mathbf{i}_e(0^+), \tag{3.2.8}
$$

where $\mathbf{Z}_e(s)$ is called the *edge impedance matrix*.

Next, we note that the KVL-equations for the independent circuits (loops) chosen may be represented in the matrix form (see Problem 2.9)

$$
\mathbf{B}\mathbf{V}_e(s) = \mathbf{0}, \tag{3.2.9}
$$

where the coefficient matrix (circuit matrix) is of order $n \times e$ and rank $n(= e - v + 1)$. Thus, one can premultiply Eq. (3.2.8) by matrix $\mathbf{B}$, set the right-hand member of the resulting equation to zero according to (3.2.9), and then solve for $\mathbf{B}\mathbf{Z}_e(s)\mathbf{I}_e(s)$ in terms of the other quantities, namely,

$$
\mathbf{B}\mathbf{Z}_e(s)\mathbf{I}_e(s) = -\mathbf{B}\{\mathbf{E}_e(s) + \frac{1}{s}\mathbf{v}_e(0^+) - \mathbf{L}_e\mathbf{i}_e(0^+)\}. \tag{3.2.10}
$$

For this example, since

$$
\mathbf{B}\mathbf{V}_e(s) = \begin{bmatrix} -1 & 1 & 0 & -1 & 0 \\ 0 & -1 & -1 & 0 & 1 \end{bmatrix} \begin{bmatrix} V_1(s) \\ V_2(s) \\ V_3(s) \\ V_4(s) \\ V_5(s) \end{bmatrix} = \mathbf{0}, \tag{3.2.11}
$$

(3.2.10) can now be written, using the matrices in (3.2.7) and (3.2.11), as follows:

$$
\begin{bmatrix} -1 & 1 & 0 & -1 & 0 \\ 0 & -1 & -1 & 0 & 1 \end{bmatrix}
\begin{bmatrix} R_1 & 0 & 0 & 0 & 0 \\ 0 & \dfrac{1}{sC_2} & 0 & 0 & 0 \\ 0 & 0 & R_3 & 0 & 0 \\ 0 & 0 & 0 & sL_4 & sM \\ 0 & 0 & 0 & sM & sL_5 \end{bmatrix}
\begin{bmatrix} I_1(s) \\ I_2(s) \\ I_3(s) \\ I_4(s) \\ I_5(s) \end{bmatrix}
$$

$$
= \begin{bmatrix} 1 & -1 & 0 & 1 & 0 \\ 0 & 1 & 1 & 0 & -1 \end{bmatrix}
\left\{
\begin{bmatrix} E_1(s) \\ 0 \\ 0 \\ 0 \\ 0 \end{bmatrix}
+ \begin{bmatrix} 0 \\ v_2(0^+) \\ 0 \\ 0 \\ 0 \end{bmatrix}
+ \frac{1}{s}
- \begin{bmatrix} 0 & 0 & 0 & 0 & 0 \\ 0 & 0 & 0 & 0 & 0 \\ 0 & 0 & 0 & 0 & 0 \\ 0 & 0 & 0 & L_4 & M \\ 0 & 0 & 0 & M & L_5 \end{bmatrix}
\begin{bmatrix} 0 \\ 0 \\ 0 \\ i_4(0^+) \\ i_5(0^+) \end{bmatrix}
\right\}.
$$

After simplification, we obtain

$$
\begin{bmatrix} -R_1 & \dfrac{1}{sC_2} & 0 & -sL_4 & -sM \\ 0 & -\dfrac{1}{sC_2} & -R_3 & sM & sL_5 \end{bmatrix}
\begin{bmatrix} I_1(s) \\ I_2(s) \\ I_3(s) \\ I_4(s) \\ I_5(s) \end{bmatrix}
= \begin{bmatrix} E_1(s) - \dfrac{1}{s}v_2(0^+) - L_4 i_4(0^+) - M i_5(0^+) \\ \dfrac{1}{s}v_2(0^+) + M i_4(0^+) + L_5 i_5(0^+) \end{bmatrix}.
$$

Next, it can be shown that, for a given network, the e (transform) edge currents $I_1(s), I_2(s), \ldots, I_e(s)$ are related to the n (transform) loop currents $I_{m1}(s), I_{m2}(s), \ldots, I_{mn}(s)$ by (see Problem 3.10)

$$
\mathbf{I}_e(s) = \mathbf{B}^T \mathbf{I}_m(s), \tag{3.2.12}
$$

where $\mathbf{B}^T$ is the transpose of the circuit matrix $\mathbf{B}$. Thus, substituting (3.2.12) into (3.2.9), we find that

$$
\mathbf{BZ}_e(s)\mathbf{B}^T \mathbf{I}_m(s) = -\mathbf{B}\{\mathbf{E}_e(s) + \frac{1}{s}\,\mathbf{v}_e(0^+) - \mathbf{L}_e \mathbf{i}_e(0^+)\} \tag{3.2.13}
$$

or

$$
\mathbf{Z}_m(s)\mathbf{I}_m(s) = -\mathbf{B}\{\mathbf{E}_e(s) + \frac{1}{s}\mathbf{v}_e(0^+) - \mathbf{L}_e \mathbf{i}_e(0^+)\}, \tag{3.2.14}
$$

where

$$\mathbf{Z}_m(s) \equiv \mathbf{B}\mathbf{Z}_e(s)\mathbf{B}^T \tag{3.2.15}$$

is called the *mesh impedance matrix* and $\mathbf{I}_m(s)$ is the vector of mesh currents having the $n(= e - v + 1)$ mesh currents as its elements. Equation (3.2.14) is known as the *mesh analysis matrix equation*, which is applicable to any linear planar network. For this example, since

$\mathbf{Z}_m(s) = \mathbf{B}\mathbf{Z}_e(s)\mathbf{B}^T$

$$= \underbrace{\begin{bmatrix} -R_1 & \dfrac{1}{sC_2} & 0 & -sL_4 & -sM \\ 0 & -\dfrac{1}{sC_2} & -R_3 & sM & sL_5 \end{bmatrix}}_{\mathbf{B}\mathbf{Z}_e(s)} \underbrace{\begin{bmatrix} -1 & 0 \\ 1 & -1 \\ 0 & -1 \\ -1 & 0 \\ 0 & 1 \end{bmatrix}}_{\mathbf{B}^T}$$

$$= \begin{bmatrix} R_1 + \dfrac{1}{sC_2} + sL_4 & -\dfrac{1}{sC_2} - sM \\ -\dfrac{1}{sC_2} - sM & \dfrac{1}{sC_2} + R_3 + sL_5 \end{bmatrix}, \tag{3.2.16}$$

we have, from (3.2.11), (3.2.14), and (3.2.16),

$$\begin{bmatrix} R_1 + \dfrac{1}{sC_2} + sL_4 & -\dfrac{1}{sC_2} - sM \\ -\dfrac{1}{sC_2} - sM & \dfrac{1}{sC_2} + R_3 + sL_5 \end{bmatrix} \begin{bmatrix} I_{m1}(s) \\ I_{m2}(s) \end{bmatrix}$$

$$= \begin{bmatrix} E_1(s) - \dfrac{1}{s}v_2(0^+) - L_4 i_4(0^+) - M i_5(0^+) \\ \dfrac{1}{s}v_2(0^+) + M i_4(0^+) + L_5 i_5(0^+) \end{bmatrix},$$

which represents the two desired loop equations in matrix form.

The following remarks are appropriate at this point.

Remark 1. Equation (3.2.13) is the matrix formulation of the mesh equations for a given *planar* linear network. It gives a concise and clear algorithmic procedure for obtaining the loop equations for the meshes of the (planar) network.

Remark 2. When the network is not planar, matrix **B** in (3.2.13) through (3.2.15) should be replaced by $\mathbf{B}_f$ (the fundamental circuit matrix) for a chosen tree and matrix $\mathbf{Z}_m(s)$ in (3.2.15) should be redefined as

$$\mathbf{Z}_l(s) \equiv \mathbf{B}_f \mathbf{Z}_e(s)\mathbf{B}_f^T, \tag{3.2.17}$$

where $\mathbf{Z}_l(s)$ is called the *loop impedance matrix*, so that (3.2.14) becomes

$$\mathbf{Z}_l(s)\mathbf{I}_c(s) = -\mathbf{B}_f\left\{\mathbf{E}_e(s) + \frac{1}{s}\mathbf{v}_e(0^+) - \mathbf{L}_e\mathbf{i}_e(0^+)\right\}. \tag{3.2.14a}$$

Equation (3.2.14a) is the *loop-analysis matrix equation* for a *general* linear (planar or nonplanar) network, and it can be shown that, in the case of a nonplanar network, the edge currents and the loop currents (that is, the fundamental circuit currents) are related by

$$\mathbf{I}_e(s) = \mathbf{B}_f^T\mathbf{I}_c(s), \tag{3.2.18}$$

where $\mathbf{I}_e(s)$ is the vector of edge currents and $\mathbf{I}_c(s)$ is the vector of chord (link) currents (which are also the *f*-circuit currents).*

Remark 3. For simple networks such as the one used in Example 3.2.2, it is evident that the loop equations may be obtained more easily by direct substitution instead of using the matrix approach. The advantages offered by the matrix formulation technique exemplified by (3.2.13) through (3.2.15) are the concise and clear representation of the formulation procedure as pointed out in Remark 1, and the facility with which matrix manipulations in general can be programmed for computers. More importantly, for a complex network, such a matrix approach offers a systematic procedure which would be less error prone than that obtained by direct substitution.

3.3 THE NODE EQUATIONS

In Section 2.6, we noted the striking similarity in form between the KVL-equations of a (planar) network N and the KCL-equations of the corresponding dual network N'. Also, by means of Example 2.6.5, we showed that the loop equations of N correspond to the node equations of N'. Thus, after our discussion on the formulation of loop equations in the preceding section, it seems logical to expect that the procedure for formulating the node equations of a given network may be similarly developed following certain "rules of duality."

In loop analysis, we began with the VCR-equation, (3.2.8); premultiplied it by the circuit matrix, **B**, which is the coefficient matrix of the KVL-matrix equation, (3.2.9); and then set it to zero. In the development of the node equation, we first rewrite (3.2.8) in such a form that the vector of edge currents $\mathbf{I}_e(s)$ are explicitly

---

* Refer to Seshu and Reed [SE 1].

expressed in terms of all other quantities in the equations. Thus, we write

$$\mathbf{I}_e(s) = \mathbf{Z}_e^{-1}(s)\{\mathbf{V}_e(s) - \mathbf{E}_e(s) + \mathbf{L}_e\mathbf{i}_e(0^+) - \frac{1}{s}\mathbf{v}_e(0^+)\}, \tag{3.3.1}$$

where $\mathbf{Z}_e^{-1}(s)$ is the inverse of $\mathbf{Z}_e(s)$.*

Next, we premultiply (3.3.1) by the incidence matrix $\mathbf{A}$ and set it to zero by the application of the KCL-equation, $\mathbf{AI}_e(s) = 0$. This results in the following expression:

$$\mathbf{AZ}_e^{-1}(s)\mathbf{V}_e(s) = \mathbf{AZ}_e^{-1}(s)\{\mathbf{E}_e(s) - \mathbf{L}_e\mathbf{i}_e(0^+) + \frac{1}{s}\mathbf{v}_e(0^+)\}. \tag{3.3.2}$$

Similar to (3.2.12) in the loop analysis, it can be shown (see Problem 3.11) that the (transform) edge voltage vector $\mathbf{V}_e(s)$ is related to the node-pair voltage vector $\mathbf{V}_n(s)$ by

$$\mathbf{V}_e(s) = \mathbf{A}^T\mathbf{V}_n(s). \tag{3.3.3}$$

In (3.3.3) $\mathbf{A}^T$ is the transpose of $\mathbf{A}$ and $\mathbf{V}_n(s)$ is a vector of the $r(= v - 1)$ independent voltages between the $v - 1$ nodes (vertices) and the reference node (arbitrarily chosen among the set of all v nodes of the given network N). Substituting (3.3.3) into (3.3.2), we find that

$$\mathbf{AZ}_e^{-1}(s)\mathbf{A}^T\mathbf{V}_n(s) = \mathbf{AZ}_e^{-1}(s)\{\mathbf{E}_e(s) - \mathbf{L}_e\mathbf{i}_e(0^+) + \frac{1}{s}\mathbf{v}_e(0^+)\} \tag{3.3.4}$$

or

$$\mathbf{Y}_n(s)\mathbf{V}_n(s) = \mathbf{AZ}_e^{-1}(s)\{\mathbf{E}_e(s) - \mathbf{L}_e\mathbf{i}_e(0^+) + \frac{1}{s}\mathbf{v}_e(0^+)\}, \tag{3.3.5}$$

where

$$\mathbf{Y}_n(s) \equiv \mathbf{AZ}_e^{-1}(s)\mathbf{A}^T \tag{3.3.6}$$

is called the *node admittance matrix*.

Equation (3.3.5), together with (3.3.6), is the desired node-analysis matrix equation which represents the r independent node equations of N.

Example 3.3.1 Consider the network N of Fig. 3.2.2 (a) used in Example 3.2.2. Let us write the node equations for N. Since there are four nodes ($v = 4$), the number r of independent node equations is $r = v - 1 = 3$. Choosing node D arbitrarily as the reference node, we write the KCL-matrix equation:

$$\mathbf{AI}_e(s) = \begin{array}{c} \\ \end{array} \begin{array}{ccccc} 1 & 2 & 3 & 4 & 5 \end{array} \\ \begin{bmatrix} -1 & 0 & 0 & 1 & 0 \\ 1 & 1 & -1 & 0 & 0 \\ 0 & -1 & 0 & -1 & -1 \end{bmatrix} \begin{bmatrix} I_1(s) \\ I_2(s) \\ I_3(s) \\ I_4(s) \\ I_5(s) \end{bmatrix} = \mathbf{0}. \tag{3.3.7}$$

---

* Here, we assume that $\mathbf{Z}_e$ is nonsingular so that its inverse exists. The case when $\mathbf{Z}_e$ is singular will be discussed later in the chapter.

From (3.2.7) in Example 3.2.2, we partition the edge impedance matrix $\mathbf{Z}_e(s)$ as follows.

$$\mathbf{Z}_e(s) = \begin{bmatrix} R_1 & 0 & 0 & 0 & 0 \\ 0 & \dfrac{1}{sC_2} & 0 & 0 & 0 \\ 0 & 0 & R_3 & 0 & 0 \\ 0 & 0 & 0 & sL_4 & sM \\ 0 & 0 & 0 & sM & sL_5 \end{bmatrix} = \begin{bmatrix} \mathbf{Z}_{11} & 0 \\ 0 & \mathbf{Z}_{22} \end{bmatrix} \qquad (3.3.8)$$

so that

$$\mathbf{Z}_e^{-1}(s) = \begin{bmatrix} \mathbf{Z}_{11}^{-1} & 0 \\ 0 & \mathbf{Z}_{22}^{-1} \end{bmatrix} = \begin{bmatrix} \dfrac{1}{R_1} & 0 & 0 & 0 & 0 \\ 0 & sC_2 & 0 & 0 & 0 \\ 0 & 0 & \dfrac{1}{R_3} & 0 & 0 \\ 0 & 0 & 0 & \dfrac{\Gamma_4}{s} & -\dfrac{\Gamma_M}{s} \\ 0 & 0 & 0 & -\dfrac{\Gamma_M}{s} & \dfrac{\Gamma_5}{s} \end{bmatrix}, \qquad (3.3.9)$$

where $\mathbf{Z}_{11}^{-1}$, $\mathbf{Z}_{22}^{-1}$, and $\mathbf{Z}_e^{-1}(s)$ in (3.3.9) are the inverses of $\mathbf{Z}_{11}$, $\mathbf{Z}_{22}$, and $\mathbf{Z}_e(s)$, respectively, as defined in (3.3.8); and

$$\Gamma_4 = \frac{L_5}{L_4 L_5 - M^2},$$

$$\Gamma_5 = \frac{L_4}{L_4 L_5 - M^2}, \qquad (3.3.10)$$

$$\Gamma_M = \frac{M}{L_4 L_5 - M^2},$$

are, respectively, the self-inverse inductance in coil 4, the self-inverse inductance in coil 5, and the mutual inverse inductance between coils 4 and 5 discussed in Chapter 1.

With the incidence matrix $\mathbf{A}$ known from (3.3.7), $\mathbf{Z}_e^{-1}(s)$ from (3.3.9) and all other quantities of (3.3.4) found from (3.2.7) in Example 3.2.2, the node-analysis matrix equation (3.3.5) can now be written, after substitution and simplification:

$$
\begin{bmatrix}
\left(\dfrac{1}{R_1} + \dfrac{\Gamma_4}{s}\right) & -\dfrac{1}{R_1} & \dfrac{1}{s}(\Gamma_M - \Gamma_4) \\[2ex]
-\dfrac{1}{R_1} & \left(\dfrac{1}{R_1} + sC_2 + \dfrac{1}{R_3}\right) & -sC_2 \\[2ex]
\dfrac{1}{s}(\Gamma_M - \Gamma_4) & -sC_2 & \left\{ sC_2 + \dfrac{1}{s}(\Gamma_4 + \Gamma_5 - 2\Gamma_M) \right\}
\end{bmatrix}
\begin{bmatrix}
V_A(s) \\[2ex]
V_B(s) \\[2ex]
V_C(s)
\end{bmatrix}
$$

$$\underbrace{\qquad\qquad\qquad\qquad\qquad\qquad\qquad\qquad\qquad}_{\mathbf{Y}_n(s)} \quad \underbrace{\qquad}_{\mathbf{V}_n(s)}$$

$$
=
\begin{bmatrix}
-\dfrac{E_1}{R_1} - \dfrac{\Gamma_4}{s}\{L_4 i_4(0^+) + M i_5(0^+)\} + \dfrac{\Gamma_M}{s}\{M i_4(0^+) + L_5 i_5(0^+)\} \\[3ex]
\dfrac{E_1}{R_1} + C_2 V_2(0^+) \\[3ex]
-C_2 V_2(0^+) - \dfrac{1}{s}(\Gamma_M - \Gamma_4)\{L_4 i_4(0^+) + M i_5(0^+)\} \\[3ex]
\qquad\qquad - \dfrac{1}{s}(\Gamma_M - \Gamma_5)\{M i_4(0^+) + L_5 i_5(0^+)\}
\end{bmatrix}
$$

$$(3.3.11)$$

It is important to note that the procedure for the node analysis outlined above is valid only when $\mathbf{Z}_e^{-1}(s)$, the inverse of the edge impedance matrix, exists. This, in turn, requires that in the given network, the following two conditions prevail so that $\mathbf{Z}_e(s)$ is a nonsingular matrix:

1. each of the current sources must be connected across a single branch so that it can be transformed into a voltage source in series with an impedance, and

2. each of the voltage sources must be in series with some impedance.

From a physical point of view, the two conditions listed above are reasonable assumptions since, in fact, any *real* voltage source indeed can always be considered as an ideal voltage source with a series impedance, and a *real* current source considered as an ideal current source with a parallel impedance. However, it is sometimes convenient to neglect the series impedance of the voltage source when its magnitude is sufficiently small in comparison with other impedances in the network. Similarly, we might wish to neglect the parallel impedance of the current source when it is sufficiently large in magnitude. Thus, the two conditions are too restrictive and they pose problems in formulating node equations. Fortunately, these difficulties can be overcome by the application of certain "source transformations" which will be the main topic of discussion in the next section.

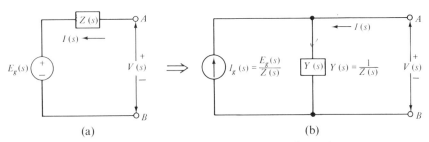

Fig. 3.4.1 Voltage-to-current source transformation.

3.4 VOLTAGE-SOURCE AND CURRENT-SOURCE TRANSFORMATIONS

Consider the simple network of Fig. 3.4.1 (a) in which $E_g(s)$ is the transform of an ideal voltage source $e_g(t)$, $Z(s)$ is the impedance in series with the source, and $V(s)$ is the transform of the terminal voltage at terminals A and B. The loop equation for this network is given by

$$V(s) = E_g(s) + I(s)Z(s), \tag{3.4.1}$$

where $I(s)$ is the current transform through $Z(s)$ from A to B. Next, consider the network of Fig. 3.4.1 (b), in which a current source (transform) $I_g(s)$ is connected in parallel with an admittance $Y(s)$. If we write the node equation at terminal A, we get

$$V(s)Y(s) = I_g(s) + I(s)$$

so that

$$V(s) = \frac{I_g(s)}{Y(s)} + \frac{I(s)}{Y(s)}. \tag{3.4.2}$$

Now let us consider the network in Fig. 3.4.1 (b) to be the equivalent of that in Fig. 3.4.1 (a) by requiring the same voltage $V(s)$ and current $I(s)$ at the terminals A–B. We see upon comparing (3.4.1) to (3.4.2), that the following conditions must be met:

$$Y(s) = \frac{1}{Z(s)}, \tag{3.4.3}$$

$$I_g(s) = E_g(s)Y(s) = \frac{E_g(s)}{Z(s)}, \tag{3.4.4}$$

which are the desired expressions for $Y(s)$ and $I_g(s)$ in the *voltage-to-current source transformation* as illustrated in Fig. 3.4.1. Rewriting (3.4.3) and (3.4.4) by solving for $Z(s)$ and $E_g(s)$, we have

$$Z(s) = \frac{1}{Y(s)}, \tag{3.4.5}$$

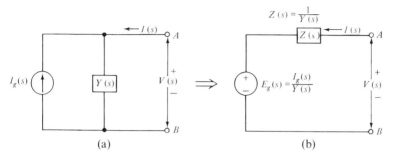

Fig. 3.4.2 Current-to-voltage transformation.

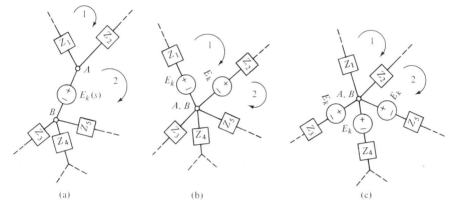

Fig. 3.4.3 Equivalent networks illustrating the E-shift transformation: (a) N_a, (b) N_b, and (c) N_c.

$$E_g(s) = I_g(s)Z(s) = \frac{I_g(s)}{Y(s)}, \qquad (3.4.6)$$

which are the desired expressions for $Z(s)$ and $E_g(s)$ in the current-to-voltage source transformation as shown in Fig. 3.4.2. With Fig. 3.4.2 and the equations, (3.4.5) and (3.4.6), all of the current sources with parallel admittances can be transformed into voltage sources with series impedances so that Condition (1) of the preceding section will be satisfied provided that all of the current sources are connected with parallel admittances. If, however, there exists one (or more) current source(s) connected across several branches or there exists one (or more) voltage source(s) without a series impedance (such that Condition 2 of Section 3.3 is violated), other source transformations must be applied to these sources so that such difficulties will be overcome and Conditions (1) and (2) will be satisfied by the resulting equivalent network.

Consider a branch, say the kth branch, of a network N_a which contains a voltage source $E_k(s)$ alone, as illustrated in Fig. 3.4.3 (a), where only the pertinent portion of N_a is shown. If we coalesce the terminals A and B of the kth branch by

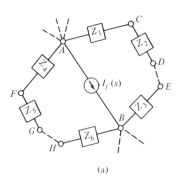

(a)

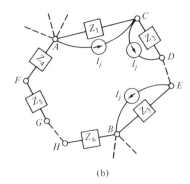

(b)

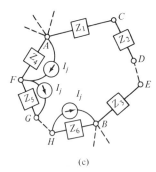

(c)

Fig. 3.4.4 Equivalent networks illustrating the I-shift transformation: (a) N_1, (b) N_2, and (c) N_3.

"pushing" the source $E_k(s)$ through node A into the other branches that are connected to A so that each of these branches will contain $E_k(s)$ in series with their respective branch impedances, the network N_b results, as shown in Fig. 3.4.3 (b). Similarly, if $E_k(s)$ is pushed through node B, instead of node A, when the two nodes are coalesced, we obtain the resultant network N_c as shown in Fig. 3.4.3 (c). The transformation of voltage sources illustrated in Fig. 3.4.3 is known as the *E-shift transformation*. That the three networks N_a, N_b, and N_c are equivalent as far as writing the loop equations is concerned can be established readily by showing that the three sets of loop equations for the three networks are the same.

Following a parallel development, the procedure for the transformation of current sources each of which is connected across several branches can also be established. Consider the network N_1 of Fig. 3.4.4 (a) in which $I_j(s)$ is such a current source. If $I_j(s)$ is transformed with respect to loop A-C-D-E-B-A so that $I_j(s)$ is removed between nodes A and B, but a parallel current source with value equal to $I_j(s)$ is added to each of the branch impedances in the loop, an equivalent network N_2 is obtained, as shown in Fig. 3.4.4 (b). Similarly, $I_j(s)$ may be transformed with respect to loop A-F-G-H-B-A [or, in fact, any other loop containing $I_j(s)$] yielding another equivalent network N_3, as depicted in Fig. 3.4.4 (c). The transformation of current sources just described is called the *I-shift transformation*.

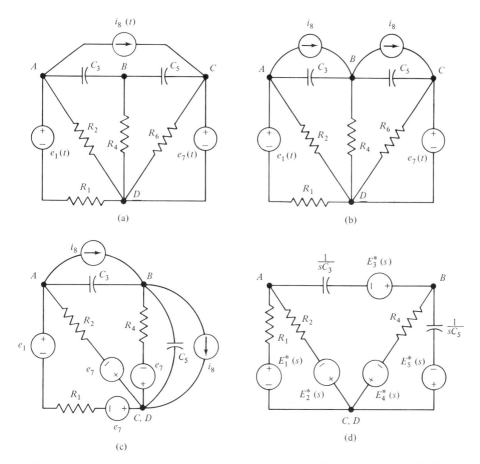

Fig. 3.4.5 Equivalent networks for Example 3.4.1: (a) N_a, (b) N_b, (c) N_c, and (d) N_d.

The fact that networks N_1, N_2, and N_3 in Fig. 3.4.4 are equivalent can be readily verified by comparing the corresponding node equations of these three networks and showing that the same set of node equations will result in each case. The following example will serve to illustrate the application of the E-shift and the I-shift transformations in connection with the formulation of node equations.

Example 3.4.1   Let us write the node equations for the network N_a of Fig. 3.4.5 (a). Our first step is to apply the I-shift transformation of the current source $i_8(t)$ with respect to loop A-B-C-A. The resultant network N_b is shown in Fig. 3.4.5 (b). Next, by the application of E-shift transformation, the voltage source, $e_7(t)$, is pushed through node D so that nodes C and D are coalesced and resistor R_6 is removed, resulting in network N_c as illustrated in Fig. 3.4.5 (c).

Finally, combining the two voltage sources $e_1(t)$ and $e_7(t)$ in the branch containing R_1 and transforming the two current sources in parallel with c_3 and c_5, respectively, we obtain the equivalent network N_d, as depicted in Fig. 3.4.5 (d) in which the transforms of the voltage sources (as functions of s) are given by

$$E_1^* = E_1(s) - E_7(s),$$

$$E_2^*(s) = E_7(s),$$

$$E_3^*(s) = \frac{1}{sC_3} I_8(s),$$

$$E_4^*(s) = E_7(s),$$

$$E_5^*(s) = \frac{1}{sC_5} I_8(s).$$

Of course the capacitors C_3 and C_5 are assumed to be initially uncharged.* It is now a simple matter to write the two node equations for N_d since both Conditions (1) and (2) as stated in Section 3.3 are satisfied. Thus, the remaining task is left to the reader as an exercise.

3.5 THE STATE EQUATIONS

In the preceding sections, we showed how a given linear network may be represented by a set of linear equations. In loop analysis, these equations are a set of $B - N + 1$ independent loop equations corresponding, for example, to the set of fundamental circuits† with respect to a chosen tree, where B is the number of branches and N is the number of nodes of the network. On the other hand, if node analysis is used, these equations are the $N - 1$ independent node equations with each equation corresponding to a node where one of the N nodes is chosen as the reference node. But in either case these network equations are formulated from the three primary systems of equations—namely, the KVL-equations, the KCL-equations, and the VCR-equations.

In this section, we shall introduce another approach to the characterization of a given network. It is known as the *state-variable* approach in which a new set of independent network equations is formulated to represent the network. This third set of network equations called the *state equations* are again obtained from the same three primary systems of equations, as in the cases of loop analysis and node analysis. But it is different from the other two sets of network equations in that some of the (state) variables may be voltages and some of them may be currents whereas in the case of loop analysis, the variables are all (loop) currents, and in node analysis, they are all (node) voltages.‡

---

* The cases with nonzero initial conditions for the capacitors and inductors will be discussed in the section on initial conditions in Chapter 4.

† Or, in case of a planar network, the set of meshes (window frames); or simply any other set of $B - N + 1$ independent loops or circuits in general.

‡ Other differences between the state-variable analysis and the loop and node analyses will be stated later in this section and in the summary of this chapter.

The concept of "state" perhaps originated from classical mechanics. The behavior of a physical system containing particles is generally characterized by the positions of the particles, denoted by the variables, say x_j, and their corresponding velocities $\dot{x}_j = dx_j/dt$ $(j = 1, \ldots, n)$. Thus, the variables x_j together with their time derivatives $\dot{x}_j$ describe the *complete state* of the physical system; and, consequently, they are called the *state variables* of the system. In simple terms, the state of a given physical system may be considered to be "the minimal amount of information necessary to characterize completely any possible future behavior of the system."*

From a mathematical point of view, the state-variable formulation of network equations may be regarded as a new application of certain existing mathematical techniques such as matrix and vector methods to manipulate a (large) number of variables (namely, the state variables) in a given physical system.

Another significant difference between the state-variable technique and the conventional loop- and node-analysis methods is that the former emphasizes the "dynamic" character of the system under study whereas the latter do not. For example, the procedure for formulating the loop or node equations of a passive network is always the same regardless of the types of elements contained in the network. On the other hand, using the state-variable approach, the quantities such as the voltage across a capacitor and the current in an inductor play an important role in the formulation of the state equations, as will be illustrated later in great detail.

When a set of ordinary differential equations representing a dynamic physical system is expressed in the form

$$\dot{x}_j = f_j(x_1, x_2, \ldots, x_n; u_1, u_2, \ldots, u_m)$$
$$(j = 1, 2, \ldots, n), \qquad\qquad (3.5.1)$$

where x_j $(j = 1, 2, \ldots, n)$ are the variables called the *state variables*, and u_k $(k = 1, 2, \ldots, m)$ are the prescribed functions of time t (inputs), the set of equations (3.5.1) is said to be in the *normal* or *standard form*. If the system is linear, the n equations in (3.5.1) can be written as

$$\dot{x}_j = \sum_{i=1}^{n} a_{ji} x_i + \sum_{k=1}^{m} b_{jk} u_k$$
$$(j = 1, 2, \ldots, n)$$

or, in matrix form,

$$\dot{\mathbf{x}} = \mathbf{A}\mathbf{x} + \mathbf{B}\mathbf{u}. \qquad\qquad (3.5.2)$$

Here the n-vector $\mathbf{x}$ is called the *state vector*, the m-vector $\mathbf{u}$ is the *input vector*, $\mathbf{A}$ is an $n \times n$-(square) matrix, simply called the $\mathbf{A}$ *matrix*, $\mathbf{B}$ is an $n \times m$-matrix known as the *distribution matrix*, and $\dot{\mathbf{x}}$ is the time derivative of $\mathbf{x}$, an n-vector, which is

---

* This simple description of the state of a system is borrowed from a paper by E. S. Kuh and R. A. Rohrer [KU 1].

defined as

$$\dot{\mathbf{x}} = \frac{d}{dt}\mathbf{x} = \frac{d}{dt}\begin{bmatrix} x_1 \\ x_2 \\ \vdots \\ x_n \end{bmatrix} = \begin{bmatrix} dx_1/dt \\ dx_2/dt \\ \vdots \\ dx_n/dt \end{bmatrix}. \tag{3.5.3}$$

These state variables may or may not be the desired output variables in the analysis. If they are the desired output variables in the problem, the solution of (3.5.2) for the state vector $\mathbf{x}$ gives the final answer to the problem. If, on the other hand, the state variables are not the output variables, another matrix equation will be needed to express the output variables in terms of the state variables and the inputs:

$$\mathbf{y} = \mathbf{Cx} + \mathbf{Du}, \tag{3.5.4}$$

where $\mathbf{y}$ is a p-vector, called the *output vector*, consisting of the p output variables $y_1, y_2, \ldots, y_p$ as its elements; $\mathbf{C}$ and $\mathbf{D}$ are matrices with (known) constant entries of order $p \times n$ and $p \times m$, respectively. In this case, the complete solution of the system consists of the solution of (3.5.2) for the state vector $\mathbf{x}$ and the determination of the output vector $\mathbf{y}$ by substituting $\mathbf{x}$ in (3.5.4).

Equations (3.5.2) and (3.5.4) are in general the two required matrix equations in the state-variable analysis of a given linear system, and (3.5.2) is called (the normal form of) the *matrix state equation* of the system.

When a set of values is assigned to the (state) variables in the state vector $\mathbf{x}$ at a given time t_0, then the state of the system at time t_0 is specified. In an electrical network, it is convenient to choose voltages across the capacitors and currents in the inductors as state variables so that $\mathbf{x}_0$ can be regarded as an *initial state vector* containing the initial conditions as elements with $t_0 = 0$. Thus, except for special cases where the network contains either all-capacitor circuits or all-inductor cutsets (or both) with voltage sources short-circuited and current sources open-circuited, the number of state variables α is given by

$$\alpha = N_c + N_l, \tag{3.5.5}$$

where N_c is the number of capacitors and N_l is the number of inductors in the network. The number, α, is also known as the *order* of the system in systems theory, or the *order of complexity* in the network in network theory. If a given network contains some all-capacitor circuits or all-inductor cutsets, or both, when the sources are removed (by short-circuiting the voltage sources and open-circuiting the current sources), the network is called a *degenerate network*. The number of state variables α for such networks is given by

$$\alpha = N_c + N_l - C_c - K_l, \tag{3.5.6}$$

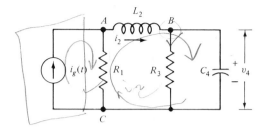

Fig. 3.6.1 An *RLC* network *N* with a current driver $i_g(t)$.

where C_c is the number of all-capacitor circuits,* K_l is the number of all-inductor cutsets,* and N_c and N_l are as defined in (3.5.5). The correctness of (3.5.6) can be readily appreciated by observing the fact that in a circuit containing n elements, only $n - 1$ element-voltages can be arbitrarily specified since the nth element-voltage must be dependent upon the other $n - 1$ element-voltages in the circuit, and can be determined from the KVL-equation for that circuit. Similarly, in a cutset (or a node, as a special case) containing m elements, only $m - 1$ element-currents can be arbitrarily specified since the mth element-current is dependent upon the other $m - 1$ element-currents and can be determined from the KCL-equation for that cutset. Thus, for each of the all-capacitor circuits (or all-inductor cutsets), it is evident that the number of independent capacitor voltages (or inductor currents) which are the state variables of the network must be reduced by one, and (3.5.6) follows.

3.6 ILLUSTRATIVE EXAMPLES

The following examples serve to illustrate some of the points about state equations discussed in the preceding section.

Example 3.6.1 We shall write the matrix state equation for the *RLC* network of Fig. 3.6.1. Since *N* contains one capacitor and one inductor, and there is no degenerate circuit or cutset in *N*, the number of state variables is given by

$$\alpha = N_c + N_l = 1 + 1 = 2.$$

They are the current i_2 in L_2 and the voltage v_4 across C_4. Thus, around the circuit containing R_1, L_2, and R_3, we write the loop equation, which is obtained from the KVL-equation around the circuit and the VCR-equations for the edges in the circuit:

$$L_2 \dot{i}_2 + v_4 + R_1(i_2 - i_g) = 0. \tag{3.6.1}$$

---

* These circuits and cutsets are similarly called *degenerate circuits* and *degenerate cutsets*, respectively, in the degenerate network.

At node B, we write the node equation, which is obtained from the KCL-equation at the node and the VCR-equations for the edges connected to that node:

$$C_4 \dot{v}_4 + \frac{v_4}{R_3} - i_2 = 0. \tag{3.6.2}$$

Rearranging the terms in (3.6.1) and (3.6.2), we find

$$\dot{i}_2 = -\frac{R_1}{L_2} i_2 - \frac{1}{L_2} v_4 + \frac{R_1}{L_2} i_g$$

$$\dot{v}_4 = \frac{1}{C_4} i_2 - \frac{1}{R_3 C_4} v_4$$

which may be written in matrix form

$$\begin{bmatrix} \dot{i}_2 \\ \dot{v}_4 \end{bmatrix} = \begin{bmatrix} -\dfrac{R_1}{L_2} & -\dfrac{1}{L_2} \\ \dfrac{1}{C_4} & -\dfrac{1}{R_3 C_4} \end{bmatrix} \begin{bmatrix} i_2 \\ v_4 \end{bmatrix} + \begin{bmatrix} \dfrac{R_1}{L_2} \\ 0 \end{bmatrix} [i_g]. \tag{3.6.3}$$

Equation (3.6.3) is the desired matrix equation in normal form. Upon comparison of (3.6.3) to (3.5.2), we find that

$$\mathbf{x} = \begin{bmatrix} x_1 \\ x_2 \end{bmatrix} = \begin{bmatrix} i_2 \\ v_4 \end{bmatrix}$$

so that

$$\dot{\mathbf{x}} = \begin{bmatrix} \dot{i}_2 \\ \dot{v}_4 \end{bmatrix}$$

and

$$\mathbf{A} = \begin{bmatrix} -\dfrac{R_1}{L_2} & -\dfrac{1}{L_2} \\ \dfrac{1}{C_4} & -\dfrac{1}{R_3 C_4} \end{bmatrix}$$

with a single input i_g so that the input vector $\mathbf{u}$ becomes a scalar:

$$\mathbf{u} = [i_g].$$

Example 3.6.2 As another example, let us obtain the matrix state equation for the network of Fig. 3.6.2.

By inspection, there are two state variables for this network due to the same reason as stated in the previous example. They are the current i_2 in L_2, and the voltage v_4 across C_4.

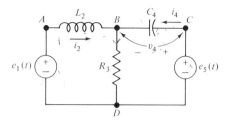

Fig. 3.6.2 An *RLC* network with two sources.

Next, we write the loop equation for the circuit containing e_1, L_2, C_4, and e_5:

$$L_2 \frac{di_2}{dt} - v_4 + e_5 - e_1 = 0 \qquad (3.6.4)$$

and the node equation for node B:

$$\frac{e_5 - v_4}{R_3} - i_2 - C_4 \frac{dv_4}{dt} = 0. \qquad (3.6.5)$$

Rewriting (3.6.4) and (3.6.5) in matrix normal form, we have

$$
\begin{bmatrix} \dot{i}_2 \\ \dot{v}_4 \end{bmatrix}
=
\begin{bmatrix} 0 & \dfrac{1}{L_2} \\ -\dfrac{1}{C_4} & -\dfrac{1}{R_3 C_4} \end{bmatrix}
\begin{bmatrix} i_2 \\ v_4 \end{bmatrix}
+
\begin{bmatrix} \dfrac{1}{L_2} & -\dfrac{1}{L_2} \\ 0 & \dfrac{1}{R_3 C_4} \end{bmatrix}
\begin{bmatrix} e_1 \\ e_5 \end{bmatrix}.
\qquad (3.6.6)
$$

By comparing (3.6.6) with (3.5.2), we have immediately the following relationships:

$$
\mathbf{x} = \begin{bmatrix} i_2 \\ v_4 \end{bmatrix}, \qquad
\mathbf{A} = \begin{bmatrix} 0 & \dfrac{1}{L_2} \\ -\dfrac{1}{C_4} & -\dfrac{1}{R_3 C_4} \end{bmatrix},
$$

$$
\mathbf{B} = \begin{bmatrix} \dfrac{1}{L_2} & -\dfrac{1}{L_2} \\ 0 & \dfrac{1}{R_3 C_4} \end{bmatrix}, \qquad \text{and} \qquad
\mathbf{u} = \begin{bmatrix} e_1 \\ e_5 \end{bmatrix}.
$$

From Examples 3.6.1 and 3.6.2, we see that the state equations of a network may be obtained from a combination of loop equations and node equations of the network. This, then, clearly indicates that the state equations are also obtained

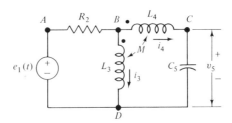

Fig. 3.6.3 An *RLC* network with mutual inductances.

from the same primary systems of equations, namely, the KVL-equations, the KCL-equations, and the VCR-equations. In the following example we shall show how the state equations may be obtained for a network with mutual magnetic coupling.

Example 3.6.3 Consider the network shown in Fig. 3.6.3. We shall choose the currents i_3 in L_3, and i_4 in L_4, and the voltage v_5 across C_5 as the three state variables. Writing the loop equation around the circuit containing e_1, R_2, and L_3, we have

$$R_2(i_3 + i_4) + L_3 \frac{di_3}{dt} + M \frac{di_4}{dt} = e_1. \qquad (3.6.7)$$

Next we write the loop equation for the circuit containing L_3, L_4, and v_5 which yields

$$-L_3 \frac{di_3}{dt} + L_4 \frac{di_4}{dt} - M \frac{di_4}{dt} + M \frac{di_3}{dt} + v_5 = 0. \qquad (3.6.8)$$

Rearranging the terms in (3.6.7) and (3.6.8) and expressing the two equations in matrix form, we have

$$\begin{bmatrix} L_3 & M \\ (L_3 - M) & (M - L_4) \end{bmatrix} \begin{bmatrix} \dfrac{di_3}{dt} \\ \dfrac{di_4}{dt} \end{bmatrix} = \begin{bmatrix} e_1 - R_2(i_3 + i_4) \\ v_5 \end{bmatrix}. \qquad (3.6.9)$$

Now, we can solve for di_3/dt and di_4/dt in (3.6.9) using the method of determinants (Cramer's rule):

$$\dot{i}_3 = \frac{di_3}{dt} = \frac{\begin{vmatrix} e_1 - R_2(i_3 + i_4) & M \\ v_5 & (M - L_4) \end{vmatrix}}{\begin{vmatrix} L_3 & M \\ (L_3 - M) & (M - L_4) \end{vmatrix}} \qquad (3.6.10)$$

$$= R_2(\Gamma_M - \Gamma_3)i_3 + R_2(\Gamma_M - \Gamma_3)i_4 + \Gamma_M v_5 + (\Gamma_3 - \Gamma_M)e_1;$$

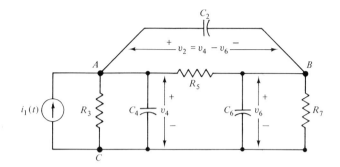

Fig. 3.6.4 A degenerate RC network with an all-C circuit.

$$\dot{i}_4 = \frac{di_4}{dt} = \frac{\begin{vmatrix} L_3 & e_1 - R_2(i_3 + i_4) \\ (L_3 - M) & v_5 \end{vmatrix}}{\begin{vmatrix} L_3 & M \\ (L_3 - M) & (M - L_4) \end{vmatrix}} \tag{3.6.11}$$

$$= R_2(\Gamma_M - \Gamma_4)i_3 + R_2(\Gamma_M - \Gamma_4)i_4 - \Gamma_4 v_5 + (\Gamma_4 - \Gamma_M)e_1;$$

where

$$\Gamma_3 = \frac{L_4}{L_3 L_4 - M^2}, \qquad \Gamma_4 = \frac{L_3}{L_3 L_4 - M^2}, \qquad \text{and} \qquad \Gamma_M = \frac{M}{L_3 L_4 - M^2}. \tag{3.6.12}$$

Combining (3.6.10) and (3.6.11) with the VCR-equation for C_5,

$$\dot{v}_5 = \frac{dv_5}{dt} = \frac{1}{C_5} i_4,$$

in a single matrix equation, we have the desired matrix state equation in normal form:

$$\begin{bmatrix} \dot{i}_3 \\ \dot{i}_4 \\ \dot{v}_5 \end{bmatrix} = \begin{bmatrix} R_2(\Gamma_M - \Gamma_3) & R_2(\Gamma_M - \Gamma_3) & \Gamma_M \\ R_2(\Gamma_M - \Gamma_4) & R_2(\Gamma_M - \Gamma_4) & -\Gamma_4 \\ 0 & \dfrac{1}{C_5} & 0 \end{bmatrix} \begin{bmatrix} i_3 \\ i_4 \\ v_5 \end{bmatrix} + \begin{bmatrix} \Gamma_3 - \Gamma_M \\ \Gamma_4 - \Gamma_M \\ 0 \end{bmatrix} [e_1]. \tag{3.6.13}$$

All of the networks in the above examples are networks which do not contain degenerate circuits or cutsets. In the next example, we shall consider a degenerate network which contains an all-capacitor circuit.

Example 3.6.4 Let us obtain the matrix state equation for the degenerate RC network shown in Fig. 3.6.4. The degenerate circuit is the one consisting of C_2, C_4, and C_6. Thus, although there are three capacitors, the number α of state variables is, according to (3.5.6),

$$\alpha = N_c + N_l - C_c - K_l = 3 + 0 - 1 - 0 = 2.$$

Also, by inspection, we see that the voltage v_2 across C_2 is the difference between the voltage v_4 across C_4 and the voltage v_6 across C_6; that is,

$$v_2 = v_4 - v_6.$$

Hence, we shall choose v_4 and v_6 as the two state variables and write the two node equations at nodes A and B as follows.

At node A: $\quad \dfrac{v_4}{R_3} + C_4 \dfrac{dv_4}{dt} + C_2 \dfrac{d}{dt}(v_4 - v_6) + \dfrac{1}{R_5}(v_4 - v_6) = i_1.$ (3.6.14)

At node B: $\quad \dfrac{v_6}{R_7} + C_2 \dfrac{d}{dt}(v_6 - v_4) + C_6 \dfrac{dv_6}{dt} + \dfrac{1}{R_5}(v_6 - v_4) = 0.$ (3.6.15)

Rearranging the terms in (3.6.14) and (3.6.15) and combining the two resulting equations in matrix form, we get

$$\begin{bmatrix} (C_2 + C_4) & -C_2 \\ -C_2 & (C_2 + C_6) \end{bmatrix} \begin{bmatrix} \dot{v}_4 \\ \dot{v}_6 \end{bmatrix} = \begin{bmatrix} -\left(\dfrac{1}{R_3} + \dfrac{1}{R_5}\right) & \dfrac{1}{R_5} \\ \dfrac{1}{R_5} & -\left(\dfrac{1}{R_5} + \dfrac{1}{R_7}\right) \end{bmatrix} \begin{bmatrix} v_4 \\ v_6 \end{bmatrix} + \begin{bmatrix} i_1 \\ 0 \end{bmatrix}$$

$$(3.6.16)$$

Finally, premultiplying (3.6.16) by the matrix

$$\begin{bmatrix} (C_2 + C_4) & -C_2 \\ -C_2 & (C_2 + C_6) \end{bmatrix}^{-1},$$

we have the desired matrix state equation in normal form:

$$\dot{\mathbf{x}} = \mathbf{Ax} + \mathbf{Bu}$$

where

$$\mathbf{x} = \begin{bmatrix} v_4 \\ v_6 \end{bmatrix}, \qquad \dot{\mathbf{x}} = \begin{bmatrix} \dot{v}_4 \\ \dot{v}_6 \end{bmatrix},$$

$$\mathbf{A} = \begin{bmatrix} \dfrac{-C_2 R_5 - C_6(R_3 + R_5)}{R_3 R_5 C_T} & \dfrac{C_6 R_7 - C_2 R_5}{R_5 R_7 C_T} \\ \dfrac{R_3 C_4 - C_2 R_5}{R_3 R_5 C_T} & \dfrac{-C_2 R_5 - C_4(R_5 + R_7)}{R_5 R_7 C_T} \end{bmatrix},$$

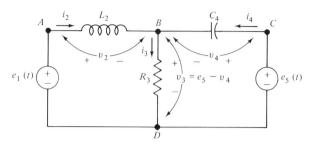

Fig. 3.6.5 The *RLC* network of Fig. 3.6.2.

$$\mathbf{B} = \begin{bmatrix} (C_2 + C_4) & -C_2 \\ \\ -C_2 & (C_2 + C_6) \end{bmatrix}^{-1} = \begin{bmatrix} \dfrac{C_2 + C_6}{C_T} & \dfrac{C_2}{C_T} \\ \\ \dfrac{C_2}{C_T} & \dfrac{C_2 + C_4}{C_T} \end{bmatrix},$$

and

$$\mathbf{u} = \begin{bmatrix} i_1 \\ 0 \end{bmatrix},$$

with $C_T = C_2 C_4 + C_2 C_6 + C_4 C_6$.

Example 3.6.5 As our final example in this section, let us consider again the network of Fig. 3.6.2. Instead of the state variables, suppose that the desired quantities in the network are the voltage v_2 across L_2, the current i_3 in R_3, and the current i_4 in C_4. The network is repeated in Fig. 3.6.5 with all the pertinent voltages and currents indicated in the figure. From Fig. 3.6.5, we write

$$v_2 = L_2 \dot{i}_2, \tag{3.6.17}$$

$$\begin{aligned} i_3 &= \frac{1}{R_3} v_3 = \frac{1}{R_3}(e_5 - v_4) \\ &= -\frac{1}{R_3} v_4 + \frac{1}{R_3} e_5, \end{aligned} \tag{3.6.18}$$

$$i_4 = C_4 \dot{v}_4. \tag{3.6.19}$$

But from (3.6.6) in Example 3.6.2, we found that

$$\dot{i}_2 = \frac{1}{L_2}(v_4 + e_1 - e_5),$$

$$\dot{v}_4 = \frac{1}{C_4}\left(-i_2 - \frac{1}{R_3} v_4 + \frac{1}{R_3} e_5 \right),$$

which, when substituted in (3.6.17) and (3.6.19), yield

$$v_2 = v_4 + e_1 - e_5, \tag{3.6.20}$$

$$i_4 = -i_2 - \frac{1}{R_3} v_4 + \frac{1}{R_3} e_5. \tag{3.6.21}$$

Combining Eqs. (3.6.18), (3.6.20), and (3.6.21) into a single matrix equation, we obtain

$$
\begin{bmatrix} v_2 \\ i_3 \\ i_4 \end{bmatrix}
=
\begin{bmatrix} 0 & 1 \\ 0 & -\dfrac{1}{R_3} \\ -1 & -\dfrac{1}{R_3} \end{bmatrix}
\begin{bmatrix} i_2 \\ v_4 \end{bmatrix}
+
\begin{bmatrix} 1 & -1 \\ 0 & \dfrac{1}{R_3} \\ 0 & \dfrac{1}{R_3} \end{bmatrix}
\begin{bmatrix} e_1 \\ e_5 \end{bmatrix},
$$

which is in the standard form as given in (3.5.4), namely,

$$\mathbf{y} = \mathbf{Cx} + \mathbf{Du}$$

with the state vector $\mathbf{x}$ and the input vector unaltered as shown in (3.6.3), and the output vector $\mathbf{y}$ and the matrices $\mathbf{C}$ and $\mathbf{D}$ given as follows:

$$
\mathbf{y} = \begin{bmatrix} v_2 \\ i_3 \\ i_4 \end{bmatrix}, \quad
\mathbf{C} = \begin{bmatrix} 0 & 1 \\ 0 & -\dfrac{1}{R_3} \\ -1 & -\dfrac{1}{R_3} \end{bmatrix}, \quad
\mathbf{D} = \begin{bmatrix} 1 & -1 \\ 0 & \dfrac{1}{R_3} \\ 0 & \dfrac{1}{R_3} \end{bmatrix}. \tag{3.6.22}
$$

3.7 ALGORITHM FOR FORMULATING STATE EQUATIONS

In the preceding section, we have seen examples illustrating how the state equations may be obtained from the KVL-, the KCL-, and the VCR-equations by the use of certain combinations of the loop and node equations. For simple networks such as those in the examples, it is rather easy to choose the right combination of loop and node equations to produce the state equations. However, when the network under study contains large numbers of circuits (loops), branches, and energy storage elements, the matter of choosing the appropriate combinations of loop and node equations by inspection for the formulation of state equations becomes difficult. For this reason, it is desirable to find a systematic procedure which will enable one to obtain the state equations of a given network step by step.

In the following, we shall develop an algorithm for formulating the state equations of a given network by the use of certain basic concepts of network topology.

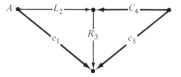

Fig. 3.7.1 The oriented graph G of N.

In a passive RLC network N, the state variables are usually taken to be the capacitor voltages and inductor currents. Thus, to obtain the state equation, we shall first select a tree T from the oriented graph G of N such that T includes (a) *all* of the voltage sources, (b) *none* of the current sources, (c) a maximum number of capacitors, and (d) a minimum number of inductors. Thus, (for easy elimination in later steps) the chord (inductor) voltages are expressed in terms of the tree-branch voltages from the fundamental circuit (KVL) equations, and the tree-branch (capacitor) currents are expressed in terms of the chord currents from the fundamental cutset (KCL) equations for the tree T. Our next step, then, is to eliminate the undesired voltages and currents (which are those voltages and currents other than the state variables and the known sources) from the VCR-equations. This elimination process is accomplished by substituting the circuit (KVL) and cutset (KCL) equations into the VCR-equations for the passive elements. The final step in the algorithm is to rearrange the terms in the equations after the elimination of undesired variables and express the resultant equations in matrix normal form, which is the matrix state equation in the form of (3.5.2).

Before we proceed to underline the necessary steps in the algorithm, we shall first demonstrate the above ideas by means of a simple example.

Example 3.7.1 Let us consider again the network N of Example 3.6.2. The state equations can be obtained step by step as follows:

Step 1. The directed graph G of N is first obtained and is shown in Fig. 3.7.1.

Step 2. Select a tree containing all of the voltage sources and the capacitor. This tree T is shown in heavy strokes in Fig. 3.7.1.

Step 3. The VCR-equations for the passive elements are given by

$$v_2 = L_2 \frac{di_2}{dt} = L_2 \dot{i}_2,$$

$$v_3 = i_3 R_3,$$

$$v_4 = \frac{1}{C_4} \int i_4 \, dt.$$

Since the state variables in this example are i_2 and v_4, the above equations can be rearranged so that time derivatives of the state variables appear on the left-hand side of the equations. Thus, we can group the above equations as follows:

Group (a):

$$\dot{i}_2 = \frac{1}{L_2} v_2, \qquad \text{or} \qquad \begin{bmatrix} \dot{i}_2 \\ \dot{v}_4 \end{bmatrix} = \begin{bmatrix} \frac{1}{L_2} & 0 \\ 0 & \frac{1}{C_4} \end{bmatrix} \begin{bmatrix} v_2 \\ i_4 \end{bmatrix}; \qquad (3.7.1)$$
$$\dot{v}_4 = \frac{1}{C_4} i_4,$$

Group (b):

$$v_3 = i_3 R_3 \qquad\qquad (3.7.2)$$

Note that the equations in Group (a) will eventually lead to the state equations after the nonstate variables v_2 and i_4 (which are those variables other than the state variables) are eliminated. The equation in Group (b) will be used later, together with KVL- and KCL-equations, for the elimination of nonstate variables.

Step 4. The f-circuit (KVL) and f-cutset (KCL) equations are given by

$$-e_1 + e_5 - v_4 + v_2 = 0,$$
$$v_3 - e_5 + v_4 = 0,$$
$$i_2 - i_3 + i_4 = 0.$$

From Step 3 the nonstate variables v_2 and i_4 are to be eliminated; therefore the above equations can be rearranged such that v_2 and i_4 appear on the left-hand side. Thus,

$$\text{Group (a):} \quad v_2 = v_4 + e_1 - e_5,$$
$$i_4 = -i_2 + i_3,$$

or

$$\begin{bmatrix} v_2 \\ i_4 \end{bmatrix} = \begin{bmatrix} 0 & 1 \\ -1 & 0 \end{bmatrix} \begin{bmatrix} i_2 \\ v_4 \end{bmatrix} + \begin{bmatrix} 1 & -1 \\ 0 & 0 \end{bmatrix} \begin{bmatrix} e_1 \\ e_5 \end{bmatrix} + \begin{bmatrix} 0 \\ i_3 \end{bmatrix}; \qquad (3.7.3)$$

$$\text{Group (b):} \quad v_3 = -v_4 + e_5. \qquad\qquad (3.7.4)$$

Remark. There are two more f-cutset (KCL) equations corresponding to the tree-branches e_1 and e_5. Since i_1 and i_5 are not asked for in this example and they are not needed to formulate the state equations, only one KCL-equation is necessary, as indicated in (3.7.4).

Step 5. Our next step is to eliminate all the nonstate variables. This can be done by first combining (3.7.2) and (3.7.4) and then substituting the resultant equation into (3.7.3)—a process which yields

$$
\begin{bmatrix} v_2 \\ i_4 \end{bmatrix} = \begin{bmatrix} 0 & 1 \\ -1 & 0 \end{bmatrix} \begin{bmatrix} i_2 \\ v_4 \end{bmatrix} + \begin{bmatrix} 1 & -1 \\ 0 & 0 \end{bmatrix} \begin{bmatrix} e_1 \\ e_5 \end{bmatrix} + \begin{bmatrix} 0 \\ -\dfrac{v_4}{R_3} + \dfrac{e_5}{R_3} \end{bmatrix}
$$

$$
= \begin{bmatrix} 0 & 1 \\ -1 & -\dfrac{1}{R_3} \end{bmatrix} \begin{bmatrix} i_2 \\ v_4 \end{bmatrix} + \begin{bmatrix} 1 & -1 \\ 0 & \dfrac{1}{R_3} \end{bmatrix} \begin{bmatrix} e_1 \\ e_5 \end{bmatrix}. \tag{3.7.5}
$$

Next, the variables v_2 and i_4 can be eliminated by substituting (3.7.5) into (3.7.1), and we obtain

$$
\begin{bmatrix} \dot{i}_2 \\ \dot{v}_4 \end{bmatrix} = \begin{bmatrix} \dfrac{1}{L_2} & 0 \\ 0 & \dfrac{1}{C_4} \end{bmatrix} \left\{ \begin{bmatrix} 0 & 1 \\ -1 & -\dfrac{1}{R_3} \end{bmatrix} \begin{bmatrix} i_2 \\ v_4 \end{bmatrix} + \begin{bmatrix} 1 & -1 \\ 0 & \dfrac{1}{R_3} \end{bmatrix} \begin{bmatrix} e_1 \\ e_5 \end{bmatrix} \right\} \tag{3.7.6}
$$

in which all the undesirable variables have been eliminated.

Step 6. Finally, combining the terms on the right-hand side of (3.7.6) and simplifying, we obtain the required matrix state equation

$$
\begin{bmatrix} \dot{i}_2 \\ \dot{v}_4 \end{bmatrix} = \begin{bmatrix} 0 & \dfrac{1}{L_2} \\ -\dfrac{1}{C_4} & \dfrac{-1}{R_3 C_4} \end{bmatrix} \begin{bmatrix} i_2 \\ v_4 \end{bmatrix} + \begin{bmatrix} \dfrac{1}{L_2} & -\dfrac{1}{L_2} \\ 0 & \dfrac{1}{R_3 C_4} \end{bmatrix} \begin{bmatrix} e_1 \\ e_5 \end{bmatrix} \tag{3.7.7}
$$

which is seen to be identical to (3.6.6), the matrix state equation for the network of Fig. 3.6.2.

Examples 3.6.2 and 3.7.1 together serve to illustrate an important fact about the algorithm for writing the state equations of a given network: the algorithm which will be stated below is not always the easiest method to use in formulating the state equations especially when the given network is a simple one. For simple *RLC* networks, it is easier to write the state equation by inspection. Thus the importance and usefulness of the algorithm lie in the fact that it provides a step-by-step systematic procedure. Consequently the algorithm lends itself readily to computer programming and constitutes a part of a comprehensive network analysis program, which is a powerful tool for solving networks and systems of high complexity as well as time-varying and nonlinear networks.

Now we are ready to summarize the steps in the algorithm as follows:

Algorithm for Formulating State Equations of an *RLC* Network *N*

Step 1. Obtain the oriented graph from *N*.

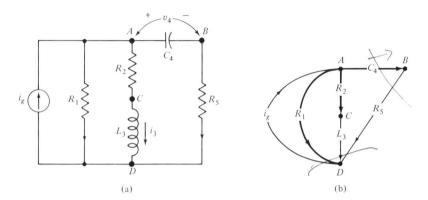

Fig. 3.7.2 (a) The RLC network N, and (b) its oriented graph G.

Step 2. Select a tree T such that it contains:

  a) all of the (known) voltage sources,
  b) none of the (known) current sources,
  c) as many capacitors as possible, and
  d) as few inductors as possible.

Step 3. Write the VCR-equations for the passive elements and separate these equations into two groups:

  a) those VCR-equations for the tree-branch capacitors and the chord inductors; and

  b) those VCR-equations for all other passive elements.

Step 4. Write the fundamental circuit (KVL) and cutset (KCL) equations and separate them in the same manner as done for the VCR-equations in Step 3 (for easy substitution in the next step for the elimination of nonstate variables).

Step 5. Eliminate all of the nonstate variables in the VCR-equations by substituting the equations in Step 4 into those in Step 3. Nonstate variables are defined as those variables that are neither state variables nor (known) sources. They are simply the tree-branch currents and chord voltages with respect to tree T.

Step 6. Rearrange the terms and express the resultant equations in the matrix normal form of (3.5.2), which is the required matrix state equation for N.*

The additional examples following will serve to illustrate the steps in the algorithm.

---

* In some cases, after rearrangement of terms, the state equations still cannot be put in the matrix normal form (3.5.2), and a modified normal form should be used in such cases, as the following example will illustrate.

Example 3.7.2 Consider the network N of Fig. 3.7.2 (a).

Step 1. The oriented graph G of N is shown in Fig. 3.7.2 (b).

Step 2. The tree T consisting of R_1, R_2, and C_4 is selected.

Step 3. The VCR-equations are

Group (a):
$$\begin{cases} \dot{i}_3 = \dfrac{1}{L_3} v_3 \\ \dot{v}_4 = \dfrac{1}{C_4} i_4 \end{cases} \Rightarrow \begin{bmatrix} \dot{i}_3 \\ \dot{v}_4 \end{bmatrix} = \begin{bmatrix} \dfrac{1}{L_3} & 0 \\ 0 & \dfrac{1}{C_4} \end{bmatrix} \begin{bmatrix} v_3 \\ i_4 \end{bmatrix}, \qquad (3.7.8)$$

Group (b):
$$\begin{cases} v_1 = R_1 i_1 \\ v_2 = R_2 i_2 \\ v_5 = R_5 i_5 \end{cases}$$

Step 4. The f-circuit (KVL) and the f-cutset (KCL) equations can be arranged so that group (a) consists of two equations with the variables v_3 and i_4 (where the variables on the right-hand side of the matrix equation in (3.7.8) now appear on the left-hand side). Thus,

Group (a): $v_3 = v_1 - v_2$ (KVL)   (3.7.9)

$\qquad\qquad\qquad\qquad i_4 = i_5$ (KCL) $V_1 = V_4 + V_5 = R_1 i_1$ (3.7.10)

$\qquad\qquad\qquad\qquad\qquad\qquad\qquad\qquad -7 \quad V_1 = V_4 + V_5 .$

Group (b): $-v_1 + v_4 + v_5 = 0$ (i) $V_4 + R_5 i_5 = R_1 \cdot$

$\qquad\qquad -i_g + i_1 + i_3 + i_5 = 0$ (ii)

$\qquad\qquad i_2 - i_3 = 0$ (iii)

Step 5. For the elimination of nonstate variables, we proceed by obtaining expressions for v_3 and i_4 in terms of state variables i_3 and v_4 and then substituting the resultant equations into Group (a) of (3.7.8) to yield the required matrix state equation. Using the VCR-equations in (3.7.8) along with (i) and (ii) from Group (b) in Step 4, we can solve for i_1 and i_5 in terms of the state variables i_3 and v_4 as well as the current source i_g. Thus,

$$i_1 = \frac{\begin{vmatrix} -v_4 & R_5 \\ -i_3 + i_g & 1 \end{vmatrix}}{\begin{vmatrix} -R_1 & R_5 \\ 1 & 1 \end{vmatrix}} = \frac{-v_4 - (-i_3 + i_g)R_5}{-(R_1 + R_5)}$$

$$= \frac{-R_5}{R_1 + R_5} i_3 + \frac{1}{R_1 + R_5} v_4 + \frac{R_5}{R_1 + R_5} i_g \qquad (3.7.11)$$

and

$$i_5 = \frac{\begin{vmatrix} -R_1 & -v_4 \\ 1 & -i_3 + i_g \end{vmatrix}}{\begin{vmatrix} -R_1 & R_5 \\ 1 & 1 \end{vmatrix}} = \frac{(i_3 - i_g)R_1 + v_4}{-(R_1 + R_5)}$$

(3.7.12)

$$= -\frac{R_1}{R_1 + R_5} i_3 + \frac{R_1}{R_1 + R_5} i_g - \frac{1}{R_1 + R_5} v_4.$$

Again using the VCR-equations in (3.7.8) together with (iii) in Group (b) of Step 4, Eq. (3.7.11), and (3.7.12), we obtain from (3.7.9) and (3.7.10) the following equations:

$$v_3 = R_1 i_1 - R_2 i_2 = R_1 i_1 - R_2 i_3$$

$$= -\frac{R_1 R_5}{R_1 + R_5} i_3 + \frac{R_1 R_5}{R_1 + R_5} i_g + \frac{R_1}{R_1 + R_5} v_4 - R_2 i_3$$

or,

$$v_3 = -\left\{ \frac{R_1(R_2 + R_5) + R_2 R_5}{R_1 + R_5} \right\} i_3 + \left(\frac{R_1}{R_1 + R_5} \right) v_4 + \left(\frac{R_1 R_5}{R_1 + R_5} \right) i_g,$$

(3.7.13)

and

$$i_4 = \left(-\frac{R_1}{R_1 + R_5} \right) i_3 - \left(\frac{1}{R_1 + R_5} \right) v_4 + \left(\frac{R_1}{R_1 + R_5} \right) i_g. \quad (3.7.14)$$

Combining (3.7.13) and (3.7.14) into one matrix equation, we have

$$\begin{bmatrix} v_3 \\ i_4 \end{bmatrix} = \begin{bmatrix} -\left(\dfrac{R_1 R_2 + R_1 R_5 + R_2 R_5}{R_1 + R_5} \right) & \left(\dfrac{R_1}{R_1 + R_5} \right) \\ -\left(\dfrac{R_1}{R_1 + R_5} \right) & -\left(\dfrac{1}{R_1 + R_5} \right) \end{bmatrix} \begin{bmatrix} i_3 \\ v_4 \end{bmatrix}$$

(3.7.15)

$$+ \begin{bmatrix} \dfrac{R_1 R_5}{R_1 + R_5} \\ \dfrac{R_1}{R_1 + R_5} \end{bmatrix} [i_g].$$

Step 6. Finally, (3.7.15) is substituted into (3.7.8) yielding, after simplification, the desired matrix state equation

$$
\begin{bmatrix} \dot{i}_3 \\ \dot{v}_4 \end{bmatrix} = \begin{bmatrix} -\dfrac{R_1R_2 + R_1R_5 + R_2R_5}{L_3(R_1 + R_5)} & \dfrac{R_1}{L_3(R_1 + R_5)} \\[2ex] -\dfrac{R_1}{C_4(R_1 + R_5)} & -\dfrac{1}{C_4(R_1 + R_5)} \end{bmatrix} \begin{bmatrix} i_3 \\ v_4 \end{bmatrix}
$$
$$
+ \begin{bmatrix} \dfrac{R_1R_5}{L_3(R_1 + R_5)} \\[2ex] \dfrac{R_1}{C_4(R_1 + R_5)} \end{bmatrix} [i_g]. \tag{3.7.16}
$$

It should be noted also that sometimes for a degenerate network, the matrix state equation cannot be put in the normal form of (3.5.2) if capacitor voltages and inductor currents are routinely chosen as state variables. Instead, the matrix state equation takes the form

$$
\dot{\mathbf{x}} = \mathbf{A}\mathbf{x} + \mathbf{B}\mathbf{u} + \mathbf{K}\dot{\mathbf{u}}. \tag{3.7.17}
$$

This can be accepted as the *modified* normal form of the state equations since it may be rewritten as

$$
\dot{\mathbf{x}} = \mathbf{A}\mathbf{x} + \mathbf{B}'\mathbf{u}, \tag{3.7.18}
$$

where

$$
\mathbf{B}' = \mathbf{B} + \mathbf{K}p \tag{3.7.19}
$$

with $\mathbf{K}$ being a matrix of constant coefficients, and

$$
p \equiv \frac{d}{dt} \tag{3.7.20}
$$

being the *linear differential operator* such that for any (matrix) function $\mathbf{F}(t)$ of time t

$$
p\mathbf{F} = \frac{d}{dt}\mathbf{F} \equiv \dot{\mathbf{F}}. \tag{3.7.21}
$$

Thus, it is evident from (3.7.9) through (3.7.21) that

$$
\mathbf{B}'\mathbf{u} = (\mathbf{B} + \mathbf{K}p)\mathbf{u} = \mathbf{B}\mathbf{u} + \mathbf{K}\dot{\mathbf{u}}
$$

and we see that (3.7.18) is identical to (3.7.17). If, however, one should insist on considering (3.5.2) as the only acceptable normal form of the state equations, the derivative term $\mathbf{K}\dot{\mathbf{u}}$ in (3.7.17) can be eliminated by redefining a new set of state variables as illustrated in the next example.

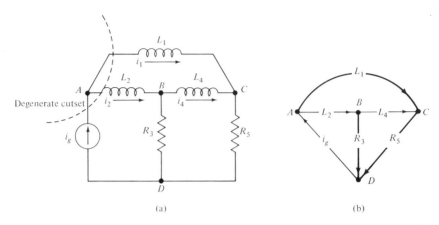

Fig. 3.7.3 (a) A degenerate RL network N with a degenerate cutset; (b) the oriented graph G of N.

Example 3.7.3 Consider the degenerate RL network N of Fig. 3.7.3 (a), the oriented graph G of which is shown in Fig. 3.7.3 (b). The tree T consisting of L_1, R_3, and R_5, is selected for the formulation of state equations.

Note that there exists a *degenerate cutset* at node A (namely a cutset consisting of only inductors and/or current sources) consisting of i_g, L_1, and L_2 as shown by the dotted line in Fig. 3.6.3 (a). We first write the VCR-equations as follows.

$$
\text{(a)}\quad
\begin{cases}
i_2 = \dfrac{1}{L_2}\, v_2, \\[2mm]
i_4 = \dfrac{1}{L_4}\, v_4;
\end{cases}
\tag{3.7.22}
$$

$$
\text{(b)}\quad
\begin{cases}
v_1 = L_1 \dot{i}_1, \\[1mm]
v_3 = R_3 i_3, \\[1mm]
v_5 = R_5 i_5.
\end{cases}
\tag{3.7.23}
$$

Next, the KVL- and KCL-equations are given by

$$
\text{(a)}\quad
\begin{cases}
v_2 = v_1 - v_3 + v_5, \\[1mm]
v_4 = v_3 - v_5;
\end{cases}
\tag{3.7.24}
$$

$$
\text{(b)}\quad
\begin{cases}
i_1 = -i_2 + i_g, \\[1mm]
i_3 = i_2 - i_4, \\[1mm]
i_5 = -i_2 + i_4 + i_g.
\end{cases}
\tag{3.7.25}
$$

Now, for the elimination of nonstate variables, we obtain from (3.7.24), (3.7.25), and (3.7.23)

$$v_2 = v_1 - v_3 + v_5$$
$$= L_1\dot{i}_1 - R_3i_3 + R_5i_5$$
$$= L_1(-\dot{i}_2 + \dot{i}_g) - R_3(i_2 - i_4) + R_5(-i_2 + i_4 + i_g)$$

or

$$v_2 = -(R_3 + R_5)i_2 + (R_3 + R_5)i_4 + R_5i_g + L_1\dot{i}_g - L_1\dot{i}_2. \qquad (3.7.26)$$

Similarly,

$$v_4 = v_3 - v_5$$
$$= R_3i_3 - R_5i_5$$
$$= R_3(i_2 - i_4) - R_5(-i_2 + i_4 + i_g),$$

or

$$v_4 = (R_3 + R_5)i_2 - (R_3 + R_5)i_4 - R_5i_g. \qquad (3.7.27)$$

Substituting (3.7.26) and (3.7.27) into (3.7.22) and simplifying, we obtain

$$\begin{bmatrix} \dot{i}_2 \\ \dot{i}_4 \end{bmatrix} = \begin{bmatrix} \dfrac{-1}{L_1 + L_2}(R_3 + R_5) & \dfrac{R_3 + R_5}{L_1 + L_2} \\ \dfrac{1}{L_4}(R_3 + R_5) & \dfrac{-1}{L_4}(R_3 + R_5) \end{bmatrix} \begin{bmatrix} i_2 \\ i_4 \end{bmatrix} + \begin{bmatrix} \dfrac{R_5}{L_1 + L_2} \\ -\dfrac{R_5}{L_4} \end{bmatrix} [i_g]$$
$$+ \begin{bmatrix} \dfrac{L_1}{L_1 + L_2} \\ 0 \end{bmatrix} [\dot{i}_g], \qquad (3.7.28)$$

which is seen to be in the form of (3.7.17) with

$$\mathbf{x} = \begin{bmatrix} i_2 \\ i_4 \end{bmatrix} \qquad \text{so that} \qquad \dot{\mathbf{x}} = \begin{bmatrix} \dot{i}_2 \\ \dot{i}_4 \end{bmatrix},$$

$$\mathbf{A} = \begin{bmatrix} \dfrac{-1}{L_1 + L_2}(R_3 + R_5) & \dfrac{1}{L_1 + L_2}(R_3 + R_5) \\ \dfrac{1}{L_4}(R_3 + R_5) & -\dfrac{1}{L_4}(R_3 + R_5) \end{bmatrix}, \qquad (3.7.29)$$

$$\mathbf{B} = \begin{bmatrix} \dfrac{R_5}{L_1 + L_2} \\ -\dfrac{R_5}{L_4} \end{bmatrix}, \qquad \mathbf{K} = \begin{bmatrix} \dfrac{L_1}{L_1 + L_2} \\ 0 \end{bmatrix}$$

and

$$\mathbf{u} = [i_g] \qquad \text{so that} \qquad \dot{\mathbf{u}} = [\dot{i}_g].$$

If we rewrite (3.7.27) by combining the last two terms of its right-hand member in light of (3.7.19) through (3.7.21), we have

$$
\begin{bmatrix} \dot{i}_2 \\ \dot{i}_4 \end{bmatrix}
=
\begin{bmatrix}
\dfrac{-1}{L_1 + L_2}(R_3 + R_5) & \dfrac{1}{L_1 + L_2}(R_3 + R_5) \\
\dfrac{1}{L_4}(R_3 + R_5) & -\dfrac{1}{L_4}(R_3 + R_5)
\end{bmatrix}
\begin{bmatrix} i_2 \\ i_4 \end{bmatrix}
$$

$$
+
\begin{bmatrix}
\left(\dfrac{R_5}{L_1 + L_2} + \dfrac{L_1}{L_1 + L_2} p \right) \\
-\dfrac{R_5}{L_4}
\end{bmatrix}
[i_g]
$$

which is in the form of (3.7.18) with

$$
\mathbf{B'} =
\begin{bmatrix}
\left(\dfrac{R_5}{L_1 + L_2} + \dfrac{L_1}{L_1 + L_2} p \right) \\
-\dfrac{R_5}{L_4}
\end{bmatrix}.
$$

Finally, if we define the new state variables as $x_1 = i_2 - \{L_1/(L_1 + L_2)\}i_g$, $x_2 = i_4$, we can express the old state variables (i_2 and i_4) in terms of the new ones (x_1 and x_2) as

$$
\begin{bmatrix} i_2 \\ i_4 \end{bmatrix}
=
\begin{bmatrix} x_1 \\ x_2 \end{bmatrix}
+
\begin{bmatrix} \dfrac{L_1}{L_1 + L_2} \\ 0 \end{bmatrix}
[i_g]. \tag{3.7.30}
$$

We can then rewrite (3.7.27) in terms of the new state variables by substituting (3.7.30) and its derivative into (3.7.28) and simplifying as

$$
\begin{bmatrix} \dot{x}_1 \\ \dot{x}_2 \end{bmatrix}
=
\begin{bmatrix}
\dfrac{-1}{(L_1 + L_2)}(R_3 + R_5) & \dfrac{R_3 + R_5}{L_1 + L_2} \\
\dfrac{1}{L_4}(R_3 + R_5) & \dfrac{-1}{L_4}(R_3 + R_5)
\end{bmatrix}
\begin{bmatrix} x_1 \\ x_2 \end{bmatrix}
$$

$$
+
\begin{bmatrix}
\dfrac{-L_1(R_3 + R_5)}{(L_1 + L_2)^2} + \dfrac{R_5}{L_1 + L_2} \\
\dfrac{L_1(R_3 + R_5)}{L_4(L_1 + L_2)} - \dfrac{R_5}{L_4}
\end{bmatrix}
[i_g]. \tag{3.7.31}
$$

$$V_4 = (R_3 + R_5)i_2 - (R_3 + R_5)i_4 - R_5 i_g$$

$$V_4 = (R_3 + R_5)\left(x_1 + \frac{L_1}{L_1 + L_2}\right)i_g - (R_3 + R_5)I_2 - R_5 - R_5 i_g$$

Equation (3.7.31) is the normal form as given by (3.5.2) with matrix $\mathbf{A}$ unchanged as given in (3.7.28) and matrix $\mathbf{B}$ to be a 2×1-matrix (or a 2-vector) as

$$\mathbf{B} = \begin{bmatrix} -\dfrac{L_1(R_3 + R_5)}{(L_1 + L_2)^2} + \dfrac{R_5}{(L_1 + L_2)} \\[3mm] \dfrac{L_1(R_3 + R_5)}{L_4(L_1 + L_2)} - \dfrac{R_5}{L_4} \end{bmatrix}.$$

3.8 SUMMARY

In the preceding sections of this chapter, we first developed the two conventional methods in network analysis and showed that both the loop and the node equations are derived from the three postulates: the VCR-equations, the KVL-equations, and the KCL-equations. The steps involved in both analysis methods can best be summarized by means of Table 3.8.1.

Table 3.8.1 not only summarizes the steps involved in the developments of the loop and node analyses but also reveals the striking parallelism between them— that is, in each step of one of the two methods, there exists a counterpart step in the other method.

Obviously in the node analysis in Table 3.8.1, it has been assumed that the inverse $\mathbf{Z}_e^{-1}(s)$ of the given network exists for the sake of brevity. In the case when $\mathbf{Z}_e^{-1}(s)$ does not exist, source transformations should be applied to modify the network as discussed in Section 3.4.

If, for a given network N, cutset analysis is used instead of node analysis, one can easily derive the cutset-analysis matrix equation following a development similar to that used in deriving (3.3.5). In fact, it can be readily shown* that in the development of node analysis, if the fundamental cutset matrix $\mathbf{Q}_f$ is used in the place of the incidence matrix $\mathbf{A}$, (3.3.5) becomes

$$\mathbf{Y}_k(s)\mathbf{V}_b(s) = \mathbf{Q}_f \mathbf{Z}_e^{-1}(s)\{\mathbf{E}_e(s) - \mathbf{L}_e \mathbf{i}_e(0^+) + \frac{1}{s}\mathbf{v}_e(0^+)\}, \qquad (3.3.5a)$$

where

$$\mathbf{Y}_k(s) \equiv \mathbf{Q}_f \mathbf{Z}_e^{-1}(s)\mathbf{Q}_f^T \qquad (3.3.6a)$$

is called the *cutset admittance matrix*. Equation (3.3.5a), together with (3.3.6a), is the desired *cutset-analysis matrix equation* which represents the n independent cutset equations of N.

In Sections 3.5 through 3.7, the state equations and their formulation were discussed. It was shown in these sections that the state equations are also derived from the three postulates. An algorithm for formulating the state equations was developed in Section 3.7, the steps of which are now summarized in Table 3.8.2.

---

* See Problem 3.21.

Table 3.8.1 A Summary of Development of Loop and Node Analyses

---

A. *Primary systems of equations (postulates)*

    1. a) The VCR-equations:

$$V_e(s) = E_e(s) + Z_e(s)I_e(s) + \frac{1}{s}v_e(0^+) - L_e i_e(0^+) \qquad (3.2.8)$$

        or

$$I_e(s) = Z_e^{-1}(s)\left\{V_e(s) - E_e(s) + L_e i_e(0^+) - \frac{1}{s}V_e(0^+)\right\} \qquad (3.3.1)$$

    2. The KVL-equations: $BV_e(s) = 0$

    3. The KCL-equations: $AI_e(s) = 0$

B. *Secondary systems of equations (auxiliary equations)*

    4. a) Mesh transformation: $I_e(s) = B^T I_m(s)$ (for planar networks only)   (3.2.12)
       b) or loop transformation: $I_e(s) = B_f^T I_c(s)$ (general)   (3.2.18)

    5. a) Node transformation: $V_e(s) = A^T V_n(s)$ (3.3.3)
       b) (or cutset transformation:

$$V_e(s) = Q_f^T V_b(s) \qquad (3.3.3a)$$

    where $V_b(s)$ is the vector of tree-branch voltages with respect to the tree, T)

Step	*Loop analysis**	*Node analysis (Assume Z_e^{-1} exists)*
a)	From (1a) and (2) obtain	From (1b) and (3) obtain
	6. $B_f Z_e I_e = -B_f\left\{E_e + \dfrac{1}{s}v_e(0^+)\right.$	8. $AZ_e^{-1}V_e = AZ_e^{-1}\left\{E_e + \dfrac{1}{s}v_e(0^+)\right.$
	$\left. - L_e i_e(0^+)\right\}.$ (3.2.8)	$\left. - L_e i_e(0^+)\right\}.$ (3.3.2)
b)	Substitute (4b) into (6) to get	Substitute (5a) into (8) to get
	7. $Z_l I_c = -B_f\left\{E_e + \dfrac{1}{s}v_e(0^+)\right.$ (3.2.14a)	9. $Y_n V_n = AZ_e^{-1}\left\{E_e + \dfrac{1}{s}v_e(0^+)\right.$ (3.3.5)
	$\left. - L_e i_e(0^+)\right\}$	$\left. - L_e i_e(0^+)\right\}$
	where	where
	$Z_l \equiv B_f Z_e B_f^T.$ (3.2.17)	$Y_n \equiv AZ_e^{-1}A^T.$ (3.3.6)

---

* In the case of mesh analysis (for planar networks only), replace B by B_f in (6) and (7). Also in Step (b), use (4a) to obtain (3.2.14) instead of using (4b) to obtain (3.2.14a).

Table 3.8.2 Algorithm for Formulating State Equations

Step	Operation
1	Obtain the oriented graph G from the given network N.
2	Select a tree T from G such that T contains: a) all of the voltage sources, b) none of the current sources, c) as many capacitors as possible, d) as few inductors as possible.
3	Write the VCR-equations for the passive elements and separate them into two groups (a) and (b) such that: Group (a) contains the VCR-equations for the tree-branch capacitors and chord inductors with respect to tree T; Group (b) contains the VCR-equations for all other passive elements.
4	Write the f-circuit (KVL) and the f-cutset (KCL) equations and separate them in the same manner as done for the VCR-equations in Step 3.
5	Eliminate all of the nonstate variables, which are those variables other than the state variables, and the (known) sources (i.e., they are the tree-branch currents and chord voltages), by substituting the equations in Step 3.
6	Rearrange and simplify the terms and express the resultant equations in the matrix normal form $$\dot{\mathbf{x}} = \mathbf{A}\mathbf{x} + \mathbf{B}\mathbf{u} \qquad (3.5.2)$$ or in the modified normal form (for some degenerate cases) $$\dot{\mathbf{x}} = \mathbf{A}\mathbf{x} + \mathbf{B'}\mathbf{u} \qquad (3.7.18)$$ or, redefine new state variables as illustrated in Example 3.7.3 so that (3.7.18) may be rewritten to be in the same form as (3.5.2) after the new state variables replace the old ones.

Finally it is worth noting that while capacitor voltages and inductor currents are usually chosen for convenience as state variables, such a choice is by no means a unique one since other choices for state variables are certainly possible. For example, linear combinations of capacitor voltages and/or inductor currents may be taken as state variables for a given network, in which case a new set of state equations will result with different $\mathbf{A}$, $\mathbf{B}$, $\mathbf{C}$, $\mathbf{D}$ matrices.*

---

* See Problem 3.20.

REFERENCES

1. S. P. Chan, *Introductory Topological Analysis of Electrical Networks*, Holt, Rinehart & Winston, 1969, Chapter 4.

2. S. Seshu and M. B. Reed, *Linear Graphs and Electrical Networks*, Addison-Wesley, 1961, Chapters 5 and 6.

3. B. C. Kuo, *Linear Networks and Systems*, McGraw-Hill, 1967, Chapters 4, 5, and 6.

4. J. B. Cruz, Jr. and M. E. Van Valkenburg, *Introductory Signals and Circuits*, Blaisdell, 1967, Chapters 5, 12, and 13.

5. C. A. Desoer and E. S. Kuh, *Basic Circuit Theory*, McGraw-Hill, 1966, Chapters 9 and 10.

6. W. A. Blackwell, *Mathematical Modeling of Physical Networks*, Macmillan, 1968, Chapters 6, 8, and 9.

7. H. E. Koenig, Y. Tokad, and H. K. Kesavan, *Analysis of Discrete Physical Systems*, McGraw-Hill, 1967, Chapters 4 and 5.

PROBLEMS

3.1 Obtain the loop equations in matrix form for the network shown in Fig. P.3.1.

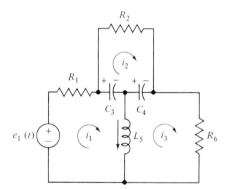

Figure P.3.1

3.2 Obtain the node equations in matrix form for the network shown in Fig. P.3.2.

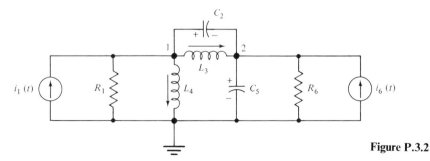

Figure P.3.2

3.3 For the coupled network of Fig. P.3.3, write the loop equations in matrix form, using the three loop currents specified in the figure. (*Hint*: Transform the parallel combination of $i_g(t)$ and R_3 into a series combination by the application of the current-to-voltage transformation.)

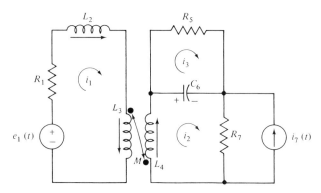

Figure P.3.3

3.4 Repeat Problem 3.1 for the network of Fig. P.3.4.

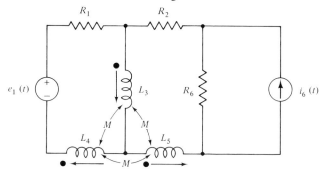

$$M_{34} = M_{35} = M_{45} = M$$

Figure P.3.4

3.5 Repeat Problem 3.2 for the same network, assuming a mutual inductance M between L_3 and L_4 with the dot of L_4 at node 1 and the dot of L_3 at node 2.

3.6 For the network shown in Fig. P.3.6, obtain the node equations by first applying the I-shift transformation to the 10-A current source, and the E-shift transformation to the 10-V voltage source, and express the node equations in matrix form.

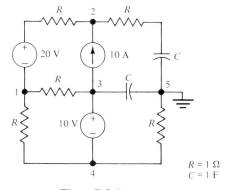

Figure P.3.6

3.7 Using both the *I*-shift and the *E*-shift transformations, obtain (a) the loop equations, and (b) the node equations for the network shown in Fig. P.3.7. Express both sets of equations in matrix form.

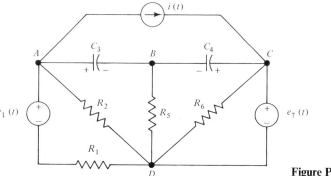

Figure P.3.7

3.8 Prove that for a given tree *T*, the edge currents in a given (connected) network can always be expressed in terms of the chord currents. That is, derive the following loop-transformation matrix equation (3.2.18):

$$\mathbf{i}_e = \mathbf{B}_f^T \mathbf{i}_c,$$

where $\mathbf{B}_f^T$ is the transpose of the fundamental circuit matrix $\mathbf{B}_f$ with respect to tree *T*.

3.9 Prove that the edge voltages in a given (connected) network can always be expressed in terms of the tree-branch voltages for a given tree *T*. That is, derive the following cutset-transformation matrix equation:

$$\mathbf{v}_e = \mathbf{Q}_f^T \mathbf{v}_b,$$

where $\mathbf{Q}_f^T$ is the transpose of the fundamental cutset matrix $\mathbf{Q}_f$ with respect to tree *T*.

3.10 Using arguments similar to those used in Problem 3.8, derive Eq. (3.2.12).

3.11 Using arguments similar to those used in Problem 3.9, derive Eq. (3.3.3).

3.12 Obtain the state equations for the network shown in Fig. P.3.12 and express the equations in standard matrix form, viz., $\dot{\mathbf{x}} = \mathbf{A}\mathbf{x} + \mathbf{B}\mathbf{u}$.

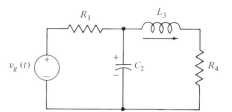

Figure P.3.12

3.13 Repeat Problem 3.12 for the network in Fig. P.3.1.

3.14 Repeat Problem 3.12 for the network in Fig. P.3.2.

3.15 Repeat Problem 3.12 for the network in Fig. P.3.3.

3.16 Applying the algorithm outlined in Section 3.7 (also see Table 3.8.2), obtain the state equations in standard matrix form, $\dot{\mathbf{x}} = \mathbf{A}\mathbf{x} + \mathbf{B}\mathbf{u}$, for the network of Fig. P.3.2.

3.17 Repeat Problem 3.16 for the network of Fig. P.3.1.

3.18 Repeat Problem 3.16 for the network shown in Fig. P.3.18.

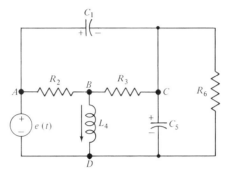

Figure P.3.18

3.19 Repeat Problem 3.16 for the network shown in Fig. P.3.19.

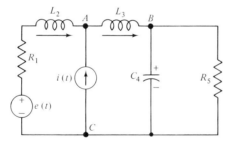

Figure P.3.19

3.20 a) Repeat Problem 3.16 for the network shown in Fig. P.3.20.

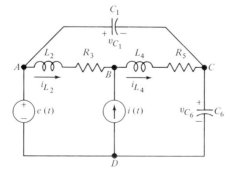

Figure P.3.20

b) Show that the choice of state variables is not unique by choosing the following new set of state variables

$$x_1 = v_{C_1} + v_{C_6}, \qquad x_2 = v_{C_1} - v_{C_6}, \qquad x_3 = i_{L_2} + i_{L_4}, \qquad x_4 = i_{L_2} - i_{L_4},$$

and then obtaining a new set of state equations (in normal form) with different $\mathbf{A}$ and $\mathbf{B}$ matrices from those obtained in (a).

3.21 Derive (3.3.5a) for the cutset analysis following a development similar to that used in deriving (3.3.5). [*Hint*: Use (3.3.3a) in Table 3.8.1.]

CHAPTER 4

NETWORK THEOREMS AND
TRANSFORM NETWORKS

4.1 INTRODUCTION

In the preceding chapter, we introduced three different sets of network equations: the loop, the node, and the state equations. Each set of these equations completely characterizes the behavior of a network. Upon solving these equations, we can easily determine the response at any point of the network.

Different methods can be used to obtain the solution of a network problem. In general, there is no rule of thumb for determining the best method of solution for a given network problem, since, depending upon the nature of the problem itself, one method may be best for one problem but not for another. Fortunately, however, some analytical means do exist and can be applied to very large classes of networks encountered in practice so that their solutions can be greatly simplified. These analytical means are given in the form of network theorems, which are the main topics of this chapter.

In the following section we shall discuss several very useful network theorems: Tellegen's theorem, the superposition theorem, the reciprocity theorem, Thévenin's theorem, Norton's theorem, and the substitution theorem. Different examples are used to demonstrate their applications. In the later part of this chapter we shall study the transformation of a network from the time domain to the frequency domain. The importance of this transformation lies in the fact that, instead of obtaining the solution of a given network problem in the time domain by solving a set of integro-differential equations, the same solution can be attained by solving a set of algebraic equations in the frequency domain.

4.2 NETWORK THEOREMS

As mentioned in the preceding section, the network theorems under consideration are very useful and are applicable to a wide spectrum of networks encountered in practice. In the subsequent presentation we shall study these network theorems separately and state the conditions under which each theorem can be applied.

Tellegen's Theorem

The first network theorem to be studied is the theorem, introduced by Tellegen,* which can be applied to any lumped-parameter network with the only restriction

---

* B. D. H. Tellegen, "A General Network Theorem, With Applications," Philips Research Reports, **7**, 1952, pp. 259–269, [TE 1].

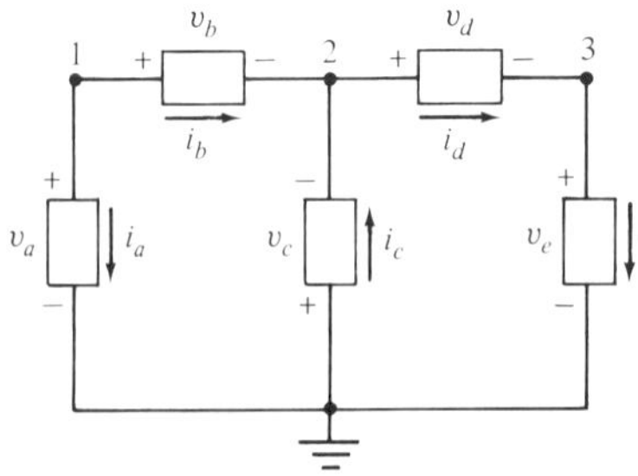

Fig. 4.2.1 Network used in Example 4.2.1.

that the two Kirchhoff's laws (KCL and KVL) be satisfied. The theorem can be stated as follows:

Theorem 4.2.1 (*Tellegen*) Let N be a lumped network of b branches and n nodes. If KVL is satisfied by the branch voltages $v_1, v_2, \ldots, v_b$ in every loop and if KCL is satisfied by the branch currents $i_1, i_2, \ldots, i_b$ at every node, then

$$\sum_{k=1}^{b} v_k i_k = 0, \tag{4.2.1}$$

where the direction of the current i_k and the polarity marks of the voltage v_k associated with the branch k follow the established convention discussed in Section 2.3.

To understand the essence of the theorem more fully before attempting to prove it, we shall consider the following example:

Example 4.2.1   Consider the network of Fig. 4.2.1. If v_1, v_2, and v_3 denote the voltages at the nodes 1, 2, and 3, respectively, then, by means of KVL, the branch voltages can be written in terms of the node voltages as

$$\begin{aligned} v_a &= v_1, & v_b &= v_1 - v_2, & v_c &= -v_2 \\ v_d &= v_2 - v_3, & v_e &= v_3. \end{aligned} \tag{4.2.2}$$

Let P be the sum of the products of all the branch voltages and their associated branch currents. Then, with the aid of (4.2.2), we find

$$
\begin{aligned}
P &\triangleq v_a i_a + v_b i_b + v_c i_c + v_d i_d + v_e i_e \\
&= (v_1)i_a + (v_1 - v_2)i_b + (-v_2)i_c + (v_2 - v_3)i_d + (v_3)i_e \tag{4.2.3} \\
&= \sum_{i=1}^{3} v_i \cdot (\text{algebraic sum of all currents leaving node } i).
\end{aligned}
$$

Observe that, in the last expression of (4.2.3), the coefficient associated with each node voltage is simply the algebraic sum of all the branch currents directed away from that node. The KCL demands that all the coefficients of the node voltages in (4.2.3) be zero, and hence

$$P = 0 \tag{4.2.4}$$

which is, of course, the consequence of Tellegen's theorem.

By means of this example it is not difficult to see that, for a lumped network of b branches and n nodes, if we denote the node-to-datum voltage by $v_{in}(i = 1, 2, \ldots, n - 1)$, then

$$\sum_{k=1}^{b} v_k i_k = v_{1n} \cdot (\text{algebraic sum of all currents leaving node 1})$$

$$+ v_{2n} \cdot (\text{algebraic sum of all currents leaving node 2})$$

$$+ \cdots$$

$$+ v_{(n-1)n} \cdot (\text{algebraic sum of all currents leaving node } n - 1)$$

which by KCL reduces to zero, or

$$\sum_{k=1}^{b} v_k i_k = 0. \tag{4.2.5}$$

In fact, an elegant proof of this theorem may be accomplished as follows. First, we observe that

$$\sum_{k=1}^{b} v_k i_k = \mathbf{I}_e^T \mathbf{V}_e,$$

with $b \equiv e$. Then, by (3.2.12) and (3.3.3), we write

$$\sum_k v_k i_k = \mathbf{I}_e^T \mathbf{V}_e = (\mathbf{I}_m^T \mathbf{B})(\mathbf{A}^T \mathbf{V}_n) = \mathbf{I}_m^T (\mathbf{A} \cdot \mathbf{B}^T)^T \mathbf{V}_n$$

which is identically zero since $\mathbf{A} \cdot \mathbf{B}^T = 0$ in light of (2.5.2).

Remarks. A careful consideration of the above development indicates the following important facts:

1. In deriving Tellegen's theorem, only two laws of Kirchhoff are used. Since the specification of the branch elements is not required, Tellegen's theorem is applicable to any network, linear or nonlinear, passive or active, time invariant or time varying so long as Kirchhoff's laws are not violated.

2. Since the branch characteristics or v-i relationships were not used in deriving Tellegen's theorem, (4.2.5) is still applicable even if the set of voltages and the corresponding set of currents are evaluated at two different time instants t_1 and $t_2(t_1 \neq t_2)$. That is,

$$\sum_{k=1}^{b} v_k(t_1) i_k(t_2) = 0. \tag{4.2.6}$$

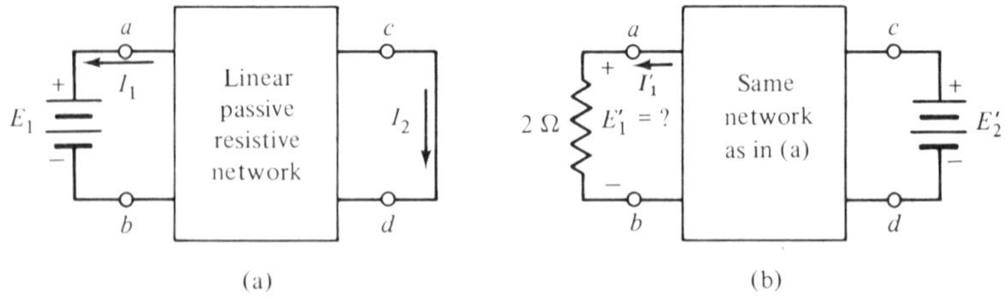

Fig. 4.2.2 Networks used in Example 4.2.2.

3. A slight generalization of the above discussion leads to an even more interesting result which can be stated as follows: Let N_1 and N_2 be two different networks of b branches and n nodes such that they have the same oriented graph G. Let $v_1, v_2, \ldots, v_b$ and $v'_1, v'_2, \ldots, v'_b$ be the branch voltages of N_1 and N_2, respectively. Thus, the corresponding branch currents of N_1 and N_2 are, respectively, given by $i_1, i_2, \ldots, i_b$ and $i'_1, i'_2, \ldots, i'_b$. By applying Tellegen's theorem, we obtain

$$\sum_{j=1}^{b} v_j(t_1) i'_j(t_2) = 0 \tag{4.2.7}$$

and

$$\sum_{j=1}^{b} v'_j(t_1) i_j(t_2) = 0, \tag{4.2.8}$$

where the time instants t_1 and t_2 may or may not be distinct.

Let us apply Tellegen's theorem to solve the following problem.

Example 4.2.2   As indicated in Fig. 4.2.2 (a), when the voltage E_1 is 20 V, the current I_1 is -10 A, and the current I_2 is 2 A. If a voltage source of $E'_2 = 10$ V is connected across terminals c–d, and a 2-Ω resistance is placed in terminals a–b as indicated in Fig. 4.2.2 (b), what voltage will occur across the 2-Ω resistance? By means of (4.2.7), we write

$$\sum_{j=1}^{b} E_j I'_j = 0, \tag{4.2.9}$$

which can be expanded as

$$E_1 I'_1 + E_2 I'_2 + \sum_{j=3}^{b} E_j I'_j = 0$$

or

$$E_1 I'_1 + E_2 I'_2 = -\sum_{j=3}^{b} R_j I_j I'_j, \tag{4.2.10}$$

where R_j is the resistance of the jth branch. In like manner, (4.2.8) can be written as

$$E_1'I_1 + E_2'I_2 = -\sum_{j=3}^{b} R_j I_j' I_j. \tag{4.2.11}$$

Equating (4.2.10) and (4.2.11), we obtain

$$E_1 I_1' + E_2 I_2' = E_1' I_1 + E_2' I_2. \tag{4.2.12}$$

Substitution of the numerical values in (4.2.12) yields

$$20 I_1' = E_1'(-10) + 10(2)$$

or

$$2 I_1' + E_1' = 2. \tag{4.2.13}$$

Equation (4.2.13), together with $I_1' = (E_1'/2)$, gives the solution

$$E_1' = 1 \text{ V}.$$

The Superposition Theorem

The usefulness of this theorem lies in dividing a complex network problem into a set of simpler problems. The response of the given network can be obtained more easily by then combining the solutions of these simpler problems. This theorem can be applied to a very large class of networks since the only restriction is that the network under consideration be linear. The network can even be time varying. Of course, the effectiveness of the theorem in terms of saving time and effort depends mainly upon the topology of the network under investigation. We shall state the theorem in the following manner:

Theorem 4.2.2 (*Superposition*) At any point of a *linear* network, the response to a number of excitations (voltage and/or current sources) acting simultaneously is equal to the algebraic sum of the responses at that point due to each excitation acting alone in the network.

Remark. The work "excitation" in the above theorem includes not only the independent sources, but also the initial condition sources. For example, a capacitor C initially charged to a voltage of V_0 volts is equivalent to the same capacitor C (initially uncharged) in series with an independent voltage source of V_0 volts. This voltage source is referred to as an initial condition source.

The proof of the theorem is immediate. For convenience, let us assume that the response is a mesh current, say $I_{mi}(s)$. Then, by means of (3.2.14), the mesh current vector $\mathbf{I}_m(s)$ is given by

$$\mathbf{I}_m(s) = -\{\mathbf{Z}_m(s)\}^{-1}\mathbf{B}\left\{\mathbf{E}_e(s) + \frac{1}{s}\mathbf{v}_e(0^+) - \mathbf{L}_e\mathbf{i}_e(0^+)\right\}. \tag{4.2.14}$$

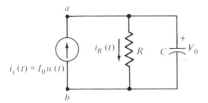

Fig. 4.2.3 An RC network.

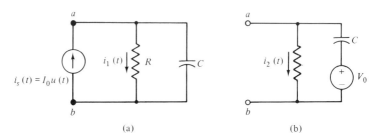

(a)   (b)

Fig. 4.2.4 Application of the superposition theorem to the network of Fig. 4.2.3.

If we let

$$[y_{ij}] \triangleq \{\mathbf{Z}_m(s)\}^{-1},\tag{4.2.15}$$

then, in terms of components, $I_{mi}(s)$ becomes

$$I_{mi}(s) = -\sum_{j=1}^{n}\sum_{k=1}^{e} y_{ij} b_{jk}\left(E_k + \frac{1}{s}v_k(0^+) - L_k i_k(0^+)\right) \qquad i = 1, 2, \ldots, n \tag{4.2.16}$$

which completes the proof.

Example 4.2.3 Determine the current $i_R(t)$ for $t > 0$ of the network shown in Fig. 4.2.3.

The current response $i_R(t)$ can be regarded as the sum of two components $i_1(t)$ and $i_2(t)$ due to sources $i_s(t)$ and V_0 acting alone as indicated in Fig. 4.2.4. Simple manipulations lead to

$$i_1(t) = I_0(1 - e^{-t/RC})u(t)$$

and

$$i_2(t) = \frac{V_0}{R} e^{-t/RC}u(t).$$

Therefore, the response current $i_R(t)$ becomes

$$i_R(t) = i_1(t) + i_2(t)$$

$$= \left[I_0(1 - e^{-t/RC}) + \frac{V_0}{R} e^{-t/RC}\right]u(t).$$

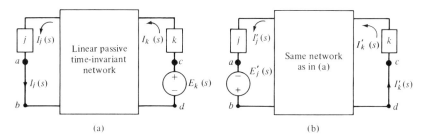

Fig. 4.2.5 Illustration of the reciprocity theorem.

The Reciprocity Theorem

A very useful property of linear time invariant passive networks is provided by the theorem of reciprocity. The property of reciprocity is powerful in the sense that it can be applied to a wide class of practical problems and is also very useful in theoretical investigations. Since the restrictions under which the theorem can be applied are not always clearly understood, it would be advantageous to itemize them before stating the theorem:

1. The network is linear and time invariant.
2. The network is initially at rest.
3. The network contains only one source which may be a voltage or current but must be an independent source.
4. The elements in the network other than the source must be restricted to R, L, C, and M only.
5. As the consequence of (3) and (4), dependent sources are not allowed in the network.

If all the conditions stated above are met, the theorem of reciprocity, which is to be stated next, can then be applied.*

Theorem 4.2.3 (*Reciprocity*) The ratio of the response transform observed at one point to the excitation transform applied at another point, is *invariant* to an interchange of positions of observation and excitation as long as the topology of the network is unaltered.

Let us verify this theorem for the special case as shown in Fig. 4.2.5 (a) in which the network is linear and time invariant with the points of observation and

---

* Actually, a linear active network *can* be reciprocal and a linear passive network *can* be nonreciprocal. For example, a linear passive network which contains (passive) gyrators (see Section 6.6) is a nonreciprocal network.

excitation indicated as j and k, respectively. Since Condition 2 above dictates that the network has zero initial conditions, the transform mesh equations for an n-mesh network are given by

$$Z_{11}(s)I_1(s) + Z_{12}(s)I_2(s) + \cdots + Z_{1j}(s)I_j(s) + \cdots + Z_{1n}(s)I_n(s) = 0$$

$$Z_{21}(s)I_1(s) + Z_{22}(s)I_2(s) + \cdots + Z_{2j}(s)I_j(s) + \cdots + Z_{2n}(s)I_n(s) = 0$$

$$\vdots$$

$$Z_{k1}(s)I_1(s) + Z_{k2}(s)I_2(s) + \cdots + Z_{kj}(s)I_j(s) + \cdots + Z_{kn}(s)I_n(s) = E_k(s) \qquad (4.2.17)$$

$$\vdots$$

$$Z_{n1}(s)I_1(s) + Z_{n2}(s)I_2(s) + \cdots + Z_{nj}(s)I_j(s) + \cdots + Z_{nn}(s)I_n(s) = 0$$

where $Z_{rs}(s)$ $(r, s = 1, 2, \ldots, n)$ are the elements of the mesh impedance matrix $\mathbf{Z}_m(s)$. If we denote the determinant of $\mathbf{Z}_m(s)$ and the (k, j)-cofactor of $\mathbf{Z}_m(s)$ by $\Delta(s)$ and $\Delta_{kj}(s)$, respectively, we can solve (4.2.17) for $I_j(s)$ by applying Cramer's rule,

$$I_j(s) = \frac{\Delta_{kj}(s)}{\Delta(s)} E_k(s), \qquad (4.2.18)$$

which yields the ratio of response to excitation

$$\frac{I_j(s)}{E_k(s)} = \frac{\Delta_{kj}(s)}{\Delta(s)}. \qquad (4.2.19)$$

Now if we place a source $E_j'(s)$ [which may not be equal to $E_k(s)$] between the terminal-pair* a–b and observe the response $I_k'(s)$ through the terminal-pair c–d as indicated in Fig. 4.2.5 (b), we find, in like manner, that

$$I_k'(s) = \frac{\Delta_{jk}(s)}{\Delta(s)} E_j'(s) \qquad (4.2.20)$$

or

$$\frac{I_k'(s)}{E_j'(s)} = \frac{\Delta_{jk}(s)}{\Delta(s)}. \qquad (4.2.21)$$

Since the network considered must contain only R, L, C, and M elements, the mesh impedance matrix $\mathbf{Z}_m(s)$ is symmetric leading to the equality

$$\Delta_{jk}(s) = \Delta_{kj}(s). \qquad (4.2.22)$$

---

* A one-terminal-pair network or one-port is a network with a pair of nodes called terminals accessible for external measurements such that the current entering one terminal is the same as that leaving the other. A detailed discussion of multi-terminal-pair networks or n-ports will be given in Chapter 6.

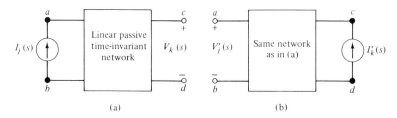

Fig. 4.2.6 Illustration of the reciprocity theorem.

By means of (4.2.19), (4.2.21), and (4.2.22), we establish the reciprocity property; that is,

$$\frac{I_j(s)}{E_k(s)} = \frac{I_k'(s)}{E_j'(s)}. \tag{4.2.23}$$

Remark. It should be emphasized that, when the excitation is a voltage and the response is a current, then the short-circuit constraints are implied at both the points of observation and excitation as evidenced in Fig. 4.2.5 (a), since an independent voltage is regarded as a generalized short circuit. After an interchange of the position of the excitation and the response as indicated in Fig. 4.2.5 (b), we still have the same short-circuit constraints at both points of observation and excitation, leaving the topology of the network invariant.

Now, let us turn to the case in which the excitation is a current and the response is a voltage as depicted in Fig. 4.2.6 (a). If an interchange of points of excitation and observation is made, we have the result as shown in Fig. 4.2.6 (b). A similar procedure on the node basis leads to

$$\frac{V_k(s)}{I_j(s)} = \frac{V_j'(s)}{I_k'(s)}. \tag{4.2.24}$$

another version of the reciprocity theorem. Again, it should be pointed out that both networks (a) and (b) of Fig. 4.2.6 have exactly the same terminal constraints; that is, open-circuit conditions at both terminal-pairs *a–b* and *c–d*. In other words, the interchange of the points of excitation and observation again leaves the topology of the network invariant.

The following simple examples will serve to further clarify the reciprocity theorem.

Example 4.2.4   Figure 4.2.7 provides a simple example in which, after an interchange of the points of excitation and observation, the topology of the network is modified and hence reciprocity does not hold. In Fig. 4.2.7 (a), the point of excitation *a–b* is constrained to a short-circuit condition while the point of observation *c–d* is constrained to an open-circuit condition. In Fig. 4.2.7 (b), *c–d* has the short-circuit constraint and *a–b* an open-circuit constraint. Therefore, the topology of the

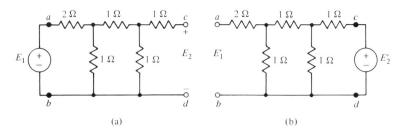

Fig. 4.2.7 Illustration of a situation to which reciprocity does not hold.

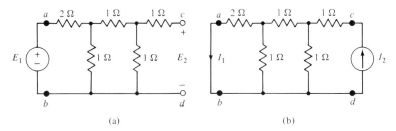

Fig. 4.2.8 Illustration of a situation in which the reciprocity theorem can be applied.

network has been modified after an interchange of the points of excitation and observation is made and as a result the reciprocity theorem does not apply. Simple calculations lead to

$$\frac{E_2}{E_1} = \frac{1}{7} \quad \text{and} \quad \frac{E_1'}{E_2'} = \frac{1}{5}. \tag{4.2.25}$$

Hence

$$\frac{E_2}{E_1} \neq \frac{E_1'}{E_2'}. \tag{4.2.26}$$

Example 4.2.5   In this example, Fig. 4.2.8 (a) is the same as that of Fig. 4.2.7 (a). However, in Fig. 4.2.8 (b) a current source I_2 is applied at c–d, and the response current I_1 is observed. An inspection reveals that in both (a) and (b) of Fig. 4.2.8 a short-circuit constraint is implied at a–b, while an open-circuit constraint is implied at c–d. Hence the reciprocity theorem holds in this case, or

$$\frac{E_2}{E_1} = \frac{I_1}{I_2} = \frac{1}{7}, \tag{4.2.27}$$

which indicates that the open-circuit voltage ratio is equal to the short-circuit current ratio.

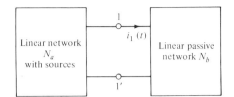

Fig. 4.2.9 Network N for development of Thévenin's theorem.

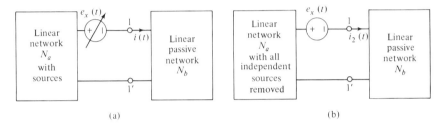

(a)   (b)

Fig. 4.2.10 Development of Thévenin's theorem.

Thévenin's and Norton's Theorems

In Section 3.4 we have seen that under certain conditions a voltage source may be transformed into a current source, and vice versa. Actually, this is a direct application of Thévenin's and Norton's theorems. If, for a given network, we wish to determine the voltage and the current of every branch, it is necessary to obtain either the node, the loop, or the state equations as discussed in Chapter 3, and then solve these equations. If, on the other hand, we are interested only in the voltage and/or the current in a particular portion of the network, we may wish to simplify the solution by reducing the entire network, excluding that part under investigation, to a simple equivalent. This reduction is accomplished by direct application of either Thévenin's or Norton's theorem. Before stating these theorems formally, we shall first consider the network N of Fig. 4.2.9, in which the two subnetworks are connected at terminals 1–1'. It is assumed that N is linear and, in addition, that no magnetic coupling is allowed between the subnetworks N_a and N_b of N. Let the current flowing through the terminal-pair 1–1' be denoted by $i_1(t)$ as indicated in the figure. This current $i_1(t)$ is, of course, the result of all the independent sources (initial condition sources are included as independent sources) acting simultaneously in N_a. In other words, $i_1(t)$ will become zero once the independent sources in N_a are removed. Next, let us insert an adjustable voltage source $e_x(t)$ between N_a and N_b resulting in a current $i(t)$ at 1–1' as depicted in Fig. 4.2.10 (a). By the superposition theorem, this current $i(t)$ is the sum of two currents: the current $i_1(t)$, which is due to the combined action of all the independent sources (including initial condition sources) acting together in N_a; and the current $i_2(t)$, caused by $e_x(t)$ acting alone with all the independent sources in N_a removed as indicated in Fig. 4.2.10 (b). The removal of an independent voltage source is

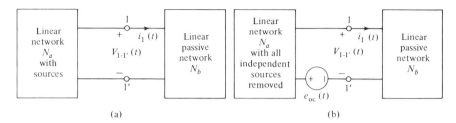

Fig. 4.2.11 (a) The original network N, and (b) its Thévenin equivalent network.

accomplished by short-circuiting the source while the removal of an independent current source is done by open-circuiting it. If the voltage source is adjusted so that the total current $i(t)$ vanishes; that is,

$$i(t) = i_1(t) + i_2(t) = 0 \qquad (4.2.28)$$

or

$$i_2(t) = -i_1(t). \qquad (4.2.29)$$

Equation (4.2.28), together with Fig. 4.2.10 (a), indicates that this value of $e_x(t)$ is simply equal to the open-circuit voltage $e_{oc}(t)$ of N_a across the terminal-pair $1-1'$, or

$$e_x(t) = e_{oc}(t) \qquad (4.2.30)$$

since $i(t) = 0$ implies that the voltage across the terminal-pair $1-1'$ is zero. Equation (4.2.29), together with Fig. 4.2.10 (b), reveals that if the polarities of $e_x(t) = e_{oc}(t)$ are reversed, the effect of N_a on terminal-pair $1-1'$ can be replaced by the equivalent network as depicted in Fig. 4.2.11. Thus, we have derived the powerful Thévenin's theorem, which can be stated as follows.

Theorem 4.2.4 (*Thévenin*) Let the linear network N be connected as shown in Fig. 4.2.11 (a) such that subnetwork N_a is composed of both independent and dependent sources, as well as linear passive elements, and subnetwork N_b is passive. In addition, it is assumed that there is no magnetic coupling between the two subnetworks. The effects of N_a upon N_b at terminal-pair $1-1'$ is equivalent to an independent voltage source $e_{oc}(t)$, which is the open-circuit voltage across terminal-pair $1-1'$, in series with the same subnetwork N_a after all the independent sources are removed as depicted in Fig. 4.2.11 (b). Furthermore, if the network N is time invariant, the Thévenin equivalent network for N_a in the s-domain is simply an independent voltage transform $E_{oc}(s) \triangleq \mathscr{L}\{e_{oc}(t)\}$ in series with an impedance $Z_0(s)$ representing the equivalent impedance looking into N_a at terminals $1-1'$ after all the independent sources in N_a are removed, as shown in Fig. 4.2.12.

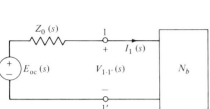

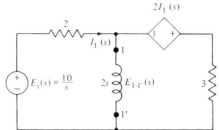

Fig. 4.2.12 The Thévenin equivalent network in the s-domain.

Fig. 4.2.13 Illustration of the application of Thévenin's theorem.

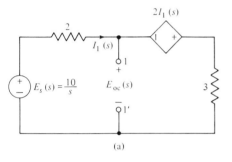

(a)

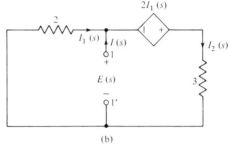

(b)

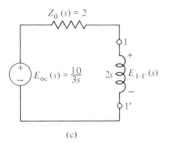

(c)

Fig. 4.2.14 Application of Thévenin's theorem to the network of Fig. 4.2.13: (a) network for calculation of $E_{oc}(s)$; (b) Network for calculation of $Z_0(s)$, and (c) the required Thévenin equivalent network

Example 4.2.6 The network of Fig. 4.2.13 contains a dependent source. Determine the Thévenin equivalent network at 1–1' with the element values as shown in the figure and then solve for the voltage across 1–1'. Assume zero initial condition.

To determine the open-circuit voltage transform $E_{oc}(s)$, we first remove the branch 2s from the network as indicated in Fig. 4.2.14 (a).

Writing the KVL-equation around the loop, we obtain

$$\frac{10}{s} + 2I_1(s) = I_1(s)(2 + 3)$$

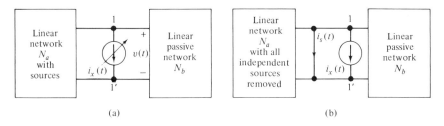

Fig. 4.2.15 Development of Norton's theorem.

or

$$I_1(s) = \frac{10}{3s}.$$

Hence

$$E_{oc}(s) = E_s(s) - 2I_1(s)$$

$$= \frac{10}{s} - 2\left(\frac{10}{3s}\right)$$

$$= \frac{10}{3s}.$$

To determine $Z_0(s)$ we remove $E_s(s)$ by short-circuiting it as described in Fig. 4.2.14 (b). An application of a voltage source $E(s)$ across 1–1' of Fig. 4.2.14 (b) gives rise to $I_1(s) = (-E(s)/2)$. Writing the KVL-equation around the loop yields the equation

$$2I_1(s) = 3I_2(s) + 2I_1(s)$$

or

$$I_2 = 0.$$

But $I(s) = I_2(s) - I_1(s) = \frac{1}{2}E(s)$, hence the impedance $Z_0(s)$ is given by

$$Z_0(s) = \frac{E(s)}{I(s)} = 2.$$

The Thévenin equivalent network at 1–1' is shown in Fig. 4.2.14 (c). Simple calculations lead to

$$E_{1-1}(s) = \frac{10}{3(1 + s)}.$$

To derive the Norton equivalent network, one needs only to realize that Thévenin's and Norton's theorems form a pair of duals. The parallel treatment

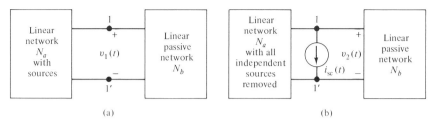

Fig. 4.2.16 Development of Norton's theorem.

would then be to insert an adjustable current source $i_x(t)$ in Fig. 4.2.9 as shown in Fig. 4.2.15 (a) and then adjust $i_x(t)$ until the voltage $v(t)$ across the terminal-pair 1–1′ is zero. Since the voltage $v(t)$ across 1–1′ is zero, a short circuit can be placed across 1–1′ without having any effect on the behavior of the network. This situation is indicated in Fig. 4.2.15 (b). The current $i_s(t)$ in the short circuit is zero and can be written, with the aid of the superposition theorem, as

$$i_s(t) = i_{sc}(t) - i_x(t) = 0 \qquad (4.2.31)$$

or

$$i_x(t) = i_{sc}(t), \qquad (4.2.32)$$

where $i_{sc}(t)$ is the current flowing through the short circuit as the result of the combined action of all the sources acting simultaneously in N_a with $i_x(t)$ removed.

Having determined the particular external current source $i_x(t)$ for $v(t) = 0$, we can write, as before,

$$v(t) = v_1(t) + v_2(t) = 0 \qquad (4.2.33)$$

or

$$v_2(t) = -v_1(t). \qquad (4.2.34)$$

In (4.2.34) $v_1(t)$ is the component of the voltage across 1–1′ due to the combined action of all the independent sources (including initial-condition sources) acting simultaneously in N_a with $i_x(t)$ removed, and the voltage $v_2(t)$ is the component of $v(t)$ due to $i_x(t) = i_{sc}(t)$ acting alone with all the independent sources in N_a set to zero. Figure 4.2.16 provides a clear picture of these two components of $v(t)$.

An inspection of (4.2.34) indicates that the two networks (a) and (b) of Fig. 4.2.16 are equivalent with respect to the terminal-pair 1–1′ provided that the direction of $i_{sc}(t)$ in (b) is reversed. This interesting result is, of course, the consequence of Norton's theorem, which can be summarized as follows.

Theorem 4.2.5 (*Norton*) Let the linear network N be connected as shown in Fig. 4.2.17 (a) such that subnetwork N_a is composed of both independent and dependent sources, as well as linear passive elements, and the subnetwork N_b is

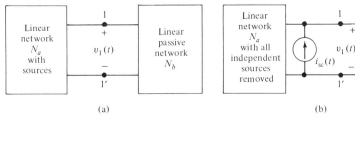

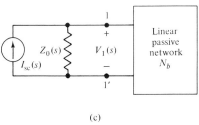

Fig. 4.2.17 (a) The original network N, (b) its Norton equivalent network in t-domain, and (c) its Norton equivalent network in the s-domain.

passive. In addition, it is assumed that there is no magnetic coupling between the subnetworks. Then the effects of N_a upon N_b at the terminal-pair 1–1′ are equivalent to an independent current source $i_{sc}(t)$, which is the short-circuit current through the terminal-pair 1–1′, in parallel with the same network N_a after all the independent sources are removed as described in Fig. 4.2.17 (b). Furthermore, if network N is time invariant and $Z_0(s)$ is the impedance representing the total impedance across the subnetwork N_a after all the independent sources are removed, the Norton equivalent network in the s-domain is that shown in Fig. 4.2.17 (c).

Example 4.2.7 Let us determine the Norton equivalent network at 1–1′ for the network of Fig. 4.2.13. The short-circuit current transform $I_{sc}(s)$ can be obtained with the aid of Fig. 4.2.18 (a). By inspection, $I_{sc}(s)$ is given by

$$I_{sc}(s) = I_1(s) - \frac{2I_1(s)}{3}$$

$$= \frac{1}{3}I_1(s)$$

$$= \frac{5}{3s}$$

and $Z_0(s)$ is, as obtained in Example 4.2.6, equal to

$$Z_0(s) = 2.$$

Hence, the Norton equivalent network is obtained as shown in Fig. 4.2.18 (b).

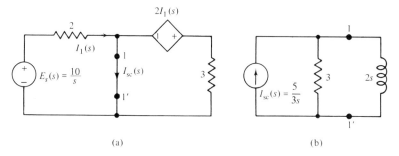

(a)   (b)

Fig. 4.2.18 Illustration of the application of Norton's theorem.

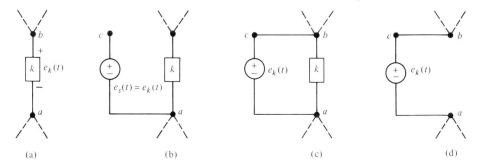

(a)   (b)   (c)   (d)

Fig. 4.2.19 Development of the substitution theorem.

The Substitution Theorem

This theorem is very general in the sense that it can be applied to any arbitrary network whether it is linear or nonlinear, time varying or time invariant. One of the important applications of this theorem is that a complicated network is first divided into a finite number of smaller sections and the solution of the original network is then obtained by combining the solutions of the smaller sections.

Consider a portion of an arbitrary network as illustrated in Fig. 4.2.19 (a). Let the voltage waveform of the kth element be denoted by $e_k(t)$. Next, let us connect an independent voltage source $e_s(t)$ to node a of the network and adjust it until it is equal to $e_k(t)$; that is, $e_s(t) = e_k(t)$. This situation is depicted in Fig. 4.2.19 (b). Now, if we connect a short circuit between nodes b and c as indicated in Fig. 4.2.19 (c), the behavior of the network will not be disturbed since nodes b and c are at the same potential. The kth element, which is in parallel with the independent voltage source $e_k(t)$, can be removed, so far as the remaining network is concerned. The net effect amounts to the replacement of the kth element by an independent voltage source of the same voltage waveform without affecting the behavior of the remaining portion of the network.

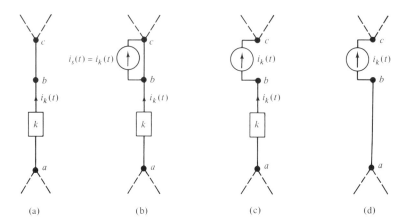

Fig. 4.2.20 Development of the substitution theorem.

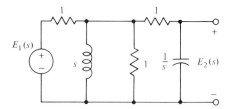

Fig. 4.2.21 Network for Example 4.2.8.

In similar manner, let us assume that the current waveform $i_k(t)$ of the kth element of an arbitrary network described in Fig. 4.2.20 (a) is known. If an independent current source with a waveform identical to that of $i_k(t)$ is connected as shown in Fig. 4.2.20 (b), the behavior of the network remains undisturbed. The KCL at node b indicates that the current in the short circuit connecting nodes b and c is zero. Hence this short circuit can be removed as depicted in Fig. 4.2.20 (c) leaving the kth element in series with the independent current source. The kth element can be removed so far as the rest of the network is concerned. A comparison between (a) and (d) of Fig. 4.2.20 reveals that the net result is the replacement of the kth element by an independent current source.

Summarizing the above discussion into a theorem, we have:

Theorem 4.2.6 (*Substitution*) A known voltage (current) can be replaced by an independent voltage (current) source.

Example 4.2.8 Let us apply the substitution theorem to determine the voltage transfer ratio $E_2(s)/E_1(s)$ of the network shown in Fig. 4.2.21.

The solution can be obtained by first dividing the network into two subnetworks, N_a and N_b, connected in cascade as indicated in Fig. 4.2.22 (a). Then we replace the subnetwork N_b by its input impedance $Z_0(s)$ at the terminal-pair a–b

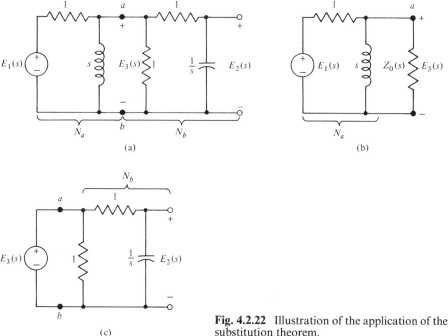

Fig. 4.2.22   Illustration of the application of the substitution theorem.

and attain the result shown in Fig. 4.2.22 (b). By inspection, this impedance $Z_0(s)$ is given by

$$Z_0(s) = \frac{1 \cdot \left(1 + \dfrac{1}{s}\right)}{1 + 1 + \dfrac{1}{s}}$$

(4.2.35)

$$= \frac{1 + s}{1 + 2s}.$$

Next, we proceed to calculate $E_3(s)$. From Fig. 4.2.22 (b), we find

$$E_3(s) = \frac{E_1(s)}{1 + \dfrac{sZ_0(s)}{s + Z_0(s)}} \cdot \frac{sZ_0(s)}{s + Z_0(s)}$$

(4.2.36)

$$= \frac{sZ_0(s)E_1(s)}{s + Z_0(s) + sZ_0(s)}.$$

Substituting (4.2.35) into (4.2.36) and then simplifying, we obtain

$$E_3(s) = \frac{s + s^2}{1 + 3s + 3s^2} E_1(s).$$

(4.2.37)

Applying the substitution theorem, the known voltage $E_3(s)$ just determined can be replaced by an independent source as depicted in Fig. 4.2.22 (c), from which the required $E_2(s)$ can be easily calculated to be

$$E_2(s) = \frac{\dfrac{1}{s}}{1 + \dfrac{1}{s}} E_3(s)$$

$$= \frac{1}{1 + s} E_3(s) \tag{4.2.38}$$

$$= \frac{s}{1 + 3s + 3s^2} E_1(s).$$

Hence the voltage-transfer function becomes

$$\frac{E_2(s)}{E_1(s)} = \frac{s}{1 + 3s + 3s^2}. \tag{4.2.39}$$

Remark. In the process of determining the solution in the above example, only one subnetwork is involved at any time. This amounts to trading a more difficult problem for two easier ones. The required solution is obtained by combining the solutions of the subnetworks.

4.3 EVALUATION OF INITIAL CONDITIONS

As discussed previously, the general solution of a set of differential equations describing the behavior of a physical system contains one or more arbitrary constants. To evaluate these constants, additional information must be either given along with the differential equations or determined directly from the physical system itself. In the case of electrical networks, if the values of inductor currents and capacitor voltages are known at a specified time, $t = t_0$, the set of arbitrary constants can be evaluated accordingly. To be more specific, let us select, for convenience, $t_0 = 0$ as the reference time at which one or more switches in a given network operate, thus upsetting the equilibrium condition of the network. The values of inductor currents and capacitor voltages at this instant of time, i.e., $t = t_0 = 0$, are usually referred to as the *initial conditions* of the network. To distinguish between the time immediately before and immediately after the operation of a switch, we designate, respectively, $t = 0^-$ and $t = 0^+$. Since we are interested in the behavior of a given network *after* the operation of the switches, we shall regard the evaluation of the voltages and currents at $t = 0^+$ as the *evaluation of initial conditions* unless stated otherwise. However, it should be noted that the values of the inductor currents and the capacitor voltages at $t = 0^+$ are closely related to their corresponding values at $t = 0^-$ as will be obvious in the following discussion.

The relationship between the values of an inductor current $i_L(t)$ immediately before and immediately after the operation of a switch can be obtained from the voltage-current relation (VCR) of an inductor, i.e.,

$$v_L(t) = L \frac{di_L(t)}{dt}$$

or

$$i_L(t) = \frac{1}{L} \int_{-\infty}^{t} v_L(x)\, dx,$$

where x is a dummy variable in the integral. Upon setting $t = 0^-$ and $t = 0^+$, the above equation yields, respectively,

$$i_L(0^-) = \frac{1}{L} \int_{-\infty}^{0^-} v_L(x)\, dx \qquad (4.3.1)$$

and

$$i_L(0^+) = \frac{1}{L} \int_{-\infty}^{0^+} v_L(x)\, dx = \frac{1}{L} \int_{-\infty}^{0^-} v_L(x)\, dx + \frac{1}{L} \int_{0^-}^{0^+} v_L(x)\, dx$$

$$= i_L(0^-) + \frac{1}{L} \int_{0^-}^{0^+} v_L(x)\, dx. \qquad (4.3.2)$$

In the last expression of (4.3.2), the integral is carried over an infinitesimally small time interval and is always equal to zero unless the integrand, $v_L(x)$, is an impulse at $x = 0$ as in the case of a degenerate network.

In a similar manner, the relationship between the values of a capacitor voltage $v_C(t)$ immediately before and immediately after the operation of a switch can be obtained by writing

$$i_C(t) = C \frac{dv_C(t)}{dt}$$

or

$$v_C(t) = \frac{1}{C} \int_{-\infty}^{t} i_C(x)\, dx.$$

After setting $t = 0^+$, we then have

$$v_C(0^+) = \frac{1}{C} \int_{-\infty}^{0^+} i_C(x)\, dx = \frac{1}{C} \int_{-\infty}^{0^-} i_C(x)\, dx + \frac{1}{C} \int_{0^-}^{0^+} i_C(x)\, dx$$

$$= v_C(0^-) + \frac{1}{C} \int_{0^-}^{0^+} i_C(x)\, dx, \qquad (4.3.3)$$

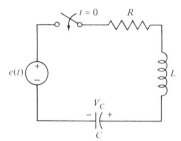

Fig. 4.3.1 Series RLC network.

with x denoting the dummy variable of integration as before. Again, in the last expression of (4.3.3), the value of the integral is zero unless $i_C(x)$ is an impulse as in the case of a degenerate network. The following example will illustrate the ideas just discussed.

Example 4.3.1 Consider the series RLC network in Fig. 4.3.1. The loop equation is given by

$$e(t) = Ri(t) + L\frac{d}{dt}i(t) + \frac{1}{C}\int i(t)\,dt. \tag{4.3.4}$$

Taking the Laplace transform of (4.3.4), we find

$$E(s) = RI(s) + L[sI(s) - i(0^+)] + \frac{1}{Cs}[I(s) + q_C(0^+)], \tag{4.3.5}$$

where

$$q_C(0^+) = \int_{-\infty}^{0^+} i(t)\,dt$$

is the charge in the capacitor at the time $t = 0^+$. Since

$$\frac{1}{C}q_C(t) = v_C(t),$$

(4.3.5) can be rewritten as

$$E(s) = \left[R + Ls + \frac{1}{Cs}\right]I(s) - Li(0^+) + \frac{1}{s}v_C(0^+). \tag{4.3.6}$$

It is clear that to solve for $I(s)$ in (4.3.6) we must first determine the values of $i(0^+)$ and $v_C(0^+)$, the initial conditions of the network. In this example, just before the switch is closed at $t = 0$, we have $i(0^-) = 0$.

In view of (4.3.2) and the fact that the network is nondegenerate, we find that $i(0^+) = i(0^-) = 0$ which implies that the current $i(t)$ is *continuous* immediately before and immediately after the switch is closed. The capacitor may or may not

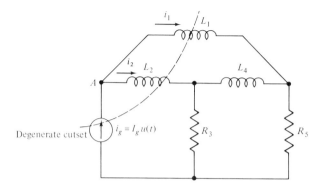

Fig. 4.3.2 Network with a
degenerate cutset.

be initially charged so that the voltage $v_C(0^-)$ at time $t = 0^-$ may or may not be
zero. But, by virtue of (4.3.3), $v_C(0^+) = v_C(0^-)$, showing that the capacitor voltage
is also *continuous* immediately before and immediately after the switch action.
Therefore, in this example, the initial conditions of the network at $t = 0^+$ are
identical to the corresponding ones evaluated at $t = 0^-$. The *continuity condition*
of the inductor currents and the capacitor voltages of a network immediately before
and immediately after the operation of a switch is generally valid *unless the given
network is degenerate* as demonstrated in the next example.

Example 4.3.2 Let us consider the network of Fig. 4.3.2, which was used in
Example 3.7.3 in Chapter 3. The degenerate cutset consists of L_1, L_2, and i_g.
Kirchhoff's current law (KCL) requires that the sum of the currents leaving node A
at any time must be equal to zero, or

$$-i_g(t) + i_1(t) + i_2(t) = 0 \tag{4.3.7}$$

which, at $t = 0^+$, becomes

$$i_1(0^+) + i_2(0^+) = I_g. \tag{4.3.8}$$

Since $i_1(0^-)$ and $i_2(0^-)$ are, in general, independent of each other (that is, they
can assume any finite values), the expression in (4.3.8) demands that the values of
the inductor currents change instantly so that KCL will not be violated. It should
be noted at this point that although we have discussed the initial conditions of a
network with only degenerate cutsets as illustrated in the above example, a similar
discussion may be given for networks containing degenerate circuits. In the
following paragraphs, we shall present a method for calculating initial conditions
at $t = 0^+$ of degenerate networks.

When the network under consideration is degenerate, the determination of
initial conditions at $t = 0^+$ can be carried out by applying the two Kirchhoff's laws
together with two physical principles—the principle of conservation of charge and
the principle of continuity of flux linkages. Stating these principles formally, we

have:

The Principle of Conservation of Charge   The total charge *transferred* into a junction of a given network at any time t is zero, i.e.,

$$\sum_k \Delta q_k = 0. \tag{4.3.9}$$

The Principle of Continuity of Flux Linkages   The summation of all the flux linkages at any time over any closed loop of a given network is continuous, i.e.,

$$\sum_k L_k i_k(t^+) - \sum_k L_k i_k(t^-) = 0, \tag{4.3.10}$$

where t^- and t^+ denote, respectively, the time instants immediately before and immediately after t.

Next, let us investigate separately how each of the two physical principles can be applied to various networks for determining the initial conditions at $t = 0^+$. First, we shall consider a network with a degenerate circuit, i.e., a loop consisting only of capacitances and/or voltage sources. Suppose that all the independent sources are finite (in magnitude) and that switching action takes place at $t = 0$. Upon examination of the v-i relations, one concludes that no charge can be instantaneously transferred at $t = 0^+$ through any of the passive elements except those capacitors forming a degenerate circuit. This can be seen by noting that the charge Δq_k being instantaneously transferred at $t = 0^+$ through the kth element is given by the relation

$$\Delta q_k = \int_{0^-}^{0^+} i_k(t)\, dt, \tag{4.3.11}$$

which is equal to zero unless $i_k(t)$ is an impulse at $t = 0$. The loop current of a degenerate circuit is indeed an impulse at $t = 0$ since the capacitor is equivalent to a short circuit at the instant of switching action. The currents in both inductors and resistors are always finite because the voltages across them are finite.

Application of (4.3.9) to various nodes will indicate that the current through a capacitor which is not an element of a degenerate circuit is finite also. In terms of Δq, (4.3.3) can be rewritten as

$$v_C(0^+) = v_C(0^-) + \frac{\Delta q}{C}. \tag{4.3.12}$$

Keeping in mind the fact that charge can be instantaneously transferred through capacitors in degenerate circuits, the initial voltages at $t = 0^+$ can be determined by applying (4.3.9) and (4.3.12) to the degenerate circuits of the network under investigation. The following example will illustrate the ideas involved.

Example 4.3.3   The network in Fig. 4.3.3(a) consists of two degenerate circuits. Since the source voltage E_0 is finite, all other voltages must be finite. Hence no charge can be instantaneously transferred through L and R and hence also through

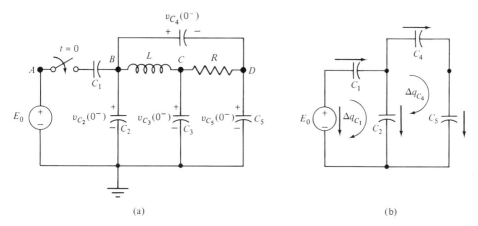

Fig. 4.3.3 (a) Network for Example 4.3.3, and (b) its reduced network.

C_3 (with the aid of (4.3.9) at node C). To determine the charge transfers Δq_k, we delete those elements not contained in degenerate circuits from the original network and assign a loop charge transfer in each of the degenerate circuits, resulting in the reduced network depicted in Fig. 4.3.3 (b).

The procedure for determining the initial conditions at $t = 0^+$ can be carried out step-by-step in the following manner:

1. For those elements with no charge transfers, the initial conditions are continuous at $t = 0$, i.e.,

$$i_L(0^+) = i_L(0^-) \qquad \text{and} \qquad v_C(0^+) = v_C(0^-).$$

These include all the elements other than those capacitors forming degenerate circuits. In this example, we have

$$i_L(0^+) = i_L(0^-) \qquad \text{and} \qquad v_{C_3}(0^+) = v_{C_3}(0^-).$$

2. For the capacitors in degenerate circuits, we proceed as follows:

a) Apply the principle of conservation of charge to the reduced network and then express the charge transfers of the individual elements in terms of the loop charge transfers. Thus, in our example, we find, in terms of the loop charge transfers, Δq_{C_1} and Δq_{C_4},

$$\Delta q_{C_1} = \Delta q_{C_1}, \qquad \Delta q_{C_4} = \Delta q_{C_4}$$

$$\Delta q_{C_2} = \Delta q_{C_1} - \Delta q_{C_4}, \qquad \Delta q_{C_5} = \Delta q_{C_4}.$$

b) Write the independent KVL-equations for the loops in the reduced network for $t = 0^+$ and then substitute (4.3.12) into these equations to yield a set of KVL-equations in terms of both the initial voltages at $t = 0^-$ and the charge transfers. In our example, we find that

$$E_0 = v_{C_1}(0^+) + v_{C_2}(0^+)$$

$$= v_{C_1}(0^-) + v_{C_2}(0^-) + \frac{\Delta q_{C_1}}{C_1} + \frac{\Delta q_{C_2}}{C_2}$$

and

$$0 = -v_{C_2}(0^+) + v_{C_4}(0^+) + v_{C_5}(0^+)$$

$$= -v_{C_2}(0^-) + v_{C_4}(0^-) + v_{C_5}(0^-) - \frac{\Delta q_{C_2}}{C_2} + \frac{\Delta q_{C_4}}{C_4} + \frac{\Delta q_{C_5}}{C_5}.$$

c) Solve for the loop charges by combining the equations in (a) and (b). We obtain, in this example,

$$E_0 - v_{C_1}(0^-) - v_{C_2}(0^-) = \left(\frac{1}{C_1} + \frac{1}{C_2}\right)\Delta q_{C_1} - \frac{1}{C_2}\Delta q_{C_4},$$

$$v_{C_2}(0^-) - v_{C_4}(0^-) - v_{C_5}(0^-) = \frac{-1}{C_2}\Delta q_{C_1} + \left(\frac{1}{C_2} + \frac{1}{C_4} + \frac{1}{C_5}\right)\Delta q_{C_4},$$

and the loop charges Δq_{C_1} and Δq_{C_4} can be solved to yield

$$\Delta q_{C_1} = \frac{K_1}{K} \qquad \text{and} \qquad \Delta q_{C_4} = \frac{K_2}{K},$$

where

$$K \triangleq \left(\frac{1}{C_1} + \frac{1}{C_2}\right)\left(\frac{1}{C_2} + \frac{1}{C_4} + \frac{1}{C_5}\right) - \frac{1}{C_2^2},$$

$$K_1 \triangleq \left(\frac{1}{C_2} + \frac{1}{C_4} + \frac{1}{C_5}\right)[E_0 - v_{C_1}(0^-) - v_{C_2}(0^-)]$$

$$+ \frac{1}{C_2}[v_{C_2}(0^-) - v_{C_4}(0^-) - v_{C_5}(0^-)]$$

$$K_2 \triangleq \left(\frac{1}{C_1} + \frac{1}{C_2}\right)[v_{C_2}(0^-) - v_{C_4}(0^-) - v_{C_5}(0^-)]$$

$$+ \frac{1}{C_2}[E_0 - v_{C_1}(0^-) - v_{C_2}(0^-)].$$

d) Finally substituting the results obtained in (c) into the expressions in (a) and then making use of (4.3.12), we obtain the required initial conditions at $t = 0^+$. Thus, in our example, we have

$$v_{C_1}(0^+) = v_{C_1}(0^-) + \frac{1}{C_1}\frac{K_1}{K},$$

$$v_{C_2}(0^+) = v_{C_2}(0^-) + \frac{1}{C_2}\left[\frac{K_1 - K_2}{K}\right],$$

$$v_{C_4}(0^+) = v_{C_4}(0^-) + \frac{1}{C_4}\frac{K_2}{K},$$

$$v_{C_5}(0^+) = v_{C_5}(0^-) + \frac{1}{C_5}\frac{K_2}{K}.$$

Next, we shall consider a network with degenerate cutsets, i.e., cutsets consisting only of inductances and/or current sources. Again, let the switching action occur at $t = 0$ and our problem is to evaluate the initial inductance currents at $t = 0^+$. An examination of the v-i relation indicates that the current in an inductor is always continuous at $t = 0$, i.e., $i(0^-) = i(0^+)$, unless the voltage across it is an impulse. This situation occurs in those inductors which, possibly together with some current sources, form an inductor-only cutset. We use $\Delta\lambda_k$ to denote the flux transfer (or more precisely flux linkage transfer) in the kth inductor at $t = 0$, and find that

$$\Delta\lambda_k = L_k[i_k(0^+) - i_k(0^-)] \tag{4.3.13}$$

or,

$$i_k(0^+) = i_k(0^-) + \frac{\Delta\lambda_k}{L_k}. \tag{4.3.14}$$

Equation (4.3.14) is identical in form to (4.3.12) as expected, since the degenerate cutset and the degenerate circuit are duals of each other. Hence, a parallel step-by-step procedure can be easily established for the determination of initial conditions at $t = 0^+$ for networks consisting of degenerate cutsets, which we now describe.

1. For those elements with no flux transfers, the initial conditions are continuous at $t = 0$, i.e.,

$$i_L(0^+) = i_L(0^-) \quad \text{and} \quad v_C(0^+) = v_C(0^-).$$

These include all the elements other than those inductors forming degenerate cutsets.

2. For those inductors forming degenerate cutsets, we proceed as follows:

a) Replace all the elements without flux transfers in the original network by *short circuits*. These elements are resistors, capacitors, voltage generators, and those inductors not contained in any of the degenerate cutsets. The resulting network is referred to as the *reduced network*.

b) Apply the principle of continuity of flux linkages to the independent loops of the reduced network, yielding a set of algebraic equations in terms of flux linkages of the individual elements.

c) Write the independent KCL-equations for the nodes in the reduced network at $t = 0^+$, and then substitute (4.3.14) into these equations.

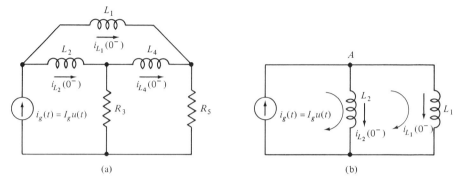

Fig. 4.3.4 (a) A degenerate network, and (b) its reduced network.

d) Solve for the flux transfers by combining the equations in (b) and (c).

e) Finally, the initial currents can be obtained by applying (4.3.14) together with the results in (d).

The following example will illustrate the ideas just described.

Example 4.3.4 Consider again the network of Fig. 4.3.2, which is repeated in Fig. 4.3.4 (a) for convenience. According to the steps underlined above, we proceed as follows:

1. Since L_4 is not in the degenerate cutset, we have immediately

$$i_{L_4}(0^+) = i_{L_4}(0^-).$$

2. a) The reduced network with R_3, L_4, and R_5 removed is drawn in Fig. 4.3.4 (b).

   b) Applying the principle of continuity of flux linkages to the reduced network, we find that

   $$\Delta\lambda_g + \Delta\lambda_2 = 0 \qquad \text{and} \qquad -\Delta\lambda_2 + \Delta\lambda_1 = 0,$$

   which is equivalent to

   $$\Delta\lambda_1 = \Delta\lambda_2 = -\Delta\lambda_g.$$

   c) Writing the KCL-equation for node A of Fig. 4.3.4 (b) yields at $t = 0^+$,

   $$I_g = i_{L_1}(0^+) + i_{L_2}(0^+)$$

   which, with the aid of (4.3.14), reduces to

   $$I_g = i_{L_1}(0^-) + i_{L_2}(0^-) + \frac{\Delta\lambda_1}{L_1} + \frac{\Delta\lambda_2}{L_2}.$$

d) Combining the equations in (b) and (c), we find that

$$\Delta\lambda_1 = \Delta\lambda_2 = \frac{1}{\left(\dfrac{1}{L_1} + \dfrac{1}{L_2}\right)} \, [I_g - i_{L_1}(0^-) - i_{L_2}(0^-)]$$

$$= \left(\frac{L_1 L_2}{L_1 + L_2}\right) [I_g - i_{L_1}(0^-) - i_{L_2}(0^-)]$$

e) Finally, by means of (4.3.14), the initial currents at $t = 0^+$ are given by

$$i_{L_1}(0^+) = i_{L_1}(0^-) + \left(\frac{L_2}{L_1 + L_2}\right) [I_g - i_{L_1}(0^-) - i_{L_2}(0^-)]$$

and

$$i_{L_2}(0^+) = i_{L_2}(0^-) + \left(\frac{L_1}{L_1 + L_2}\right) [I_g - i_{L_1}(0^-) - i_{L_2}(0^-)].$$

Remark. If the network under investigation consists of both degenerate circuits as well as degenerate cutsets, then two reduced networks will result—one corresponds to the degenerate circuits and the other to the degenerate cutsets. The two procedures underlined above can be applied separately to the reduced networks so that the complete set of initial conditions can be determined.

4.4 TRANSFORM NETWORKS

In Chapter 3, the matrix formulation of network equations in the *s*-domain was obtained by first writing the *t*-domain equations and then taking the Laplace transform of these equations. Sometimes it is more convenient to determine these *s*-domain network equations directly from the *transform network*, which is the equivalent of the original network in the *s*-domain. Thus, in this section, we shall discuss the *s*-domain representation of a given network using the concept of transform impedances and admittances. More precisely, for each of the elements in the given network, there is a corresponding transform *immittance** (either impedance or admittance) in the transform network. We shall first derive the transform-impedance representation of the resistor, the inductor, and the capacitor with the initial conditions taken into account. Then we shall show how the transform network can be obtained.

As we recall, the transform impedance of a passive element is defined as the ratio of the transform voltage across the element to the transform current through that element with the initial condition set to zero. Thus, using the VCR-equations for the resistor, the inductor, and the capacitor, the corresponding transform

---

* The term "immittance" is derived from the words "*im*pedance" and "ad*mittance*."

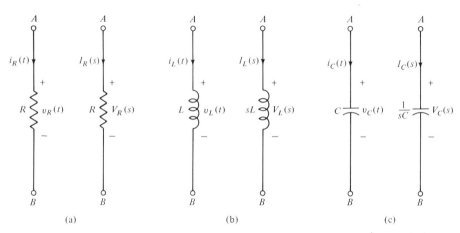

Fig. 4.4.1 Equivalent representations in the t- and s-domains for (a) a resistor, (b) an inductor, and (c) a capacitor.

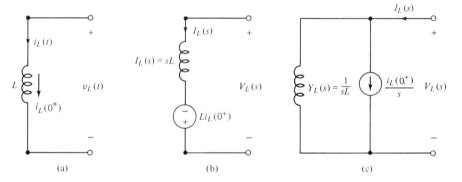

Fig. 4.4.2 Equivalent representation of an inductor with an initial current (a) in the t-domain, (b) in the s-domain with the voltage-generator equivalent, and (c) in the s-domain with the current-generator equivalent.

impedances can be easily shown to be, respectively,

$$Z_R(s) = R, \qquad Z_L(s) = sL, \qquad \text{and} \qquad Z_C(s) = \frac{1}{sC}. \qquad (4.4.1)$$

The equivalent representations of these elements in both the t- and the s-domains are depicted in Fig. 4.4.1.

Next, we shall obtain the s-domain representation of an inductor L shown in Fig. 4.4.2 (a) with an initial current $i_L(0^+)$. Taking the Laplace transform of the

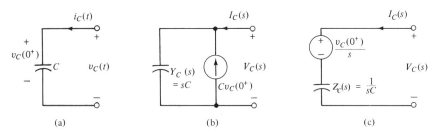

Fig. 4.4.3 Equivalent representations of an initially charged capacitor: (a) An initially charged capacitor in t-domain, (b) s-domain equivalent with a current source, and (c) s-domain equivalent with a voltage source.

VCR-equation for the inductor, namely,

$$v_L(t) = L\frac{d}{dt}i_L(t), \tag{4.4.2}$$

yields

$$V_L(s) = L[sI_L(s) - i_L(0^+)]$$
$$= sLI_L(s) - Li_L(0^+). \tag{4.4.3}$$

It is evident from (4.4.3) that an inductor L with an initial current $i_L(0^+)$ is equivalent to the same inductor with zero initial current connected in series with an independent voltage source of magnitude $Li_L(0^+)$. This situation is shown in Fig. 4.4.2 (b). Another s-domain representation can be obtained by solving for the transform current $I_L(s)$ in (4.4.3). Thus

$$I_L(s) = \frac{V_L(s)}{sL} + \frac{i_L(0^+)}{s}, \tag{4.4.4}$$

which indicates that an inductor L with an initial current $i_L(0^+)$ is equivalent to the same inductor with zero initial current connected in parallel with an independent current source of magnitude $i_L(0^+)/s$. This situation is depicted in Fig. 4.4.2 (c). Note that the directions of both the equivalent voltage and current sources are dictated by that of the initial current.

In like manner, taking the Laplace transform of the VCR-equation for an initially charged capacitor, i.e.,

$$i_C(t) = C\frac{dv_C(t)}{dt}, \tag{4.4.5}$$

yields

$$I_C(s) = sCV_C(s) - Cv_C(0^+) \tag{4.4.6}$$

or, equivalently,

$$V_C(s) = \frac{1}{sC} I_C(s) + \frac{v_C(0^+)}{s}. \qquad (4.4.7)$$

By means of (4.4.6) and (4.4.7), one obtains the equivalent representations for an initially charged capacitor as described in Fig. 4.4.3.

Again we should observe that the directions of both the current and the source in (b) and (c) of Fig. 4.4.3 are dictated by that of the initial voltage of the capacitor in Fig. 4.4.3 (a).

Having discussed the s-domain equivalents of the individual network elements, our next task is to show how a transform network can be obtained from a given electrical network. For convenience we shall divide networks into two classes: (1) networks without mutual inductances, and (2) networks with mutual inductances.

Networks without Mutual Inductances

For a network without mutual inductances a transform network can be obtained by replacing each passive element with its s-domain equivalent and each source by its transform. In general, if the loop analysis is used, the s-domain equivalent with a voltage source for every reactive element is usually preferred in obtaining an equivalent network, since with this choice the total number of loops will be smaller. On the other hand, if the node analysis is used, it would be more advantageous to apply the s-domain equivalents with current sources for all the reactive elements as the following examples will illustrate.

Example 4.4.1 Consider the network of Fig. 4.4.4. Our problem is to determine the mesh currents $i_1(t)$ and $i_2(t)$.

Using the voltage-source equivalents for both the inductor and the capacitor, the transform network is drawn in Fig. 4.4.5. The transform mesh equations can be easily shown to be

$$(R_1 + R_3 + sL_2)I_1(s) - R_3I_2(s) = E_1(s) + L_2i_{L_2}(0^+) \qquad (4.4.8)$$

$$-R_3I_1(s) + \left(R_3 + \frac{1}{sC_4}\right)I_2(s) = \frac{-v_{C_4}(0^+)}{s}. \qquad (4.4.9)$$

Needless to say, the two equations (4.4.8) and (4.4.9) are the Laplace transforms of the corresponding integro-differential equations for the two meshes in Fig. 4.4.4. Thus, instead of writing first the integro-differential equations (loop or node equations) for a given network in the t-domain and then taking the Laplace transform, we can obtain the transform equations directly with the aid of a transform network.

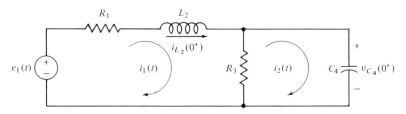

Fig. 4.4.4 Network for Example 4.4.4.

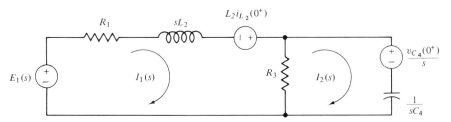

Fig. 4.4.5 Transform network for the network in Fig. 4.4.4 using voltage-source equivalent.

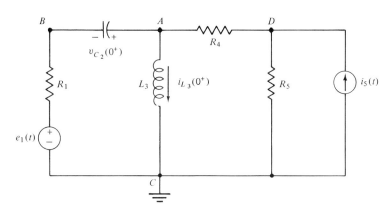

Fig. 4.4.6 Network for Example 4.4.2.

Example 4.4.2 Draw the transform network appropriate for writing the node equations of the network in Fig. 4.4.6. Using the current-source equivalents for both the capacitor and the inductor and replacing the independent voltage source $e_1(t)$ by the equivalent current source, the transform network is drawn in Fig. 4.4.7. Note that admittances are used for all the elements in the figure. Denoting the

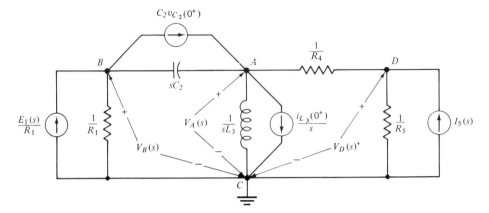

Fig. 4.4.7 Transform network for the network in Fig. 4.4.6.

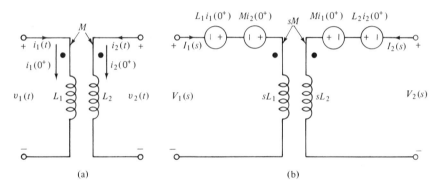

Fig. 4.4.8 (a) A pair of coupled coils, and (b) its s-domain equivalent.

node-to-datum voltages as $V_A(s)$, $V_B(s)$, and $V_D(s)$ as indicated, the network equations for nodes A, B, and D can be arranged into the following matrix form:

$$
\underbrace{\begin{bmatrix} \left(\dfrac{1}{R_4} + C_2 s + \dfrac{1}{L_3 s}\right) & -(C_2 s) & -\left(\dfrac{1}{R_4}\right) \\[2mm] -(C_2 s) & \left(\dfrac{1}{R_1} + C_2 s\right) & 0 \\[2mm] -\left(\dfrac{1}{R_4}\right) & 0 & \left(\dfrac{1}{R_4} + \dfrac{1}{R_5}\right) \end{bmatrix}}_{\mathbf{Y}_n(s)}
\begin{bmatrix} V_A(s) \\[2mm] V_B(s) \\[2mm] V_D(s) \end{bmatrix}
=
\begin{bmatrix} C_2 v_{C_2}(0^+) - \dfrac{1}{s} i_{L_3}(0^+) \\[2mm] \dfrac{1}{R_1} E_1(s) - C_2 v_{C_2}(0^+) \\[2mm] I_5(s) \end{bmatrix}
$$

$$(4.4.10)$$

where $\mathbf{Y}_n(s)$ is, of course, the node admittance matrix of the network.

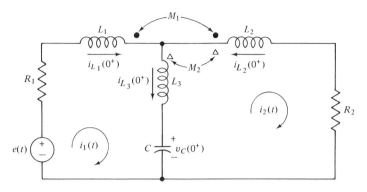

Fig. 4.4.9 Network for Example 4.4.3.

Networks with Mutual Inductances

When a network contains magnetically coupled coils, the voltage across one coil is affected by the time rate of change of the current through the other coil. For the coupled coils of Fig. 4.4.8 (a), the voltages $v_1(t)$ and $v_2(t)$ are, respectively, given by

$$v_1(t) = L_1 \frac{di_1(t)}{dt} + M \frac{di_2(t)}{dt} \tag{4.4.11}$$

and

$$v_2(t) = M \frac{di_1(t)}{dt} + L_2 \frac{di_2(t)}{dt}. \tag{4.4.12}$$

Taking the Laplace transform of these equations, we find, respectively,

$$V_1(s) = sL_1 I_1(s) + sM I_2(s) - L_1 i_1(0^+) - M i_2(0^+) \tag{4.4.13}$$

and

$$V_2(s) = sM I_1(s) + sL_2 I_2(s) - M i_1(0^+) - L_2 i_2(0^+). \tag{4.4.14}$$

From (4.4.13) and (4.4.14), the s-domain equivalent for the coupled coils can be drawn as shown in Fig. 4.4.8 (b).

An examination of this figure points out, as before, the close relationship between the polarity marks of an equivalent voltage source representing the effect of the initial current on a coil and the direction of the initial current itself. For example, the voltage source $L_1 i_1(0^+)$ in Fig. 4.4.8 (b) will *tend to cause* a current to flow in the same direction as that of $i_1(0^+)$ in Fig. 4.4.8 (a), i.e., into the coil through the dotted terminal of the coil. Similar observation can be made for the other voltage sources.

Having observed the close relationship between the polarity marks of a voltage source in the transform network and the direction of the initial current in the

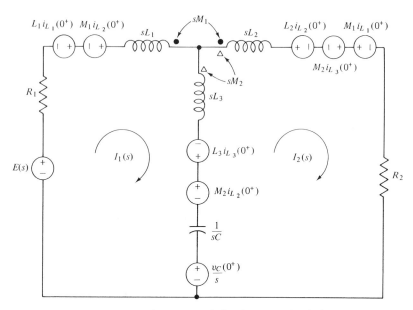

Fig. 4.4.10 Transform network for the network of Fig. 4.4.9.

original network, we should be able to draw the transform network by inspection—regardless of whether the given network consists of coupled coils as the following example will illustrate.

Example 4.4.3 We shall proceed to draw the transform network for the network of Fig. 4.4.9. Consider coil L_1 (i.e., the coil designated by L_1). Since this coil is coupled only to coil L_2 and not to coil L_3, there will be two voltage sources of magnitudes $L_1 i_{L_1}(0^+)$ and $M_1 i_{L_2}(0^+)$ connected in series with the impedance sL_1 in the transform network. The polarity marks of the voltage source $L_1 i_{L_1}(0^+)$ must be assigned in such a way that it will tend to cause a current to flow in the same direction as that of $i_{L_1}(0^+)$, i.e., *leaving* the dotted terminal of coil L_1. In like manner, since the initial current $i_{L_2}(0^+)$ is leaving the dotted terminal of coil L_2, the polarity marks of the corresponding voltage source $M_1 i_{L_2}(0^+)$, must be designated so that the source will tend to send a current flowing out of the dotted terminal of coil L_1.

Next, consider coil L_2. Since it is coupled to both coil L_1 and coil L_3, there will be three voltage sources in series with the impedance sL_2 in the transform network with magnitudes $L_2 i_{L_2}(0^+)$, $M_2 i_{L_3}(0^+)$, and $M_1 i_{L_1}(0^+)$. Since the currents $i_{L_1}(0^+)$ and $i_{L_2}(0^+)$ are flowing out of the dotted terminals of their coils while $i_{L_3}(0^+)$ is entering the dotted terminal of coil L_3, the polarity marks of $M_2 i_{L_3}(0^+)$ must be opposite to that of both $L_2 i_{L_2}(0^+)$ and $M_1 i_{L_1}(0^+)$. A similar consideration for coil L_3 leads to the transform network of Fig. 4.4.10. Once the transform network is drawn, the loop equations can be written down immediately by inspection. How-

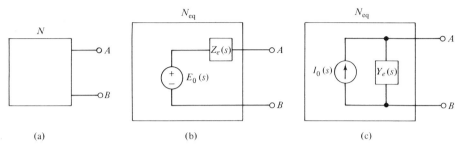

Fig. 4.5.1 (a) A two-terminal network N, (b) the Thévenin equivalent N_{eq} of N, and (c) the Norton equivalent N_{eq} of N.

ever, in the case of writing the node equations for a network with mutual inductances, this "generator" representation of initial conditions becomes rather involved, especially when the network contains more than two mutual inductances, and hence will not be discussed here. The interested reader is referred to the literature elsewhere.*

4.5 SUMMARY

In Section 4.2 we introduced several network theorems: Tellegen's theorem, the superposition theorem, the reciprocity theorem, Thévenin's theorem, Norton's theorem, and the substitution theorem. These theorems are quite general and can be applied to very large classes of networks encountered in practice so that their solutions can be greatly simplified.

Both Thévenin's and Norton's theorems apply to linear networks containing sources and passive elements. The Thévenin and the Norton equivalents of a two-terminal network N are shown in Fig. 4.5.1. In this figure $E_0(s)$ is the open-circuit voltage transform across the terminals of N, $I_0(s)$ is the transform of the short-circuit current from terminal A to terminal B of N with the two terminals coalesced, and $Z_e(s)$ and $Y_e(s)$ are respectively the input impedance and input admittance of N at the terminals with all the *independent* sources and initial conditions of N reduced to zero.

The remaining theorems are summarized in Table 4.5.1.

In Section 4.3 we discussed the methods for determining the initial conditions. The transform immittance for individual elements was discussed in Section 4.4 with the equivalent representation of the elements R, L, and C in both the t-domain and the s-domain summarized in Table 4.5.2. By using the equivalent representation for each element in a given network with no mutual inductances, we showed how to construct a transform network from which the transform equations (loop

---

* For example, see [CR 1].

or node equations) can be written directly. The method for constructing transform networks was then extended to include networks with magnetic couplings.

Table 4.5.1 A Summary of Network Theorems

Theorem	Assumptions	Statement
Tellegen	Lumped elements which obey KCL and KVL.	$\sum\limits_{k=1}^{b} v_k i_k = 0$
Superposition	Linear network.	The response to a number of excitations acting simultaneously is equal to the algebraic sum of the responses to the individual excitations acting alone.
Reciprocity	Linear, passive, bilateral elements with zero initial condition. One source only.	The ratio of the response transform to the excitation transform is invariant upon the interchange of the positions of excitation and observation provided that the topology of the network is unaltered.
Substitution	None.	A known voltage (current) can be replaced by an independent voltage (current) source.

Table 4.5.2 Equivalent Representation of R, L, and C

t-domain	s-domain

t-domain	s-domain	
	With voltage source equivalent	With current source equivalent
$i_L(t)$ A + L $i_L(0^+)$ $v_L(t)$ − B	$I_L(s)$ A $Li_L(0^+)$ + $V_L(s)$ $V_0(s)$ $Z_L(s) = Ls$ − B	$I_L(s)$ A $Y_L(s) = \dfrac{1}{Ls}$ $\dfrac{i_L(0^+)}{s}$ $V_L(s)$ − B
$i_C(t)$ A + C $v_C(0^+)$ $v_C(t)$ − B	$I_C(s)$ A + $\dfrac{v_C(0^+)}{s}$ $V_C(s)$ $Z_C(s) = \dfrac{1}{Cs}$ − B	$I_C(s)$ A + $Y_C(s) = Cs$ $Cv_C(0^+)$ $V_C(s)$ − B

REFERENCES

1. M. E. Van Valkenburg, *Network Analysis*, 2nd Ed., Prentice-Hall, 1964, Chapters 5 and 9.

2. C. A. Desoer and E. S. Kuh, *Basic Circuit Theory*, McGraw-Hill, 1967, Chapter 14.

3. E. J. Craig, *Laplace and Fourier Transforms for Electrical Engineers*, Holt, Rinehart & Winston, 1964, Chapter 4.

4. B. C. Kuo, *Linear Networks and Systems*, McGraw-Hill, 1967, Chapter 9.

5. G. I. Atabekov, *Linear Network Theory*, translated by J. Yeoman, Pergamon Press, 1965, Chapter 4.

6. R. Rudenberg, *Transient Performance of Electric Power System*, McGraw-Hill, 1950, Chapter 37.

PROBLEMS

4.1 Use Tellegen's theorem to verify the reciprocity property of (4.2.23).

4.2 The following measurements are made on a linear resistive network shown in Fig. P.4.2(a): $E = 20$ V, $I_1 = 10$ A, and $I_2 = 2$ A. If the current I'_1 in Fig. P.4.2(b) is to be equal to 4 A, what value must E'_2 have?

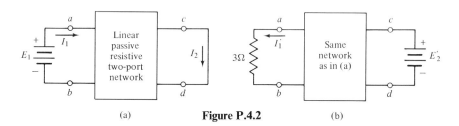

(a)   **Figure P.4.2**   (b)

4.3 The network of Fig. P.4.3 is initially at rest. At $t = 0$, the switch S is suddenly closed. Determine the voltage across the terminal-pair a–b as a function of time for $t > 0$. (*Hint*: Use the substitution theorem.)

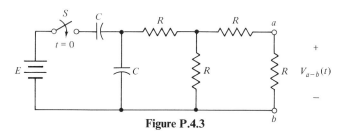

Figure P.4.3

4.4 As indicated in Fig. P.4.4(a), when the voltage E_1 is 20 V, the current I_1 is 5 A, and the current I_2 is 2 A. If a voltage source of $E'_2 = 30$ V is connected across terminals c–d and a 2-Ω resistor is placed in terminals a–b as indicated in Fig. P.4.4(b), determine the Norton equivalent network for the terminals a–b of Fig. P.4.4(b), and then calculate the current in the 2-Ω resistor.

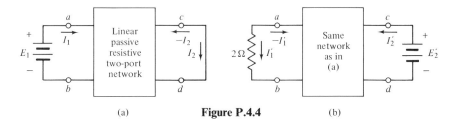

(a)   **Figure P.4.4**   (b)

4.5 Determine the Thévenin equivalent network for the determination of the current i_3 in the resistor R_3 in Fig. P.4.5.

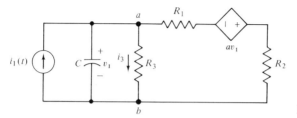

Figure P.4.5

4.6 The network of Fig. P.4.6 contains a dependent source. For zero initial conditions, determine

a) the Thévenin equivalent network at a–b, and
b) the Norton equivalent network at c–d.

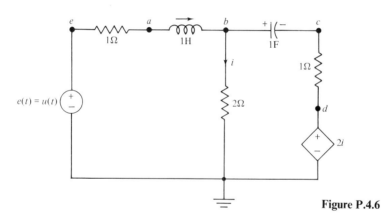

Figure P.4.6

4.7 For the network of Fig. P.4.7, determine the Norton equivalent network for the determination of the current i in the 1-Ω resistor.

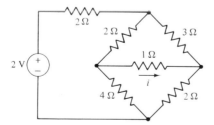

Figure P.4.7

4.8 The network of Fig. P.4.8 is initially at rest. For the element values given, with $e_1(t) = u(t)$, determine the initial conditions at $t = 0^+$.

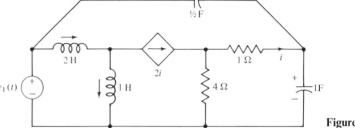

Figure P.4.8

4.9 The network of Fig. P.4.9 has been in steady state with the switch S open. At $t = 0$, S is suddenly closed. Determine the initial conditions at $t = 0^+$.

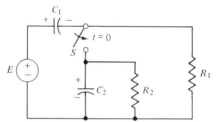

Figure P.4.9

4.10 For the element values given, with $e(t) = 10\ u(t)$, determine the initial values of the network shown in Fig. P.4.10.

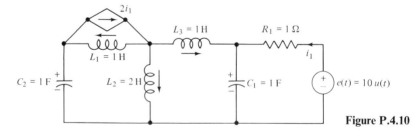

Figure P.4.10

4.11 Obtain a transform network for the network shown in Fig. P.4.11.

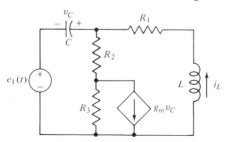

Figure P.4.11

4.12 Repeat Problem 4.11 for the network of Fig. P.4.9 for $t > 0$.

4.13 Refer to the network of Fig. P.4.13. A steady state has been reached with the switch S in position a. At $t = 0$, S is thrown to position b. Draw a transform network for $t > 0$, and then write the mesh equations in terms of the mesh currents shown.

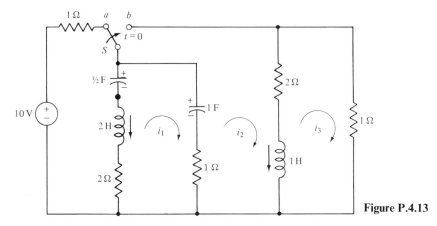

Figure P.4.13

4.14 Draw a transform network for the network of Fig. P.4.14, and then write the transform loop equations. Let M_{ij} be the mutual inductance between L_i and L_j $(i, j = 1, 2, 3)$.

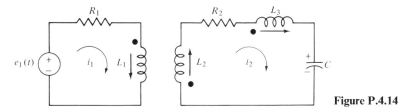

Figure P.4.14

4.15 Determine a transform network for the network of Fig. P.4.6 and then write the node equations.

CHAPTER 5

RESPONSES OF LINEAR NETWORKS

5.1 INTRODUCTION

In Chapter 3, we discussed different methods of formulating network equations which are a set of simultaneous, linear integro-differential equations in the time domain. In mathematical terms, the *general* (*complete*) *solution* to such a system of equations consists of two parts, namely the *complementary function* and *the particular integral.* The *complementary function* is the general solution of the homogeneous system (with the forcing functions set to zero) and is also known as the *zero-input solution* of the given network. The particular integral is the particular solution of the nonhomogeneous system with all arbitrary constants set to zero and is also called the *zero-state solution* of the network.* Thus, the *complete solution* is the sum of the zero-input and the zero-state solutions. We shall see that the zero-input solution is closely related to the element values as well as to the topology of the network itself, whereas the zero-state solution is the direct consequence of the application of energy sources (inputs) to the system.

Sometimes, the complete response of a network is, by convention, divided into: (a) the *transient response*, and (2) the *steady-state response.** Thus, we can regard the complete response as being composed of the transient and steady-state responses.

In the subsequent sections of this chapter, we shall discuss the responses of linear networks in both the time and the frequency domains using loop and node equations as well as the representations of network functions together with their properties. The standard methods for solving matrix state equations will be discussed in Chapter 7.

5.2 SOLUTIONS OF LOOP AND NODE EQUATIONS BY LAPLACE TRANSFORMATION

Network equations can be written either in the time-domain (*t*-domain) or in the frequency domain (*s*-domain) as discussed in Chapters 3 and 4. If the equations are given in the *t*-domain, a convenient method of obtaining the solution may be

---

* The existence of the steady-state solution of a network will be discussed in detail in Section 5.6.

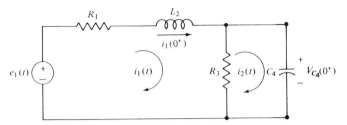

Fig. 5.2.1 The network of Fig. 4.4.4.

described by the following three steps:

a) Take the Laplace transforms of these equations.

b) Solve these transform equations for the desired quantities (currents and/or voltages).

c) Take the inverse Laplace transforms of these quantities to determine t-domain solutions.

Let us illustrate the above procedure by means of the examples that follow.

Example 5.2.1 Consider again the network of Fig. 4.4.4, which is repeated here for convenience in Fig. 5.2.1. The mesh equations for this network are:

$$(R_1 + R_3)i_1 + L_2 \frac{di_1}{dt} - R_3 i_2 = e_1(t)$$

$$- R_3 i_1 + R_3 i_2 + \frac{1}{C_4} \int i_2 \, dt = 0. \tag{5.2.1}$$

The corresponding set of transform equations is given by

$$(R_1 + R_3 + sL_2)I_1(s) - R_3 I_2(s) = E_1(s) + L_2 i_1(0^+)$$

$$- R_3 I_1(s) + \left(R_3 + \frac{1}{sC_4}\right) I_2(s) = - \frac{v_c(0^+)}{s}. \tag{5.2.2}$$

In matrix form, the set becomes

$$\begin{bmatrix} R_1 + R_3 + sL_2 & -R_3 \\ -R_3 & R_3 + \dfrac{1}{sC_4} \end{bmatrix} \begin{bmatrix} I_1(s) \\ I_2(s) \end{bmatrix} = \begin{bmatrix} E_1(s) + L_2 i_1(0^+) \\ -\dfrac{v_c(0^+)}{s} \end{bmatrix}. \tag{5.2.3}$$

We shall proceed to determine the mesh currents $i_1(t)$ and $i_2(t)$ for $t > 0$ with the following set of data:

$$R_1 = 1\Omega, \qquad L_2 = 2\mathrm{H}, \qquad R_3 = 3\Omega, \qquad C_4 = \tfrac{1}{4}\mathrm{F};$$

$$i_1(0^+) = 2\,\mathrm{A}, \qquad v_c(0^+) = 5\,\mathrm{V}, \qquad e_1(t) = 10\,u(t)\,\mathrm{V}. \tag{5.2.4}$$

Substituting the numerical values into (5.2.3), we find

$$
\begin{bmatrix} 4 + 2s & -3 \\ -3 & 3 + \dfrac{4}{s} \end{bmatrix} \begin{bmatrix} I_1(s) \\ I_2(s) \end{bmatrix} = \begin{bmatrix} \dfrac{10}{s} + 4 \\ -\dfrac{5}{s} \end{bmatrix},
$$
(5.2.5)

which can be solved to yield

$$
\begin{bmatrix} I_1(s) \\ I_2(s) \end{bmatrix} = \begin{bmatrix} 4 + 2s & -3 \\ -3 & 3 + \dfrac{4}{s} \end{bmatrix}^{-1} \begin{bmatrix} \dfrac{10}{s} + 4 \\ -\dfrac{5}{s} \end{bmatrix}
$$

$$
= \frac{s}{6s^2 + 11s + 16} \begin{bmatrix} \dfrac{3s + 4}{s} & 3 \\ 3 & 2s + 4 \end{bmatrix} \begin{bmatrix} \dfrac{4s + 10}{s} \\ -\dfrac{5}{s} \end{bmatrix}
$$

$$
= \begin{bmatrix} \dfrac{12s^2 + 31s + 40}{6s(s^2 + 1.835s + 2.67)} \\ \dfrac{2s + 10}{6(s^2 + 1.835s + 2.67)} \end{bmatrix}.
$$
(5.2.6)

The corresponding inverse Laplace transforms are given by

$$
\begin{bmatrix} i_1(t) \\ i_2(t) \end{bmatrix} = \begin{bmatrix} 2.5\, u(t) - 1.98\, e^{-0.918t} \cos(1.26t - 8°)\, u(t) \\ 1.13\, e^{-0.918t} \cos(1.26t - 72.8°)\, u(t) \end{bmatrix}.
$$
(5.2.7)

An examination of (5.2.7) indicates that both $i_1(t)$ and $i_2(t)$ contain an exponentially decaying term which is the transient portion of the solution while the steady-state value of $i_1(t)$ is $2.5\, u(t)$ and that of $i_2(t)$ is zero.

Example 5.2.2   The mesh equation for the network of Fig. 5.2.2 is

$$
L\frac{di}{dt} + Ri = e_s(t).
$$
(5.2.8)

If $e_s(t) = E\, u(t)$, the transform equation of (5.2.8) is given by

$$
(sL + R)I(s) = \frac{E}{s} + LI_0
$$

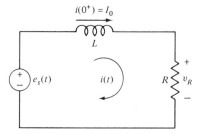

$i(0^+) = I_0$

L

$e_s(t)$ $i(t)$ $R \gtrless v_R$ $+$ $-$

Fig. 5.2.2 An RL network.

which can be solved for $I(s)$, or

$$I(s) = \frac{E}{s(sL + R)} + \frac{LI_0}{sL + R}$$

$$= \frac{E}{R}\left(\frac{1}{s} - \frac{1}{s + \frac{R}{L}}\right) + \frac{I_0}{s + \frac{R}{L}}.$$
(5.2.9)

Taking the inverse transform of (5.2.9), we obtain

$$i(t) = \frac{E}{R}(1 - e^{-(R/L)t})\, u(t) + I_0 e^{-(R/L)t}\, u(t).$$
(5.2.10)

If $v_R(t)$, the voltage across the resistor, is taken as the response of the network, then we have from (5.2.10):

$$v_R(t) = E(1 - e^{-(R/L)t})\, u(t) + I_0 R e^{-(R/L)t}\, u(t).$$
(5.2.11)

An inspection of (5.2.11) indicates that if $I_0 = 0$, we find

$$v_R(t)\Big|_{I_0 = 0} = E(1 - e^{-(R/L)t})\, u(t),$$
(5.2.12)

which is the zero-state response of the network because, in obtaining (5.2.12), the initial value I_0 of the *state* variable $i(t)$ (current in an inductor) was set to *zero*. On the other hand, if $E = 0$, we find that

$$v_R(t)\Big|_{E = 0} = I_0 R e^{-(R/L)t}\, u(t),$$
(5.2.13)

which is the zero-input response, since $E = 0$ implies that the *input* $e_s(t)$ is equal to *zero* for any value of t.

Example 5.2.2 clearly indicates that the complete response is the sum of the zero-state response and the zero-input response, as is always true for all linear networks. It should be emphasized that, for a linear network, the zero-state response is always a linear function of the input whereas the zero-input response is always a linear function of the initial state.

If (5.2.10) is rearranged as

$$i(t) = \left(I_0 - \frac{E}{R}\right)e^{-(R/L)t}\,u(t) + \frac{E}{R}\,u(t), \tag{5.2.14}$$

then the first term, a decaying exponential, is called the *transient response*; the second is called the *steady-state response*. This example illustrates that the transient response of a network, in general, depends on both the initial conditions as well as the input, whereas the steady-state response is a function of the input only.

The method of solving the node equations is identical to that of solving the loop equations since both the node and the loop equations have exactly the same mathematical form. If the behavior of a network is described by the state equations, several methods are available for the solution of these equations as will be discussed in Chapter 7.

5.3 NETWORK FUNCTIONS

In this section we shall discuss two types of network functions, namely, driving-point and transfer functions. These functions relate the transform quantities (voltage and current transforms) in the s-domain. A driving-point function can be expressed as a ratio of the transform voltage and the transform current associated with a specified pair of nodes called terminals of a given network. On the other hand, a transfer function relates two transform quantities associated with two different pairs of terminals in the network. We have seen, in Chapter 4, that the transform impedance (or simply impedance) of a two-terminal passive element (i.e., R, L, or C) is defined as the ratio of the transform voltage $V(s)$ across the element to the transform current $I(s)$ through the element with the initial condition set to zero. This concept of impedance can be applied equally well to two-terminal networks or one-ports in terms of driving-point functions. The *driving-point* (or simply *input*) *impedance* $Z_d(s)$ of a linear two-terminal network containing no *independent* sources is given by

$$Z_d(s) = \frac{V_1(s)}{I_1(s)}, \tag{5.3.1}$$

where $V_1(s)$ and $I_1(s)$ are, respectively, the transforms of the voltage and the current at the terminals with the network in zero-state (zero-initial conditions). Similarly, the reciprocal of $Z_d(s)$ is defined as the *driving-point admittance* $Y_d(s)$ of the network. That is,

$$Y_d(s) = \frac{1}{Z_d(s)} = \frac{I_1(s)}{V_1(s)}. \tag{5.3.2}$$

It is important to note that in the definitions of driving-point functions (impedance or admittance) only independent sources must be excluded in the two-terminal network. That is, the network may contain dependent sources as the following example will illustrate.

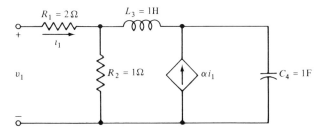

Fig. 5.3.1 Network with a dependent source.

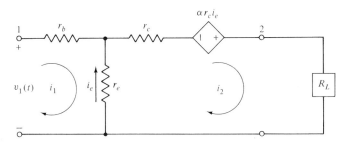

Fig. 5.3.2 The equivalent network of a grounded emitter transistor terminated in R_L.

Example 5.3.1 Figure 5.3.1 shows a network containing a dependent current source, αi_1, which is controlled by the current i_1 with an amplification factor α. We can show that

$$Z_d(s) = \frac{3s^2 + 2s + (\alpha + 3)}{s^2 + s + 1}$$

is the driving-point impedance of this network.*

For simple networks, *immittance* functions can usually be obtained by inspection. As an example, for a passive-ladder type two-terminal network consisting of only R, L, C elements such as the one in Fig. 4.2.21, the driving-point impedance may be obtained by combining the series and parallel branches (starting from the far right end of the network) into a single impedance. For more complicated networks, the Y–Δ or Δ–Y transformation may be helpful in the determination. A systematic procedure for determining the driving-point immittance would be to write either the transform loop or node equations and then solve for the quantities to form the required function. The next example will illustrate this point.

Example 5.3.2 The network of Fig. 5.3.2 is the small-signal equivalent circuit of a grounded-emitter transistor terminated in a resistive load R_L. If we assign the

---

* This is left as an exercise.

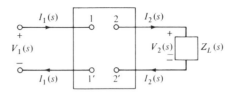

Fig. 5.3.3 Network for the definitions of the transfer functions.

loop currents i_1 and i_2 as shown in the figure, the transform loop equations can be shown to be

$$(r_b + r_e)I_1(s) - r_e I_2(s) = V_1(s) \tag{5.3.3}$$

$$-r_e I_1(s) + (r_e + r_c + R_L)I_2(s) = \alpha r_c I_e(s). \tag{5.3.4}$$

Since $I_e(s) = I_2(s) - I_1(s)$, (5.3.4) can be rewritten as

$$-[r_e - \alpha r_c]I_1(s) + [r_e + R_L + r_c(1 - \alpha)]I_2(s) = 0. \tag{5.3.5}$$

Solving (5.3.3) and (5.3.5) for $I_1(s)$, we find that

$$I_1(s) = \frac{[r_e + R_L + r_c(1 - \alpha)]V_1(s)}{r_b[r_e + R_L + r_c(1 - \alpha)] + r_e(r_c + R_L)}. \tag{5.3.6}$$

Finally, $Z_d(s)$ is given by

$$Z_d(s) = \frac{V_1(s)}{I_1(s)} = r_b + \frac{r_e(r_c + R_L)}{r_e + R_L + r_c(1 - \alpha)}.$$

While a driving-point function relates the current and the voltage transforms associated with a single pair of terminals of a network, a transfer function relates two transform quantities associated with two different pairs of terminals of a network. Since there are four transform quantities associated with the two pairs of terminals, eight different ratios of these transform quantities can be obtained. However, four of these ratios are merely the inverses of the other four. Thus, we shall define only four transfer functions.

Refer to the network of Fig. 5.3.3 where the two pairs of terminals, 1–1' and 2–2', are of interest and the assigned voltage and current directions are as depicted in the figure. The transfer functions of the four-terminal (two-port) network are tabulated in Table 5.3.1.

As in the case of driving-point functions, the definitions of the transfer functions are based on the following assumptions:

1. The network contains no independent sources; and
2. The network is in zero-state.

An inspection of Table 5.3.1 indicates that each of the transfer functions is defined as a ratio of two transform quantities with respect to different terminal-

Table 5.3.1

Symbol and definition	Name of the function
$Z_{21} = \dfrac{V_2(s)}{I_1(s)}$	Transfer impedance
$Y_{21} = \dfrac{I_2(s)}{V_1(s)}$	Transfer admittance
$G_{21} = \dfrac{V_2(s)}{V_1(s)}$	Voltage ratio transfer function $\checkmark$
$\alpha_{21} = \dfrac{I_2(s)}{I_1(s)}$	Current ratio transfer function

pairs. The first subscript of the symbol of each transfer function refers to the quantity in the numerator (output) of the defining ratio while the other subscript refers to that in the denominator (input).

The procedure for determining the transfer functions of a four-terminal network is the same as that for determining the driving-point functions, as the next example will illustrate.

Example 5.3.3 Let us determine the current ratio transfer function $\alpha_{21}(s)$ for the network of Fig. 5.3.2. From the loop equations (5.3.3) and (5.3.5) one can solve for $I_2(s)$

$$I_2(s) = \frac{[r_e - \alpha r_c]V_1(s)}{r_b[r_e + R_L + r_c(1 - \alpha)] + r_e(r_c + R_L)}. \tag{5.3.7}$$

The ratio of (5.3.7) to (5.3.6) yields

$$\alpha_{21}(s) = \frac{I_2(s)}{I_1(s)} = \frac{r_e - \alpha r_c}{r_e + R_L + r_c(1 - \alpha)}$$

which is also known as the *current gain* of the network.

As will be discussed later in the chapter, network functions play an important role in obtaining network responses. Before discussing the methods for determining these responses, we shall first study how a network function can be represented as well as some of its properties.

5.4 REPRESENTATION OF NETWORK FUNCTIONS

The network functions obtained in the last section are real rational functions of the complex frequency variable s; that is, they are ratios of polynomials in s with real

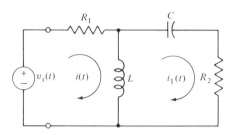

Fig. 5.4.1 Network for Example 5.4.1.

coefficients.* In this section, we shall show that all the network functions of a linear passive network can always be expressed as a real rational function in s.

We shall begin by considering a linear network in which any two quantities $x(t)$ and $y(t)$ (one may be a current and the other a voltage referring to the same or different points of the network, or they may be both voltages or currents in different points of the network) are always related by a differential equation of the form

$$\left(a_n \frac{d^n}{dt^n} + a_{n-1} \frac{d^{n-1}}{dt^{n-1}} + \cdots + a_1 \frac{d}{dt} + a_0\right) y(t)$$

$$= \left(b_m \frac{d^m}{dt^m} + b_{m-1} \frac{d^{m-1}}{dt^{m-1}} + \cdots + b_1 \frac{d}{dt} + b_0\right) x(t), \quad (5.4.1)$$

where the constants $a_1, a_2, \ldots, a_n$, and $b_1, b_2, \ldots, b_m$ depend on the element values as well as the structure of the network. The differential equation (5.4.1) is obtained from the Kirchhoff-law equations and the branch voltage-current relations of the network. By means of the following example, we shall illustrate the method of writing the differential equation which relates any two variables of a network.

Example 5.4.1 For the network of Fig. 5.4.1, the differential equation relating the source voltage $v_s(t)$ and the current $i(t)$ can be obtained by first writing the mesh equations

$$v_s = R_1 i + L \frac{di}{dt} - L \frac{di_1}{dt}$$

$$= (R_1 + Lp)i - Lpi_1, \quad (5.4.2)$$

$$0 = -L \frac{di}{dt} + L \frac{di_1}{dt} + R_2 i_1 + \frac{1}{C} \int i_1 \, dt$$

$$= -Lpi + \left(Lp + R_2 + \frac{1}{Cp}\right) i_1, \quad (5.4.3)$$

---

* In the trivial case of a purely resistive network, the function becomes a constant. Examples of this are the functions I_2/V_1 and α_{12} in Example 5.3.3.

where p and $1/p$ are, respectively, the *differential* and *integral operators* defined by

$$pF(t) = \frac{d}{dt} F(t) \tag{5.4.4}$$

and

$$\frac{1}{p} F(t) = \int F(t)\, dt \tag{5.4.5}$$

for any function $F(t)$ of the variable t. It should be noted that in terms of the differential and integral operators, the mesh equations (5.4.2) and (5.4.3) become simply algebraic equations. Thus, the current response $i(t)$ can be found by eliminating $i_1(t)$ between (5.4.2) and (5.4.3):

$$i(t) = \frac{\begin{vmatrix} v_s & -Lp \\[2mm] 0 & Lp + R_2 + \dfrac{1}{Cp} \end{vmatrix}}{\begin{vmatrix} R_1 + Lp & -Lp \\[2mm] -Lp & Lp + R_2 + \dfrac{1}{Cp} \end{vmatrix}} = \frac{\left(Lp + R_2 + \dfrac{1}{Cp}\right) v_s(t)}{(R_1 + Lp)\left(Lp + R_2 + \dfrac{1}{Cp}\right) - L^2 p^2}$$

$$= \frac{\left(Lp + R_2 + \dfrac{1}{Cp}\right) v_s(t)}{(LR_2 + LR_1)p + \left(R_1 R_2 + \dfrac{L}{C}\right) + \dfrac{R_1}{Cp}}$$

$$= \frac{\left(Lp^2 + R_2 p + \dfrac{1}{C}\right) v_s(t)}{(LR_2 + LR_1)p^2 + \left(R_1 R_2 + \dfrac{L}{C}\right)p + \dfrac{R_1}{C}},$$

or

$$\left[(LR_2 + LR_1)p^2 + \left(R_1 R_2 + \frac{L}{C}\right)p + \frac{R_1}{C}\right] i(t) = \left[Lp^2 + R_2 p + \frac{1}{C}\right] v_s(t). \tag{5.4.6}$$

Equation (5.4.6) corresponds to the differential equation

$$\left[(LR_2 + LR_1)\frac{d^2}{dt^2} + \left(R_1 R_2 + \frac{L}{C}\right)\frac{d}{dt} + \frac{R_1}{C}\right] i(t)$$

$$= \left[L\frac{d^2}{dt^2} + R_2 \frac{d}{dt} + \frac{1}{C}\right] v_s(t). \tag{5.4.7}$$

Thus, (5.4.7) is the desired differential equation which relates the two quantities $v_s(t)$ and $i(t)$ of the network of Fig. 5.4.1.

Taking the Laplace transform of (5.4.1) with all the initial conditions set to zero, we find

$$(a_n s^n + a_{n-1} s^{n-1} + \cdots + a_1 s + a_0) Y(s)$$
$$= (b_m s^m + b_{m-1} s^{m-1} + \cdots + b_1 s + b_0) X(s) \qquad (5.4.8)$$

from which the function relating the transformed variables $X(s)$ and $Y(s)$ of the network is given by

$$H(s) \triangleq \frac{X(s)}{Y(s)} = \frac{a_n s^n + a_{n-1} s^{n-1} + \cdots + a_1 s + a_0}{b_m s^m + b_{m-1} s^{m-1} + \cdots + b_1 s + b_0}. \qquad (5.4.9)$$

Since each of the a and b constants is made up of a sum of products of element values, the coefficients of both numerator and denominator polynomials of $H(s)$ must be real constants. Denoting $p(s)$ and $q(s)$ as the numerator and denominator polynomials, $H(s)$ can be expressed in factored form as

$$H(s) = \frac{p(s)}{q(s)} = K \frac{(s - z_1)(s - z_2) \cdots (s - z_n)}{(s - p_1)(s - p_2) \cdots (s - p_m)} = K \frac{\prod_{i=1}^{n} (s - z_i)}{\prod_{j=1}^{m} (s - p_j)}, \qquad (5.4.10)$$

where $K = a_n/b_m$ is the scale factor, $z_i(i = 1, \ldots, n)$ is called the ith *zero* of $H(s)$, and $p_j(j = 1, \ldots, m)$ is called the jth *pole* of $H(s)$. From (5.4.10) it is evident that the zeros of $H(s)$ are simply the roots of $p(s) = 0$ whereas the poles of $H(s)$ are the roots of $q(s) = 0$. If, in (5.4.10), all the zeros and poles are *distinct*, then they are referred to as *simple* zeros and poles; otherwise, $H(s)$ can be written as

$$H(s) = K \frac{(s - z_1)^{k_1}(s - z_2)^{k_2} \cdots (s - z_r)^{k_i}}{(s - p_1)^{l_1}(s - p_2)^{l_2} \cdots (s - p_t)^{l_j}}, \qquad (5.4.11)$$

where $r \leq n$, $k_1 + k_2 + \cdots + k_i = n$, $t \leq m$, and $l_1 + l_2 + \cdots + l_j = m$. In (5.4.11), z_1 is called a zero of $H(s)$ of *order* k_1.* Similarly, z_2 is a zero [of $H(s)$] of order k_2, p_1 is a pole [of $H(s)$] of order l_1, p_2 is a pole of order l_2, and so forth.

An inspection of (5.4.10) or (5.4.11) indicates that the network function is completely specified once the poles and zeros, together with the scale factor K, are known. Therefore the locations of the poles and the zeros of a network function play a very important role in network analysis. To investigate further how the locations of poles and zeros affect the behavior of a network function, we shall classify $H(s)$ into two classes according to whether the two quantities $x(t)$ and $y(t)$ are referred to the same point of the network: (1) driving-point immittance (impedance or admittance) functions and (2) transfer functions.

---

* The *order* of a zero (pole) is also known as the *multiplicity* of that zero (pole). Thus, z_1 may also be described as a zero of multiplicity k_1.

Driving-Point Immittance Functions

If, in (5.4.1), both the excitation and the response of a network are referred to the same terminal-pair, then one must be a voltage and the other must be a current. Without loss of generality, we shall assume for convenience that $x(t) = e(t)$ is an excitation and $y(t) = i(t)$ is a response so that (5.4.1) becomes

$$\left(a_n \frac{d^n}{dt^n} + a_{n-1} \frac{d^{n-1}}{dt^{n-1}} + \cdots + a_1 \frac{d}{dt} + a_0\right)i(t)$$

$$= \left(b_m \frac{d^m}{dt^m} + b_{m-1} \frac{d^{m-1}}{dt^{m-1}} + \cdots + b_1 \frac{d}{dt} + b_0\right)e(t). \tag{5.4.12}$$

The Laplace transform of (5.4.12) yields the impedance function*

$$Z_d(s) = \frac{E(s)}{I(s)} = \frac{a_n s^n + a_{n-1}s^{n-1} + \cdots + a_1 s + a_0}{b_m s^m + b_{m-1}s^{m-1} + \cdots + b_1 s + b_0}. \tag{5.4.13}$$

The transient response is simply the solution of the homogeneous equation

$$\left(a_n \frac{d^n}{dt^n} + a_{n-1} \frac{d^{n-1}}{dt^{n-1}} + \cdots + a_1 \frac{d}{dt} + a_0\right)i(t) = 0, \tag{5.4.14}$$

which is obtained from (5.4.12) with $e(t) = 0$ for all t. Assuming that the solution of (5.4.14) is of the form

$$i(t) = Ae^{st}, \tag{5.4.15}$$

(5.4.15), after substitution into (5.4.14), leads to

$$(a_n s^n + a_{n-1}s^{n-1} + \cdots + a_1 s + a_0)Ae^{st} = 0. \tag{5.4.16}$$

A nontrivial solution for $i(t)$ demands that

$$(a_n s^n + a_{n-1}s^{n-1} + \cdots + a_1 s + a_0) = 0, \tag{5.4.17}$$

which is, of course, the characteristic equation of (5.4.14). Let $z_1, z_2, \ldots, z_n$ be the roots of (5.4.17), which can be recognized as the *zeros* of $Z_d(s)$. Then the natural (transient) response is given by†

$$i(t) = A_1 e^{z_1 t} + A_2 e^{z_2 t} + \cdots + A_n e^{z_n t} \tag{5.4.18}$$

with the constants $A_1, A_2, \ldots, A_n$ to be determined by the initial conditions of the network. Since $e(t) = 0$ corresponds to the short-circuit condition, the zeros of the impedance function $Z_d(s)$ can be interpreted as the *short-circuit complex natural frequencies* of the network.

---

* The same analysis applies to the admittance function $Y_d(s)$, since one is simply the reciprocal of the other.

† Here, for convenience, the roots $z_1, z_2, \ldots, z_n$ are assumed to be distinct.

If the excitation is the current $i(t)$, then the transient response of the network is the solution of

$$\left(b_m \frac{d^m}{dt^m} + b_{m-1} \frac{d^{m-1}}{dt^{m-1}} + \cdots + b_1 \frac{d}{dt} + b_0 \right) e(t) = 0 \qquad (5.4.19)$$

which is obtained from (5.4.12) with $i(t) = 0$. An assumed nontrivial solution of the form

$$e(t) = Be^{st} \qquad (5.4.20)$$

requires that

$$b_m s^m + b_{m-1} s^{m-1} + \cdots + b_1 s + b_0 = 0. \qquad (5.4.21)$$

If the roots of the characteristic equation (5.4.21) are denoted by $p_1, p_2, \ldots, p_m$, which are simply the poles of $Z_d(s)$ in (5.4.13), then the natural (transient) response $e(t)$ is given by

$$e(t) = B_1 e^{p_1 t} + B_2 e^{p_2 t} + \cdots + B_m e^{p_m t}, \qquad (5.4.22)$$

where the constants $B_1, B_2, \ldots, B_m$ are again determined by the initial conditions of the network. Since $i(t) = 0$ corresponds to the open-circuit condition, the poles of the impedance function $Z_d(s)$ can be regarded as the *open-circuit complex natural frequencies* of the network.

Since the natural response of a *passive* physical network must decrease to zero as a function of t as the energy originally stored in the network dissipates gradually through the resistive elements in the form of heat, this leads to the conclusion that the poles and zeros of $Z_d(s)$ must have *negative real* parts. However, in the special case of a lossless system in which the stored energy is conserved, the poles and zeros of $Z_d(s)$ must be purely imaginary numbers and they must be simple.*

Since the behavior of a network function is determined by the poles and zeros together with the scale factor K, it is sometimes convenient to represent the given function pictorially by plotting the locations of the *critical frequencies* (zeros and poles) of the function in the s-plane called the *pole-zero plot*. The following example will illustrate the steps required for making the plot.

Example 5.4.2   Obtain the pole-zero plot for the function

$$F(s) = \frac{3s^2 + 3s - 6}{s^4 + 2s^3 + 5s^2 + 8s + 4}.$$

Expressing $F(s)$ in factored form, we find that

$$F(s) = 3 \frac{(s - 1)(s + 2)}{(s + j2)(s - j2)(s + 1 + j1)(s + 1 - j1)}.$$

---

* The reader is invited to prove it as an exercise.

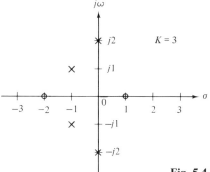

Fig. 5.4.2 The pole-zero plot of $F(s)$ in Example 5.4.2.

Obviously, the poles occur at $-j2$, $+j2$, $-1 - j1$, and $-1 + j1$, and the zeros are located at 1 and -2. Using the symbols $\times$ and $\circ$ to designate the poles and zeros, respectively, the pole-zero plot is drawn as indicated in Fig. 5.4.2.

It should be noted that, in addition to the zeros indicated in Fig. 5.4.2, there is a zero of order 2 (counted as two zeros) at $s = \infty$ since the numbers of poles and zeros for any network function are always equal when all the poles and zeros including those at $s = 0$ and $s = \infty$ are taken into consideration.* Furthermore, it is interesting to point out that, in the above example, one of the zeros has a positive real part, and hence the given function $F(s)$ cannot be a driving-point immittance function of a passive network. Next, we shall investigate the behavior of a transfer function.

Transfer Functions

If, in (5.4.1), the excitation and the response are in different points of a network, e.g., if the excitation is in terminal-pair "1" while the response is considered to be in terminal-pair "2," then, for convenience, we can rewrite (5.4.1) as follows:

$$\left(a_n \frac{d^n}{dt^n} + a_{n-1} \frac{d^{n-1}}{dt^n} + \cdots + a_1 \frac{d}{dt} + a_0 \right) e_s(t)$$

$$= \left(b_m \frac{d^m}{dt^m} + b_{m-1} \frac{d^{m-1}}{dt^{m-1}} + \cdots + b_1 \frac{d}{dt} + b_0 \right) v(t), \qquad (5.4.23)$$

where, without loss of generality, both the excitation and the response are assumed to be voltages with $e_s(t)$ as excitation. Taking the Laplace transform of (5.4.23) *with initial conditions set to zero* and defining the transfer function $H(s)$ as the ratio of the transformed response $V(s)$ to the transformed excitation $E_s(s)$, we find that

$$H(s) \triangleq \frac{V(s)}{E_s(s)} = \frac{a_n s^n + a_{n-1} s^{n-1} + \cdots + a_1 s + a_0}{b_m s^m + b_{m-1} s^{m-1} + \cdots + b_1 s + b_0}. \qquad (5.4.24)$$

---

* The reader is invited to verify this statement as an exercise.

In factored form, (5.2.24) becomes

$$H(s) = K \frac{(s - z_1)(s - z_2) \cdots (s - z_n)}{(s - p_1)(s - p_2) \cdots (s - p_m)}, \tag{5.4.25}$$

where $K \triangleq a_n/b_m$.

If the excitation $e_s(t)$ is set to zero, then the short-circuit voltage response $v(t)$ is the solution of the homogeneous equation

$$\left(b_m \frac{d^m}{dt^m} + b_{m-1} \frac{d^{m-1}}{dt^{m-1}} + \cdots + b_1 \frac{d}{dt} + b_0 \right) v(t) = 0. \tag{5.4.26}$$

The natural (transient) response in this case is then given by

$$v(t) = C_1 e^{p_1 t} + C_2 e^{p_2 t} + \cdots + C_m e^{p_m t}, \tag{5.4.27}$$

where $C_1, C_2, \ldots, C_m$ are constants to be determined by applying initial conditions and $p_1, p_2, \ldots, p_m$ are the roots of the characteristic equation

$$b_m s^m + b_{m-1} s^{m-1} + \cdots + b_1 s + b_0 = 0. \tag{5.4.28}$$

Clearly, these characteristic values $p_1, p_2, \ldots, p_m$ are the poles of the transfer function $H(s)$ defined in (5.4.24). Again, for a passive physically realizable network, the transient response must decrease to zero as a function of time because the energy originally stored in the system must dissipate gradually through the resistive elements in the form of heat. This fact leads to the conclusion that the *poles of a transfer function must have negative real parts* (an equivalent statement would be that the poles must lie on the left-half of the complex s-plane). However, in the limiting case of a lossless system, the poles can lie on the $j\omega$-axis of the s-plane, but they must be simple. On the other hand, if the response $v(t)$ is set to zero, (5.4.23) becomes

$$\left(a_n \frac{d^n}{dt^n} + a_{n-1} \frac{d^{n-1}}{dt^{n-1}} + \cdots + a_1 \frac{d}{dt} + a_0 \right) e_s(t) = 0 \tag{5.4.29}$$

which amounts to a particular linear combination of the excitation function $e_s(t)$ and its derivatives of different orders such that the response of the network is zero. Since the roots of the characteristic equation associated with (5.4.29) are the zeros of the transfer function $H(s)$, the above argument leads to the conclusion that *no restrictions can be placed on the zero locations of the transfer function of a general passive network*. In fact, *they can lie anywhere in the complex s-plane with any order of multiplicity*, as the following example will illustrate.

Example 5.4.3 Refer to the network of Fig. 5.4.3. The voltage ratio

$$H(s) \triangleq \frac{V_o(s)}{V_i(s)}$$

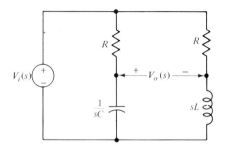

Fig. 5.4.3 Illustration of zeros of transmission.

can be readily shown to be

$$H(s) = \frac{V_o(s)}{V_i(s)}$$

$$= \frac{R(1 - s^2 LC)}{(1 + sCR)(R + sL)}. \tag{5.4.30}$$

From (5.4.30), it is evident that the zeros of $H(s)$ occur at $s = \pm 1/\sqrt{LC}$, viz., one on the left-half and the other on the right-half of the s-plane.

Remark. Since the zeros of $H(s)$ are the complex frequencies at which the network response is zero, they are sometimes referred to as the *zeros of transmission.*

The above discussion on the locations of zeros and poles of both driving-point immittances and transfer functions of a *passive* network leads to the following observations:

a) The poles and zeros of both driving-point immittances and transfer functions are either real or they occur in conjugate pairs since the coefficients of both the numerator and the denominator polynomials of an immittance or a transfer function are made up of some combinations of element values which are real and positive.

b) All poles and zeros of driving-point immittance functions have negative-real parts. This restriction also applies to the poles of a transfer function but the zeros can lie anywhere in the s-plane.

c) In the special case of a lossless system, the poles and zeros of an immittance function lie on the $j\omega$-axis and must be of simple order. This rule applies to the poles of a transfer function as well.

There are, of course, other properties of network functions in addition to these listed above; the interested reader is referred to literature elsewhere for a more detailed discussion on this topic.*

---

* See, for example, Van Valkenburg, [VA 1].

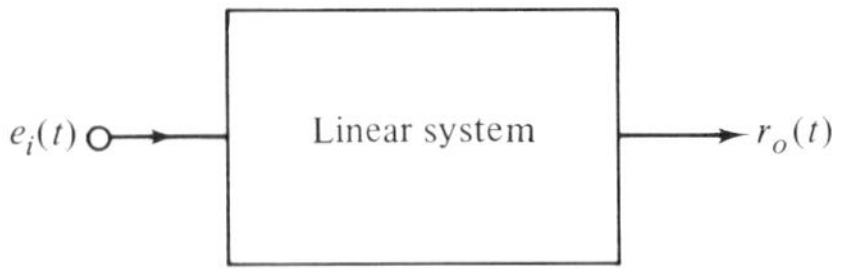

Fig. 5.5.1 A network considered as a single-input–single-output system.

5.5 NETWORK RESPONSES

In Section 5.2, the method of solving the loop or node equation by Laplace transformation was discussed. The solution of a system of loop or node equations together with the set of VCR-equations enables one to determine the responses at all points of the network. However, in most cases, we are not interested in determining the voltage and the current in every branch of the network. Instead, a network is often considered as a single-input–single-output system as depicted in Fig. 5.5.1, where $e_i(t)$ is the voltage or current excitation applied at some point of the network, $r_o(t)$ is the voltage or current response taken at the same or any other point of the network, and where the objective is to determine the zero-state response to various types of excitation. To this end, it would be most appropriate to use the transfer function $H(s)$ since it relates the input and the output directly in the following form:

$$H(s) \triangleq \frac{R_o(s)}{E_i(s)}, \tag{5.5.1}$$

where, of course, $R_o(s)$ and $E_i(s)$ are, respectively, the response and excitation transforms. This definition of transfer function in the system point of view is more general in the sense that it covers all the four transfer functions defined in Table 5.3.1. Thus, for example, if $R_o(s) = V_2(s)$ and $E_i(s) = I_1(s)$, then (5.5.1) becomes

$$H(s) = \frac{V_2(s)}{I_1(s)}$$

which is simply the transfer impedance defined in Section 5.3. From (5.5.1) the transform output $R_o(s)$ is given by

$$R_o(s) = H(s)E_i(s). \tag{5.5.2}$$

Equation (5.5.2) indicates that once $H(s)$ is specified, $R_o(s)$ can be obtained for any input function as long as it is Laplace transformable, and that taking the inverse transform of $R_o(s)$ yields the required *zero-state* response.

A physical interpretation of the transfer function $H(s)$ is possible. If the input $e_i(t)$ is a unit impulse, that is,

$$e_i(t) = \delta(t) \tag{5.5.3}$$

and hence $E_i(s) = 1$, then (5.5.2) becomes

$$R_o(s) = H(s). \tag{5.5.4}$$

In other words, the transfer function $H(s)$ is simply the Laplace transform of the *impulse response denoted by h(t)** (i.e., the response to the input of a unit impulse) of the system, or

$$r_o(t) = h(t).$$ (5.5.5)

For any other input, the zero-state response $r_o(t)$ is obtained by taking the inverse transform of (5.5.2)†

$$r_o(t) = \int_0^t h(\tau)e_i(t - \tau)\, d\tau$$

$$= \int_0^t e_i(\tau)h(t - \tau)\, d\tau, \qquad \text{for } t > 0.$$ (5.5.6)

It should be noted that (5.5.6), a convolution integral, can also be derived directly in the time-domain by using the properties of linearity and time-invariance.‡ It can be shown § that the zero-state response $r_o(t)$ of a system at time t caused by the input $e_i(t)$ applied at t_0 is given by

$$r_o(t) = \int_{t_0}^t e_i(\tau)h(t - \tau)\, d\tau$$

$$= \int_0^{t-t_0} h(\tau)e_i(t - \tau)\, d\tau, \qquad \text{for } t \geq t_0,$$ (5.5.7)

which, of course, reduces to (5.5.6) when $t_0 = 0$. The usefulness of (5.5.6) or (5.5.7) lies in the fact that, in many cases, both the impulse response $h(t)$ and the input $e_i(t)$ to a given system exist only in graphical form as a result of experimental measurements, so that analytical expressions, especially in closed forms, are not obtainable for $H(s)$ or $E_i(s)$ and hence the Laplace transform method cannot be used for obtaining the solution. Furthermore, even if the analytical expressions of both $h(t)$ and $e_i(t)$ are available, the Laplace transform method may become extremely difficult to apply for systems of moderate complexity. On the other hand, the solution of a convolution integral can be readily obtained without any difficulty by the use of a digital computer even for complicated systems.

Let us now illustrate some of the ideas discussed in this section by considering the following examples.

Example 5.5.1 Consider the parallel *RC* network *N* of Fig. 5.5.2. We shall proceed to determine the impulse response $h(t)$ with $i_1(t)$ and $v_2(t)$ taken to be the input and the output, respectively. The transfer function $H(s)$ in this case becomes

$$H(s) = \frac{V_2(s)}{I_1(s)} = \frac{1}{s + \frac{1}{2}}.$$ (5.5.8)

---

* In some literature, $h(t)$ is also called the Green's function.
† See Appendix A.2.
‡ See Problem 5.21.
§ See Problem 5.22.

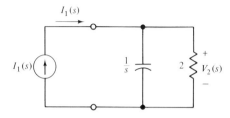

Fig. 5.5.2 Parallel RC network N.

Hence the impulse response $h(t)$ is given by

$$h(t) = \mathscr{L}^{-1}\{H(s)\} = e^{-(1/2)t}\,u(t). \tag{5.5.9}$$

Example 5.5.2   Let us determine by two different methods the zero-state response of the network used in Example 5.5.1 to a unit-step input.

Method A: The Laplace transform method.   Since $i_1(t) = u(t)$, we have

$$I_1(s) = \frac{1}{s}. \tag{5.5.10}$$

Substituting (5.5.10) into (5.5.8), we find that

$$V_2(s) = \frac{1}{s(s + \frac{1}{2})} = \frac{2}{s} - \frac{2}{s + \frac{1}{2}}$$

and hence that

$$v_2(t) = 2(1 - e^{-(1/2)t})u(t). \tag{5.5.11}$$

Method B: The impulse response method.   We obtain $v_2(t)$ directly in the t-domain by using the convolution integral (5.5.6). With $h(t)$ given in (5.5.9), the first integral of (5.5.6) can be written as

$$
\begin{aligned}
v_2(t) &= \int_0^t h(\tau)u(t - \tau)\,d\tau \\
&= \int_0^t e^{-(1/2)\tau}u(\tau)u(t - \tau)\,d\tau \\
&= \int_0^t e^{-(1/2)\tau}\,d\tau \\
&= \left. \frac{e^{-(1/2)\tau}}{-\frac{1}{2}} \right|_0^t \\
&= 2(1 - e^{-(1/2)t})u(t),
\end{aligned}
\tag{5.5.12}
$$

which is, of course, identical to (5.5.11).

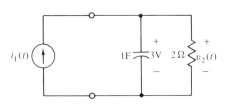

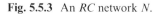

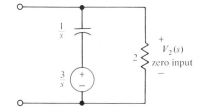

Fig. 5.5.3 An RC network N.

Fig. 5.5.4 Transform network N' of N of Fig. 5.5.2 with $I_1(s)$ removed.

It should be emphasized that the solution obtained by the impulse response method is actually the zero-state response of the system. Therefore, if the given system is not in zero-state, the zero-input response must be determined first and then added to the zero-state response to obtain the complete response as the following example will illustrate.

Example 5.5.3 Consider the network N of Fig. 5.5.3, which is the same as that of Fig. 5.5.1 with a three-volt initial voltage across the capacitor as indicated in the figure. Let us determine the unit-step response $v_2(t)$. From Example 5.5.2, the zero-state response was found to be

$$v_2(t) \underset{\text{zero-state}}{} = 2(1 - e^{-(1/2)t})u(t). \tag{5.5.13}$$

To determine the zero-input response, we remove the current source $i_1(t)$ from N and obtain the transform network N' of the resulting network as shown in Fig. 5.5.4. From N' we find

$$V_2(s) \underset{\text{zero-input}}{} = \left(\frac{2}{2 + (1/s)}\right)\left(\frac{3}{s}\right) = \frac{3}{s + (1/2)} \tag{5.5.14}$$

and hence

$$v_2(t) \underset{\text{zero-input}}{} = 3e^{-(1/2)t}u(t). \tag{5.5.15}$$

Finally by virtue of the principle of superposition, we add (5.5.13) and (5.5.15) to yield the unit-step response of the network

$$v_2(t) = 2(1 - e^{-(1/2)t})u(t) + 3e^{-(1/2)t}u(t)$$

$$= (2 + e^{-(1/2)t})u(t). \tag{5.5.16}$$

5.6 SINUSOIDAL STEADY-STATE ANALYSIS

All excitations are either periodic or nonperiodic in nature. A nonperiodic waveform is one in which we cannot find a finite repetition period; otherwise it is a periodic function. Periodic waveforms can be further classified as either sinusoidal or nonsinusoidal. In this section, we are mainly concerned with networks excited by sinusoidal inputs since they play a very important role in science and engineering. The study of sinusoidal steady-state responses of linear networks is of great importance, because most physical systems in our daily life such as electrical apparatus, lights, etc., use sinusoidal sources. Furthermore, responses of a linear network to excitations of various forms can be readily derived from the sinusoidal steady-state response of the network by invoking the principle of superposition.

The Complete Response and the Sinusoidal Steady-State Response

As discussed in the preceding sections, the response $y(t)$ of a linear network to an excitation $x(t)$ is related by a linear differential equation of the form

$$a_n \frac{d^n y}{dt^n} + a_{n-1} \frac{d^{n-1} y}{dt^{n-1}} + \cdots + a_1 \frac{dy}{dt} + a_0 y$$

$$= b_m \frac{d^m x}{dt^m} + b_{m-1} \frac{d^{m-1} x}{dt^{m-1}} + \cdots + b_1 \frac{dx}{dt} + b_0 x. \quad (5.6.1)$$

If $x(t)$ is sinusoidal, the response $y(t)$ of (5.6.1) can be regarded as the response of the network to a sum of m sinusoidal excitations since the derivative of a sinusoid is still a sinusoid of the same frequency. Thus, the response $y(t)$ can be obtained by first finding the response of the network to each of the sinusoidal excitations and then adding these m individual responses.

Hence, it is sufficient to consider, without loss of generality, the response $y(t)$ of a linear network to a single sinusoidal input $x(t)$. Such a response can be described by the following differential equation:

$$a_n \frac{d^n y}{dt^n} + a_{n-1} \frac{d^{n-1} y}{dt^{n-1}} + \cdots + a_1 \frac{dy}{dt} + a_0 y = x(t). \quad (5.6.2)$$

As discussed in previous sections, the *complete* response $y(t)$ is given by

$$y(t) = y_h(t) + y_p(t), \quad (5.6.3)$$

where $y_p(t)$, the particular solution, is a sinusoidal function of the same frequency as the input, and $y_h(t)$, the solution of the homogeneous equation, is given by

$$y_h(t) = c_1 e^{s_1 t} + c_2 e^{s_2 t} + \cdots + c_n e^{s_n t} = \sum_{i=1}^{n} c_i e^{s_i t}. \quad (5.6.4)$$

Here, in (5.6.4), the natural frequencies $s_1, s_2, \ldots, s_n$ are assumed to be distinct and the constants $c_1, c_2, \ldots, c_n$ are to be determined by the initial conditions. As discussed in Section 5.4, the natural response $y_h(t)$ of a passive physical network must decrease to zero as $t \to \infty$ since the energy originally stored in the network must dissipate gradually through the resistive elements in the form of heat. The sinusoidal steady-state response of the network is simply the sinusoidal part of the complete response—the particular solution $y_p(t)$. However, in the special case of a lossless passive system, the natural frequencies in (5.6.4) are simple and lie on the

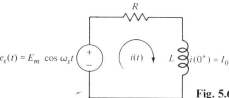

Fig. 5.6.1 Series RL network with sinusoidal input.

$j\omega$-axis of the s-plane. The natural response will consist of at least a term of the form

$$c_i \cos(\omega_0 t + \phi_i) \tag{5.6.5}$$

corresponding to a natural frequency $s_i = j\omega_0$ with ω_0 not equal to the frequency ω_s of the sinusoidal excitation. In this case, the sinusoidal steady-state response can still be defined as the part of the complete response having the same frequency as the input, which is again the particular solution. If, on the other hand, $\omega_0 = \omega_s$ the complete response will contain a term of the form

$$ct \cos(\omega_0 t + \phi) \tag{5.6.6}$$

which increases in magnitude without bound as $t \to \infty$. Under this situation, the sinusoidal steady-state response cannot be defined at all. The same conclusion can be drawn for a linear network containing dependent sources, in which the natural frequencies may have positive real parts. We shall illustrate the above ideas by investigating the following examples.

Example 5.6.1 The differential equation for the current $i(t)$ of the series RL network excited by a sinusoidal voltage (Fig. 5.6.1) is given by

$$L\frac{di}{dt} + Ri = e_s(t). \tag{5.6.7}$$

Taking the Laplace transform of this equation, we obtain

$$(Ls + R)I(s) - E_s(s) + LI_0, \tag{5.6.8}$$

which can be solved to yield

$$I(s) = \frac{E_s(s)}{Ls + R} + \frac{LI_0}{Ls + R}$$

$$= \frac{E_m s}{(s^2 + \omega_s^2)(Ls + R)} + \frac{I_0}{s + (R/L)} \tag{5.6.9}$$

since

$$E_s(s) = \frac{E_m s}{s^2 + \omega_s^2}.$$

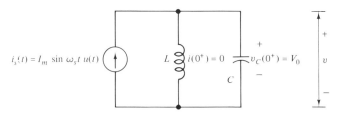

Fig. 5.6.2 LC network.

With the aid of partial fraction expansions, the inverse transform $i(t)$ can be readily shown to be

$$i(t) = \left[I_0 - \frac{E_m R}{R^2 + (\omega_s L)^2} \right] e^{-(R/L)t} + \frac{E_m}{\sqrt{R^2 + (\omega_s L)^2}} \cos\left(\omega_s t - \tan^{-1}(\omega_s L/R)\right),$$

$$\text{for } t > 0. \tag{5.6.10}$$

It is clear in (5.6.10) that the natural frequency of the network is $s = -(R/L)$, a negative real number and hence the sinusoidal steady-state response can be obtained from (5.6.10) as

$$i_{ss}(t) = \frac{E_m}{\sqrt{R^2 + (\omega L)^2}} \cos\left(\omega_s t - \tan^{-1} \frac{\omega_s L}{R}\right), \qquad t > 0. \tag{5.6.11}$$

Example 5.6.2 As another example, let us consider the LC network of Fig. 5.6.2. For simplicity, at $t = 0^+$, the capacitor is assumed to be charged to V_0 volts while the inductor is assumed to be initially de-energized. The differential equation for the voltage $v(t)$ is

$$i_s(t) = \frac{1}{L} \int_0^t v \, dt + C \frac{dv}{dt} \tag{5.6.12}$$

and the corresponding transform is given by

$$\frac{I_m \omega_s}{s^2 + \omega_s^2} = \left[\frac{1}{Ls} + Cs \right] V(s) - CV_0. \tag{5.6.13}$$

Solving (5.6.13) for $V(s)$, we find

$$V(s) = \frac{\dfrac{I_m \omega_s s}{C}}{(s^2 + \omega_s^2)\left(s^2 + \dfrac{1}{LC}\right)} + \frac{V_0 s}{\left(s^2 + \dfrac{1}{LC}\right)}. \tag{5.6.14}$$

Let us consider the following two cases:

Case 1. If $\omega_s^2 \neq (1/LC)$, (5.6.14) can be simplified to

$$V(s) = \frac{\left\{\dfrac{I_m\omega_s Ls}{1 - \omega_s^2 LC}\right\}}{s^2 + \omega_s^2} - \frac{\left\{\dfrac{I_m\omega_s Ls}{1 - \omega_s^2 LC}\right\}}{s^2 + \dfrac{1}{LC}} + \frac{V_0 s}{s^2 + \dfrac{1}{LC}}$$

with the corresponding inverse transform given by

$$v(t) = \frac{I_m\omega_s L}{(1 - \omega_s^2 LC)}\cos \omega_s t - \frac{I_m\omega_s L}{1 - \omega_s^2 LC}\cos\frac{1}{\sqrt{LC}}t + V_0\cos\frac{1}{\sqrt{LC}}t,$$

$$\text{for } t > 0. \qquad (5.6.15)$$

An inspection of (5.6.15) shows that only the first of the three sinusoidal terms has the same frequency as the source current, and hence it is taken to be the sinusoidal steady-state response, i.e.,

$$v_{ss}(t) = \frac{I_m\omega_s L}{(1 - \omega_s^2 LC)}\cos \omega_s t, \qquad t > 0. \qquad (5.6.16)$$

Case 2. If $\omega_s^2 = (1/LC)$, (5.6.14) reduces to

$$V(s) = \frac{\left\{\dfrac{I_m\omega_s}{C}\right\}s}{(s^2 + \omega_s^2)^2} + \frac{V_0 s}{(s^2 + \omega_s^2)}, \qquad (5.6.17)$$

and the corresponding inverse transform becomes

$$v(t) = \frac{I_m}{2C}t \sin \omega_s t + V_0 \cos \omega_s t, \qquad t > 0. \qquad (5.6.18)$$

In such a case, the sinusoidal steady-state does not exist at all since the magnitude of the first term of the complete response $v(t)$ in (5.6.18) increases without bound as $t \to \infty$.

Representation of Sinusoids by Phasors

In Example 5.6.1 or 5.6.2 the Laplace transform method was used in determining the complete response of a given network. If only the sinusoidal steady-state response is required, a simpler method using phasors may be applied to obtain the solution. The term sinusoid includes any waveform which is sinusoidal in nature—any sine or cosine function with an arbitrary phase angle. In general, a sinusoid,

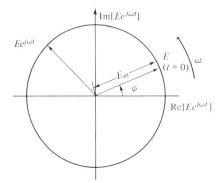

Fig. 5.6.3 Representation of $Ee^{j\omega t}$.

say a voltage $e(t)$, of angular frequency ω can be written as

$$e(t) = E_m \cos(\omega t + \phi)$$
$$= \text{Re}\,[E_m e^{j(\omega t + \phi)}] \qquad (5.6.19)$$
$$= \text{Re}\,[\mathbf{E}e^{j\omega t}],$$

where "Re" denotes the *real part of* the function and

$$\mathbf{E} \triangleq E_m e^{j\phi} \qquad (5.6.20)$$

is defined as the *phasor* representing the sinusoid in (5.6.19).

In the same manner, we may write

$$e(t) = E_m \sin(\omega t + \phi)$$
$$= \text{Im}\,[E_m e^{j(\omega t + \phi)}] \qquad (5.6.21)$$
$$= \text{Im}\,[\mathbf{E}e^{j\omega t}],$$

where "Im" denotes the *imaginary part of* the function. Thus

$$\mathbf{E}e^{j\omega t} = E_m e^{j\phi} e^{j\omega t}$$
$$= E_m e^{j(\omega t + \phi)}$$
$$= E_m \cos(\omega t + \phi) + jE_m \sin(\omega t + \phi) \qquad (5.6.22)$$
$$= \text{Re}[\mathbf{E}e^{j\omega t}] + j\,\text{Im}[\mathbf{E}e^{j\omega t}].$$

The term $\mathbf{E}e^{j\omega t}$ can be interpreted as a phasor (vector) rotating in the counterclockwise direction with an angular velocity ω as indicated in Fig. 5.6.3. It should be noted that the projection of this rotating phasor at any time t on the horizontal axis is $E_m \cos(\omega t + \phi)$, while the projection on the vertical axis is $E_m \sin(\omega t + \phi)$.

Next, let us consider a network in which the source is an exponential function and the response is required for steady-state operation. To be specific, let it be a network consisting of a resistor in parallel with a capacitor as depicted in Fig. 5.6.4. The differential equation for the voltage $v(t)$ across the parallel combination

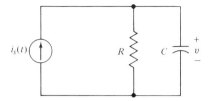

Fig. 5.6.4 Simple RC network.

is given by

$$i_s(t) = Gv + C\frac{dv}{dt}. \tag{5.6.23}$$

If the source current is an exponential function, i.e.,

$$i_s(t) = \mathbf{I}e^{j\omega t}, \tag{5.6.24}$$

the voltage response will also be of the same form, or

$$v(t) = \mathbf{V}e^{j\omega t}. \tag{5.6.25}$$

Substituting (5.6.24) and (5.6.25) into (5.6.23), we get

$$\mathbf{I}e^{j\omega t} = (G + j\omega C)\mathbf{V}e^{j\omega t} \tag{5.6.26}$$

which may be solved to yield

$$\mathbf{V} = \frac{\mathbf{I}}{(G + j\omega C)}. \tag{5.6.27}$$

Hence, from (5.6.25),

$$v(t) = \frac{\mathbf{I}}{(G + j\omega C)}e^{j\omega t}. \tag{5.6.28}$$

It should be noted that the two quantities $\mathbf{V}$ and $\mathbf{I}$ in the above example are related by the driving-point admittance of the network with the variable s replaced by $j\omega$, or

$$\mathbf{V} = \frac{\mathbf{I}}{Y(j\omega)}, \tag{5.6.29}$$

where $Y(j\omega) = G + j\omega C$.

If we assume that $\mathbf{I} = I_m e^{j\phi}$, then (5.6.24) can be written as

$$i_s(t) = I_m e^{j\phi} e^{j\omega t}$$

$$= I_m e^{j(\omega t + \phi)}$$

$$= I_m \cos(\omega t + \phi) + jI_m \sin(\omega t + \phi). \tag{5.6.30}$$

In exactly the same manner, the steady-state response $v(t)$ in (5.6.28) can also be

Fig. 5.6.5 Single-input–single-output linear system.

put into the same form

$$v(t) = \frac{I_m e^{j\phi} e^{j\omega t}}{(G + j\omega C)}$$

$$= \frac{I_m e^{j\phi} e^{j\omega t}}{\sqrt{G^2 + (\omega C)^2} \; e^{j\tan^{-1}(\omega C/G)}}$$

$$= \frac{I_m}{\sqrt{G^2 + (\omega C)^2}} \; e^{j(\omega t + \phi - \tan^{-1}(\omega C/G))}$$

$$= \frac{I_m}{\sqrt{G^2 + (\omega C)^2}} \cos\left(\omega t + \phi - \tan^{-1}\frac{\omega C}{G}\right)$$

$$+ j \frac{I_m}{\sqrt{G^2 + (\omega C)^2}} \; \sin\left(\omega t + \phi - \tan^{-1}\frac{\omega C}{G}\right). \tag{5.6.31}$$

With the knowledge that if the input to a linear network is a "real" sinusoid, the response will also be a "real" sinusoid, one may wonder whether the real part (imaginary part) of $v(t)$ in (5.6.31) is the steady-state response of the network to the real part (imaginary part) of the input current $i_s(t)$ in (5.6.30). By invoking the principle of linearity, it is not difficult to see that this is indeed the case.

Next, let us turn our attention to a more general problem. Refer to the single-input–single-output system as shown in Fig. 5.6.5. Let the behavior of the system be described by the linear differential equation

$$a_n \frac{d^n y}{dt^n} + a_{n-1} \frac{d^{n-1} y}{dt^{n-1}} + \cdots + a_1 \frac{dy}{dt} + a_0 y = x(t). \tag{5.6.32}$$

If the input $x(t)$ is given in the form of sine or cosine function, i.e.,

$$x(t) = X_m \cos(\omega t + \phi) \tag{5.6.33}$$

or

$$x(t) = X_m \sin(\omega t + \phi), \tag{5.6.34}$$

then the steady-state response $y(t)$ can be obtained by the following procedure. Instead of substituting (5.6.33) or (5.6.34) into (5.6.32), we first determine the input phasor $\mathbf{X}$ which from (5.6.33) or (5.6.34) can be written as

$$\mathbf{X} = X_m e^{j\phi}. \tag{5.6.35}$$

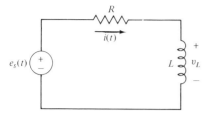

Fig. 5.6.6 Series RL network driven by a sinusoidal voltage.

In the same manner, let us introduce the output phasor as

$$\mathbf{Y} = Y_m e^{j\psi}. \tag{5.6.36}$$

Next, substituting $x(t) = \mathbf{X}e^{j\omega t}$ and $y(t) = \mathbf{Y}e^{j\omega t}$ into (5.6.32), we obtain

$$[a_n(j\omega)^n + a_{n-1}(j\omega)^{n-1} + \cdots + a_1(j\omega) + a_0]\mathbf{Y}e^{j\omega t} = \mathbf{X}e^{j\omega t}$$

or

$$\mathbf{Y} = \frac{\mathbf{X}}{a_n(j\omega)^n + a_{n-1}(j\omega)^{n-1} + \cdots + a_1(j\omega) + a_0}. \tag{5.6.37}$$

Substituting (5.6.35) into (5.6.37) and simplifying, we find

$$\mathbf{Y} = \frac{X_m \exp\left\{ j\left[\phi - \tan^{-1}\dfrac{\omega(a_1 - a_3\omega^2 + \cdots)}{a_0 - a_2\omega^2 + \cdots} \right] \right\}}{\{[a_0 + a_2(j\omega)^2 + \cdots]^2 + \omega^2[a_1 + a_3(j\omega)^2 + \cdots]^2\}^{1/2}}. \tag{5.6.38}$$

Comparing (5.6.36) and (5.6.38), we obtain

$$Y_m = \frac{X_m}{\{[a_0 + a_2(j\omega)^2 + \cdots]^2 + \omega^2[a_1 + a_3(j\omega)^2 + \cdots]^2\}^{1/2}} \tag{5.6.39}$$

and

$$\psi = \phi - \tan^{-1}\left[\frac{\omega(a_1 - a_3\omega^2 + \cdots)}{a_0 - a_2\omega^2 + \cdots} \right]. \tag{5.6.40}$$

Finally, if $x(t) = X_m \cos(\omega t + \phi)$, the steady-state response $y(t)$ is given by

$$y(t) = \mathrm{Re}[\mathbf{Y}e^{j\omega t}] \tag{5.6.41}$$

with $\mathbf{Y}$ given in (5.6.38). On the other hand, if $x(t) = X_m \sin(\omega t + \phi)$, the corresponding steady-state response is

$$y(t) = \mathrm{Im}[\mathbf{Y}e^{j\omega t}]. \tag{5.6.42}$$

Let us illustrate the above procedure with the specific examples that follow.

Example 5.6.3 Consider the series RL network as shown in Fig. 5.6.6. Let the input be the sinusoidal source

$$e_s(t) = E_m \sin(\omega t + \phi) = \text{Im}[\mathbf{E}e^{j\omega t}], \tag{5.6.43}$$

where $\mathbf{E} \triangleq E_m e^{j\phi}$. We wish to determine the response current $i(t)$ in the steady state. Using the procedure outlined above, the equation

$$L\frac{di}{dt} + Ri = \mathbf{E}e^{j\omega t}$$

has the solution

$$\mathbf{I}e^{j\omega t} = \frac{\mathbf{E}}{(R + j\omega L)}e^{j\omega t}. \tag{5.6.44}$$

Since the source voltage is expressed in terms of a sine function, the steady-state current $i(t)$ can be obtained by using (5.6.42), or

$$
\begin{aligned}
i(t) &= \text{Im}\left[\frac{\mathbf{E}}{R + j\omega L}e^{j\omega t}\right] \\
&= \text{Im}\left[\frac{E_m}{\sqrt{R^2 + (\omega L)^2}}e^{j(\omega t + \phi - \tan^{-1}(\omega L/R))}\right] \\
&= \frac{E_m}{\sqrt{R^2 + (\omega L)^2}}\sin(\omega t + \phi - \tan^{-1}(\omega L/R)). \tag{5.6.45}
\end{aligned}
$$

Example 5.6.4 Consider again the series RL network of Fig. 5.6.6. Assume that the source voltage is a sinusoidal excitation of the form

$$
\begin{aligned}
e_s(t) &= E_m \cos(\omega t + \phi) \\
&= \text{Re}[\mathbf{E}e^{j\omega t}] \tag{5.6.46}
\end{aligned}
$$

and that we wish to determine the steady-state voltage v_L across the inductor. The differential equation in this case is

$$L\frac{dv_L}{dt} + Rv_L = L\frac{de_s}{dt}. \tag{5.6.47}$$

With $e_s(t) = \mathbf{E}e^{j\omega t}$, (5.6.47) has the solution

$$\mathbf{V}_L e^{j\omega t} = \frac{j\omega L\mathbf{E}e^{j\omega t}}{(R + j\omega L)}. \tag{5.6.48}$$

Since the excitation is given in terms of a cosine function, the response $v_L(t)$ in the steady state is then

$$v_L(t) = \text{Re}[\mathbf{V}_L e^{j\omega t}]$$

$$= \text{Re}\left[\frac{j\omega L \mathbf{E} e^{j\omega t}}{R + j\omega L}\right]. \tag{5.6.49}$$

After simple algebraic manipulations, we obtain

$$v_L(t) = \frac{\omega L E_m}{\sqrt{R^2 + (\omega L)^2}} \cos\left(\omega t + \phi + \frac{\pi}{2} - \tan^{-1}\frac{\omega L}{R}\right). \tag{5.6.50}$$

It should be noted that in each of the above examples the first step in determining the steady-state response is to find the output phasor in terms of the input phasor. The next step is to take either the real part or the imaginary part of the product of the output phasor and $e^{j\omega t}$, according to whether the excitation is expressed in terms of a cosine or a sine function, respectively. In determining the output phasor, it is necessary, in the first place, to set up the differential equation for the output variable which may be time-consuming even for networks of moderate complexity. Fortunately, a simpler method is available for the determination of the output phasor. Instead of writing the differential equation for the output variable, one may proceed to determine the transfer function which is, of course, the ratio of the output transform to the input transform. In Example 5.6.1, it can be readily shown that the transfer function $H(s)$ is given by

$$H(s) = \frac{I(s)}{E_s(s)} = \frac{1}{R + sL}, \tag{5.6.51}$$

while the ratio of the phasor current to the phasor voltage can be obtained from (5.6.44):

$$\frac{\mathbf{I}}{\mathbf{E}} = \frac{1}{R + j\omega L}. \tag{5.6.52}$$

Comparing (5.6.51) with (5.6.52), we find that the ratio of the output phasor to the input phasor is equal to the transfer function with s replaced by $j\omega$. In like manner, the transfer function and the ratio of the output phasor to the input phasor in Example 5.6.4 are, respectively, given by

$$H(s) = \frac{V_L(s)}{E_s(s)} = \frac{sL}{R + sL} \tag{5.6.53}$$

and

$$\frac{\mathbf{V}_L}{\mathbf{E}} = \frac{j\omega L}{R + j\omega L}. \tag{5.6.54}$$

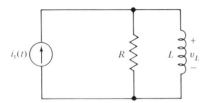

Fig. 5.6.7 Parallel RL network with sinusoidal current excitation.

Again we find

$$\frac{\mathbf{V}_L}{\mathbf{E}} = H(s)\bigg|_{s=j\omega}.\tag{5.6.55}$$

It should be emphasized that the close relationship between the transfer function and the ratio of the output phasor to the input phasor of a given network is by no means a coincidence. Indeed, if we consider the phasor quantities for the individual elements (i.e., R, L, C), we find

$$\mathbf{V}_R = R\mathbf{I}, \qquad \mathbf{V}_L = j\omega L\mathbf{I}, \qquad \mathbf{V}_C = \frac{1}{j\omega C}\mathbf{I}.\tag{5.6.56}$$

If, on the other hand, the transform VCR-equations for the same elements are written down for comparison, we get

$$V_R(s) = RI(s), \qquad V_L(s) = sLI(s), \qquad V_C(s) = \frac{1}{sC}I(s).\tag{5.6.57}$$

An examination of (5.6.56) and (5.6.57) immediately indicates that the driving-point immittances in sinusoidal steady state can be obtained from the driving-point immittances in the s-domain as special cases by replacing s by $j\omega$.

In summary, the steady-state response $y(t)$ of a given network excited by a sinusoidal input $x(t)$ can be determined by means of the following steps.

Step 1. Determine the input phasor **X**.

Step 2. Determine the transfer function

$$H(s) = \frac{Y(s)}{X(s)},$$

which, in the case of one-terminal-pair networks or one-ports, becomes an immittance function in the s-domain.

Step 3. The output phasor **Y** can be obtained from Step 2 by the following relation

$$\mathbf{Y} = H(j\omega)\mathbf{X}.\tag{5.6.58}$$

Step 4. Form the product $\mathbf{Y}e^{j\omega t}$.

Step 5. If the excitation is expressed in terms of a cosine (sine) function, then the steady-state solution $y(t)$ is

$$y(t) = \mathrm{Re}\,[\mathbf{Y}e^{j\omega t}] \qquad (y(t) = \mathrm{Im}\,[\mathbf{Y}e^{j\omega t}]). \tag{5.6.59}$$

Example 5.6.5   The network of Fig. 5.6.7 is excited by a current source of the form

$$i_s(t) = I_m \cos(\omega t + \phi). \tag{5.6.60}$$

We wish to determine the voltage $v_L(t)$ in the steady state. Using the procedure outlined above, the solution is obtained in the following steps.

Step 1. $\mathbf{I} = I_m e^{j\phi}$

Step 2.

$$H(s) = \frac{V_L(s)}{I_s(s)} = Z_d(s) = \frac{RsL}{R + sL} \tag{5.6.61}$$

Step 3.

$$\mathbf{V}_L = \frac{j\omega LR}{R + j\omega L}\,\mathbf{I}$$

Step 4.

$$\mathbf{V}_L e^{j\omega t} = \frac{j\omega LR}{R + j\omega L}\,\mathbf{I}e^{j\omega t}$$

$$= \frac{j\omega LR}{R + j\omega L}\,I_m e^{j\phi} e^{j\omega t}$$

Step 5.

$$v_L(t) = \mathrm{Re}\left[\frac{j\omega LR}{R + j\omega L}\,I_m e^{j\phi} e^{j\omega t}\right]$$

$$= \mathrm{Re}\left[\frac{\omega LRI_m}{\sqrt{R^2 + (\omega L)^2}}\,e^{j(\omega t + \phi + (\pi/2) - \tan^{-1}(\omega L/R))}\right]$$

$$= \frac{\omega LRI_m}{\sqrt{R^2 + (\omega L)^2}}\cos\left(\omega t + \phi + \frac{\pi}{2} - \tan^{-1}\frac{\omega L}{R}\right)$$

Phasor Diagram

To acquire a geometric picture about the relationships between the voltages and currents of a given network, we introduce the *phasor diagram*. The phasor diagram is constructed by representing each sinusoidal voltage and current by a phasor having a length equal to the maximum amplitude of the sinusoid and having an angle of inclination with respect to the positive real axis equal to the angle of the

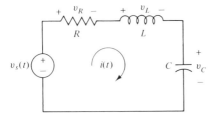

Fig. 5.6.8 Series *RLC* network.

equivalent cosine function of the given sinusoid at $t = 0$. Thus, the first step in constructing the phasor diagram is to convert all sinusoids into equivalent cosine functions. Let us demonstrate the method of constructing the phasor diagram by means of an example that follows.

Example 5.6.6   Refer to the network of Fig. 5.6.8. If the applied voltage is taken as reference, that is, the voltage phasor $\mathbf{V}_s = V_m e^{j0}$ is drawn in the complex plane along the positive real axis, then various phasor quantities are given by

$$\mathbf{I} = \frac{\mathbf{V}_s}{R + j\left(\omega L - \dfrac{1}{\omega C}\right)} = \frac{V_m}{|Z|} e^{-j\theta_z},$$

$$\mathbf{V}_R = \mathbf{I}R = \frac{V_m R}{|Z|} e^{-j\theta_z},$$

$$\mathbf{V}_L = j\omega L\mathbf{I} = \frac{V_m \omega L}{|Z|} e^{j((\pi/2) - \theta_z)},$$

$$\mathbf{V}_C = \frac{1}{j\omega C}\mathbf{I} = \frac{V_m}{\omega C|Z|} e^{-j((\pi/2) + \theta_z)},$$

where

$$|Z| = \sqrt{R^2 + \left(\omega L - \frac{1}{\omega C}\right)^2}$$

and

$$\theta_z = \tan^{-1}\left\{\frac{\omega L - \dfrac{1}{\omega C}}{R}\right\}.$$

Assuming that the network is *inductive*, that is, $\theta_z > 0$, the phasor diagram is constructed as indicated in Fig. 5.6.9.

It should be noted that any other phasor quantity instead of $\mathbf{V}_s$ could have been used as reference for the construction of the phasor diagram. The relative

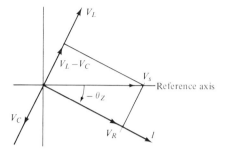

Fig. 5.6.9 Complete phasor diagram for network of Example 5.6.6.

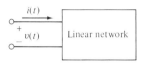

Fig. 5.6.10 Linear one-port.

positions of all the phasors remain exactly the same except that the complete diagram would have been rotated with the reference phasor lying on the positive real axis.

Power, Average Power, Power Factor and Complex Power

When a network is made up solely of inductors and capacitors, it is frequently referred to as a *lossless network*, since both inductors and capacitors are regarded as *storage elements*, where energy is being stored in their associated magnetic and electric fields. If some resistive elements are present, then the network is termed *dissipative* because energy is being consumed in the network through the resistive elements in the form of heat.

In this section it is intended to derive some of the power expressions which are commonly used when the network is operated in the sinusoidal steady state. Such studies are important, because, from the economic point of view, a good network design may enable the system to transfer energy efficiently from the source to the load. Efficient energy transfer means economy and this cost-saving aspect is one of many important factors to be considered in engineering design problems.

Consider the one-port with the voltage and current directions as shown in Fig. 5.6.10. The instantaneous power dissipated in the network is

$$p(t) = v(t)i(t) \tag{5.6.62}$$

and the corresponding energy absorbed by the network from time t_1 to t_2 is given by

$$W = \int_{t_1}^{t_2} p(t)\, dt = \int_{t_1}^{t_2} v(t)i(t)\, dt. \tag{5.6.63}$$

If the network is operated in the sinusoidal steady state, the voltage and the current take the following forms:

$$v(t) = V_m \cos(\omega t + \phi), \tag{5.6.64}$$

$$i(t) = I_m \cos(\omega t + \psi). \tag{5.6.65}$$

Then from (5.6.62), the instantaneous power dissipated in the network is

$$p(t) = v(t)i(t)$$

$$= V_m I_m \cos(\omega t + \phi) \cos(\omega t + \psi)$$

$$= \tfrac{1}{2} V_m I_m \cos(\phi - \psi) + \tfrac{1}{2} V_m I_m \cos(2\omega t + \phi + \psi), \tag{5.6.66}$$

where, in obtaining the last expression of (5.6.66), the trigonometric identity

$$\cos A \cos B = \tfrac{1}{2}[\cos(A - B) + \cos(A + B)]$$

has been applied. If we denote the average power by P_{ave}, then from (5.6.66) we get

$$P_{\text{ave}} = \tfrac{1}{2} V_m I_m \cos(\phi - \psi). \tag{5.6.67}$$

In terms of phasor quantities $\mathbf{V} = V_m e^{j\phi}$ and $\mathbf{I} = I_m e^{j\psi}$, the driving-point impedance of the one-port is

$$Z = \frac{\mathbf{V}}{\mathbf{I}} = \frac{V_m}{I_m} e^{j(\phi - \psi)}. \tag{5.6.68}$$

From (5.6.68), it is obvious that $\phi - \psi$ is simply the impedance angle or

$$\theta_Z = \phi - \psi. \tag{5.6.69}$$

Knowing the fact that in the sinusoidal steady-state analysis, the amplitude and the effective (rms) value of a sinusoid are related by the factor of $\sqrt{2}$, that is,

$$V_{\text{eff}} = \frac{V_m}{\sqrt{2}} \qquad \text{or} \qquad I_{\text{eff}} = \frac{I_m}{\sqrt{2}}, \tag{5.6.70}$$

the average power can be rewritten in terms of effective values as

$$P_{\text{ave}} = \frac{V_m}{\sqrt{2}} \frac{I_m}{\sqrt{2}} \cos(\phi - \psi)$$

$$= V_{\text{eff}} I_{\text{eff}} \cos \theta_Z. \tag{5.6.71}$$

In (5.6.71), $\cos \theta_Z$ is defined as the *power factor* or simply *p.f.* The usual convention is that the power factor is said to be *lagging* if the current *lags* the voltage, and *leading* if the current *leads* the voltage.

Next, let us rearrange the terms in $p(t)$ in the following manner: From (5.6.66), we find (with $\phi = 0$ for convenience)*

$$
\begin{aligned}
p(t) &= \tfrac{1}{2}V_m I_m \cos\theta_Z + \tfrac{1}{2}V_m I_m \cos(2\omega t - \theta_Z) \\
&= V_{\text{eff}}I_{\text{eff}}\cos\theta_Z + V_{\text{eff}}I_{\text{eff}}\cos\theta_Z \cos 2\omega t + V_{\text{eff}}I_{\text{eff}}\sin\theta_Z \sin 2\omega t \\
&= V_{\text{eff}}I_{\text{eff}}\cos\theta_Z(1 + \cos 2\omega t) + V_{\text{eff}}I_{\text{eff}}\sin\theta_Z \sin 2\omega t.
\end{aligned}
\tag{5.6.72}
$$

The first term on the right-hand side of (5.6.72) can be interpreted as the power dissipated in the resistive elements of the one-port, which oscillates between the values of zero and $2V_{\text{eff}}I_{\text{eff}}\cos\theta_Z$ at twice the frequency of the applied source with an average value of $P_{\text{ave}} = V_{\text{eff}}I_{\text{eff}}\cos\theta_Z$. The second term can be considered as the power borrowed from the source by the reactive elements for one quarter of a cycle of the applied source and then returned to the source in the next quarter cycle. This borrowed power oscillates between the extremes $\pm V_{\text{eff}}I_{\text{eff}}\sin\theta_Z$ at twice the frequency of the applied source and is averaged to zero.

It is convenient to define this peak value $V_{\text{eff}}I_{\text{eff}}\sin\theta_Z$ as the *reactive power*, or

$$
Q \triangleq V_{\text{eff}}I_{\text{eff}}\sin\theta_Z \text{ vars.}
\tag{5.6.73}
$$

As indicated in (5.6.73), the unit for the reactive power is the volt-ampere reactive or vars. If the network is inductive, that is, if $(\pi/2) \geq \theta_Z > 0$, then the reactive power Q is positive. If, on the other hand, the network is capacitive, that is, if $-(\pi/2) \leq \theta_Z < 0$, then the reactive power Q is negative. Of course, when the network is purely resistive, that is, $\theta_Z = 0$, the reactive power is zero.

It is possible to relate both the average power and the reactive power together by one simple expression. In terms of effective phasor quantities $\mathbf{V}_e$ and $\mathbf{I}_e$, we find from (5.6.64) and (5.6.65) that

$$
\mathbf{V}_e = \frac{V_m}{\sqrt{2}} e^{j\phi}
\tag{5.6.74}
$$

and

$$
\mathbf{I}_e = \frac{I_m}{\sqrt{2}} e^{j\psi}.
\tag{5.6.75}
$$

We define the *complex* (or phasor) *power* as

$$
\begin{aligned}
\mathbf{S} &\triangleq \mathbf{V}_e\mathbf{I}_e^* \\
&= \tfrac{1}{2}V_m I_m e^{j(\phi - \psi)} \\
&= \tfrac{1}{2}V_m I_m e^{j\theta_Z} \\
&= V_{\text{eff}}I_{\text{eff}} e^{j\theta_Z},
\end{aligned}
\tag{5.6.76}
$$

---

* Without loss of generality, the arbitrary voltage phase angle ϕ is set to zero for convenience since only the impedance angle θ_z affects the average power.

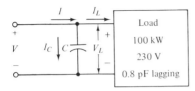

Fig. 5.6.11 Network for power correction.

where $\mathbf{I}_e^*$ is defined as the complex conjugate of $\mathbf{I}_e$. Expressing (5.6.76) in trigono-metric form, we obtain

$$S = V_{\text{eff}}I_{\text{eff}} \cos \theta_Z + jV_{\text{eff}}I_{\text{eff}} \sin \theta_Z$$
$$= P_{\text{ave}} + jQ. \tag{5.6.77}$$

The magnitude of the complex power

$$|S| = \sqrt{P_{\text{ave}}^2 + Q^2} = V_{\text{eff}}I_{\text{eff}} \text{ VA} \tag{5.6.78}$$

is sometimes referred to as the *apparent power* measured in volt-amperes or simply VA. A number of expressions for the power quantities are possible. If we express the driving-point impedance Z as

$$Z = |Z|e^{j\theta_Z} = R + jX, \tag{5.6.79}$$

with $R = |Z| \cos \theta_Z$ and $X = |Z| \sin \theta_Z$, then we can write different expressions for P_{ave} and Q: Thus,

$$P_{\text{ave}} = V_{\text{eff}}I_{\text{eff}} \cos \theta_Z$$
$$= I_{\text{eff}}^2 R$$
$$= \text{Re}[S] \tag{5.6.80}$$

and

$$Q = V_{\text{eff}}I_{\text{eff}} \sin \theta_Z$$
$$= I_{\text{eff}}^2 X$$
$$= \text{Im}[S]. \tag{5.6.81}$$

Example 5.6.7 An industrial load such as shown in Fig. 5.6.11 is rated at 100 kW, 220 V (effective value) at a power factor of 0.8 lagging. It is intended to install a bank of parallel capacitors represented by C in the figure so that the resultant voltage $\mathbf{V}$ and the current $\mathbf{I}$ through the combined load are in phase. Determine the required capacitive reactance.

Let the effective values of the load voltage and the load current be denoted by V_L and I_L, respectively. Then

$$P_{\text{ave}} = V_L I_L \cos \theta_Z = 100 \times 10^3 \text{W}.. \tag{5.6.82}$$

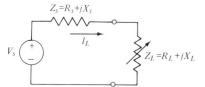

Fig. 5.6.12 Network illustrating power transfer from source to load.

But the reactive power Q_L of the load is given by

$$Q_L = V_L I_L \sin \theta_Z$$
$$= P_{\text{ave}} \tan \theta_Z$$
$$= 100 \times 10^3 \tan 36.9°$$
$$= 75 \times 10^3 \text{ vars.} \tag{5.6.83}$$

Note that Q_L is positive because of the lagging power factor at the load. In order that the resultant voltage **V** and the current **I** through the combined load be in phase, the total reactive power Q_T must be zero. This implies that the reactive power of the combined capacitance C must be

$$Q_c = -Q_L = -75 \times 10^3 \text{ vars.} \tag{5.6.84}$$

But

$$Q_c = I_c^2 X_c = \left(\frac{V_L}{X_c}\right)^2 X_c = \frac{V_L^2}{X_c}. \tag{5.6.85}$$

Therefore

$$X_c = \frac{V_L^2}{Q_c} = \frac{220^2}{-75 \times 10^3} = -0.645\Omega. \tag{5.6.86}$$

Example 5.6.8 Refer to the network in Fig. 5.6.12. The *effective* phasor of the source voltage and its internal impedance are denoted by $\mathbf{V}_s$ and Z_s, respectively. The problem is to determine, at a specified source frequency ω, the value of the load impedance so that it will draw the maximum power from the source. It is assumed that both R_L and X_L of the load Z_L can be varied individually.

If we denote the effective phasor of the load current by $\mathbf{I}_L$, then

$$\mathbf{I}_L = \frac{\mathbf{V}_s}{Z_s + Z_L} \tag{5.6.87}$$

and the power delivered to the load is given by

$$P_L = |\mathbf{I}_L|^2 R_L$$
$$= \frac{|\mathbf{V}_s|^2 R_L}{(R_s + R_L)^2 + (X_s + X_L)^2}. \tag{5.6.88}$$

First, R_L is considered to be a constant and the value of X_L is sought so that the average power P_L is maximized. Differentiating (5.6.88) with respect to X_L, we find that

$$\frac{\partial P_L}{\partial X_L} = \frac{-2|\mathbf{V}_s|^2 R_L(X_s + X_L)}{[(R_s + R_L)^2 + (X_s + X_L)^2]^2}. \tag{5.6.89}$$

Setting $(\partial P_L/\partial X_L) = 0$, we find that

$$X_L = -X_s. \tag{5.6.90}$$

Substituting this value of X_L into (5.6.88) yields

$$P_L = \frac{|\mathbf{V}_s|^2 R_L}{(R_s + R_L)^2}. \tag{5.6.91}$$

Next, we vary R_L by differentiating (5.6.91):

$$\frac{\partial P_L}{\partial R_L} = |\mathbf{V}_s|^2 \frac{[(R_s + R_L)^2 - 2(R_s + R_L)R_L]}{[(R_s + R_L)^2]^2}. \tag{5.6.92}$$

The condition $(\partial P_L/\partial R_L) = 0$ demands that

$$R_L = R_s. \tag{5.6.93}$$

Combining both conditions (5.6.90) and (5.6.93), we conclude that the maximum power will enter the load if

$$Z_L = Z_s^* = R_s - jX_s. \tag{5.6.94}$$

When this condition is met, the load is said to be *conjugately matched* with the source, and the maximum power entering the load is given by

$$P_{\substack{\text{ave} \\ \text{max}}} = \frac{1}{4} \frac{|\mathbf{V}_s|^2}{R_L}. \tag{5.6.95}$$

Summarizing the above result into a theorem, we obtain *the maximum power transfer theorem*: For the network of Fig. 5.6.12, maximum power transfer is accomplished when the load impedance Z_L is equal to the complex conjugate of Z_s; that is, $Z_L = Z_s^*$.

5.7 FREQUENCY RESPONSE PLOTS

As mentioned previously, a sinusoidal excitation is one of the most important types of excitation functions in engineering applications. The behavior of a network function as a function of ω is of prime interest to the network designer since specifications from which a network is to be designed are usually given in terms of magnitude and phase as functions of frequency. There are different methods for obtaining the frequency response of a network function $H(j\omega)$. For example, the frequency response of $H(j\omega)$ is completely determined if both the magnitude and

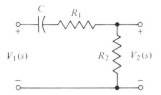

Fig. 5.7.1   Network for Example 5.7.1.

the phase of $H(j\omega)$ as functions of ω are known. On the other hand, the same information can be obtained if both the real and the imaginary parts of $H(j\omega)$ are specified. Still another possibility of representing the frequency response is to plot both the magnitude and the phase of the network function on the complex plane with the frequency ω as the parameter. In what follows, we shall discuss each of these plots in some detail.

Magnitude and Phase Plots

Let $H(s)$ be a network function which may be either a driving-point or a transfer function. If $s = j\omega$, then $H(j\omega)$ can be expressed in the form

$$H(j\omega) = |H(j\omega)|e^{j\phi(\omega)}, \tag{5.7.1}$$

where $|H(j\omega)|$ is the magnitude of $H(j\omega)$ and $\phi(\omega)$ is the phase. In simple cases, it is quite possible to obtain the magnitude and the phase plots by inspection. However, for network functions of moderate complexity, it may be necessary to obtain these plots by point-by-point calculations even though graphical construction can be utilized to facilitate the calculations.

Example 5.7.1   Consider the two-port network of Fig. 5.7.1 with the voltage-ratio transfer function $H(s)$ given by

$$H(s) = \frac{R_2}{R_2 + R_1 + (1/sC)} = \frac{sCR_2}{1 + sC(R_1 + R_2)}. \tag{5.7.2}$$

In sinusoidal steady state, we find

$$H(j\omega) = \frac{j\omega CR_2}{1 + j\omega C(R_1 + R_2)} \tag{5.7.3}$$

with the corresponding magnitude and phase, respectively, given by

$$|H(j\omega)| = \frac{\omega CR_2}{\sqrt{1 + \omega^2 C^2(R_1 + R_2)^2}} \tag{5.7.4}$$

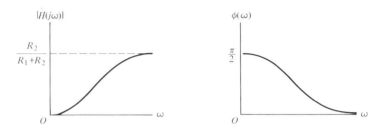

Fig. 5.7.2 The variation of magnitude and phase for the transfer function of Fig. 5.7.1.

and

$$\phi(\omega) = \frac{\pi}{2} - \tan^{-1}[\omega C(R_1 + R_2)]. \tag{5.7.5}$$

The plots of $|H(j\omega)|$ and $\phi(\omega)$ versus ω are sketched in Fig. 5.7.2.

It should be noted that both $|H(j\omega)|$ and $\phi(\omega)$ of the given voltage-ratio transfer function can be determined graphically by first plotting the zeros and the poles of $H(s)$ on the complex plane. Thus, from (5.7.2), we write

$$H(s) = \frac{R_2}{R_1 + R_2} \frac{s}{s + \dfrac{1}{C(R_1 + R_2)}} \tag{5.7.6}$$

from which the zero and the pole, together with the scale factor $K = R_2/(R_1 + R_2)$, are plotted in Fig. 5.7.3.

If we denote by **A** and **B** the vectors originating from the zero and the pole of $H(s)$ which are directed to the point $s = j\omega$ as indicated in Fig. 5.7.3, then

$$\mathbf{A} = j\omega + \frac{1}{C(R_1 + R_2)} \tag{5.7.7}$$

and

$$\mathbf{B} = j\omega. \tag{5.7.8}$$

An examination of (5.7.6) indicates that for a given value of ω, we have

$$|H(j\omega)| = \frac{K|\mathbf{B}|}{|\mathbf{A}|} \tag{5.7.9}$$

and

$$\phi(\omega) = \phi_1 - \phi_2 = \frac{\pi}{2} - \tan^{-1}[\omega C(R_1 + R_2)]. \tag{5.7.10}$$

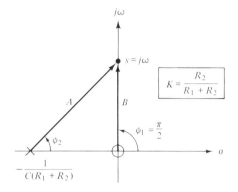

Fig. 5.7.3 Graphical determination of magnitude and phase plots of $H(s)$.

If the point s is allowed to vary along the positive $j\omega$-axis, the variation of the magnitude and the phase of $H(j\omega)$ can be obtained by first measuring the magnitudes and the phase angles of the two vectors **A** and **B** and then making use of (5.7.9) and (5.7.10). This same method can be applied equally well to any other transfer function.

Polar Plot

The frequency response of a network function can also be represented by plotting both the real and the imaginary parts of a given function on the complex plane with the frequency ω as a parameter. If we write

$$H(j\omega) = R(\omega) + jX(\omega), \tag{5.7.11}$$

then, for every value of ω, $H(j\omega)$ can be represented as a point in the complex plane. As ω is varied, the point representing $H(j\omega)$ traces a curve in the complex plane, which is referred to as the *locus* of the transfer function. The following example will illustrate some of the ideas involved.

Example 5.7.2 Consider the RC network of Fig. 5.7.4. The voltage-ratio transfer function is

$$H(s) = \frac{V_2(s)}{V_1(s)} = \frac{(1/sC)}{R + (1/sC)} = \frac{1}{1 + sCR}. \tag{5.7.12}$$

For sinusoidal steady state, $s = j\omega$, and $H(j\omega)$ becomes

$$H(j\omega) = \frac{1}{1 + j\omega CR} = \frac{1}{1 + (\omega CR)^2} - j\frac{\omega CR}{1 + (\omega CR)^2}. \tag{5.7.13}$$

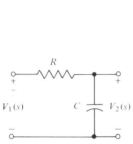

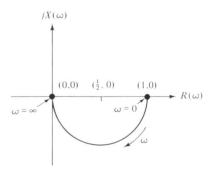

Fig. 5.7.4 Network for Example 5.7.2.   Fig. 5.7.5 Locus of $H(j\omega)$ in Example 5.7.2.

Denoting the real and the imaginary parts of $H(j\omega)$ by $R(\omega)$ and $X(\omega)$, respectively, we have from (5.7.13)

$$R(\omega) = \frac{1}{1 + (\omega CR)^2} \qquad (5.7.14)$$

and

$$X(\omega) = \frac{-\omega CR}{1 + (\omega CR)^2}. \qquad (5.7.15)$$

Combining (5.7.14) and (5.7.15) by eliminating ω, we find that

$$R^2(\omega) + X^2(\omega) = R(\omega) \qquad (5.7.16)$$

which can be simplified to be

$$(R(\omega) - \tfrac{1}{2})^2 + X^2(\omega) = (\tfrac{1}{2})^2. \qquad (5.7.17)$$

It is obvious that (5.7.17) is the equation of a circle centered at $(\tfrac{1}{2}, 0)$ with radius $\tfrac{1}{2}$. The *locus* for $H(\omega)$ is plotted in Fig. 5.7.5.

As indicated in the figure, the locus is a semi-circle with the direction of ω increasing in the clockwise direction. An examination of (5.7.13) indicates that for all positive values of ω, the imaginary part $X(\omega)$ is always negative and hence the locus must be a semi-circle located in the fourth quadrant.

It should be emphasized that only in very simple cases can the locus be obtained analytically. For more complicated networks, point-by-point plots may be required even though asymptotes are usually very helpful in sketching the loci.

Bode's Diagrams

If the logarithmic plots of the magnitude and the phase of a given transfer function $H(j\omega)$ are made on a logarithmic frequency scale, they can usually be approximated quite closely by a set of straight line segments. Consequently, point-by-point

calculations can be avoided since the logarithm of a function converts the operations of multiplication and division into simple addition and subtraction.

If the symbol "ln" is used to represent the natural logarithm, then the natural
logarithm of the transfer function

$$H(j\omega) = |H(j\omega)|e^{j\phi(\omega)} \tag{5.7.18}$$

is the complex function

$$\ln H(j\omega) = \ln |H(j\omega)| + j\phi(\omega). \tag{5.7.19}$$

The first term on the right-hand side of (5.7.19) is referred to as the *gain* measured
in *nepers*. In a similar manner, if the symbol "log" is used to indicate the common
logarithm, then we find

$$\log H(j\omega) = \log|H(j\omega)| + \log e^{j\phi(\omega)}$$

$$= \log|H(j\omega)| + j\,0.434\,\phi(\omega). \tag{5.7.20}$$

The *gain in* dB (decibels) is defined as

$$\text{Lm } H(j\omega) \triangleq 20\log|H(j\omega)| \text{ dB}. \tag{5.7.21}$$

It should be noted that the conversion from nepers to dB or vice versa is given
by the equation:

$$\text{Number of dB} = 8.68 \times \text{number of nepers} \tag{5.7.22}$$

Given a transfer function $H(j\omega)$, the *Bode diagrams* are the plots of both the log
magnitude, Lm $H(j\omega)$, and phase, $\phi(\omega)$, as functions of ω on a logarithmic frequency scale. As will be seen in the following discussion, the log magnitude and
phase diagrams may be obtained easily by means of straight-line asymptotic
approximations.

In general, a transfer function consists of the following four basic types of
factors:

1. a constant K (5.7.23)
2. a $j\omega$ factor   (5.7.24)
3. a first-order factor of the form $(1 + j\omega T)$ (5.7.25)
4. a second-order factor of the form $[1 + 2\zeta(j\omega)T + (j\omega)^2 T^2]$ (5.7.26)

The curves of log magnitude and phase versus the log frequency can be
approximated very easily for each of the above factors. A linear combination of
these curves yields the required Bode plots of a given transfer function. In the
following pages, we shall discuss the asymptotic approximations to each of these
four factors.

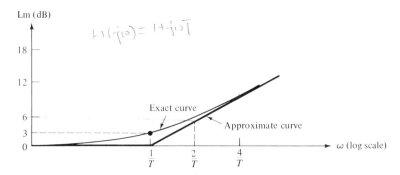

Fig. 5.7.6 Log magnitude of $(1 + j\omega T)$ and its asymptotes.

1. *The constant K.* Since the log of a constant is again a constant, we find that

$$\text{Lm } K = 20 \log K \text{ dB} \qquad (5.7.27)$$

is a horizontal line, independent of the frequency ω.

2. *The $j\omega$ factor.* The log magnitude of this factor is

$$\text{Lm}(j\omega) = 20 \log \omega \text{ dB}. \qquad (5.7.28)$$

When plotted on log frequency scale, this curve is a straight line. The slope of this straight line is found by noting that the line increases 6 dB every time the frequency is doubled, or increases 20 dB when ω is increased ten-fold. This means that the straight line, going through the 0 dB point at $\omega = 1$, has a slope of 6 dB/octave or 20 dB/decade. The phase of this factor is a constant and is equal to $(\pi/2)$.

3. *The $(1 + j\omega T)$ factor.* The log magnitude of this factor is

$$\text{Lm}(1 + j\omega T) = 20 \log|1 + j\omega T| = 10 \log[1 + \omega^2 T^2]. \qquad (5.7.29)$$

For very small ω; that is, $\omega T \ll 1$,

$$\text{Lm}(1 + j\omega T) \cong 20 \log 1 = 0 \text{ dB}. \qquad (5.7.30)$$

For very large ω; that is, $\omega T \gg 1$,

$$\text{Lm}(1 + j\omega T) \cong 10 \log[\omega^2 T^2] = 20 \log \omega T \text{ dB} \qquad (5.7.31)$$

which is a straight line going through the 0 dB point at $\omega = (1/T)$ with a slope of 6 dB/octave or 20 dB/decade. The two asymptotes described by (5.7.30) and (5.7.31) are drawn in Fig. 5.7.6, where the frequency $\omega = (1/T)$ at which the two asymptotes meet is defined as the *corner frequency* or the *break frequency*. At $\omega T = 1$, (5.7.29) becomes

$$\text{Lm}(1 + j\omega T) = 10 \log 2 \cong 3 \text{ dB} \qquad (5.7.32)$$

from which, together with the two asymptotes, the exact curve of $\text{Lm}(1 + j\omega T)$ is shown in Fig. 5.7.6.

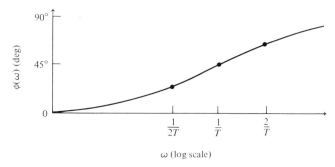

Fig. 5.7.7 Phase of $(1 + j\omega T)$.

To determine the phase of $(1 + j\omega T)$, we write

$$\phi(\omega) = \tan^{-1}\omega T. \qquad (5.7.33)$$

With the help of Table 5.7.1, $\phi(\omega)$ is drawn in Fig. 5.7.7.

Table 5.7.1

ω	$\phi(\omega) = \tan^{-1} T$
0	0
$1/2T$	26.6°
$1/T$	45°
$2T$	63.4°
∞	90°

4. *The factor* $[1 + 2\zeta(j\omega)T + (j\omega)^2 T^2]$. The log magnitude and the phase of this factor are given, respectively, by

$$\text{Lm}[1 + 2\zeta(j\omega)T + (j\omega)^2 T^2] = 10 \log[(1 - \omega^2 T^2)^2 + (2\zeta\omega T)^2] \quad (5.7.34)$$

and

$$\phi(\omega) = \tan^{-1}\left(\frac{2\zeta\omega T}{1 - \omega^2 T^2}\right). \qquad (5.7.35)$$

For small ω, that is, $\omega T \ll 1$, (5.7.34) becomes

$$\text{Lm}[1 + 2\zeta(j\omega)T + (j\omega)^2 T^2] \cong 0\,\text{dB} \qquad (5.7.36)$$

which is, of course, the low-frequency asymptote. In like manner, for large ω, that

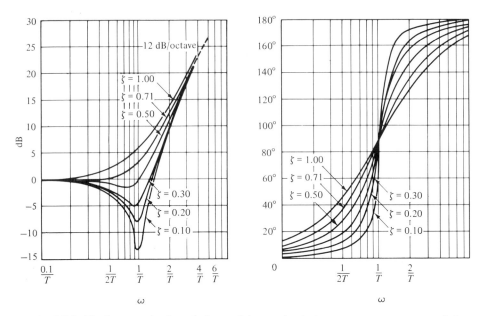

Fig. 5.7.8 The log magnitude and phase of the quadratic factor $1 + 2\zeta(j\omega)T + (j\omega)^2 T^2$.

is, $\omega T \gg 1$, (5.7.34) reduces to

$$\text{Lm}[1 + 2\zeta(j\omega)T + (j\omega)^2 T^2] \cong 40 \log \omega T \text{ dB} \qquad (5.7.37)$$

which is a straight line going through the 0 dB point at the corner frequency $\omega = (1/T)$ with a slope of 12 dB/octave or 40 dB/decade. At $\omega = (1/T)$, we find from (5.7.34)

$$\text{Lm}[1 + 2\zeta(j\omega)T + (j\omega)^2 T^2] = 20 \log 2\zeta \text{ dB}. \qquad (5.7.38)$$

The plots of the log magnitude and the phase of this quadratic factor are shown in Fig. 5.7.8.

Example 5.7.3 The transfer function of a linear system is given by

$$H(j\omega) = \frac{K}{(j\omega)(1 + j\omega T)}. \qquad (5.7.39)$$

The log magnitude and the phase of this transfer function are readily found to be

$$\text{Lm } H(j\omega) = \text{Lm } K - \text{Lm}(j\omega) - \text{Lm}(1 + j\omega T)$$

$$= 20 \log K - 20 \log \omega - 10 \log[1 + (\omega T)^2] \qquad (5.7.40)$$

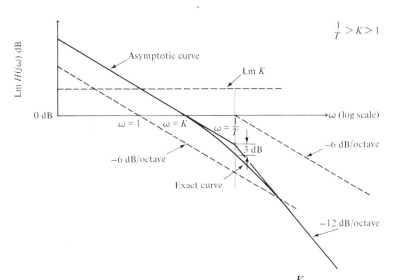

Fig. 5.7.9 Log magnitude plot for $H(j\omega) = \dfrac{K}{j\omega(1 + j\omega T)}$.

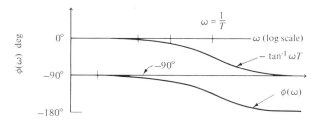

Fig. 5.7.10 Phase plot for $H(j\omega) = \dfrac{K}{j\omega(1 + j\omega T)}$.

and

$$\phi(\omega) = -90° - \tan^{-1}\omega T, \tag{5.7.41}$$

respectively.

An examination of (5.7.40) shows that the log magnitude of the transfer function is the sum of three terms. Figure 5.7.9 indicates how these terms are added together to yield the required log magnitude curve. The phase of the transfer function is drawn separately in Fig. 5.7.10.

Example 5.7.4 As another example, consider the transfer function

$$H(j\omega) = \frac{(1 + j\omega T_1)}{(1 + j\omega T_2)(1 + j\omega T_3)} \tag{5.7.42}$$

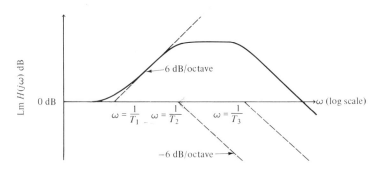

Fig. 5.7.11   Log magnitude plot of (5.7.42).

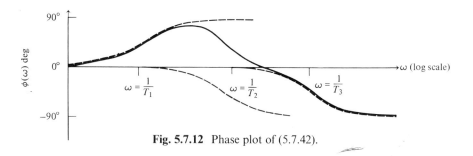

Fig. 5.7.12   Phase plot of (5.7.42).

with $T_1 > T_2 > T_3$. Then the log magnitude and the phase of this transfer function are

$$\text{Lm } H(j\omega) = \text{Lm}(1 + j\omega T_1) - \text{Lm}(1 + j\omega T_2) - \text{Lm}(1 + j\omega T_3)$$

$$= 10 \log[1 + (\omega T_1)^2] - 10 \log[1 + (\omega T_2)^2] - 10 \log[1 + (\omega T_3)^2] \quad (5.7.43)$$

and

$$\phi(\omega) = \tan^{-1}\omega T_1 - \tan^{-1}\omega T_2 - \tan^{-1}\omega T_3, \quad (5.7.44)$$

respectively, with the corresponding curves drawn in Fig. 5.7.11 and Fig. 5.7.12.

5.8 NETWORK ANALYSIS BY FOURIER SERIES

In the preceding sections of this chapter, we have restricted ourselves, in the steady-state analysis of networks, to periodic inputs which are sinusoidal in nature. This is due to the fact that in our universe the sinusoidal waveform is one that occurs frequently in our immediate environment. As a matter of fact, most of the power systems in the world today are producing sinusoidal waveforms. The methods we have developed for solving network problems with sinusoidal steady-state inputs

are very efficient in the sense that a general step-by-step procedure can be applied
to obtain the solution of any network problem with sinusoidal excitation. In this
section we shall discuss how these very same methods can be used to attack problems with nonsinusoidal steady-state periodic inputs.

   If a sectionally continuous function $f(t)$ is periodic with period $t = 2\pi/\omega$, then
it can always be resolved into a *Fourier series* of the form

$$f(t) = C_0 + C_1 \cos(\omega t + \phi_1) + C_2 \cos(2\omega t + \phi_2) + \cdots$$
$$+ C_n \cos(n\omega t + \phi_n) + \cdots, \qquad (5.8.1)$$

where C_0 is the dc-component of the wave, $C_1 \cos(\omega t + \phi_1)$ is called the *fundamental*, and $C_n \cos(n\omega t + \phi_n)$ is referred to as the nth *harmonic* of the function $f(t)$.
The amplitude C_n and the phase ϕ_n of the nth harmonic are, respectively,

$$C_0 = a_0,$$
$$C_n^2 = a_n^2 + b_n^2, \qquad n = 1, 2, \ldots, \qquad (5.8.2)$$

and

$$\phi_n = \tan^{-1}\left(\frac{-b_n}{a_n}\right), \qquad n = 1, 2, \ldots, \qquad (5.8.3)$$

with a_n and b_n being the *Fourier coefficients* defined by

$$a_0 = \frac{1}{T}\int_0^T f(t)\, dt, \qquad (5.8.4)$$

$$a_n = \frac{2}{T}\int_0^T f(t) \cos n\omega t\, dt, \qquad n = 1, 2, \ldots, \qquad (5.8.5)$$

and

$$b_n = \frac{2}{T}\int_0^T f(t) \sin n\omega t\, dt, \qquad n = 1, 2, \ldots. \qquad (5.8.6)$$

   In the Fourier representation (5.8.1), it should be noted that the function $f(t)$
is completely specified once the set of numbers C_n and ϕ_n for $n = 0, 1, \ldots$ is known.
In the following discussion we shall assume that the network under consideration
is excited by a periodic signal which can be represented by a Fourier series.

Steady-State Response to Nonsinusoidal Periodic Signals

When the input to a network is periodic, we can express it in the form of a Fourier
series such as the one shown in (5.8.1). If the input is a voltage source, then its
Fourier series representation can be regarded as a set of sinusoidal voltage sources
connected in series. If, on the other hand, the input is a current source, then its
Fourier series representation can be interpreted as a set of sinusoidal current
sources connected in parallel. Since the network under consideration is assumed
to be a linear network, its steady-state response can be obtained by invoking the

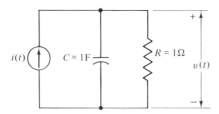

Fig. 5.8.1 Parallel RC network with current input.

principle of superposition. It is accomplished by first finding the steady-state response of the network when acted upon individually by each of the harmonic terms of the Fourier series and then adding these individual responses to form the complete steady-state solution. Since all the individual components of the source are sinusoids, the individual responses can be calculated by the methods developed in the preceding sections of this chapter.

Example 5.8.1 Consider the parallel RC network of Fig. 5.8.1. Let the input current be a square wave as drawn in Fig. 5.8.2. The problem is to determine the steady-state response $v(t)$ across the parallel combination. The Fourier series representation of the current waveform is readily calculated to be

$$i(t) = \frac{4}{\pi}\left(\cos t - \frac{\cos 3t}{3} + \frac{\cos 5t}{5} - \cdots\right)$$

$$= \frac{4}{\pi}\sum_{k=0}^{\infty}(-1)^k\frac{\cos(2k+1)t}{2k+1}. \tag{5.8.7}$$

The amplitude and the phase of the nth harmonic are given by the phasor representation

$$I(n) = \frac{4}{n\pi}e^{j[(n-1)/2]\pi}, \qquad n = 1, 3, 5, \dots. \tag{5.8.8}$$

Denoting the nth harmonic impedance of the parallel RC combination by $Z(n)$, then

$$Z(n) = \frac{R\dfrac{1}{jn\omega C}}{R + \dfrac{1}{jn\omega C}}$$

$$= \frac{1}{1 + jn}$$

$$= \frac{1}{\sqrt{1 + n^2}}e^{-j\tan^{-1}n}, \qquad n = 1, 3, 5, \dots, \tag{5.8.9}$$

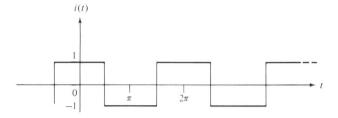

Fig. 5.8.2 Square wave current waveform.

where the values of the elements as well as $\omega = 1$ have been substituted to obtain the last expression of (5.8.9). Thus, the nth harmonic of the voltage $v(t)$ is given by

$$E(n) = I(n)Z(n) = \frac{4}{n\pi\sqrt{1 + n^2}}\, e^{j\,((n-1)/2)\pi - \tan^{-1}n}, \qquad n = 1, 3, 5, \ldots \qquad (5.8.10)$$

Finally, the steady-state response $v(t)$ becomes

$$v(t) = \sum_{n=1,3,5}^{\infty} \frac{4}{n\pi\sqrt{1 + n^2}}\, \cos\left(nt + \frac{n-1}{2}\pi - \tan^{-1}n\right)$$

$$= \sum_{k=0}^{\infty} \frac{4}{(2k + 1)\pi\sqrt{1 + (2k + 1)^2}}\, \cos[(2k + 1)t + k\pi - \tan^{-1}(2k + 1)].$$
$$(5.8.11)$$

As indicated in (5.8.11), the steady-state response is made up of an infinite number of sinusoidal components which become unrealistic in actual applications. Fortunately, in practice, most periodic input waveforms can be approximated very accurately by the first few terms in the Fourier series. Of course, the number of terms retained depends on the nature of the problem as well as the degree of accuracy required. Thus, for instance, the magnitude of the fifth harmonic of the input current in Example 5.8.1 amounts to twenty percent of that of the fundamental, whereas the magnitude of the fifth harmonic of the output voltage $v(t)$ is equal to only about four percent of that of the fundamental. In other words, the combined action of the resistor and the capacitor serves as a filter with the effect of attenuating or suppressing the magnitudes of the higher harmonic terms. With the above argument, it is not difficult to convince ourselves that the steady-state response $v(t)$ of the above example can be approximated with a high degree of accuracy by the first three nonzero terms of the series.

Effective Values and Power Products

Let us recall that the effective or rms value of a periodic function $f(t)$ is defined by

$$F_{\text{eff}} = \left[\frac{1}{T}\int_0^T f^2(t)\, dt\right]^{1/2}. \qquad (5.8.12)$$

If $f(t)$ is periodic, then it can be represented by a Fourier series. Using the exponential form, we get

$$f(t) = \sum_{n=-\infty}^{\infty} D_n e^{jn\omega t}. \tag{5.8.13}$$

To determine the effective value, we substitute (5.8.13) into (5.8.12) yielding

$$F_{\text{eff}}^2 = \frac{1}{T} \int_0^T \sum_{n=-\infty}^{\infty} \sum_{m=-\infty}^{\infty} D_n D_m e^{j(n+m)\omega t}\, dt. \tag{5.8.14}$$

By means of the orthogonality condition

$$\int_0^T e^{j(n+m)\omega t}\, dt = \begin{cases} 0, & n+m \neq 0; \\ T, & n+m = 0; \end{cases} \tag{5.8.15}$$

expression (5.8.14) reduces to

$$F_{\text{eff}}^2 = \sum_{n=-\infty}^{\infty} D_n D_n^* = D_0^2 + 2\sum_{n=1}^{\infty} D_n D_{-n} = D_0^2 + 2\sum_{n=1}^{\infty} |D_n|^2, \tag{5.8.16}$$

where D_n^* and $|D_n|$ represent, respectively, the complex conjugate and the absolute value of D_n. The result expressed in (5.8.16) is known as the *Parseval relation*. If, on the other hand, the periodic function is represented in the form of (5.8.1), that is,

$$f(t) = C_0 + C_1 \cos(\omega t + \phi_1) + C_2 \cos(2\omega t + \phi_2) + \cdots$$
$$+ C_n \cos(n\omega t + \phi_n) + \cdots, \tag{5.8.17}$$

then the corresponding Parseval relation can be written as

$$F_{\text{eff}}^2 = C_0^2 + \sum_{n=1}^{\infty} \left(\frac{C_n}{\sqrt{2}}\right)^2 \tag{5.8.18}$$

which indicates that *the square of the effective value of a periodic function $f(t)$ is equal to the sum of the squares of its dc-component C_0 and the effective values of the harmonic components $C_n/\sqrt{2}$ of the function.*

Example 5.8.2   The effective value of the current waveform described by

$$i(t) = 50 + 100 \cos t + 150 \cos 2t$$

is

$$I_{\text{eff}} = \left[50^2 + \left(\frac{100}{\sqrt{2}}\right)^2 + \left(\frac{150}{\sqrt{2}}\right)^2 \right]^{1/2} = 137 \text{ A}.$$

In like manner, if the terminal voltage and the current of a one-port are described by

$$v(t) = \sum_{n=-\infty}^{\infty} D_n e^{jn\omega t} \tag{5.8.19}$$

and

$$i(t) = \sum_{m=-\infty}^{\infty} G_n e^{jm\omega t}, \tag{5.8.20}$$

then the average power delivered to the network can be readily shown to be

$$P_{\text{ave}} = \frac{1}{T}\int_0^T v(t)i(t)\,dt = \sum_{n=-\infty}^{\infty} D_n G_n^*, \tag{5.8.21}$$

where G_n^* is the complex conjugate of the nth harmonic G_n. If both the voltage and the current are expressed in the following forms,

$$v(t) = V_{\text{dc}} + \sum_{n=-\infty}^{\infty} V_n \cos(n\omega t + \phi_n) \tag{5.8.22}$$

and

$$i(t) = I_{\text{dc}} + \sum_{m=1}^{\infty} I_m \cos(m\omega t + \psi_m), \tag{5.8.23}$$

respectively, then the average power delivered to the one-port is given by

$$P_{\text{ave}} = P_0 + \sum_{n=1}^{\infty} P_n, \tag{5.8.24}$$

where

$$P_0 = V_{\text{dc}} I_{\text{dc}} \tag{5.8.25}$$

and

$$P_n = \tfrac{1}{2} V_n I_n \cos(\phi_n - \psi_n)$$
$$= V_{n_{\text{eff}}} I_{n_{\text{eff}}} \cos \theta_n \tag{5.8.26}$$

with

$$\theta_n \triangleq \phi_n - \psi_n.$$

Example 5.8.3 The terminal voltage and the current of a one-port are given, respectively, by the expressions

$$v(t) = 10 + 15\cos t + 20\cos 2t \text{ V}$$

and

$$i(t) = 5 + 30\cos(t - 60°) + 40\cos 3t \text{ A}.$$

Determine the average power delivered to the network. Here, $P_0 = V_{\text{dc}} I_{\text{dc}}$ $= 10\,(5) = 50$ watts. Since $V_1 = 15$ V, $I_1 = 30$ A, and $\theta_1 = 60°$, $P_1 = 15 \times 30$ $\cos 60° = 112.5$ watts. Because $I_2 = 0$, we have $P_2 = 0$. Also $P_3 = 0$, since $V_3 = 0$. Therefore, $P_{\text{ave}} = P_0 + P_1 = 162.5$ watts.

An examination of (5.8.24) reveals the very important property: In determining the total average power delivered to a network, the principle of superposition can be applied to find the power components due to each of the harmonics. The sum of these components gives the total average power. This property is very interesting, since, in general, the power is not a linear function of the voltage or the current; yet the orthogonality condition (5.8.15) permits the use of the superposition principle to calculate the total average power.

5.9 SUMMARY

In the preceding sections of this chapter, we have discussed in detail the responses of linear networks. The complete response of a given network can be either regarded as (1) the sum of the transient and the steady-state responses, or (2) the sum of the zero-state and the zero-input responses. If the network is stable, then the transient response exists and is equal to the part of the complete response, which vanishes as $t \to \infty$ when the network is subject to a sinusoidal input. The remaining part of the complete response is, of course, the steady-state response of the network. The zero-input response is the portion of the complete response when the energy sources (inputs) in the system are removed, while the zero-state response is the part of the complete response when the energy originally stored in the system is set to zero.

As discussed in Section 5.4, any two quantities of a linear network can always be related by a linear differential equation with constant coefficients whose Laplace transform (with all the initial conditions set to zero) yields either a transfer function for a two-port network or a driving-point function for a one-port. In other words, both driving-point and transfer functions can always be represented as real rational functions.

In the sinusoidal steady-state analysis of networks, the conventional method of phasors is used. The phasor method is not only very efficient in determining the sinusoidal steady-state response, but is also applicable in obtaining the non-sinusoidal steady-state response when the network is excited by a periodic forcing function. Of course, in the latter case, the principle of superposition must be applied by first resolving the periodic function into a Fourier series.

As will be seen in Chapter 11, the frequency response plots discussed in Section 5.7 are directly applicable in determining the stability of a given system.

REFERENCES

1. M. E. Van Valkenburg, *Network Analysis*, 2nd Ed., Prentice-Hall, 1964, Chapters 9, 10, 11, 12, 13, and 15.

2. C. A. Desoer and E. S. Kuh, *Basic Circuit Theory*, McGraw-Hill, 1967, Chapters 5 and 6.

3. E. A. Guillemin, *Introductory Circuit Theory*, John Wiley & Sons, Inc., 1953, Chapters 5, 6, 7, 8, and 9.

4. F. F. Kuo, *Network Analysis and Synthesis*, John Wiley & Sons, Inc., 2nd Ed., 1966, Chapter 8.

5. E. Brenner and M. Javid, *Analysis of Electric Circuits*, 2nd Ed., McGraw-Hill, 1967, Chapters 9 and 10.

PROBLEMS

5.1 The network shown in Fig. P.5.1 has been in steady state with the switch S in position a. At $t = 0$, S is thrown from position a to position b. Determine the current $i(t)$ for $t > 0$ by means of the Laplace transformation method.

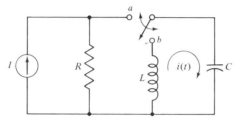

Figure P.5.1

5.2 The network of Fig. P.5.2 contains a dependent source. For zero initial conditions, determine the transform equations for the meshes as shown and then solve for $i_2(t)$ for $t > 0$.

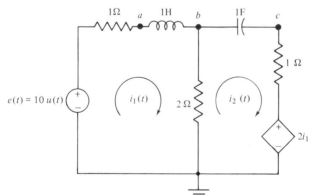

Figure P.5.2

5.3 Write the transform node equations for the network of Fig. P.5.3 and then solve for $v_a(t)$ for $t > 0$, i.e., the voltage at node a with respect to ground. Assume zero initial conditions.

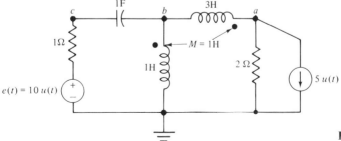

Figure P.5.3

5.4 Repeat Problem 5.3 for the network of Fig. P.5.2 for $t > 0$.

5.5 Refer to the network of Fig. P.5.5. A steady state has been reached with the switch S in position a. At $t = 0$, S is thrown from a to b. For the element values given and with $V = 10$ V, determine $v_1(t)$ for $t > 0$ by first writing the transform loop equations. Verify your result by the node analysis method.

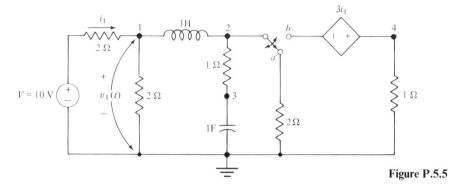

Figure P.5.5

5.6 The impedance $Z(s)$ of the network shown in Fig. P.5.6(a) has the pole-zero distribution sketched in Fig. P.5.6(b). If $Z(0) = \frac{1}{2}\Omega$, find the values for L_1, L_2, M, C, and R.

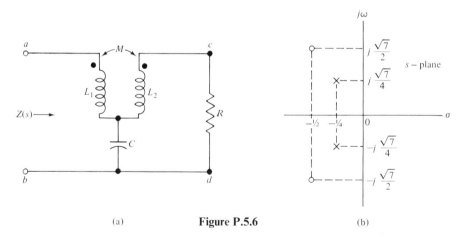

(a)   **Figure P.5.6**   (b)

5.7 Consider the network of Fig. P.5.6(a). If a current source $I_s(s)$ is connected across terminal-pair a–b such that the current flows into node a and the voltage response $V_{cd}(s)$ is defined as the voltage across the terminal-pair c–d with positive polarity taken at node c,

a) obtain an expression for the transfer function $H(s) \triangleq \dfrac{V_{cd}(s)}{I_s(s)}$, and then

b) make a pole-zero plot of $H(s)$ in the s-plane for the following set of element values:
$L_1 = 2H$, $L_2 = 8H$, $M = 4H$, $R = 4\Omega$, and $C = \frac{1}{4}F$.

5.8 Repeat Problem 5.7(a) for the network of Fig. P.5.8.

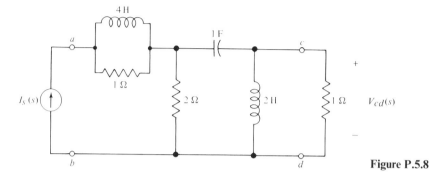

Figure P.5.8

5.9 Suppose that, in the network of Fig. P.5.8, a voltage source $V_{cd}(s)$ is placed across the terminal-pair c–d and the short-circuit current $I_s(s)$ in the terminal-pair a–b is defined as the current response, resulting in the network of Fig. P.5.9. Determine an expression for the transfer function

$$H_1(s) = \frac{I_s(s)}{V_{cd}(s)}.$$

Can any relationship be established between the transfer function $H(s)$ in Problem 5.8 and $H_1(s)$ just found? Explain.

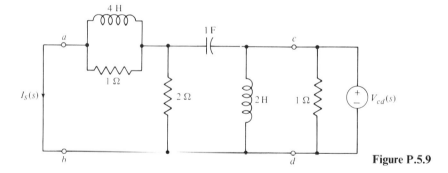

Figure P.5.9

5.10 Determine the zero-input response $v(t)$ for the network of Fig. P.5.10 for $t > 0$.

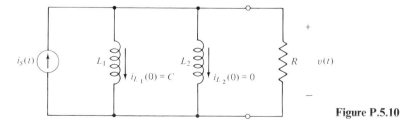

Figure P.5.10

5.11 Refer to the network of Fig. P.5.10. Let the current source $i_s(t)$ be given by

$$i_s(t) = A\delta(t) + Bu(t - t_0),$$

where A, B, and t_0 are constants. Determine

a) the zero-state response $v(t)$ of the network and
b) the complete response $v(t)$.

5.12 The network shown in Fig. P.5.12 is the small signal incremental model of a transistor amplifier. Determine

a) the transfer function

$$H(s) = \frac{V_2(s)}{V_s(s)}$$

and

b) the corresponding impulse response $h(t)$ for the following set of element values: $R_1 = 2\Omega$, $R_1 = 1\Omega$, $C_1 = C_2 = 1F$, $G_3 = 2\mho$, and $R_3 = 2\Omega$.

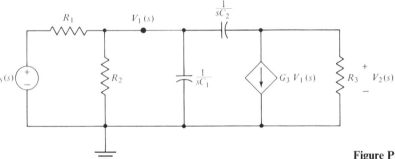

Figure P.5.12

5.13 Determine the zero-state voltage response $v_2(t)$ of the network in Fig. P.5.12 to an input $v_s(t)$ of the form $V_s(t) = 10\,[u(t) - u(t - 1)]$.

5.14 The network of Fig. P.5.14 is operating in the sinusoidal steady state with $e_s(t) = 5\cos 2t$. Given that $L = 1H$, $R_1 = R_2 = 2\Omega$, and $C = \frac{1}{2}F$, determine the current $i_c(t)$.

5.15 Refer to the network of Fig. P.5.14. For the same information as that given in Problem 5.14, determine the node voltage $v_a(t)$.

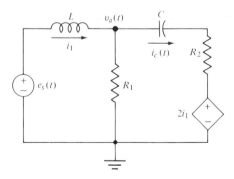

Figure P.5.14

5.16 The network of Fig. P.5.16 is adjusted so that $R^2 = (L/C)$. If the sinusoidal steady state has to be attained for the network with $e_s(t) = 5 \cos 2t$, determine the voltage $v_0(t)$ for all time.

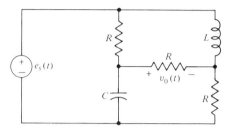

Figure P.5.16

5.17 For the network of Fig. P.5.17, determine both the polar plot and the Bode diagrams for the transfer function

$$H(s) = \frac{E_2(s)}{E_1(s)}.$$

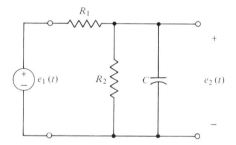

Figure P.5.17

5.18 Repeat Problem 5.17 for the network of Fig. P.5.18.

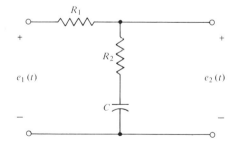

Figure P.5.18

5.19 The voltage source $e_1(t)$ in the network of Fig. P.5.18 is given by

$$e_1(t) = 10 \cos 10t + 20 \cos 20t.$$

Determine the steady-state response $e_2(t)$.

5.20 The voltage and the current of a certain one-port network are known to be, respectively,

$$v(t) = 10 + 20\cos 10t + 30\cos 30t \text{ V}$$

and

$$i(t) = 5 + 10\cos(10t + 30^0) + 15\cos 20t + 10\cos(30t - 60^0) \text{ A}.$$

Determine the average power P_{ave} consumed in the network.

CHAPTER 6

TWO-PORTS, N-PORTS, AND N-TERMINAL NETWORKS

6.1 INTRODUCTION

Consider an arbitrary network N with a finite number of nodes, t of which are accessible for making external measurements or connections as shown in Fig. 6.1.1. Then those t external nodes of N are called the *terminals* of the network. Since at least two terminals are needed for making any external connection of N to another network such as a voltage (or current) source, a measuring instrument (such as a voltmeter or an ammeter), or a load of some kind, these t terminals are often grouped in pairs. If, for example, a voltage source is connected across a pair of terminals, say k and k', of N causing a current to flow in the network, then the current $i_k(t)$ flowing *into* terminal k must necessarily be the same as the current $i_{k'}(t)$ flowing *out* of terminal k', as depicted in Fig. 6.1.2. Thus, we have

$$i_k(t) = i_{k'}(t). \tag{6.1.1}$$

Under this condition, the terminal-pair k–k' is said to constitute a *port* (which signifies a port of entry). Now if the t terminals of N are grouped into n terminal-pairs (with $n = t/2$, assuming t to be even) such that for each terminal-pair the current flowing into one terminal is equal to that flowing out of the other as illustrated in Fig. 6.1.3, then N is called an *n-port*. Thus, an n-port is an n-terminal-pair network such that (6.1.1) is satisfied for $k = 1, 2, \ldots, n$.

When a network has only two terminals accessible for external connections it is referred to as a *one-port*, which is, of course, a special case of an n-port with $n = 1$. Some of the properties of the general passive one-port as well as the two (driving-point) immittance functions associated with it have been discussed in the

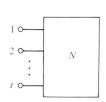

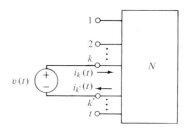

Fig. 6.1.1 A t-terminal network.

Fig. 6.1.2 Illustration of a port in a network.

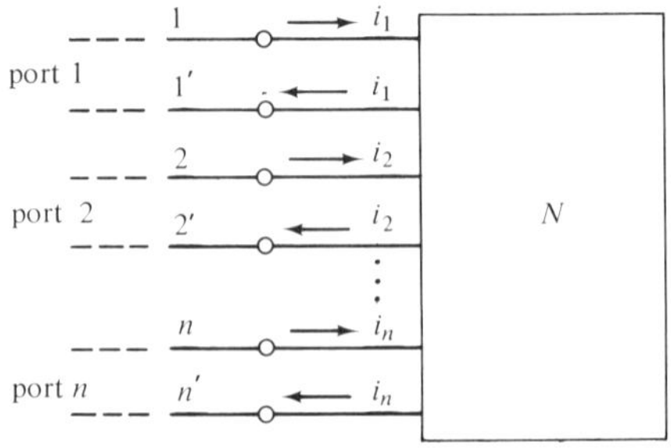

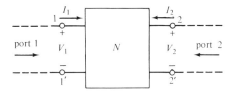

Fig. 6.1.3 An *n*-port network.

Fig. 6.2.1 The general linear two-port *N*.

preceding chapters. In the next section, we shall discuss, in great detail, properties and applications of the general two-port which is perhaps the most important class of networks in the study of network analysis and design.

6.2 THE LINEAR TWO-PORT

Consider the general linear two-port *N* shown in Fig. 6.2.1, which may be a passive or an active two-port with no independent sources. The two terminal-pairs are identified as 1–1′ and 2–2′, and are referred to as ports 1 and 2, respectively. Note that both currents I_1 and I_2 are assigned to flow *into* the *unprimed* terminals (and out from the primed terminals) and both voltages V_1 and V_2 are assigned to have their positive (+) polarity marks at the unprimed terminals of *N*, as a matter of convention.

There are four variables of concern in a general two-port; namely, the two voltages V_1 and V_2 and the two currents I_1 and I_2 measurable at both ports of the network. Thus, to characterize the two-port, one needs to relate the four variables by means of certain equations. However, only two of these four variables are independent since the specification of any two variables completely determines the remaining two.* In other words, when any two of the four variables are chosen

---

* If two sources are applied to the network at both ports, representing two independent variables, then the remaining two must be dependent variables, since they are the responses to the sources. If, on the other hand, only one source is applied at one port, and a load *Z* is connected at the other, then there still can be only two independent variables, since one of the two variables at the output port (where the load is connected) must be dependent upon the other variable at the same port as both are related by the equation $V = IZ$.

as *independent* variables, the other two may be expressed as *dependent* variables in terms of the two chosen variables by means of two independent equations. Since there are six possible combinations of four quantities taken two at a time, it is evident that there are at least six ways of characterizing a general two-port.* For example, if we choose I_1 and I_2 as independent variables, we may express V_1 and V_2 in the following equations

(a) $$V_1 = z_{11}I_1 + z_{12}I_2,$$

(6.2.1)

(b) $$V_2 = z_{21}I_1 + z_{22}I_2.$$

In matrix form $\mathbf{V} = \mathbf{Z}\mathbf{I}$, the set of equations (6.2.1) becomes

$$\begin{bmatrix} V_1 \\ V_2 \end{bmatrix} = \begin{bmatrix} z_{11} & z_{12} \\ z_{21} & z_{22} \end{bmatrix} \begin{bmatrix} I_1 \\ I_2 \end{bmatrix},$$

(6.2.2)

where the z_{ij}'s are the coefficients which form a set of *network parameters* that completely characterize the two-port, and can be determined from the network by making external measurements of the two voltages and the two currents at the terminals. Hence, for each choice of independent and dependent variables, a unique set of network parameters are determined as a characterization of the two-port provided that this particular set exists for the given network. In the next section, we shall define the six sets of network parameters.

6.3 TWO-PORT NETWORK PARAMETERS

Open-Circuit Impedances, or the z_{ij}-Parameters

Refer to the general two-port of Fig. 6.2.1. If a current source I_1 is connected to port 1, and port 2 is open-circuited, then current I_2 is equal to zero and the two equations in (6.2.1) become

$$V_1 = z_{11}I_1 \qquad (I_2 = 0)$$
$$V_2 = z_{21}I_1 \qquad (I_2 = 0)$$

from which we obtain

(a) $$z_{11} = \left. \frac{V_1}{I_1} \right|_{I_2 = 0}$$

(6.3.1)

(b) $$z_{21} = \left. \frac{V_2}{I_1} \right|_{I_2 = 0}.$$

---

* Actually there are more than six ways of characterizing networks. For example, scattering parameters are used to describe networks at high frequencies or distributed networks. In fact, the topic of scattering parameters will be discussed in a separate section, as the concept is quite different from that used for the six conventional sets of parameters presented in this section.

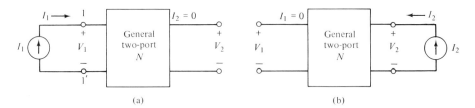

Fig. 6.3.1 Connections for determining (a) z_{11} and z_{21}, and (b) z_{12} and z_{22}.

Similarly, if port 1 is open-circuited, and a current source I_2 is connected to port 2, we have from (6.2.1)

$$V_1 = z_{12}I_2 \qquad (I_1 = 0)$$
$$V_2 = z_{22}I_2 \qquad (I_1 = 0)$$

which yield directly

(a)
$$z_{12} = \left.\frac{V_1}{I_2}\right|_{I_1=0}$$

(6.3.2)

(b)
$$z_{22} = \left.\frac{V_2}{I_2}\right|_{I_1=0}.$$

Since the four parameters z_{11}, z_{12}, z_{21}, and z_{22} are ratios of voltage to current under certain open-circuit conditions as expressed in the four defining equations in (6.3.1) and (6.3.2), they are referred to as the *open-circuit impedances* of the two-port N. The corresponding matrix **Z** in (6.2.2) is called the *open-circuit impedance matrix* of N. The conditions under which the four parameters may be determined are depicted in Fig. 6.3.1.

Short-Circuit Admittances, or the y_{ij}-Parameters

If the two voltages V_1 and V_2 in the two-port N of Fig. 6.2.1 are chosen to be independent variables, then the currents I_1 and I_2 may be expressed in terms of V_1 and V_2; namely,

(a)
$$I_1 = y_{11}V_1 + y_{12}V_2,$$

(b)
$$I_2 = y_{21}V_1 + y_{22}V_2.$$

(6.3.3)

In matrix form $\mathbf{I} = \mathbf{YV}$, we have

$$\begin{bmatrix} I_1 \\ I_2 \end{bmatrix} = \begin{bmatrix} y_{11} & y_{12} \\ y_{21} & y_{22} \end{bmatrix} \begin{bmatrix} V_1 \\ V_2 \end{bmatrix},$$

(6.3.4)

where the coefficients y_{ij} are called the *short-circuit admittances* of N, and the matrix **Y** which contains these four coefficients is known as the *short-circuit admittance matrix*. The defining equations for the short-circuit admittances can be

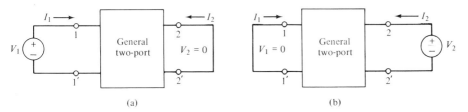

Fig. 6.3.2 Connections for determining (a) y_{11} and y_{21}, and (b) y_{12} and y_{22}.

readily obtained by short-circuiting one of the two ports with the other port driven by a voltage source as illustrated in Fig. 6.3.2. By setting the two voltages in (6.3.3) to zero one at a time, one immediately obtains the following desired expressions:

(a) $$y_{11} = \left.\frac{I_1}{V_1}\right|_{V_2=0},$$

(b) $$y_{21} = \left.\frac{I_2}{V_1}\right|_{V_2=0},$$

                                                          (6.3.5)

(c) $$y_{12} = \left.\frac{I_1}{V_2}\right|_{V_1=0},$$

(d) $$y_{22} = \left.\frac{I_2}{V_2}\right|_{V_1=0}.$$

Transmission a_{ij}-Parameters

When the independent variables are V_2 and I_2 and the remaining two variables V_1 and I_1 are expressed in terms of them, we have the following matrix equation

$$\begin{bmatrix} V_1 \\ I_1 \end{bmatrix} = \begin{bmatrix} a_{11} & a_{12} \\ a_{21} & a_{22} \end{bmatrix} \begin{bmatrix} V_2 \\ -I_2 \end{bmatrix}, \qquad (6.3.6)$$

where the coefficients a_{11}, a_{12}, a_{21}, and a_{22} are called the *transmission a_{ij}-param-eters* of the two-port. The defining expressions for them are easily obtained as

(a) $$a_{11} = \left.\frac{V_1}{V_2}\right|_{I_2=0},$$

(b) $$a_{21} = \left.\frac{I_1}{V_2}\right|_{I_2=0},$$

                                                           (6.3.7)

(c) $$a_{12} = \left.\frac{V_1}{-I_2}\right|_{V_2=0},$$

(d) $$a_{22} = \left.\frac{I_1}{-I_2}\right|_{V_2=0}.$$

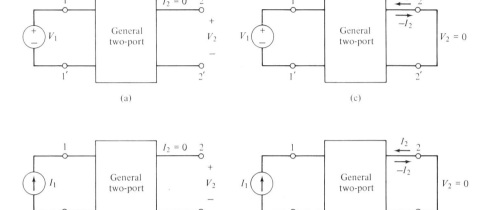

Fig. 6.3.3 Connections for determining (a) a_{11}, (b) a_{21}, (c) a_{12}, and (d) a_{22}.

The necessary connections for determining these a_{ij}-parameters are shown in Fig. 6.3.3. It should be noted that the a_{ij}-parameters are also known as the *ABCD-parameters* where $A = a_{11}$, $B = a_{12}$, $C = a_{21}$, and $D = a_{22}$; that is

$$\begin{bmatrix} a_{11} & a_{12} \\ a_{21} & a_{22} \end{bmatrix} = \begin{bmatrix} A & B \\ C & D \end{bmatrix}. \tag{6.3.8)*}$$

Transmission b_{ij}-Parameters

If the variables V_1 and I_1 at port 1 are chosen as the independent variables, we may express V_2 and I_2 at port 2 in terms of them as

$$\begin{bmatrix} V_2 \\ I_2 \end{bmatrix} = \begin{bmatrix} b_{11} & b_{12} \\ b_{21} & b_{22} \end{bmatrix} \begin{bmatrix} V_1 \\ -I_1 \end{bmatrix}, \tag{6.3.9}$$

where the b_{ij}'s are referred to as the *transmission b_{ij}-parameters* and are defined by

(a) $\quad b_{11} = \dfrac{V_2}{V_1}\Bigg|_{I_1 = 0}$,

(b) $\quad b_{21} = \dfrac{I_2}{V_1}\Bigg|_{I_1 = 0}$,

$$\tag{6.3.10}$$

(c) $\quad b_{12} = \dfrac{V_2}{-I_1}\Bigg|_{V_1 = 0}$,

(d) $\quad b_{22} = \dfrac{I_2}{-I_1}\Bigg|_{V_1 = 0}$.

---

* They are also called the *chain* parameters.

Similar to the a_{ij}'s, these b_{ij}-parameters are also known as the $\mathscr{ABCD}$-parameters where $\mathscr{A} = b_{11}$, $\mathscr{B} = b_{12}$, $\mathscr{C} = b_{21}$, and $\mathscr{D} = b_{22}$, or

$$\begin{bmatrix} b_{11} & b_{12} \\ b_{21} & b_{22} \end{bmatrix} = \begin{bmatrix} \mathscr{A} & \mathscr{B} \\ \mathscr{C} & \mathscr{D} \end{bmatrix}.$$

(6.3.11)

Hybrid h_{ij}-Parameters

Sometimes, for a given two-port, it is convenient to express the voltage V_1 at port 1 and the current I_2 at port 2 in terms of the current I_1 at port 1 and the voltage V_2 at port 2. That is, in matrix form:

$$\begin{bmatrix} V_1 \\ I_2 \end{bmatrix} = \begin{bmatrix} h_{11} & h_{12} \\ h_{21} & h_{22} \end{bmatrix} \begin{bmatrix} I_1 \\ V_2 \end{bmatrix},$$

(6.3.12)

where I_1 and V_2 are the independent variables, V_1 and I_2 the dependent ones, and the h_{ij}'s are called the *hybrid h_{ij}-parameters*, defined by the following expressions:

(a) $h_{11} = \dfrac{V_1}{I_1}\bigg|_{V_2=0}$,   (b) $h_{21} = \dfrac{I_2}{I_1}\bigg|_{V_2=0}$,

(6.3.13)

(c) $h_{12} = \dfrac{V_1}{V_2}\bigg|_{I_1=0}$ (d) $h_{22} = \dfrac{I_2}{V_2}\bigg|_{I_1=0}$.

Hybrid g_{ij}-Parameters

Finally, as the last of the six possible ways of characterizing the general two-port, one might wish to express I_1 and V_2 in terms of V_1 and I_2; namely

$$\begin{bmatrix} I_1 \\ V_2 \end{bmatrix} = \begin{bmatrix} g_{11} & g_{12} \\ g_{21} & g_{22} \end{bmatrix} \begin{bmatrix} V_1 \\ I_2 \end{bmatrix},$$

(6.3.14)

where the g_{ij}'s are known as the *hybrid g_{ij}-parameters* and are defined by

(a) $g_{11} = \dfrac{I_1}{V_1}\bigg|_{I_2=0}$,   (b) $g_{21} = \dfrac{V_2}{V_1}\bigg|_{I_2=0}$,

(6.3.15)

(c) $g_{12} = \dfrac{I_1}{I_2}\bigg|_{V_1=0}$,   (d) $g_{22} = \dfrac{V_2}{I_2}\bigg|_{V_1=0}$.

which completes the introduction of the six sets of network parameters. The following example will illustrate how these different sets of parameters may be obtained from a given two-port.

Example 6.3.1   Consider the passive two-port N of Fig. 6.3.4. The two loop equations are given by

$$V_1 = (Z_1 + Z_3 + Z_4)I_1 + Z_3I_2,$$
$$V_2 = Z_3I_1 + (Z_2 + Z_3 + Z_5)I_2.$$

(6.3.16)

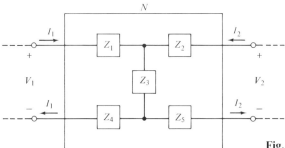

Fig. 6.3.4 A passive two-port *N*.

In matrix form $\mathbf{V} = \mathbf{Z}\mathbf{I}$, we have

$$\begin{bmatrix} V_1 \\ V_2 \end{bmatrix} = \begin{bmatrix} z_{11} & z_{12} \\ z_{21} & z_{22} \end{bmatrix} \begin{bmatrix} I_1 \\ I_2 \end{bmatrix}, \qquad (6.3.17)$$

where

$$\mathbf{Z} = \begin{bmatrix} z_{11} & z_{12} \\ z_{21} & z_{22} \end{bmatrix} = \begin{bmatrix} Z_1 + Z_3 + Z_4 & Z_3 \\ Z_3 & Z_2 + Z_3 + Z_5 \end{bmatrix} \qquad (6.3.18)$$

is, of course, the open-circuit impedance matrix of *N*.

Next, solving (6.3.16) for I_1 and I_2 using Cramer's rule, we obtain the matrix equation $\mathbf{I} = \mathbf{Y}\mathbf{V}$:

$$\begin{bmatrix} I_1 \\ I_2 \end{bmatrix} = \begin{bmatrix} \dfrac{\Delta_{11}}{\Delta_z} & \dfrac{\Delta_{21}}{\Delta_z} \\[2mm] \dfrac{\Delta_{12}}{\Delta_z} & \dfrac{\Delta_{22}}{\Delta_z} \end{bmatrix} \begin{bmatrix} V_1 \\ V_2 \end{bmatrix}, \qquad (6.3.19)$$

where Δ_z is the determinant of $\mathbf{Z}$, namely

$$\begin{aligned}
\Delta_z &\equiv \det Z \\
&= z_{11}z_{22} - z_{12}z_{21} \\
&= (Z_1 + Z_3 + Z_4)(Z_2 + Z_3 + Z_5) - Z_3^2 \qquad (6.3.20) \\
&= Z_1Z_2 + Z_1Z_3 + Z_1Z_5 + Z_2Z_3 + Z_2Z_4 + Z_3Z_4 + Z_3Z_5 + Z_4Z_5.
\end{aligned}$$

Also, the Δ_{ij}'s are the (i,j)-cofactors of $\Delta_z (i, j = 1, 2)$, which are given by

$$\begin{aligned}
\Delta_{11} &= z_{22} = Z_2 + Z_3 + Z_5, & \Delta_{21} &= -z_{12} = -Z_3, \\
\Delta_{12} &= -z_{21} = -Z_3, & \Delta_{22} &= z_{11} = Z_1 + Z_3 + Z_4.
\end{aligned} \qquad (6.3.21)$$

Hence, we obtain the short-circuit admittance matrix:

$$
\mathbf{Y} = \begin{bmatrix} y_{11} & y_{12} \\ \\ y_{21} & y_{22} \end{bmatrix} = \begin{bmatrix} \dfrac{\Delta_{11}}{\Delta_z} & \dfrac{\Delta_{21}}{\Delta_z} \\ \\ \dfrac{\Delta_{12}}{\Delta_z} & \dfrac{\Delta_{22}}{\Delta_z} \end{bmatrix} \tag{6.3.22}
$$

with Δ_z and its cofactors defined by (6.3.20) and (6.3.21), respectively. From (6.3.17) through (6.3.22), it should be obvious that $\mathbf{Y} = \mathbf{Z}^{-1}$ which is true for a general two-port provided that $\mathbf{Z}^{-1}$ exists.

If we now open-circuit port 2 (so that $I_2 = 0$), then the two equations in (6.3.16) become

$$
\begin{aligned}
V_1 &= (Z_1 + Z_3 + Z_4)I_1 \\
V_2 &= Z_3 I_1
\end{aligned} \tag{6.3.23}
$$

which according to (6.3.7) give rise to

$$
\begin{aligned}
a_{11} &= \left.\frac{V_1}{V_2}\right|_{I_2=0} = \frac{Z_1 + Z_3 + Z_4}{Z_3} \\
a_{21} &= \left.\frac{I_1}{V_2}\right|_{I_2=0} = \frac{1}{Z_3}.
\end{aligned} \tag{6.3.24}
$$

Next, with port 2 short-circuited (i.e., $V_2 = 0$), we obtain from (6.3.19)

$$
\begin{aligned}
I_1 &= \frac{\Delta_{11}}{\Delta_z} V_1 \qquad (V_2 = 0) \\
I_2 &= \frac{\Delta_{12}}{\Delta_z} V_1 \qquad (V_2 = 0)
\end{aligned} \tag{6.3.25}
$$

which, together with (6.3.7) and (6.3.21), yield

$$
\begin{aligned}
a_{12} &= \left.\frac{V_1}{-I_2}\right|_{V_2=0} = -\frac{\Delta_z}{\Delta_{12}} = \frac{\Delta_z}{Z_3} \\
a_{22} &= \left.\frac{I_1}{-I_2}\right|_{V_2=0} = -\frac{\Delta_{11}}{\Delta_{12}} = \frac{Z_2 + Z_3 + Z_5}{Z_3},
\end{aligned} \tag{6.3.26}
$$

where Δ_z is expressed in (6.3.20). Substituting (6.3.24) and (6.3.26), we find

$$
\begin{bmatrix} a_{11} & a_{12} \\ a_{21} & a_{22} \end{bmatrix} \equiv \begin{bmatrix} A & B \\ C & D \end{bmatrix} = \begin{bmatrix} \dfrac{Z_1 + Z_3 + Z_4}{Z_3} & \dfrac{\Delta_z}{Z_3} \\ \\ \dfrac{1}{Z_3} & \dfrac{Z_2 + Z_3 + Z_5}{Z_3} \end{bmatrix}, \tag{6.3.27}
$$

which, of course, defines the a_{ij}'s or the $ABCD$-parameters.

Similarly, by open-circuiting and then short-circuiting port 1, the b_{ij}- or the $\mathscr{ABCD}$-parameters may be determined from (6.3.16), (6.3.19), and (6.3.21) together with the defining expressions in (6.3.10). They are now given in matrix form:

$$
\begin{bmatrix} b_{11} & b_{12} \\ \\ b_{21} & b_{22} \end{bmatrix} \equiv \begin{bmatrix} \mathscr{A} & \mathscr{B} \\ \\ \mathscr{C} & \mathscr{D} \end{bmatrix} = \begin{bmatrix} \dfrac{Z_2 + Z_3 + Z_5}{Z_3} & \dfrac{\Delta_z}{Z_3} \\ \\ \dfrac{1}{Z_3} & \dfrac{Z_1 + Z_3 + Z_4}{Z_3} \end{bmatrix}, \quad (6.3.28)
$$

where Δ_z is defined in (6.3.20) as before.

For the determination of the hybrid h_{ij}-parameters, we simply set $V_2 = 0$ in (6.3.19) and solve for h_{11} and h_{21} according to (6.3.13). We then set $I_1 = 0$ in (6.3.16) and solve for h_{12} and h_{22}. The results are expressed in matrix form:

$$
\begin{bmatrix} h_{11} & h_{12} \\ \\ h_{21} & h_{22} \end{bmatrix} = \begin{bmatrix} \dfrac{\Delta_z}{Z_2 + Z_3 + Z_5} & \dfrac{Z_3}{Z_2 + Z_3 + Z_5} \\ \\ \dfrac{-Z_3}{Z_2 + Z_3 + Z_5} & \dfrac{1}{Z_2 + Z_3 + Z_5} \end{bmatrix}, \quad (6.3.29)
$$

where Δ_z is again given in (6.3.20).

Finally, the hybrid g_{ij}-parameters are determined in a similar manner except that now $I_2 = 0$ in (6.3.16) in solving for g_{11} and g_{21}, and $V_1 = 0$ in (6.3.19) in solving for g_{12} and g_{22}. The results are given by

$$
\begin{bmatrix} g_{11} & g_{12} \\ \\ g_{21} & g_{22} \end{bmatrix} = \begin{bmatrix} \dfrac{1}{Z_1 + Z_3 + Z_4} & \dfrac{-Z_3}{Z_1 + Z_3 + Z_4} \\ \\ \dfrac{Z_3}{Z_1 + Z_3 + Z_4} & \dfrac{\Delta_z}{Z_1 + Z_3 + Z_4} \end{bmatrix}, \quad (6.3.30)
$$

with Δ_z defined in (6.3.20).

From the example above, it is evident that the six sets of network parameters are interrelated and once one set of parameters is known the other five may be determined from it through the corresponding interrelationships. For example, by examining (6.3.17) through (6.3.22), it is readily seen that

$$
\begin{bmatrix} y_{11} & y_{12} \\ \\ y_{21} & y_{22} \end{bmatrix} = \begin{bmatrix} \dfrac{z_{22}}{\Delta_z} & \dfrac{-z_{12}}{\Delta_z} \\ \\ \dfrac{-z_{21}}{\Delta_z} & \dfrac{z_{11}}{\Delta_z} \end{bmatrix}, \quad (6.3.31)
$$

where

$$
\Delta_z = z_{11}z_{22} - z_{12}z_{21}. \quad (6.3.32)
$$

Table 6.3.1 Interrelationships Among Two-Port Parameters.

Type	Immittance parameters $[z_{ij}]$	Immittance parameters $[y_{ij}]$	Transmission parameters $[a_{ij}]$	Transmission parameters $[b_{ij}]$	Hybrid parameters $[h_{ij}]$	Hybrid parameters $[g_{ij}]$
$[z_{ij}]$	$\begin{matrix} z_{11} & z_{12}\\ z_{21} & z_{22}\end{matrix}$	$\begin{matrix} \frac{y_{22}}{\Delta_y} & \frac{-y_{12}}{\Delta_y}\\ \frac{-y_{21}}{\Delta_y} & \frac{y_{11}}{\Delta_y}\end{matrix}$	$\begin{matrix} \frac{a_{11}}{a_{21}} & \frac{\Delta_a}{a_{21}}\\ \frac{1}{a_{21}} & \frac{a_{22}}{a_{21}}\end{matrix}$	$\begin{matrix} \frac{b_{22}}{b_{21}} & \frac{1}{b_{21}}\\ \frac{\Delta_b}{b_{21}} & \frac{b_{11}}{b_{21}}\end{matrix}$	$\begin{matrix} \frac{\Delta_h}{h_{22}} & \frac{h_{12}}{h_{22}}\\ \frac{-h_{21}}{h_{22}} & \frac{1}{h_{22}}\end{matrix}$	$\begin{matrix} \frac{1}{g_{11}} & \frac{-g_{12}}{g_{11}}\\ \frac{g_{21}}{g_{11}} & \frac{\Delta_g}{g_{11}}\end{matrix}$
$[y_{ij}]$	$\begin{matrix} \frac{z_{22}}{\Delta_z} & \frac{-z_{12}}{\Delta_z}\\ \frac{-z_{21}}{\Delta_z} & \frac{z_{11}}{\Delta_z}\end{matrix}$	$\begin{matrix} y_{11} & y_{12}\\ y_{21} & y_{22}\end{matrix}$	$\begin{matrix} \frac{a_{22}}{a_{12}} & \frac{-\Delta_a}{a_{12}}\\ \frac{-1}{a_{12}} & \frac{a_{11}}{a_{12}}\end{matrix}$	$\begin{matrix} \frac{b_{11}}{b_{12}} & \frac{-\Delta_b}{b_{12}}\\ \frac{-1}{b_{12}} & \frac{b_{22}}{b_{12}}\end{matrix}$	$\begin{matrix} \frac{1}{h_{11}} & \frac{-h_{12}}{h_{11}}\\ \frac{h_{21}}{h_{11}} & \frac{\Delta_h}{h_{11}}\end{matrix}$	$\begin{matrix} \frac{\Delta_g}{g_{22}} & \frac{g_{12}}{g_{22}}\\ \frac{-g_{21}}{g_{22}} & \frac{1}{g_{22}}\end{matrix}$
$[a_{ij}]$	$\begin{matrix} \frac{z_{11}}{z_{21}} & \frac{\Delta_z}{z_{21}}\\ \frac{1}{z_{21}} & \frac{z_{22}}{z_{21}}\end{matrix}$	$\begin{matrix} \frac{-y_{22}}{y_{21}} & \frac{-1}{y_{21}}\\ \frac{-\Delta_y}{y_{21}} & \frac{-y_{11}}{y_{21}}\end{matrix}$	$\begin{matrix} a_{11} & a_{12}\\ a_{21} & a_{22}\end{matrix}$	$\begin{matrix} \frac{b_{22}}{\Delta_b} & \frac{b_{12}}{\Delta_b}\\ \frac{b_{21}}{\Delta_b} & \frac{b_{11}}{\Delta_b}\end{matrix}$	$\begin{matrix} \frac{-\Delta_h}{h_{21}} & \frac{-h_{11}}{h_{21}}\\ \frac{-h_{22}}{h_{21}} & \frac{-1}{h_{21}}\end{matrix}$	$\begin{matrix} \frac{1}{g_{21}} & \frac{g_{22}}{g_{21}}\\ \frac{g_{11}}{g_{21}} & \frac{\Delta_g}{g_{21}}\end{matrix}$
$[b_{ij}]$	$\begin{matrix} \frac{z_{22}}{z_{12}} & \frac{\Delta_z}{z_{12}}\\ \frac{1}{z_{12}} & \frac{z_{11}}{z_{12}}\end{matrix}$	$\begin{matrix} \frac{-y_{11}}{y_{12}} & \frac{-1}{y_{12}}\\ \frac{-\Delta_y}{y_{12}} & \frac{-y_{22}}{y_{12}}\end{matrix}$	$\begin{matrix} \frac{a_{22}}{\Delta_a} & \frac{a_{12}}{\Delta_a}\\ \frac{a_{21}}{\Delta_a} & \frac{a_{11}}{\Delta_a}\end{matrix}$	$\begin{matrix} b_{11} & b_{12}\\ b_{21} & b_{22}\end{matrix}$	$\begin{matrix} \frac{1}{h_{12}} & \frac{h_{11}}{h_{12}}\\ \frac{h_{22}}{h_{12}} & \frac{\Delta_h}{h_{12}}\end{matrix}$	$\begin{matrix} \frac{-\Delta_g}{g_{12}} & \frac{-g_{22}}{g_{12}}\\ \frac{-g_{11}}{g_{12}} & \frac{-1}{g_{12}}\end{matrix}$
$[h_{ij}]$	$\begin{matrix} \frac{\Delta_z}{z_{22}} & \frac{z_{12}}{z_{22}}\\ \frac{-z_{21}}{z_{22}} & \frac{1}{z_{22}}\end{matrix}$	$\begin{matrix} \frac{1}{y_{11}} & \frac{-y_{12}}{y_{11}}\\ \frac{y_{21}}{y_{11}} & \frac{\Delta_y}{y_{11}}\end{matrix}$	$\begin{matrix} \frac{a_{12}}{a_{22}} & \frac{\Delta_a}{a_{22}}\\ \frac{-1}{a_{22}} & \frac{a_{21}}{a_{22}}\end{matrix}$	$\begin{matrix} \frac{b_{12}}{b_{11}} & \frac{-\Delta_b}{b_{11}}\\ \frac{-1}{b_{11}} & \frac{b_{21}}{b_{11}}\end{matrix}$	$\begin{matrix} h_{11} & h_{12}\\ h_{21} & h_{22}\end{matrix}$	$\begin{matrix} \frac{g_{22}}{\Delta_g} & \frac{-g_{12}}{\Delta_g}\\ \frac{-g_{21}}{\Delta_g} & \frac{g_{11}}{\Delta_g}\end{matrix}$
$[g_{ij}]$	$\begin{matrix} \frac{1}{z_{11}} & \frac{-z_{12}}{z_{11}}\\ \frac{z_{21}}{z_{11}} & \frac{\Delta_z}{z_{11}}\end{matrix}$	$\begin{matrix} \frac{\Delta_y}{y_{22}} & \frac{y_{12}}{y_{22}}\\ \frac{-y_{21}}{y_{22}} & \frac{1}{y_{22}}\end{matrix}$	$\begin{matrix} \frac{a_{21}}{a_{11}} & \frac{-\Delta_a}{a_{11}}\\ \frac{1}{a_{11}} & \frac{a_{12}}{a_{11}}\end{matrix}$	$\begin{matrix} \frac{b_{21}}{b_{22}} & \frac{-1}{b_{22}}\\ \frac{\Delta_b}{b_{22}} & \frac{b_{12}}{b_{22}}\end{matrix}$	$\begin{matrix} \frac{h_{22}}{\Delta_h} & \frac{-h_{12}}{\Delta_h}\\ \frac{-h_{21}}{\Delta_h} & \frac{h_{11}}{\Delta_h}\end{matrix}$	$\begin{matrix} g_{11} & g_{12}\\ g_{21} & g_{22}\end{matrix}$
Δ_k	$\Delta_z = z_{11}z_{22} - z_{12}z_{21}$	$\Delta_y = y_{11}y_{22} - y_{12}y_{21}$	$\Delta_a = a_{11}a_{22} - a_{12}a_{21}$	$\Delta_b = b_{11}b_{22} - b_{12}b_{21}$	$\Delta_h = h_{11}h_{22} - h_{12}h_{21}$	$\Delta_g = g_{11}g_{22} - g_{12}g_{21}$

Similarly, other interrelationships between two sets of parameters may be obtained. These interrelationships are now summarized in Table 6.3.1.

Remarks. The following remarks serve to clarify some of the important concepts in the study of the six sets of network parameters introduced in this section.

1.   The definitions of the z_{ij}-, y_{ij}-, h_{ij}-, and g_{ij}-parameters as given by (6.3.1) and (6.3.2), (6.3.5), (6.3.13), and (6.3.15), respectively, can be used to measure these parameters directly from the physical network; they can also be computed from the network model by applying the specified constraints as defined by the expressions cited above. However, if the definitions as given by (6.3.7) and (6.3.10) are used to find the transmission (a_{ij}- and b_{ij}-) parameters, one would encounter difficulty because of conflicting constraint requirements. For example, take the case of $a_{11} \triangleq V_1/V_2$ with $I_2 = 0$. If V_2 is considered as the excitation, how can one set $I_2 = 0$ at the same time? These difficulties can be resolved if the a_{ij}- and b_{ij}-parameters are considered as *inverse network functions*. Thus, to compute a_{11}, one applies V_1 with $I_2 = 0$ (output open-circuited) and evaluates V_2. Then the *inverse* of the network function $V_2/V_1|_{I_2=0}$ yields a_{11}. For this reason, the a_{ij} and b_{ij} are referred to as *inverse* network functions whereas the other four sets are all (ordinary) network functions as discussed in Chapter 5.

2.   From Table 6.3.1, the following facts can be readily observed:

a)   For a general linear two-port, we define

$$\Delta_k = k_{11}k_{22} - k_{12}k_{21} \quad (k = z, y, a, b, h, \text{ or } g) \tag{6.3.33}$$

b)   For a given linear two-port, it is possible that a certain set of parameters cannot be determined from another set. For example, if $b_{21} = 0$ for some two-port, then the z_{ij}-parameters cannot be expressed in terms of the b_{ij}'s using the expressions in Table 6.3.1. As another example, if $\Delta_z = 0$ for a particular two-port, then the y_{ij}-parameters cannot be determined from the z_{ij}'s since they will be undefined in terms of the z_{ij}-parameters.

In the next section, various equivalent representations for a general linear two-port will be discussed.

6.4 EQUIVALENT NETWORKS FOR A LINEAR TWO-PORT

We showed in the preceding section that the terminal behavior of a general linear two-port N such as shown in Fig. 6.2.1 can be described by two independent equations. Furthermore, we showed that in terms of the open-circuit impedances z_{ij}, these two independent equations are given by (6.2.1) or, in matrix notation, by (6.2.2). In the two equations of (6.2.1), the two currents I_1 and I_2 are the independent variables, and the two voltages V_1 and V_2 are the dependent ones. In the first equation, let us consider the term $z_{12}I_2$ as a dependent voltage source with a value of $z_{12}I_2$ volts, and the term $z_{11}I_1$ as the voltage drop across the impedance z_{11} due to the current I_1. Then the equation may be interpreted as the loop equation

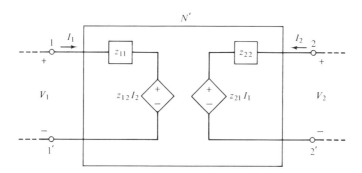

Fig. 6.4.1 Equivalent network N' of a linear two-port N of Fig. 6.2.1, using the z_{ij} parameters.

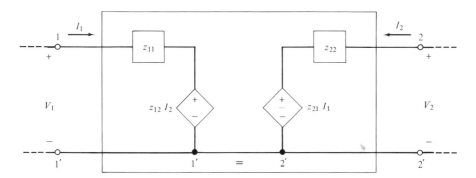

Fig. 6.4.2 Equivalent network of a grounded two-port.

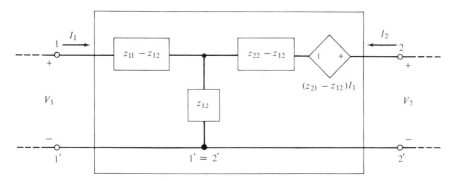

Fig. 6.4.3 Single-voltage-source equivalent of a grounded two-port.

written for the loop consisting of the dependent voltage source $z_{12}I_2$ connected in series with the impedance z_{11} so that the loop current is I_1 and the voltage across the series combination is V_1. Similar interpretation can be made for the second equation of (6.2.1). Thus, an equivalent network N' of N can be obtained in terms

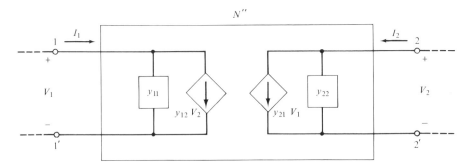

Fig. 6.4.4 Two-current-source model of N of Fig. 6.2.1 using the y_{ij} parameters.

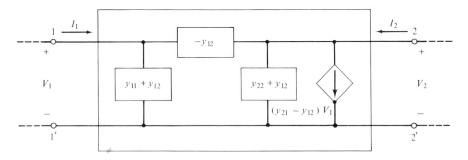

Fig. 6.4.5 Single-current-source model for a grounded two-port.

of the z_{ij}-parameters as depicted in Fig. 6.4.1. Here, by equivalence, we mean that both N and N' are described by the same set of equations, viz., (6.2.1), using the z_{ij}-parameters. This equivalent network or *model* N' is known as the *two-voltage-source equivalent* of the general linear two-port N, and is one of the most general representations of N.

If N is a *grounded* (or *common-ground*) two-port, that is, terminals 1′ and 2′ of N in Fig. 6.2.1 are the same terminal (the common ground), then the equivalent N' of N, illustrated in Fig. 6.4.1, becomes that shown in Fig. 6.4.2, which, as can be easily shown,* can be further reduced to a single-voltage-source model as depicted in Fig. 6.4.3.

In a dual development, one can obtain the *two-current-source equivalent* N'' for the general linear two-port N of Fig. 6.2.1, as shown in Fig. 6.4.4. When N is a grounded two-port, with terminals 1′ and 2′ coalesced, the corresponding equivalent may be reduced to a single-current-source model, as illustrated in Fig. 6.4.5.

Remarks. At this point, a few remarks are in order.

1. In comparing the two-current-source model N'' of N, shown in Fig. 6.4.4, to the

---

* See Problem 6.4.

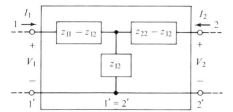

Fig. 6.4.6 Equivalent T of a linear grounded, reciprocal two-port.

two-voltage-source model N' of the same two-port N, shown in Fig. 6.4.1, it is evident that one is the dual of the other. For example, to obtain N'' from N', one needs only to replace the impedances by the (corresponding) admittances, the currents by the (corresponding) voltages, and the parallel connections by the (corresponding) series connections (so that loop equations become node equations). Similarly, duality properties may also be applied in determining N' from N''.

2. In obtaining the single-voltage-source equivalent, shown in Fig. 6.4.3, from that of Fig. 6.4.2 for a grounded linear two-port, the term $z_{12}I_1$ has been added to, and, at the same time, subtracted from, the right-hand side of (6.2.1 b), giving rise to

$$V_2 = z_{12}I_1 + z_{22}I_2 + (z_{21} - z_{12})I_1 \tag{6.4.1}$$

from which a dependent voltage source has been used to represent the term $(z_{21} - z_{12})I_1$ in the loop associated with the loop current I_2. The other elements in that model can be readily identified by inspection of (6.2.1) and (6.4.1) along with Fig. 6.4.3. A similar development may now be applied in obtaining the single-current-source model for a grounded two-port, shown in Fig. 6.4.5, from (6.4.3) by adding the term $y_{12}V_1$ to, and subtracting it from, the second of the two equations. Or, alternatively, one may obtain this equivalent directly from its z_{ij}-parameter counterpart, viz., the single-voltage-source model shown in Fig. 6.4.3, by applying the duality properties so that the two series arms become two shunt arms, the shunt arm becomes a series arm (so that the passive T subnetwork in Fig. 6.4.3 becomes the passive π subnetwork in Fig. 6.4.4), and the series dependent voltage source becomes the shunt dependent current source.

3. Obviously, the two single-source models in Figs. 6.4.3 and 6.4.5, as well as that in Fig. 6.4.2, are valid only for linear grounded two-ports. If, for any linear two port, terminals $1'$ and $2'$ are not at the same potential, the two-source models shown in Figs. 6.4.1 and 6.4.4 must be used.

4. When the given two-port N is a linear reciprocal network, the two *open-circuit transfer impedances* z_{12} and z_{21} are equal,* viz.,

$$z_{12} = z_{21}. \tag{6.4.2}$$

As a result, the dependent voltage source $(z_{21} - z_{12})I_1$ in Fig. 6.4.3 is reduced to zero, and the single-voltage-source equivalent becomes the familiar *equivalent T,*

---

* A direct consequence of the reciprocity theorem discussed in Chapter 4.

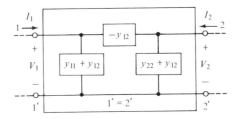

Fig. 6.4.7 Equivalent π of a linear grounded, reciprocal two-port.

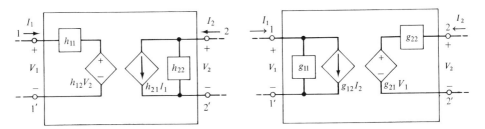

Fig. 6.4.8 A two-source equivalent for N of Fig. 6.2.1 based on the h_{ij} parameters.

Fig. 6.4.9 A two-source equivalent for N of Fig. 6.2.1 based on the g_{ij} parameters.

as shown in Fig. 6.4.6. In terms of the y_{ij}-parameters, under the assumption of a linear reciprocal two-port, we write

$$y_{12} = y_{21} \tag{6.4.3}$$

and the grounded single-current-source model shown in Fig. 6.4.5 is reduced to the *equivalent* π as depicted in Fig. 6.4.7.

5. Other equivalent networks for a linear two-port are based on the hybrid parameters and may be obtained by representing various terms in the two corresponding independent equations relating the four variables involved. For example, it can be shown that, in terms of the h_{ij}-parameters, a two-source model for the linear two-port N of Fig. 6.2.1 is that shown in Fig. 6.4.8, and a two-source model for the same two-port N based on the g_{ij}-parameters is the one illustrated in Fig. 6.4.9. The detailed development of these two models, as well as others, is left to the problems.* In the case of the transmission parameters, although they are very useful in the analysis of two-ports that are connected in *cascade* (or in *chain*, giving rise to the name *chain* parameters of the a_{ij}'s), the equivalent networks based on either the a_{ij}- or the b_{ij}-parameters are usually of little interest. This is a result of the fact that both sets of parameters are mixtures of immittances and voltage and current ratios which are *dimensionless* as evident by inspection of (6.3.7) and (6.3.10).

---

* See Problem 6.7.

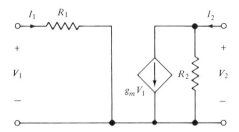

Fig. 6.4.10 Model of a common-emitter transistor amplifier.

6. It should be emphasized also that the six sets of two-port parameters discussed form three sets of duals; namely

$$[y_{ij}] = [z_{ij}]^{-1}, \qquad [g_{ij}] = [h_{ij}]^{-1}, \qquad [b_{ij}] = [a_{ij}]^{-1}, \tag{6.4.4}$$

provided, of course, that the inverses exist for the two-port under study.

The following example will serve to illustrate some of the points discussed above.

Example 6.4.1 The incremental model of a common-emitter transistor amplifier is shown in Fig. 6.4.10. The h_{ij}-parameters can be readily determined to be:

$$h_{11} = \left.\frac{V_1}{I_1}\right|_{V_2=0} = R_1,$$

$$h_{12} = \left.\frac{V_1}{V_2}\right|_{I_1=0} = 0,$$

$$h_{21} = \left.\frac{I_2}{I_1}\right|_{V_2=0} = g_m R_1, \tag{6.4.5}$$

$$h_{22} = \left.\frac{I_2}{V_2}\right|_{I_1=0} = \frac{1}{R_2}.$$

Of course, the h_{ij}-parameter found in (6.4.5) can be obtained just as easily: (1) by comparing Fig. 6.4.10 with the equivalent network of a two-port in terms of the hybrid h_{ij}-parameters shown in Fig. 6.4.8 noting that the dependent voltage source $h_{12}V_2$ is short-circuited since $h_{12} = 0(V_1 = 0$ for $I_1 = 0)$; and (2) by comparing the dependent current sources in both figures, $h_{21}I_1 = g_mV_1$, so that

$$h_{21} = g_m \frac{V_1}{I_1} = g_m R_1$$

since $V_1 = R_1I_1$ from Fig. 6.4.10.

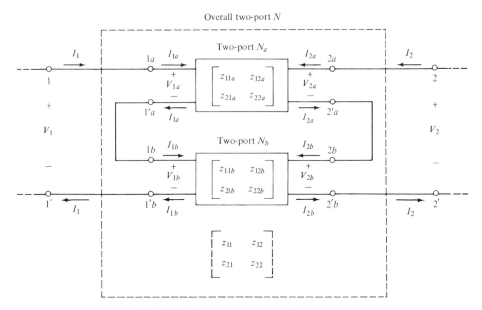

Fig. 6.5.1 Series connection of two-ports.

From Example 6.4.1, it is evident that the transmission b_{ij}-parameters for the common-emitter amplifier shown in Fig. 6.4.10 *cannot* be obtained from its h_{ij}-parameters using the expressions in Table 6.3.1. This is due to the fact that the common denominator in the expressions involved is equal to zero for the given network, viz., $h_{12} = 0$, so that the b_{ij}'s cannot be expressed in terms of the h_{ij}-parameters. In fact, by inspection of Fig. 6.4.10 and the defining equations in (6.3.10), we readily see that the b_{ij}-parameters *do not exist* at all for the common-emitter amplifier under consideration, since $I_1 = 0$ implies $V_1 = 0$ also.

6.5 INTERCONNECTIONS OF TWO-PORTS

Series Connection

Consider a linear two-port N consisting of two linear two-ports, N_a and N_b, interconnected as shown in Fig. 6.5.1. The current I_{1a} flowing into terminal $1a$ at port $1a$ of N_a is equal to the current I_{1b} flowing into terminal $1b$ at port $1b$ of N_b. The situation is similar for the currents I_{2a} and I_{2b} at the other port of the two subnetworks N_a and N_b of N. That is,

$$\begin{bmatrix} I_1 \\ I_2 \end{bmatrix} = \begin{bmatrix} I_{1a} \\ I_{2a} \end{bmatrix} = \begin{bmatrix} I_{1b} \\ I_{2b} \end{bmatrix}. \tag{6.5.1}$$

Also, by inspection of Fig. 6.5.1, the terminal voltages of N_a, N_b, and N may be related by

$$\begin{bmatrix} V_1 \\ V_2 \end{bmatrix} = \begin{bmatrix} V_{1a} \\ V_{2a} \end{bmatrix} + \begin{bmatrix} V_{1b} \\ V_{2b} \end{bmatrix}. \tag{6.5.2}$$

When the two subnetworks N_a and N_b (both are two-ports, of course) of N are interconnected as depicted in Fig. 6.5.1 such that their terminal voltages and currents satisfy both (6.5.1) and (6.5.2), they are said to be *connected in series*.

By (6.2.1), we write for N_a

$$\begin{bmatrix} V_{1a} \\ V_{2a} \end{bmatrix} = \begin{bmatrix} z_{11a} & z_{12a} \\ z_{21a} & z_{22a} \end{bmatrix} \begin{bmatrix} I_{1a} \\ I_{2a} \end{bmatrix}, \tag{6.5.3}$$

and for N_b,

$$\begin{bmatrix} V_{1b} \\ V_{2b} \end{bmatrix} = \begin{bmatrix} z_{11b} & z_{12b} \\ z_{21b} & z_{22b} \end{bmatrix} \begin{bmatrix} I_{1b} \\ I_{2b} \end{bmatrix}. \tag{6.5.4}$$

Thus, combining (6.5.1) through (6.5.4) for the two-port N (which contains N_a and N_b), we get

$$\begin{aligned} \begin{bmatrix} V_1 \\ V_2 \end{bmatrix} &= \begin{bmatrix} V_{1a} \\ V_{2a} \end{bmatrix} + \begin{bmatrix} V_{1b} \\ V_{2b} \end{bmatrix} \\[2mm] &= \left\{ \begin{bmatrix} z_{11a} & z_{12a} \\ z_{21a} & z_{22a} \end{bmatrix} + \begin{bmatrix} z_{11b} & z_{12b} \\ z_{21b} & z_{22b} \end{bmatrix} \right\} \begin{bmatrix} I_1 \\ I_2 \end{bmatrix} \\[2mm] &= \begin{bmatrix} z_{11a} + z_{11b} & z_{12a} + z_{12b} \\ z_{21a} + z_{21b} & z_{22a} + z_{22b} \end{bmatrix} \begin{bmatrix} I_1 \\ I_2 \end{bmatrix}. \end{aligned} \tag{6.5.5}$$

Upon comparison of (6.5.5) and (6.2.1), we find

$$\begin{bmatrix} z_{11} & z_{12} \\ z_{21} & z_{22} \end{bmatrix} = \begin{bmatrix} z_{11a} + z_{11b} & z_{12a} + z_{12b} \\ z_{21a} + z_{21b} & z_{22a} + z_{22b} \end{bmatrix} \tag{6.5.6}*$$

or, in compact matrix notation,

$$[z_{ij}] = [z_{ija}] + [z_{ijb}] \qquad (i, j = 1, 2) \tag{6.5.7}$$

which implies that *when N_a and N_b are connected in series, the sum of their z_{ij}'s is equal to the corresponding values of the z_{ij}'s of the overall two-port N.*

---

* Expression (6.5.6) is valid only if the constraints in (6.5.1) and (6.5.2) are satisfied. If not, an ideal transformer must be inserted in port 2 of either N_a or N_b in Fig. 6.5.1 to make the series connection possible. This point will be discussed in Section 6.9.

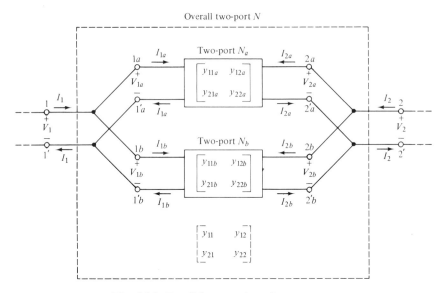

Fig. 6.5.2 Parallel connection of two-ports.

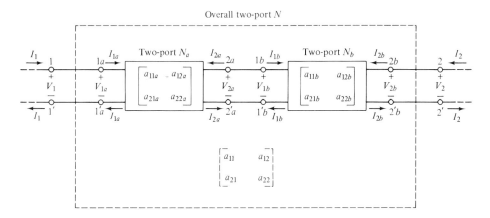

Fig. 6.5.3 Cascade connection of two-ports.

Parallel Connection

If the two subnetworks N_a and N_b of the linear two-port N are interconnected as shown in Fig. 6.5.2 such that the terminal voltages and currents are described by

$$\begin{bmatrix} V_1 \\ V_2 \end{bmatrix} = \begin{bmatrix} V_{1a} \\ V_{2a} \end{bmatrix} = \begin{bmatrix} V_{1b} \\ V_{2b} \end{bmatrix} \tag{6.5.8}$$

and

$$\begin{bmatrix} I_1 \\ I_2 \end{bmatrix} = \begin{bmatrix} I_{1a} \\ I_{2a} \end{bmatrix} + \begin{bmatrix} I_{1b} \\ I_{2b} \end{bmatrix}, \tag{6.5.9}$$

then N_a and N_b are said to be *connected in parallel*.

In light of (6.3.3), we write for N_a and N_b

$$\begin{bmatrix} I_{1k} \\ I_{2k} \end{bmatrix} = \begin{bmatrix} y_{11k} & y_{12k} \\ y_{21k} & y_{22k} \end{bmatrix} \begin{bmatrix} V_{1k} \\ V_{2k} \end{bmatrix} \qquad (k = a, b). \tag{6.5.10}$$

Hence, making use of (6.5.8) through (6.5.10), we write for N

$$\begin{aligned} \begin{bmatrix} I_1 \\ I_2 \end{bmatrix} &= \begin{bmatrix} I_{1a} \\ I_{2a} \end{bmatrix} + \begin{bmatrix} I_{1b} \\ I_{2b} \end{bmatrix} \\[2mm]
&= \left\{ \begin{bmatrix} y_{11a} & y_{12a} \\ y_{21a} & y_{22a} \end{bmatrix} + \begin{bmatrix} y_{11b} & y_{12b} \\ y_{21b} & y_{22b} \end{bmatrix} \right\} \begin{bmatrix} V_1 \\ V_2 \end{bmatrix} \\[2mm]
&= \begin{bmatrix} y_{11a} + y_{11b} & y_{12a} + y_{12b} \\ y_{21a} + y_{21b} & y_{22a} + y_{22b} \end{bmatrix} \begin{bmatrix} V_1 \\ V_2 \end{bmatrix} \end{aligned} \tag{6.5.11}$$

which, when compared with (6.3.3), yields

$$\begin{bmatrix} y_{11} & y_{12} \\ y_{21} & y_{22} \end{bmatrix} = \begin{bmatrix} y_{11a} + y_{11b} & y_{12a} + y_{12b} \\ y_{21a} + y_{21b} & y_{22a} + y_{22b} \end{bmatrix} \tag{6.5.12}*$$

or

$$[y_{ij}] = [y_{ija}] + [y_{ijb}] \qquad (i, j = 1, 2). \tag{6.5.13}$$

Equation (6.5.13) indicates that, *when N_a and N_b are connected in parallel, the sum of their y_{ij}'s is equal to the y_{ij}'s of the overall two-port N.*

Cascade Connection

A third possible interconnection of two-ports is the so-called cascade connection which is illustrated in Fig. 6.5.3. In this figure, the two two-ports N_a and N_b are the subnetworks of N *connected in cascade* such that the following constraints are satisfied:

$$\begin{bmatrix} V_1 \\ I_1 \end{bmatrix} = \begin{bmatrix} V_{1a} \\ I_{1a} \end{bmatrix}, \tag{6.5.14}$$

---

* Again, we assume conditions (6.5.8) and (6.5.9) are satisfied. If not, an ideal transformer must be used to make the parallel connection possible. See Section 6.9.

$$\begin{bmatrix} V_{2a} \\ -I_{2a} \end{bmatrix} = \begin{bmatrix} V_{1b} \\ I_{1b} \end{bmatrix},$$ (6.5.15)

and

$$\begin{bmatrix} V_{2b} \\ I_{2b} \end{bmatrix} = \begin{bmatrix} V_2 \\ I_2 \end{bmatrix}.$$ (6.5.16)

By (6.3.6), we write for N_a and N_b

$$\begin{bmatrix} V_{1k} \\ I_{1k} \end{bmatrix} = \begin{bmatrix} a_{11k} & a_{12k} \\ a_{21k} & a_{22k} \end{bmatrix} \begin{bmatrix} V_{2k} \\ -I_{2k} \end{bmatrix} \qquad (k = a, b).$$ (6.5.17)

Thus, combining (6.5.14) through (6.5.17), we have for N

$$\begin{bmatrix} V_1 \\ I_1 \end{bmatrix} = \begin{bmatrix} V_{1a} \\ I_{1a} \end{bmatrix} = \begin{bmatrix} a_{11a} & a_{12a} \\ a_{21a} & a_{22a} \end{bmatrix} \begin{bmatrix} V_{2a} \\ -I_{2a} \end{bmatrix}$$

$$= \begin{bmatrix} a_{11a} & a_{12a} \\ a_{21a} & a_{22a} \end{bmatrix} \begin{bmatrix} V_{1b} \\ I_{1b} \end{bmatrix}$$

$$= \begin{bmatrix} a_{11a} & a_{12a} \\ a_{21a} & a_{22a} \end{bmatrix} \left\{ \begin{bmatrix} a_{11b} & a_{12b} \\ a_{21b} & a_{22b} \end{bmatrix} \begin{bmatrix} V_{2b} \\ -I_{2b} \end{bmatrix} \right.$$

$$= \left\{ \begin{bmatrix} a_{11a} & a_{12a} \\ a_{21a} & a_{22a} \end{bmatrix} \begin{bmatrix} a_{11b} & a_{12b} \\ a_{21b} & a_{22b} \end{bmatrix} \right\} \begin{bmatrix} V_2 \\ -I_2 \end{bmatrix}$$ (6.5.18)

which, upon comparing with (6.3.6), gives

$$\begin{bmatrix} a_{11} & a_{12} \\ a_{21} & a_{22} \end{bmatrix} = \begin{bmatrix} a_{11a} & a_{12a} \\ a_{21a} & a_{22a} \end{bmatrix} \begin{bmatrix} a_{11b} & a_{12b} \\ a_{21b} & a_{22b} \end{bmatrix},$$ (6.5.19)

or simply

$$[a_{ij}] = [a_{ija}][a_{ijb}] \qquad (i, j = 1, 2).$$ (6.5.20)

The latter expression implies that *when N_a and N_b are connected in cascade, the product of their a_{ij}-matrices is equal to the a_{ij}-matrix of the overall two-port N.*

Series-Parallel and Parallel-Series Connections

If two two-ports N_a and N_b are interconnected in such a way that two of their available ports are connected in series and the other two are connected in parallel, then N_a and N_b are said to be connected in *series-parallel* or in *parallel-series*. These two possible connections of two-ports are shown in Figs. 6.5.4 and 6.5.5. In Fig. 6.5.4, port $1a$–$1'a$ of N_a is connected in series with port $1b$–$1'b$ of N_b (so that

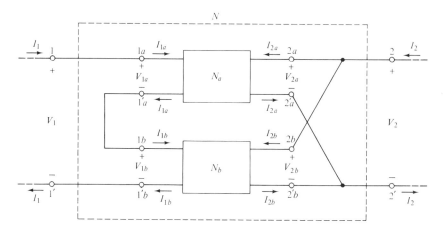

Fig. 6.5.4 Series-parallel connection of two-ports.

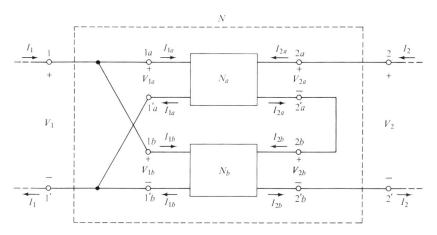

Fig. 6.5.5 Parallel-series connection of two-ports.

$I_1 = I_{1a} = I_{1b}$ and $V_1 = V_{1a} + V_{1b}$*), whereas port $2a$–$2'a$ of N_a is connected in parallel with port $2b$–$2'b$ of N_b (so that $V_2 = V_{2a} = V_{2b}$ and $I_2 = I_{2a} + I_{2b}$*) thus formulating the series-parallel connections of the two two-ports. In Fig. 6.5.5, N_a and N_b are connected in parallel-series in the sense that port $1a$–$1'a$ of N_a and port $1b$–$1'b$ are connected in parallel, whereas port $2a$–$2'a$ of N_a and port $2b$–$2'b$ of N_b are connected in series (so that $V_1 = V_{1a} = V_{1b}$, $I_1 = I_{1a} + I_{1b}$; and $I_2 = I_{2a} = I_{2b}$, $V_2 = V_{2a} + V_{2b}$*).

---

* Sometimes, for certain network structures, these conditions cannot be satisfied when a particular connection of N_a and N_b is made. Then an ideal transformer must be inserted at port 2 of either N_a or N_b to ensure satisfaction of these terminal conditions. This point, as well as others, will be discussed in Section 6.9.

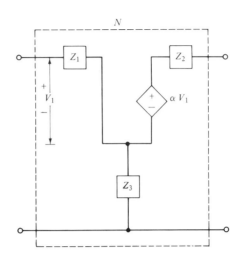

Fig. 6.5.6   Linear two-port N for Example 6.5.1.

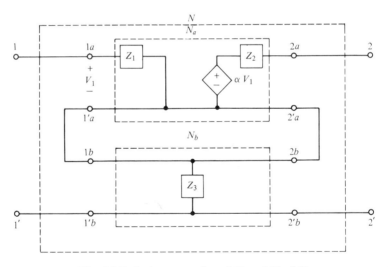

Fig. 6.5.7   Series connection of N_a and N_b of N.

It can be readily established that for the series-parallel connection shown in Fig. 6.5.4, the h_{ij}-matrix of the overall two-port N is related to the h_{ij}-matrices of its subnetworks N_a and N_b by*

$$[h_{ij}] = [h_{ija}] + [h_{ijb}].\qquad(6.5.21)$$

Similarly, it is also an easy matter to show that for the parallel-series connection

---

* The derivations of (6.5.21) and (6.5.22) are left as problems. See Problems 6.9 and 6.10.

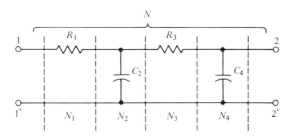

Fig. 6.5.8 *RC* ladder for Example 6.5.2.

depicted in Fig. 6.5.5, the g_{ij}-matrix of the overall two-port N and the g_{ij}-matrices of its subnetworks N_a and N_b are related by

$$[g_{ij}] = [g_{ija}] + [g_{ijb}].$$ (6.5.22)

Example 6.5.1   Figure 6.5.6 shows a linear grounded two-port N which consists of a dependent voltage source αV_1 and three passive elements $Z_1, Z_2,$ and Z_3. To obtain the z_{ij}-parameters of N we consider the two subnetworks N_a and N_b of N such that the series connection of the two subnetworks yields the overall two-port N as illustrated in Fig. 6.5.7. Then, by inspection, we obtain the z_{ij}-matrices of N_a and N_b, which are given by

$$[z_{ija}] = \begin{bmatrix} Z_1 & 0 \\ \alpha Z_1 & Z_2 \end{bmatrix}$$ (6.5.23)

and

$$[z_{ijb}] = \begin{bmatrix} Z_3 & Z_3 \\ Z_3 & Z_3 \end{bmatrix},$$ (6.5.24)

respectively. It is worth noting that N_a is essentially the same network as that of Fig. 6.4.10 used in Example 6.4.1, after replacing the series branch of the dependent voltage source and the impedance Z_2 by its Norton equivalent. Thus, substituting (6.5.23) and (6.5.24) into (6.5.7), we have the z_{ij}-matrix for N, viz.,

$$[z_{ij}] = [z_{ija}] + [z_{ijb}]$$
$$= \begin{bmatrix} Z_1 + Z_3 & Z_3 \\ \alpha Z_1 + Z_3 & Z_2 + Z_3 \end{bmatrix}$$ (6.5.25)

which is the desired result.

Example 6.5.2 The *RC* two-port N shown in Fig. 6.5.8 is known as a *ladder network*. Suppose that we wish to characterize this *RC* ladder by a set of parameters which are now to be determined. If we divide the network N into four sections $N_1, N_2, N_3,$ and N_4, as indicated by the vertical dotted lines in the figure, then each section may be considered as a two-port. All four sections connected in cascade constitute the given ladder. Furthermore, each series section (R) is characterized by the a_{ij}-matrix

$$[a_{ij}]_R = \begin{bmatrix} 1 & R \\ 0 & 1 \end{bmatrix}$$

and each shunt section (C) by

$$[a_{ij}]_C = \begin{bmatrix} 1 & 0 \\ sC & 1 \end{bmatrix}$$

Thus, it is evident that the a_{ij}-matrix of the RC ladder is the product of the a_{ij} matrices of the four sections, viz.,

$$[a_{ij}]_N = \begin{bmatrix} 1 & R_1 \\ 0 & 1 \end{bmatrix}_{N_1} \begin{bmatrix} 1 & 0 \\ sC_2 & 1 \end{bmatrix}_{N_2} \begin{bmatrix} 1 & R_3 \\ 0 & 1 \end{bmatrix}_{N_3} \begin{bmatrix} 1 & 0 \\ sC_4 & 1 \end{bmatrix}_{N_4}$$

$$= \begin{bmatrix} 1 + sR_1C_2 & R_1 \\ sC_2 & 1 \end{bmatrix} \begin{bmatrix} 1 + sR_3C_4 & R_3 \\ sC_4 & 1 \end{bmatrix}$$

or

$$[a_{ij}]_N = \begin{bmatrix} R_1C_2R_3C_4s^2 + (R_1C_2 + R_3C_4 + R_1C_4)s + 1 & R_1C_2R_3s + (R_1 + R_3) \\ C_2R_3C_4s^2 + (C_2 + C_4)s & C_2R_3s + 1 \end{bmatrix}$$

which is the desired result.

Example 6.5.3 Consider the two "simple-T" two-ports N_a and N_b shown in Fig. 6.5.9 (a). It can be easily shown that the open-circuit impedances of N_a and N_b are related to their corresponding impedance branches by

$$z_{11k} = Z_{1k} + Z_{3k},$$

$$z_{22k} = Z_{2k} + Z_{3k}, \tag{6.5.26}*$$

$$z_{12k} = z_{21k} = Z_{3k},$$

with $k = a$ for N_a and $k = b$ for N_b. Thus, if N_a and N_b are interconnected in series such that terminals $1'a$ and $1b$, as well as terminals $2'a$ and $2b$, are joined together as shown in Fig. 6.5.9 (b), then according to (6.5.7), the open-circuit impedances for the overall two-port N is given by

$$z_{11} = z_{11a} + z_{11b} = Z_{1a} + Z_{1b} + Z_{3a} + Z_{3b},$$

$$z_{22} = z_{22a} + z_{22b} = Z_{2a} + Z_{2b} + Z_{3a} + Z_{3b}, \tag{6.5.27}$$

$$z_{12} = z_{12a} + z_{12b} = Z_{3a} + Z_{3b} = z_{21}.$$

---

* See Problem 6.1(c).

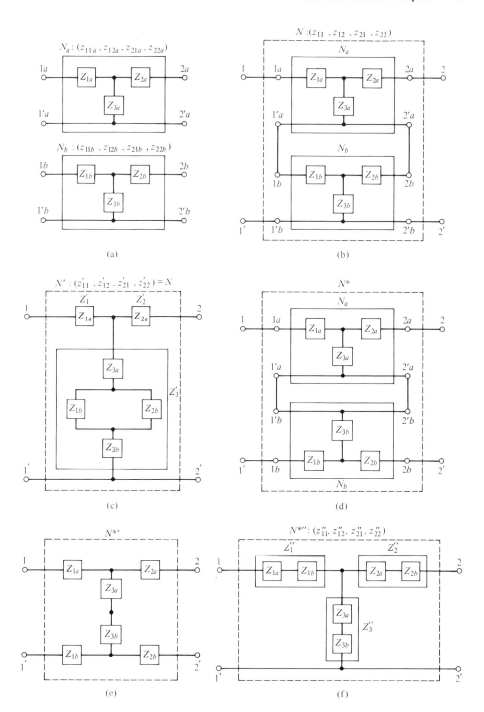

Fig. 6.5.9 Illustrations for Example 6.5.3.

However, by inspection, it is evident that the entire network N of Fig. 6.5.9 (b) can be redrawn to yield N' as illustrated in Fig. 6.5.9 (c) from which the open-circuit impedances can be written directly, using (6.5.26):

$$z'_{11} = Z'_1 + Z'_3 = Z_{1a} + Z_{3a} + Z_{3b} + \frac{Z_{1b}Z_{2b}}{Z_{1b} + Z_{2b}},$$

$$z'_{22} = Z'_2 + Z'_3 = Z_{2a} + Z_{3a} + Z_{3b} + \frac{Z_{1b}Z_{2b}}{Z_{1b} + Z_{2b}}, \qquad (6.5.28)$$

$$z'_{12} = z'_{21} = Z'_3 = Z_{3a} + Z_{3b} + \frac{Z_{1b}Z_{2b}}{Z_{1b} + Z_{2b}}.$$

An inspection of (6.5.27) and (6.5.28) immediately reveals the contradiction between the two sets of values for the z_{ij}-parameters of the *same* two-port N (or N'). This contradiction clearly indicates that (6.5.7) cannot be applied in obtaining the z_{ij}-parameters for the overall two-port N resulting from the series connection of the two simple-T two-ports N_a and N_b shown in Fig. 6.5.9 (b).

On the other hand, if the series connection between N_a and N_b is made in such a way that terminals $1'a$ and $1'b$, as well as terminals $2'a$ and $2'b$, are coalesced to yield the overall two-port N^* as illustrated in Fig. 6.5.9 (d), then the four terminals $1'a$, $1'b$, $2'a$, and $2'b$ are at the same potential. Thus N^* can be simplified into $N^{*'}$ as shown in Fig. 6.5.9 (e) which is seen to be equivalent to the simple-T two-port $N^{*''}$ as depicted in Fig. 6.5.9 (f). Using (6.5.26), the z_{ij}-parameters for $N^{*''}$ are found to be

$$z''_{11} = Z''_1 + Z''_3 = Z_{1a} + Z_{1b} + Z_{3a} + Z_{3b},$$

$$z''_{22} = Z''_2 + Z''_3 = Z_{2a} + Z_{2b} + Z_{3a} + Z_{3b}, \qquad (6.5.29)$$

$$z''_{12} = z''_{21} = Z_{3a} + Z_{3b},$$

which are seen to be identical to the z_{ij}-parameters of the two-port N shown in Fig. 6.5.9 (b) as stated in (6.5.27). This implies that for the series connection of N_a and N_b shown in Fig. 6.5.9 (d), expression (6.5.7) may be applied to determine the z_{ij}-parameters of the overall two-port N^* in terms of the z_{ij}-parameters of N_a and N_b.

Example 6.5.3 serves to demonstrate a very important fact, namely, *expressions (6.5.7), (6.5.13), (6.5.20), (6.5.21) and (6.5.22) are valid for the corresponding connections of two two-ports only if the terminal constraints under which these connections are made are satisfied.* For example, in the case of a series connection, the constraints, $I_1 = I_{1a} = I_{1b}$ and $I_2 = I_{2a} = I_{2b}$, as stated in (6.5.1) must be satisfied for (6.5.7) to be valid. Thus, in Example 6.5.3, the contradiction which appeared during the calculation of the z_{ij}-parameters of N for the series connection of N_a and N_b shown in Fig. 6.5.9 (b) simply indicates that the constraints stated in (6.5.1) were not satisfied!

In general, a test can be performed to determine if the terminal constraints for a particular connection are satisfied. In the case of a series connection, the test

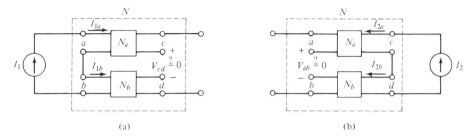

(a)   (b)

Fig. 6.5.10 The Brune test for the series connection.

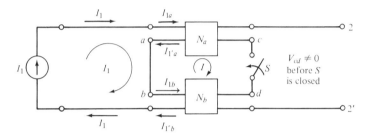

Fig. 6.5.11 Illustration of a violation of terminal constraints.

can be described as follows. First, as shown in Fig. 6.5.10 (a), the series connection at port 1 of both two-ports N_a and N_b is made by joining together terminals a and b, and a current source I_1 is used to excite the overall two-port N, with port 2 of N open-circuited. Next, measure the open-circuit voltage V_{cd} between terminals c and d at port 2. If $V_{cd} = 0$, then it is certain that there will be no current flowing after terminals c and d are coalesced and the terminal constraint $I_{1a} = I_{1b} = I_1$ will be satisfied since port 2 is open-circuited ($I_2 = 0$). Finally repeat the same procedure after the port conditions are interchanged as depicted in Fig. 6.5.10 (b) to see if the constraint $I_{2a} = I_{2b} = I_2$ is satisfied. If $V_{ab} = 0$ before terminals a and b are coalesced, then $I_{2a} = I_{2b} = I_2$ after the short-circuit connection between a and b is made, and the test is complete. Hence (6.5.7) will be valid for the series connection made by coalescing terminals a and b and likewise for terminals c and d.

If, on the other hand, during the test, either (1) $V_{cd} \neq 0$ or $V_{ab} \neq 0$, or (2) both $V_{cd} \neq 0$ and $V_{ab} \neq 0$, then the test fails (1) at one of the two ports, or (2) at both ports; in neither case can (6.5.7) be applied. To see why the terminal constraints are violated when one of the two voltages, say V_{cd}, is not zero, we refer to Fig. 6.5.11. If $V_{cd} \neq 0$ before terminals c and d are coalesced, then the two terminals are not at the same potential, and when the short-circuit connection between c and d is made, there exists a circulating current I in the loop a-c-d-b-a, which is the current flowing from c to d. Thus, by inspection, we write

$$I_{1a} = I_{1'b} = I_1$$
$$I_{1b} = I_{1'a} = I_1 - I. \tag{6.5.30}$$

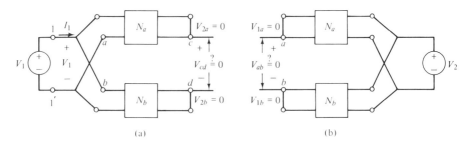

Fig. 6.5.12 The Brune test for the parallel connection.

From (6.5.30), it is evident that if $I \neq 0$ (which must be the case if $V_{cd} \neq 0$ before the two terminals c and d are coalesced), then $I_{1a} \neq I_{1b}$, which reveals the violation of the terminal constraint at port 1.

Similar testing procedures can be developed for the parallel, the series-parallel, and the parallel-series connections. (These tests are sometimes referred to as the *Brune tests*.*) For example, a test for the parallel connection is illustrated in Fig. 6.5.12 which is self-explanatory. The justification of this test as well as the derivations of the Brune tests for the series-parallel and the parallel-series connections of two two-ports are left as exercises.†

6.6 TWO-PORT DEVICES

So far, we have studied in some detail the characterization of a general linear two-port in terms of various sets of network parameters, the equivalent networks or models based on these parameters, and the different interconnections that can be made between two given two-ports. In this section, we shall discuss three important ideal two-port devices, namely; (a) the ideal transformer, (b) the gyrator, and (c) the negative immittance converter (NIC), which are, among others, fundamental building blocks in the development of network theory, both in analysis and synthesis.

The Ideal Transformer

Consider the general two-coil transformer and its schematic diagram shown in Fig. 6.6.1. The equations relating the voltages and currents are given in matrix form by

$$\begin{bmatrix} V_1 \\ V_2 \end{bmatrix} = \begin{bmatrix} sL_1 & sM \\ sM & sL_2 \end{bmatrix}\begin{bmatrix} I_1 \\ I_2 \end{bmatrix},$$ (6.6.1)

---

* Named after their inventor, O. Brune.
† See Problems 6.12 and 6.13.

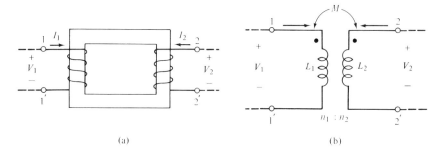

Fig. 6.6.1 (a) A two-coil transformer, and (b) its schematic diagram.

where the mutual inductance M is related to the two self inductances L_1 and L_2 by

$$M = k\sqrt{L_1 L_2} \qquad (6.6.2)$$

with k defined as the *coefficient of coupling* between the two coils, $k \leq 1$. For the limiting case, when $k = 1$, the transformer is called a *perfectly coupled* (or simply *perfect*) transformer.

It can be shown* that if a load Z_L is connected at one of the two ports, say port 2, of the transformer (so that the terminal constraint at port 2 is $V_2 = -I_2 Z_L$) the following relations hold for the *perfect* transformer.

$$\frac{I_2}{I_1} = \frac{-s\sqrt{L_1 L_2}}{sL_2 + Z_L} \qquad (6.6.3)$$

$$\frac{V_2}{V_1} = \sqrt{\frac{L_2}{L_1}} = \frac{n_2}{n_1} = \frac{1}{n}, \qquad (6.6.4)$$

where n is known as the *turns ratio* of the transformer. An examination of (6.6.4) reveals the fact that the voltage ratio of a perfect transformer under load Z_L is independent of Z_L.

Suppose that we make L_1 and L_2 very large so that the magnitude of their corresponding impedances are much greater than that of the load Z_L while at the same time the ratio L_2/L_1 remains finite; that is,

$$L_1 \to \infty, \qquad L_2 \to \infty, \qquad L_1/L_2 = \text{finite}. \qquad (6.6.5)$$

Then we see from (6.6.3) that the current ratio I_2/I_1 under these conditions is also independent of Z_L; namely,

$$\frac{I_2}{I_1} = -\sqrt{\frac{L_1}{L_2}} = -n. \qquad (6.6.6)$$

A perfect transformer satisfying (6.6.4) through (6.6.6) is called an *ideal transformer*

---

* See Problem 6.14.

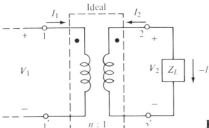

Fig. 6.6.2 An ideal transformer with a load of Z_L.

which may be characterized by

$$\begin{bmatrix} V_1 \\ I_1 \end{bmatrix} = \begin{bmatrix} n & 0 \\ 0 & \dfrac{1}{n} \end{bmatrix} \begin{bmatrix} V_2 \\ -I_2 \end{bmatrix}. \tag{6.6.7}$$

From (6.6.7) it is evident that the a_{ij}-matrix of an ideal transformer is

$$[a_{ij}] = \begin{bmatrix} n & 0 \\ 0 & \dfrac{1}{n} \end{bmatrix}. \tag{6.6.8}$$

Since $a_{12} = a_{21} = 0$, it is evident in light of Table 6.3.1, that the z_{ij}- and the y_{ij}-parameters do not exist for the ideal transformer.

Example 6.6.1 Let us obtain the input impedance Z_{in} of the network consisting of an ideal transformer with a turns ratio n loaded by an impedance Z_L as shown in Fig. 6.6.2. From this figure, we see that the terminal constraint at the load is

$$V_2 = -I_2 Z_L,$$

which together with (6.6.7) gives

$$Z_{in} = \frac{V_1}{I_1} = \frac{nV_2}{-(1/n)I_2}$$

$$= n^2 \left(\frac{V_2}{-I_2} \right) = n^2 Z_L. \tag{6.6.9}$$

Equation (6.6.9) indicates an interesting property of the ideal transformer: *it serves as a positive impedance converter* which presents an input impedance Z_{in} at the input port, directly proportional to the load Z_L. Another property of the ideal transformer is its isolating property; that is, it serves to ensure that the port conditions are met when the transformer is used in making certain interconnections between two-ports. The discussion of these properties will be presented in Section 6.10.

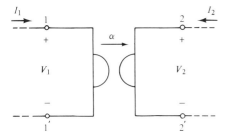

Fig. 6.6.3 A symbol for the gyrator.

The Gyrator

Like the ideal transformer, the gyrator* is an ideal two-port coupling device, shown in Fig. 6.6.3, which is characterized by

$$\begin{bmatrix} V_1 \\ V_2 \end{bmatrix} = \begin{bmatrix} 0 & -\alpha \\ \alpha & 0 \end{bmatrix} \begin{bmatrix} I_1 \\ I_2 \end{bmatrix}, \tag{6.6.10}$$

where α is a real constant called the *gyration ratio*.† The z_{ij}-matrix of the gyrator as obtained directly from (6.6.10) is

$$[z_{ij}] = \begin{bmatrix} 0 & -\alpha \\ \alpha & 0 \end{bmatrix}. \tag{6.6.11}‡}$$

It is easy to show that when the gyrator is terminated in an impedance Z_L at port 2, the input impedance Z_{in} is given by

$$Z_{\text{in}} = \frac{\alpha^2}{Z_L} \tag{6.6.12}$$

which implies that the gyrator may be viewed as an *impedance inverter*. That is, it transforms an impedance Z_L into α^2/Z_L, so that for $\alpha = 1$, it provides impedance inversion.

Example 6.6.2 Determine the input impedance of the network N, consisting of two gyrators N_a and N_b which have gyration ratios of 2 and 1, respectively, and which are connected in cascade, terminated in a resistive load Z_L as depicted in Fig. 6.6.4.

---

* First introduced by B. D. H. Tellegen in 1948. See [TE 2].
† Other names include the *gyrostatic coefficient* and *gyrating impedance*.
‡ Here, in (6.6.11), we note that $z_{12} = -\alpha$ and $z_{21} = \alpha$. However, if $z_{12} = -\alpha_1$, $g_{21} = \alpha_2$, and $\alpha_1 \neq \alpha_2$, then the total power delivered can become negative for certain specified excitations, and the gyrator becomes an active, nonreciprocal two-port known as an *active gyrator*. For a detailed discussion of this topic, refer to T. Yanagisawa and Y. Kawashima, [YA 1].

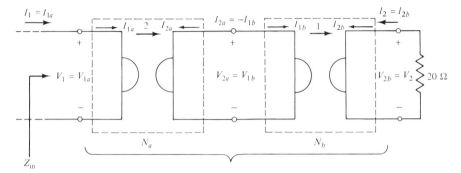

Fig. 6.6.4 Network N for Example 6.6.2.

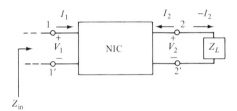

Z_{in}

Fig. 6.6.5 The *NIC* representation.

Using (6.6.10), we have

$$V_{1a} = -2I_{2a}, \qquad V_{2a} = 2I_{1a}, \qquad V_{1b} = -I_{2b}, \qquad V_{2b} = I_{1b},$$

from which we form the ratios V_{1a}/V_{2b} and I_{1a}/I_{2b}, leading to the following equations for N:

$$V_1 = 2V_2, \qquad I_1 = -\tfrac{1}{2}I_2.$$

Thus,

$$Z_{in} = \frac{V_1}{I_1} = (2)^2\left(\frac{V_2}{-I_2}\right) = 4 \times 20 = 80\Omega.$$

From Example 6.6.2, it is evident that two gyrators connected in cascade constitute an equivalent ideal transformer (with a turns ratio equal to 2), which presents an input impedance Z_{in} (of 80Ω) directly proportional to the load Z_L (of 20Ω).

The Negative-Immittance Converter (NIC)

A *negative-immittance converter* (NIC) is a two-port device which converts an impedance Z_L to its own negative. That is, with reference to Fig. 6.6.5,

$$Z_{in} = -Z_L \qquad\qquad (6.6.13)$$

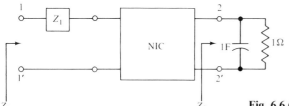

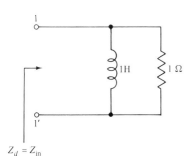

Fig. 6.6.6 Network N for Example 6.6.3.

Fig. 6.6.7 RL one-port realizing Z_{in} of (6.6.17).

or

$$\frac{V_1}{I_1} = -\left(\frac{V_2}{-I_2}\right), \tag{6.6.14}$$

Expression (6.6.14) leads to the following a_{ij}-characterization of the NIC:

$$\begin{bmatrix} V_1 \\ I_1 \end{bmatrix} = \begin{bmatrix} k & 0 \\ 0 & -k \end{bmatrix} \begin{bmatrix} V_2 \\ -I_2 \end{bmatrix} \tag{6.6.15}$$

or

$$\begin{bmatrix} V_1 \\ I_1 \end{bmatrix} = \begin{bmatrix} -k & 0 \\ 0 & k \end{bmatrix} \begin{bmatrix} V_2 \\ -I_2 \end{bmatrix}, \tag{6.6.16}$$

where k $(k > 0)$ is known as the *conversion ratio* of the device. When an NIC satisfies (6.6.15), it is termed a *current NIC* or *INIC* since, as indicated in the second equation of (6.6.15), the current direction is reversed. Similarly, if the NIC satisfies (6.6.16), it is called a *voltage NIC* or *VNIC* because of the fact that the voltage is reversed, as suggested in the first equation of (6.6.16).

The following examples suggest some applications of the NIC.

Example 6.6.3 Calculate the input impedance of the network N shown in Fig. 6.6.6. The impedance Z_L connected to port 2 of N is

$$Z_L = \frac{1}{s + 1}.$$

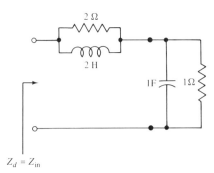

Fig. 6.6.8 *RLC* one-port realizing Z_{in} as its driving-point impedance.

Thus using (6.6.13), we find that the input impedance at port 1 of N for $Z_1 = 1$ is

$$Z_{\text{in}} = 1 - \frac{1}{s+1} = \frac{s}{s+1} \qquad (6.6.17)$$

which is also the driving-point impedance Z_d of the *RL* one-port depicted in Fig. 6.6.7.

Example 6.6.4 Refer to the same network N used in Example 6.6.3 as shown in Fig. 6.6.6. If $Z_1 = 2$, we have

$$Z_{\text{in}} = 2 - \frac{1}{s+1} = \frac{2s+1}{s+1}$$

$$= \frac{2s}{s+1} + \frac{1}{s+1}$$

$$= \frac{1}{\dfrac{1}{2} + \dfrac{1}{2s}} + \frac{1}{s+1}$$

which is the driving-point impedance Z_d of the *RLC* one-port shown in Fig. 6.6.8. From Examples 6.6.3 and 6.6.4, it is clear that certain *RL* and *RLC* types of networks may be realized by networks consisting of NIC's and *RC* elements alone. This is a significant advantage in active network synthesis since networks without inductances are attractive to the network designer where weight and size must be minimized.*

Remarks. We have in this section discussed three ideal two-port devices. The ideal transformer is a two-port device which: (1) is passive, since $V_1 I_1 + V_2 I_2 = 0$, as is evident by inspection of (6.6.7); (2) is reciprocal, since $a_{11}a_{22} - a_{21}a_{12} = 1$†; and (3) may be considered as a *positive impedance converter*, as illustrated in

---

* Refer to S. K. Mitra [MI 1], K. L. Su [SU 1], P. M. Chirlian [CH 4], R. W. Newcomb [NE 1], and L. P. Huelsman [HU 4].
† See Section 6.10.

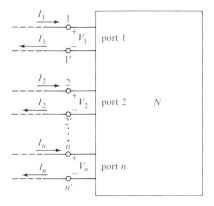

Fig. 6.7.1 A linear *n*-port *N*.

Example 6.6.1.   The gyrator is a two-port device which: (1) is passive, since $V_1 I_1 + V_2 I_2 = 0$ in light of (6.6.10); (2) is nonreciprocal, since $z_{12} \neq z_{21}$ as indicated in (6.6.11); (3) may be interpreted as an *impedance inverter* as evidenced in (6.6.12); or (4) when two of them are connected in cascade, may serve as an equivalent of an ideal transformer as demonstrated in Example 6.6.2. The NIC is an active,[*] nonreciprocal two-port device which, in addition to the applications illustrated in Examples 6.6.3 and 6.6.4, is a useful tool in active network synthesis for realizing many new circuit elements because of its ability to realize negative inductance and negative capacitance. The interested reader is referred to literature elsewhere[†] as the coverage of these topics is beyond the scope of this text.

6.7 THE GENERAL *N*-PORT

We shall now extend the theory developed for the linear two-port to the general linear *n-terminal-pair network*, or *n-port N* shown in Fig. 6.7.1, in which the current $I_k(s)$ entering terminal k is equal to that leaving terminal k' at port k ($k = 1, 2, \ldots, n$). Also the voltage $V_k(s)$ across the two terminals of port k is always *assigned* to be positive at the unprimed terminal; namely, terminal k.

The *n* equations relating the terminal voltages and currents at the *n* ports of *N* are given by

$$
\begin{aligned}
V_1 &= z_{11} I_1 + z_{12} I_2 + \cdots + z_{1n} I_n, \\
V_2 &= z_{21} I_1 + z_{22} I_2 + \cdots + z_{2n} I_n, \\
&\;\;\vdots \\
V_n &= z_{n1} I_1 + z_{n2} I_2 + \cdots + z_{nn} I_n.
\end{aligned}
\tag{6.7.1}
$$

---

[*] This is left as a problem. See Problem 6.24.
[†] See L. P. Huelsman [HU 4], K. L. Su [SU 1], S. K. Mitra [MI 1], for example.

Expression (6.7.1) may be written in matrix form $\mathbf{V} = \mathbf{ZI}$,

$$\begin{bmatrix} V_1 \\ V_2 \\ \vdots \\ V_n \end{bmatrix} = \begin{bmatrix} z_{11} & z_{12} \cdots z_{1n} \\ z_{21} & z_{22} \cdots z_{2n} \\ \vdots & \vdots & \vdots \\ z_{n1} & z_{n2} \cdots z_{nn} \end{bmatrix} \begin{bmatrix} I_1 \\ I_2 \\ \vdots \\ I_n \end{bmatrix}, \tag{6.7.2}$$

where the $n \times n$ matrix $\mathbf{Z} = [z_{ij}]$ is called the z_{ij}-*matrix* or the *open-circuit-impedance matrix* and the z_{ij}'s are referred to as the *open-circuit impedances*, or simply the z_{ij}-*parameters*. When $i = j$, z_{ii} is called the *open-circuit driving-point impedance at port i*; and for $i \neq j$, z_{ij} is referred to as the *open-circuit transfer impedance from port j to port i*. The general expression for determining any z_{ij} is found from (6.7.1) or (6.7.2) to be

$$z_{ij} = \left. \frac{V_i}{I_j} \right|_{I_k = 0 \quad (k \neq j)}. \tag{6.7.3}$$

For example, to determine the open-circuit driving-point impedance z_{11} of a linear n-port, we have

$$z_{11} = \left. \frac{V_1}{I_1} \right|_{I_2 = I_3 = \cdots = I_n = 0}. \tag{6.7.4}$$

Expression (6.7.4) implies that, to find z_{11}, one may excite the n-port by a (known) current source $I_1(s)$ at port 1 with all other ports open-circuited, and then measure the voltage $V_1(s)$ across terminals 1 and 1′. Then the ratio V_1/I_1 is the desired z_{11}. Similarly, for the determination of z_{ij}, the n-port is excited by a current source I_j at port j and the open-circuit voltage V_i is measured at port i with all the other ports open-circuited, yielding the desired ratio as indicated in (6.7.3). The connections described above for determining z_{ij} are shown in Fig. 6.7.2.

If the n equations relating the terminal voltages and currents at the n ports of N are written in such a way that the currents are expressed in terms of the voltages, we have the matrix formulation $\mathbf{I} = \mathbf{YV}$,

$$\begin{bmatrix} I_1 \\ I_2 \\ \vdots \\ I_n \end{bmatrix} = \begin{bmatrix} y_{11} & y_{12} \cdots y_{1n} \\ y_{21} & y_{22} \cdots y_{2n} \\ \vdots & \vdots & \vdots \\ y_{n1} & y_{n2} \cdots y_{nn} \end{bmatrix} \begin{bmatrix} V_1 \\ V_2 \\ \vdots \\ V_n \end{bmatrix}, \tag{6.7.5}$$

where the $n \times n$-matrix $\mathbf{Y} = [y_{ij}]$ is known as the y_{ij}-*matrix* or the *short-circuit-admittance matrix*, and the y_{ij}'s are termed the *short-circuit admittances* or simply the y_{ij}-*parameters*. When $i = j$, y_{ii} is called the *short-circuit driving-point admittance at port i*; and for $i \neq j$, y_{ij} is known as the *short-circuit transfer admittance from port j to port i*. The general expression for determining any y_{ij} is obtained from

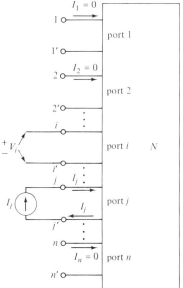

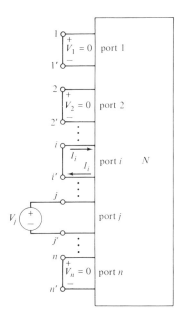

Fig. 6.7.2 Connections for determining z_{ij} of the n-port N.

Fig. 6.7.3 Connections for determining y_{ij} of the n-port N.

(6.7.5) as

$$y_{ij} = \frac{I_i}{V_j}\bigg|_{V_k = 0 \ (k \neq j)} . \tag{6.7.6}$$

Note that (6.7.6) may be obtained directly from (6.7.3) by replacing each symbol by its dual, while the subscripts remain unchanged:

$$z_{ij} \to y_{ij}, \qquad V_i \to I_i, \qquad I_j \to V_j, \qquad I_k \to V_k.$$

This implies that the necessary connections for determining y_{ij} may also be obtained from Fig. 6.7.3 by replacing the current source I_j by the voltage source V_j at port j, replacing the open-circuit voltage V_i by the short-circuit current I_i at the port i, and by short-circuiting all other previously open-circuited ports resulting in the connections shown in Fig. 6.7.3.

Example 6.7.1 Let us determine the z_{ij}-matrix of the RL three-port shown in Fig. 6.7.4. By inspection, we write

$$z_{11} = s + \frac{1(2 + 1)}{1 + (2 + 1)} = s + \frac{3}{4} = z_{22}$$

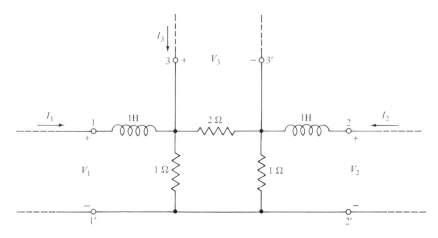

Fig. 6.7.4 A passive three-port.

and

$$z_{33} = \frac{2(1 + 1)}{2 + (1 + 1)} = \frac{4}{4} = 1.$$

For determination of $z_{21}(= z_{12})$, port 1 is excited by a current source I_1, and ports 2 and 3 are open-circuited. The (open-circuit) terminal voltage at port 2 is easily shown to be

$$V_2 = V_{22'} = (\tfrac{1}{4}I_1)(1) = \tfrac{1}{4}I_1$$

so that

$$z_{21} = \frac{V_2}{I_1}\bigg|_{I_2 = I_3 = 0} = \frac{\tfrac{1}{4}I_1}{I_1} = \frac{1}{4} = z_{12}.$$

To calculate z_{31} $(= z_{13})$, with port 1 still excited by the current source I_1 and ports 2 and 3 open-circuited, we now determine the (open-circuit) terminal voltage at port 3 to be

$$V_3 = V_{33'} = (\tfrac{1}{4}I_1)(2) = \tfrac{1}{2}I_1$$

so that

$$z_{31} = \frac{V_3}{I_1}\bigg|_{I_2 = I_3 = 0} = \frac{\tfrac{1}{2}I_1}{I_1} = \frac{1}{2} = z_{13}.$$

To determine z_{32} $(= z_{23})$, port 2 is excited by a current source I_2, and ports 1 and 3 are open-circuited. The (open-circuit) terminal voltage at port 3 is given by

$$V_3 = V_{33'} = (-\tfrac{1}{4}I_2)(2) = -\tfrac{1}{2}I_2$$

so that

$$z_{32} = \left. \frac{V_3}{I_2} \right|_{I_1 = I_3 = 0} = \frac{-\frac{1}{2}I_2}{I_2} = -\frac{1}{2} = z_{23}.$$

Finally, the desired open-circuit impedance matrix is given by

$$\mathbf{Z} = [z_{ij}] = \begin{bmatrix} s + \frac{3}{4} & \frac{1}{4} & \frac{1}{2} \\ \frac{1}{4} & s + \frac{3}{4} & -\frac{1}{2} \\ \frac{1}{2} & -\frac{1}{2} & 1 \end{bmatrix}. \tag{6.7.7}$$

Example 6.7.2   We shall obtain the y_{ij}-matrix of the same *RL* three-port used in Example 6.7.1. By inspection of Fig. 6.7.4, we have

$$y_{11} = y_{22} = \frac{1}{s + \frac{\frac{1}{2}s}{s + \frac{1}{2}}} = \frac{s + \frac{1}{2}}{s(s + 1)}$$

$$y_{33} = \frac{1}{2} + \frac{1}{\dfrac{s}{s + 1} + \dfrac{s}{s + 1}} = \frac{1}{2} + \frac{s + 1}{2s} = 1 + \frac{1}{2s}.$$

To determine $y_{21} (= y_{12})$, with $V_2 = V_3 = 0$, we write

$$I_1 = y_{11}V_1 = \frac{s + \frac{1}{2}}{s(s + 1)} V_1.$$

The voltage $V_{31'}$ across the 1–Ω resistor between terminals 3 and 1′ is

$$V_{31'} = V_1 - sI_1$$

$$= V_1 - \frac{s + \frac{1}{2}}{s + 1} V_1$$

$$= \frac{\frac{1}{2}}{s + 1} V_1$$

(which is also equal to the voltage $V_{3'2'}$ since $V_{33'} = V_3 = 0$) so that the short-circuit current through the 1-H inductor at port 2 is given by

$$-I_2 = \frac{V_{31'}}{s}$$

$$= \frac{1}{2s(s + 1)} V_1.$$

Hence

$$y_{21} = \left. \frac{I_2}{V_1} \right|_{V_2 = V_3 = 0} = \frac{-1}{2s(s + 1)} = y_{12}.$$

Similarly, it can be easily shown that

$$y_{31} = \left.\frac{I_3}{V_1}\right|_{V_2 = V_3 = 0} = -\frac{1}{2s} = y_{13}$$

and

$$y_{32} = \left.\frac{I_3}{V_2}\right|_{V_1 = V_3 = 0} = \frac{1}{2s} = y_{23}.$$

Consequently, the desired y_{ij}-matrix is

$$\mathbf{Y} = [y_{ij}] = \begin{bmatrix} \dfrac{s + \frac{1}{2}}{s(s + 1)} & -\dfrac{1}{2s(s + 1)} & -\dfrac{1}{2s} \\[2ex] -\dfrac{1}{2s(s + 1)} & \dfrac{s + \frac{1}{2}}{s(s + 1)} & \dfrac{1}{2s} \\[2ex] -\dfrac{1}{2s} & \dfrac{1}{2s} & 1 + \dfrac{1}{2s} \end{bmatrix}. \qquad (6.7.8)$$

It should be pointed out that the short-circuit admittance matrix $\mathbf{Y}$ obtained in Example 6.7.2 can be determined directly from the open-circuit impedance matrix $\mathbf{Z}$ found in Example 6.7.1 through the relationship

$$\mathbf{Y} = \mathbf{Z}^{-1}. \qquad (6.7.9)$$

Since it is a straightforward problem of matrix inversion, the verification of (6.7.9) using (6.7.7) and (6.7.8) is left as an exercise.*

It should now be evident that other *n*-port parameters, namely the hybrid parameters (the h_{ij}'s and the g_{ij}'s) and the transmission parameters (the a_{ij}'s and the b_{ij}'s), can also be obtained by extending the corresponding defining equations for the two-port following a similar development as that for the z_{ij}'s or y_{ij}'s discussed above. The interested reader is referred to other literature.†

6.8 THE *N*-TERMINAL NETWORK

So far we have discussed how a multi-terminal network may be characterized as an *n*-port such that the current entering a terminal of a given port is equal to that leaving the other terminal of the same port. However, in certain cases, the method of port designation may not be used to characterize a given multi-terminal network as the following example will illustrate.

---

* See Problem 6.21.
† For the hybrid and the transmission-parameters, the input and the output variables should first be specified. See P. M. Chirlian [CH 4], Chapter 3; or L. P. Huelsman [HU 2], Chapter 3, for a detailed discussion.

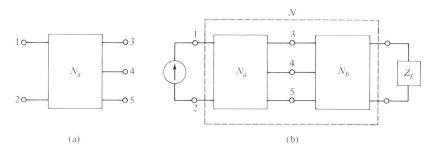

Fig. 6.8.1 A case where the port characterization cannot be applied.

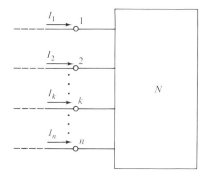

Fig. 6.8.2 An *n*-terminal network *N*.

Example 6.8.1 Consider the network N_a of Fig. 6.8.1 (a), which consists of five terminals 1 through 5. If N_a is taken to be a subnetwork of another network N as depicted in Fig. 6.8.1 (b), it is evident that the port characterization may not be applied to N_a since its terminals cannot be paired off in such a manner that the current entering one terminal of each pair is the same as that leaving the other.

This above example clearly points out the need for methods of characterizing a multi-terminal network other than the port designation as discussed in the preceding sections.

Consider a linear network N consisting of n terminals as shown in Fig. 6.8.2. Such a network is called an *n-terminal network*. If the KCL is applied to the entire network N, we conclude that

$$\sum_{k=1}^{n} I_k = 0 \qquad (6.8.1)$$

and no other conclusions can be drawn about N. If we choose an arbitrary *reference* (datum) node 0 of zero potential outside the network N, so that the voltage $V_k(s)$ between terminal k ($k = 1, 2, \ldots, n$) and the datum node 0 has its positive polarity mark at k (Fig. 6.8.3), then the terminal voltages and currents of

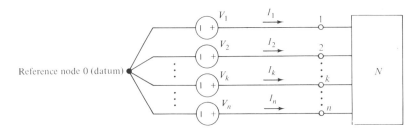

Fig. 6.8.3 An *n*-terminal network described by (6.8.2).

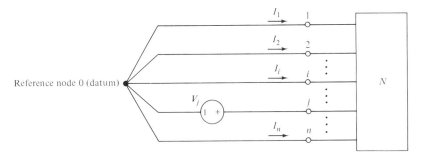

Fig. 6.8.4 Connections for determining y_{ij}.

N can be characterized by the matrix equation

$$
\begin{bmatrix} I_1 \\ I_2 \\ \vdots \\ I_n \end{bmatrix} = \begin{bmatrix} y_{11} & y_{12} \cdots y_{1n} \\ y_{21} & y_{22} \cdots y_{2n} \\ \vdots & \vdots \quad \vdots \\ y_{n1} & y_{n2} \cdots y_{nn} \end{bmatrix} \begin{bmatrix} V_1 \\ V_2 \\ \vdots \\ V_n \end{bmatrix}, \tag{6.8.2}
$$

where the $n \times n$-matrix, $[y_{ij}]$, is called the *indefinite admittance matrix*, and the (i, j) element y_{ij} of this matrix is found from (6.8.2) to be

$$
y_{ij} = \left. \frac{I_i}{V_j} \right|_{V_k = 0 \ (k \neq j)}. \tag{6.8.3}
$$

Expression (6.8.3) implies that y_{ij} may be determined by short-circuiting all terminal voltages V_k (with $k \neq j$) and measuring the short-circuit current entering terminal *i* so as to form the desired ratio I_i/V_j. The connection for determining y_{ij} is illustrated in Fig. 6.8.4. From this figure it is clear that (6.8.1) is the KCL-equation written at the datum node 0. Making use of this equation and adding the *n* equations in (6.8.2), we find

$$
\left(\sum_{i=1}^{n} y_{i1} \right) V_1 + \left(\sum_{i=1}^{n} y_{i2} \right) V_2 + \cdots + \left(\sum_{i=1}^{n} y_{in} \right) V_n = 0 \tag{6.8.4}
$$

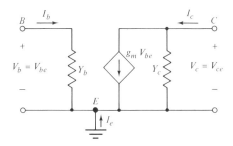

Fig. 6.8.5 Model of a common-emitter transistor amplifier.

which leads to the conclusion that

$$\sum_{i=1}^{n} y_{i1} = \sum_{i=1}^{n} y_{i2} = \cdots = \sum_{i=1}^{n} y_{in} = 0 \tag{6.8.5}$$

since the voltages $V_1, V_2, \ldots, V_n$ are arbitrary.

Equation (6.8.5) indicates that *the sum of the elements in every column of the indefinite admittance matrix* $[y_{ij}]$ *is equal to zero.* Next, by increasing the potential of every terminal of N by an equal amount, say ΔV, and recognizing that the currents in N will remain unaltered, it can be shown that

$$\sum_{i=1}^{n} y_{1i} \Delta V = \sum_{i=1}^{n} y_{2i} \Delta V$$

$$= \cdots = \sum_{i=1}^{n} y_{ni} \Delta V = 0 \tag{6.8.6}^*$$

which implies that the sum of the elements in every row of $[y_{ij}]$ *is equal to zero also.*

Other properties of $[y_{ij}]$ include that all the (first-order) cofactors Δ_{jk} of the determinant Δ of $[y_{ij}]$ are equal and that $[y_{ij}]$ is symmetrical if N is a passive reciprocal network; that is, $y_{jk} = y_{kj}$ for all j and k for a passive reciprocal N. The proofs of these properties are left as problems.†

Example 6.8.2 Consider the network shown in Fig. 6.8.5, which is essentially the same as the common emitter transistor amplifier model of Fig. 6.4.10 used in Example 6.4.1. The matrix equation based on the y_{ij}-parameters for this model is easily shown to be

$$\begin{bmatrix} I_b \\ I_c \end{bmatrix} = \begin{bmatrix} Y_b & 0 \\ g_m & Y_c \end{bmatrix} \begin{bmatrix} V_b \\ V_c \end{bmatrix}. \tag{6.8.7}$$

Now if we choose a reference node 0 as the zero potential outside the amplifier as shown in Fig. 6.8.6, we have the modified matrix equation, with the help of (6.8.5)

---

* See Problem 6.22.

† See Problem 6.23.

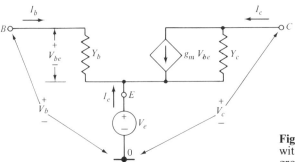

Fig. 6.8.6 The modified network with the emitter terminal E ungrounded giving rise to (6.8.8).

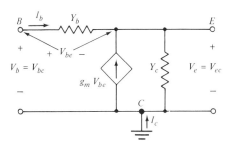

Fig. 6.8.7 The grounded-collector model of a transistor amplifier.

and (6.8.6):

$$\begin{bmatrix} I_b \\ I_c \\ I_e \end{bmatrix} = \begin{bmatrix} Y_b & 0 & -Y_b \\ g_m & Y_c & -(g_m + Y_c) \\ -(Y_b + g_m) & -Y_c & Y_b + g_m + Y_c \end{bmatrix} \begin{bmatrix} V_b \\ V_c \\ V_e \end{bmatrix}. \quad (6.8.8)$$

Thus we can readily obtain the matrix equation for the grounded-collector model of the transistor amplifier depicted in Fig. 6.8.7 by deleting the second equation from (6.8.8):

$$\begin{bmatrix} I_b \\ I_e \end{bmatrix} = \begin{bmatrix} Y_b & -Y_b \\ -(Y_b + g_m) & Y_b + g_m + Y_c \end{bmatrix} \begin{bmatrix} V_b \\ V_e \end{bmatrix}. \quad (6.8.9)$$

Example 6.8.2 clearly illustrates some interesting applications of the indefinite-admittance matrix. Still other applications of this may be found in the literature elsewhere.*

6.9 THE SCATTERING PARAMETERS

One-Port Network Relations

Since their introduction in about 1920, the scattering parameters have become increasingly important in modern electrical engineering. An understanding of

---

* See S. K. Mitra [MI 1], Chapter 4; and L. P. Huelsman [HU 2], Chapter 3, for example.

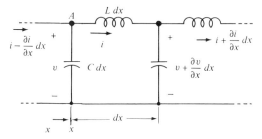

Fig. 6.9.1 An infinitesimal section of distributed network.

these parameters is essential for design of most high-frequency components such as filters, antennas, and related structures. At frequencies much above 100 MHz, it becomes difficult to *measure* conventional z_{ij}, y_{ij}, or h_{ij} parameters across a band of frequencies for certain devices. For example, this is the reason why most transistors are characterized in terms of scattering parameters for applications above 100 MHz.

To motivate our discussion of scattering parameters, we illustrate the concept of waves traveling in a lossless distributed network. Figure 6.9.1 depicts an infinitesimal section of a lossless network consisting of a series inductor, L henries per unit length, and a shunt capacitor, C farads per unit length. The section is considered to be an infinitesimal length dx long. At the input, we have voltage v and current i, and at the output $v + (\partial v/\partial x)\, dx$ and $i + (\partial i/\partial x)\, dx$, respectively. We should note that v and i are functions of time t and distance x.

Writing the KVL-equation around the loop, we have

$$-v + (L\, dx)\frac{\partial i}{\partial t} + v + \frac{\partial v}{\partial x}\, dx = 0.$$

Application of KCL at node A gives

$$-\left(i - \frac{\partial i}{\partial x}\, dx\right) + (C\, dx)\frac{\partial v}{\partial t} + i = 0.$$

These equations can be simplified to yield

$$\text{(a)} \quad \frac{\partial v}{\partial x} = -L\frac{\partial i}{\partial t} \qquad \text{(b)} \quad \frac{\partial i}{\partial x} = -C\frac{\partial v}{\partial t}. \qquad (6.9.1)$$

Equations (6.9.1a) and (6.9.1b) indicate, respectively, the manner in which the instantaneous voltage v and current i vary along the distributed network.

To obtain a solution for the voltage, we take the partial derivative of the first equation of (6.9.1) with respect to the variable x

$$\frac{\partial^2 v}{\partial x^2} = -L\frac{\partial}{\partial x}\left(\frac{\partial i}{\partial t}\right). \qquad (6.9.2a)$$

We may now eliminate the current from this expression by taking the partial derivative of (6.9.1b) with respect to time to obtain

$$\frac{\partial^2 i}{\partial t \partial x} = -C \frac{\partial^2 v}{\partial t^2}$$

which is then substituted into (6.9.2a) yielding

$$\frac{\partial^2 v}{\partial x^2} = LC \frac{\partial^2 v}{\partial t^2}. \tag{6.9.2b}$$

Note that the order of differentiation is assumed to be immaterial so that

$$\frac{\partial^2 i}{\partial t \partial x} = \frac{\partial^2 i}{\partial x \partial t}.$$

Recalling the fact that v and i are both functions of t and x, we assume a sinusoidal steady-state solution of the form $v(x, t) = \sqrt{2}V(x)e^{j\omega t}$, which can be substituted into (6.9.2b) to yield

$$\frac{d^2 V(x)}{dx^2} = LC[(j\omega)^2]V(x) \triangleq -k_1^2 V(x), \tag{6.9.2c}$$

where

$$k_1^2 = -(j\omega)^2 LC = \omega^2 LC.$$

The differential equation (6.9.2c) possesses the well-known general solution

$$V(x) = V_a e^{-jk_1 x} + V_b e^{jk_1 x}. \tag{6.9.3}$$

By discussion of Section 5.6 on phasors, we may write the voltage as a function of t and x as follows:

$$v(x, t) = \sqrt{2}V_a \cos(\omega t - k_1 x) + \sqrt{2}V_b \cos(\omega t + k_1 x).$$

It can be easily shown* that the first term of $v(x, t)$ represents a wave traveling to the right (see Fig. 6.9.1 for the orientation of x) as time t (the parameter) increases, and that the second term represents a wave traveling to the left.

Returning to (6.9.3), we write

$$V = V(x) = a + b, \tag{6.9.4a}$$

where we have made the substitutions

$$a = V_a e^{-jk_1 x}$$

to represent the phasor of a wave traveling to the right and

$$b = V_b e^{+jk_1 x}$$

to represent the phasor of a wave traveling to the left.

---

* See Problem 6.25.

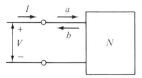

Fig. 6.9.2 One-port network.

By following the same argument given above for the general one-port in Fig. 6.9.2, it may be shown* that the current I is given by

$$I = I(x)$$

$$= \frac{1}{k_2}(a - b),$$

(6.9.4b)

where k_2 is equal to $\sqrt{L/C}$ and has units of an impedance.

Thus, the voltage $V = a + b$ and current $I = (1/k_2)(a - b)$ may be represented by a superposition of an *incident wave a* traveling to the right, and a *reflected wave b* traveling to the left.

Solving for the waves a and b in terms of voltage and current yields

$$a = \frac{1}{2}(V + I)$$

$$b = \frac{1}{2}(V - I),$$

where k_2 has arbitrarily been set equal to unity for convenience. Thus, we have obtained solutions for the waves a and b in terms of voltages and currents at various points of the network. We may consider that the network may be described by either the voltage V and current I at the port terminals (e.g., $V = ZI$ where Z is said to describe the one-port), or by its incident and reflected waves a and b, respectively. In the latter case, the relationship between b and a is defined by

$$b = sa,$$

(6.9.5)

where s is a parameter, called the *scattering parameter* of the one-port, which characterizes the network (to be explained further later). Thus, in general, we may consider the waves a and b to be a linear combination of V and I as follows:

$$a = \tfrac{1}{2}(k_V V + k_I I)$$

$$b = \tfrac{1}{2}(k_V V - k_I I).$$

(6.9.6)

Now consider the one-port to be driven by a generator E which has an *internal impedance z* as shown in Fig. 6.9.3.

---

* See Problem 6.26.

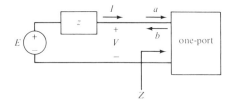

Fig. 6.9.3 One-port network driven by a generator with internal impedance z.

For reasons which will become evident in the discussion of (6.9.9), we define

$$k_V = \frac{1}{\sqrt{z}}$$

and

$$k_I = \sqrt{z}.$$

Hence (6.9.6) may be rewritten as

$$a = \frac{1}{2}(k_V V + k_I I) = \frac{1}{2}\left(\frac{V}{\sqrt{z}} + I\sqrt{z}\right)$$

$$b = \frac{1}{2}(k_V V - k_I I) = \frac{1}{2}\left(\frac{V}{\sqrt{z}} - I\sqrt{z}\right)$$
(6.9.7)

and solving for $V/\sqrt{z}$ and $I\sqrt{z}$, we obtain

$$\frac{V}{\sqrt{z}} = a + b$$

$$I\sqrt{z} = a - b.$$
(6.9.8)

Upon comparison with (6.9.4a) and (6.9.4b), (6.9.8) implies that we have *normalized* the voltage and current to yield the above values.

To better understand the physical significance of the waves a and b, we compute the average power delivered to the one-port for real frequencies ($j\omega$), assuming, for the moment, that z is real:†

$$P_{ave} = \text{Re}\,[VI^*] = \text{Re}\left[\left(\frac{V}{\sqrt{z}}\right)(\sqrt{z}I)^*\right]$$

$$= \text{Re}\,[(a + b)(a - b)^*]$$

$$= \text{Re}\,[aa^* - ab^* + a^*b - bb^*]$$
(6.9.9)

$$= aa^* - bb^*$$

$$= |a|^2 - |b|^2.$$

----

† See Problem 6.27.

Thus, the average power delivered to the network is given by the difference of a term due to the incident wave and a term due to the reflected wave; hence, we make the logical definitions that $|a|^2$ represents the incident power to the network and $|b|^2$ represents the power reflected from the network.

To conserve this property for the case when z is complex, we must slightly modify the definitions given in (6.9.7).†

Assuming $r = \text{Re } z > 0$, we define

$$a = \frac{1}{2}\left[\frac{V}{\sqrt{r}} + \frac{Iz}{\sqrt{r}}\right]$$

$$b = \frac{1}{2}\left[\frac{V}{\sqrt{r}} - \frac{Iz^*}{\sqrt{r}}\right]$$
(6.9.10a)

from which we may obtain‡

$$V = \left[\frac{z^*a}{\sqrt{r}} + \frac{zb}{\sqrt{r}}\right]$$

$$I = \left[\frac{a}{\sqrt{r}} - \frac{b}{\sqrt{r}}\right].$$
(6.9.10b)

A simple calculation for power§ now yields the same result for power entering the one-port as was given in (6.9.9).

It was mentioned previously that the one-port could be described by the parameter s which was defined by the relationship $b = sa$ (analogous to $V = ZI$).

Thus

$$s = \frac{b}{a} = \frac{V - Iz^*}{V + Iz},$$
(6.9.11)

where (6.9.10a) has been used.

Dividing the numerator and denominator of the right-hand expression of (6.9.11) by I and identifying $Z = V/I$ as the input impedance of the one-port, we obtain

$$s = \frac{Z - z^*}{Z + z}.$$
(6.9.11a)

Solving for Z, we obtain

$$Z = \frac{z^* + zs}{1 - s} = (z^* + zs)(1 - s)^{-1}.$$
(6.9.11b)

---

† Refer to D. C. Youla [YO 1].
‡ See Problem 6.28.
§ See Problem 6.29.

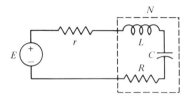

Fig. 6.9.4 Network N driven by a generator E with real internal impedance R.

For the special case where z is real, (6.9.11b) becomes

$$Z = z \frac{(1 + s)}{1 - s} = z^{1/2}(1 - s)^{-1}(1 + s)z^{1/2}. \qquad (6.9.11c)$$

These expressions give the relationship between the impedance characterization and scattering parameter description of the one-port. It is important to note that when the generator internal impedance z is real, the scattering parameter s is termed the *reflection coefficient* of the network.

Example 6.9.1 The application of (6.9.11a) will be illustrated by writing an expression for the scattering parameter of the network N shown in Fig. 6.9.4 where $z = r$ is a real number; hence, we are actually computing the *reflection coefficient* of the network.

First, we write

$$Z = R + j\omega L + \frac{1}{j\omega C} = \frac{(j\omega)^2 LC + 1 + j\omega RC}{j\omega C}.$$

From (6.9.11a), and noting that $z = r = z^*$ since z is real, we have

$$s = \frac{-\omega^2 LC + 1 + j\omega RC - r(j\omega C)}{-\omega^2 LC + 1 + j\omega RC + r(j\omega C)}$$

$$= \frac{(1 - \omega^2 LC) + j\omega C(R - r)}{(1 - \omega^2 LC) + j\omega C(R + r)}.$$

From the discussion of Section 5.6, in (5.6.94), we noted that, when the network is conjugately matched with the generator; that is, $Z = z^*$, then (6.9.11a) becomes

$$s = 0.$$

Note that, for this *matched case*, the power relation of (6.9.9) becomes

$$P_{\text{ave}} = |a|^2 - |b|^2$$

$$= |a|^2,$$

where we have used the fact that $b = sa = 0$. But from (5.6.95) in Chapter 5, under this condition, we find that

$$P_{ave} = \text{the available power from the generator};$$

therefore

$$P_{ave} = \frac{1}{4}\frac{|E|^2}{r},$$

where we have made the identifications $V_s = E$ and $R_L = r = \text{Re}(z)$ due to the condition for matching. Thus $P_{ave} = |a|^2$ implies that

$$|a|^2 = \frac{|E|^2}{4r}. \tag{6.9.12}$$

It should be reiterated at this point that the units of $|a|^2$ and $|b|^2$ are watts.

The above discussion permits the following interpretation of the waves a and b: the voltage source E sends a wave with power $|a|^2$ to the one-port, and this is independent of the input impedance seen. If the input impedance of the one-port is equal to the conjugately matched value, then all this generated power is absorbed by the one-port (i.e., the power delivered is equal to the maximum power).

If the input impedance is not equal to the matched value, then the power absorbed by the network will be $|a|^2 - |b|^2$, and thus the one-port reflects a power equal to $|b|^2 = |sa|^2$.

If the one-port network is lossless (having only capacitive- and inductive-type elements), we know that the total average power absorbed is equal to zero. Hence (6.9.9) becomes

$$0 = |a|^2 - |b|^2$$

$$= aa^* - bb^*.$$

Therefore

$$\begin{aligned} a^*a - b^*b &= a^*a - (sa)^*sa \\ &= a^*a - a^*s^*sa \\ &= a^*(1 - s^*s)a = 0. \end{aligned}$$

Consequently

$$s^*s = 1 \qquad \text{or} \qquad |s|^2 = 1, \tag{6.9.13}$$

and

$$|s| = 1. \tag{6.9.14}$$

This relation is analogous to the impedance relationship for a lossless network; that is,

$$Z + Z^* = 0.$$

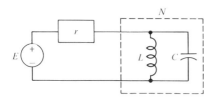

Fig. 6.9.5 A lossless network driven by a voltage source.

Thus, the s parameter of a lossless network must always satisfy the condition stated in (6.9.14).

Example 6.9.2 We shall now demonstrate that the s-parameter of the network N shown in Fig. 6.9.5 satisfies the relation of (6.9.14).

The input impedance of the network is

$$Z = \cfrac{1}{\cfrac{1}{j\omega L} + j\omega C}$$

$$= \frac{j\omega L}{1 - \omega^2 LC},$$

and we note that $Z + Z^* = 0$, and $z = r = z^*$.

From (6.9.11a)

$$s = \frac{Z - r}{Z + r} = \frac{j\omega L - (1 - \omega^2 LC)r}{j\omega L + (1 - \omega^2 LC)r}$$

$$= \frac{-r(1 - \omega^2 LC) + j\omega L}{r(1 - \omega^2 LC) + j\omega L}$$

and

$$|s| = \left[\frac{(r(1 - \omega^2 LC))^2 + (\omega L)^2}{(r(1 - \omega^2 LC))^2 + (\omega L)^2} \right]^{1/2} = 1.$$

For a general passive network, the total average power absorbed is always greater than or, in the lossless case, equal to zero. Hence, the above relationship may be written as

$$P_{ave} = |a|^2 - |b|^2 \geq 0$$

$$a^*(1 - s^*s)a \geq 0$$

which yields the following condition on the s-parameter of a passive one-port:

$$1 \geq |s|^2.$$

This in turn implies that

$$|s| \leq 1, \tag{6.9.15a}$$

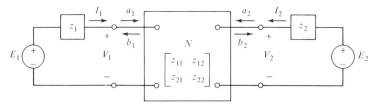

Fig. 6.9.6 A two-port network N driven by voltage generators.

and thus we say that the magnitude of the s-parameter of a passive one-port is bounded by unity.

Example 6.9.3 We previously computed the s-parameter for the lossy network of Example 6.9.1 to be

$$s = \frac{(1 - \omega^2 LC) + j\omega C(R - r)}{(1 - \omega^2 LC) + j\omega C(R + r)}.$$

Thus, we have

$$|s| = \left[\frac{(1 - \omega^2 LC)^2 + (\omega C(R - r))^2}{(1 - \omega^2 LC)^2 + (\omega C(R + r))^2}\right]^{1/2} \leq 1 \qquad (6.9.15b)$$

since it is always true that, in (6.9.15b), $\omega C(R - r) \leq \omega C(R + r)$.

The Two-Port Scattering Matrix

In this paragraph we develop the two-port scattering matrix, and derive the fundamental relationship between the power absorbed by a passive network and its scattering matrix. Subsequently, the constraints imposed by passivity and losslessness are developed. Finally, after obtaining the interrelationship between scattering and impedance matrices, we derive the reciprocity relations for the two-port scattering matrix.

Consider the two-port network N shown in Fig. 6.9.6. The network is driven by generators E_1 and E_2 having internal impedances z_1 and z_2, respectively.

The incident and reflected waves at each port are defined similarly to those of (6.9.10):

$$a_1 = \frac{1}{2\sqrt{r_1}}(V_1 + z_1 I_1), \qquad (6.9.16a)$$

$$b_1 = \frac{1}{2\sqrt{r_1}}(V_1 - z_1^* I_1), \qquad (6.9.16b)$$

$$a_2 = \frac{1}{2\sqrt{r_2}}(V_2 + z_2 I_2), \qquad (6.9.16c)$$

$$b_2 = \frac{1}{2\sqrt{r_2}}(V_2 - z_2^* I_2), \qquad (6.9.16d)$$

where $r_i = \mathrm{Re}(z_i)(i = 1, 2)$.

The *scattering matrix* $\mathbf{S}$ of the two-port is defined by the matrix relationship

$$\begin{bmatrix} b_1 \\ b_2 \end{bmatrix} = \begin{bmatrix} s_{11} & s_{12} \\ s_{21} & s_{22} \end{bmatrix} \begin{bmatrix} a_1 \\ a_2 \end{bmatrix}. \tag{6.9.17a}$$

Writing the vectors

$$\mathbf{a} = \begin{bmatrix} a_1 \\ a_2 \end{bmatrix}$$

and

$$\mathbf{b} = \begin{bmatrix} b_1 \\ b_2 \end{bmatrix}$$

and using the following definition

$$\mathbf{S} = \begin{bmatrix} s_{11} & s_{12} \\ s_{21} & s_{22} \end{bmatrix},$$

we may write (6.9.17a) as

$$\mathbf{b} = \mathbf{Sa}.$$

The parameters s_{ij} are called the *scattering parameters* of the two-port, and they are defined by the relationships

$$s_{11} = \left. \frac{b_1}{a_1} \right|_{a_2 = 0}, \qquad s_{12} = \left. \frac{b_1}{a_2} \right|_{a_1 = 0},$$

$$s_{21} = \left. \frac{b_2}{a_1} \right|_{a_2 = 0}, \qquad s_{22} = \left. \frac{b_2}{a_2} \right|_{a_1 = 0}.$$

Example 6.9.4 For the two-port shown in Fig. 6.9.6, with $z_1 = r_1$ and $z_2 = r_2$ both being real numbers, we shall compute the scattering parameters of the network. Using the defining expressions at port 1 above with (6.9.16a) and (6.9.16b), we have

$$s_{11} = \left. \frac{b_1}{a_1} \right|_{a_2 = 0} = \frac{\dfrac{1}{2\sqrt{r_1}}(V_1 - z_1^* I_1)}{\dfrac{1}{2\sqrt{r_1}}(V_1 + z_1 I_1)} = \frac{V_1 - z_1^* I_1}{V_1 + z_1 I_1}.$$

Dividing numerator and denominator by I_1 yields

$$s_{11} = \frac{Z_1 - z_1^*}{Z_1 + z_1},$$

where $Z_1 = V_1/I_1$.

A similar computation at port 2 will give

$$s_{22} = \frac{Z_2 - z_2^*}{Z_2 + z_2},$$

where $Z_2 = V_2/I_2$.

To calculate s_{21}, we note that

$$s_{21} = \left.\frac{b_2}{a_1}\right|_{a_2 = 0},$$

but $a_2 = 0$ implies that

$$\frac{V_2}{\sqrt{r_2}} + \frac{z_2 I_2}{\sqrt{r_2}} = 0.$$

Hence $V_2 + r_2 I_2 = 0$ since $z_2 = r_2$ for this example. Next, we may write $V_2/I_2 = -r_2$ which from the figure implies that $E_2 = 0$, and consequently,

$$b_2 = \frac{1}{2\sqrt{r_2}}(V_2 - z_2^* I_2) = \frac{1}{2\sqrt{r_2}}(V_2 - z_2 I_2)$$

$$= \frac{V_2}{\sqrt{r_2}}.$$

Now a_1 is the incident wave under conditions of no reflection so that $s_{11} = 0$; thus

$$\frac{V_1}{I_1} = z_1^* = z_1 \qquad \text{and} \qquad a_1 = \frac{1}{2\sqrt{r_1}}(V_1 + z_1 I_1) = \frac{1}{2\sqrt{r_1}} 2V_1.$$

With this identification, we can see from the figure that

$$V_1 = \frac{E_1}{2z_1} \cdot z_1 = \frac{E_1}{2}.$$

Substituting this identity into the relation for a_1, we have

$$a_1 = \frac{1}{2\sqrt{r_1}} 2V_1 = \frac{1}{\sqrt{r_1}} \frac{E_1}{2}.$$

Therefore

$$s_{21} = \left.\frac{b_2}{a_1}\right|_{a_2 = 0} = \frac{\dfrac{V_2}{\sqrt{r_2}}}{\dfrac{E_1}{2\sqrt{r_1}}} = \frac{2V_2}{E_1}\sqrt{\frac{r_1}{r_2}}.$$

The reader should demonstrate by a similar argument that

$$s_{12} = \frac{2V_1}{E_2}\sqrt{\frac{r_2}{r_1}}.$$

We will now proceed with the discussion of some fundamental properties of the scattering parameters. Solving (6.9.16) for voltage and current in terms of incident and reflected waves, one obtains

$$V_1 = \frac{1}{\sqrt{r_1}} (a_1 z_1^* + b_1 z_1)$$

(6.9.17b)†

$$I_1 = \frac{1}{\sqrt{r_1}} (a_1 - b_1).$$

Consequently, the total average power P_1 entering port 1 can be shown to be‡

$$P_1 = \text{Re}[V_1 I_1^*]$$

$$= \text{Re}[|a_1|^2 - |b_1|^2 + \text{Im}(a_1^* b_1) - \text{Im } z_1 |a_1 - b_1|^2]$$

$$= |a_1|^2 - |b_1|^2.$$

(6.9.18)

Similarly, at port 2, we have

$$P_2 = |a_2|^2 - |b_2|^2.$$

Thus, the total average power absorbed by the network can be written in matrix form as follows:

$$P_{\text{tot}} = P_1 + P_2$$

$$= |a_1|^2 + |a_2|^2 - |b_1|^2 - |b_2|^2.$$

(6.9.19)

Defining a matrix $\mathbf{a}^+$, called the tranjugate (meaning *tran*spose of the con*jugate*) of $\mathbf{a}$, by the relationship

$$\mathbf{a}^+ = \begin{bmatrix} a_1 \\ a_2 \end{bmatrix}^+ = \begin{bmatrix} a_1^* \\ a_2^* \end{bmatrix}^T = [a_1^*, a_2^*],$$

we may write

$$\mathbf{a}^+ \mathbf{a} = [a_1^*, a_2^*] \begin{bmatrix} a_1 \\ a_2 \end{bmatrix} = a_1^* a_1 + a_2^* a_2$$

and

$$\mathbf{b}^+ \mathbf{b} = [b_1^*, b_2^*] \begin{bmatrix} b_1 \\ b_2 \end{bmatrix} = b_1^* b_1 + b_2^* b_2.$$

Utilizing these identities, we may then put (6.9.19) in the following form:

$$P_{\text{tot}} = \mathbf{a}^+ \mathbf{a} - \mathbf{b}^+ \mathbf{b}.$$

(6.9.20a)

---

† See Problem 6.28.
‡ See Problem 6.29.

Now

$$\mathbf{b} = \mathbf{S}\mathbf{a}$$

and

$$\mathbf{b}^+ = [\mathbf{S}\mathbf{a}]^+ = [\mathbf{S}*\mathbf{a}*]^T = (\mathbf{a}*)^T(\mathbf{S}*)^T = \mathbf{a}^+\,\mathbf{S}^+.$$

Thus,

$$P_{\text{tot}} = \mathbf{a}^+\,\mathbf{a} - \mathbf{b}^+\,\mathbf{b} = \mathbf{a}^+\,\mathbf{a} - \mathbf{a}^+\,\mathbf{S}^+\mathbf{S}\mathbf{a}.$$

Factoring out an $\mathbf{a}^+$ on the left, we obtain

$$P_{\text{tot}} = \mathbf{a}^+\,(\mathbf{a} - \mathbf{S}^+\mathbf{S}\mathbf{a}), \qquad (6.9.20\text{b})$$

but

$$\mathbf{a} = \begin{bmatrix} a_1 \\ a_2 \end{bmatrix} = \begin{bmatrix} 1 & 0 \\ 0 & 1 \end{bmatrix}\begin{bmatrix} a_1 \\ a_2 \end{bmatrix} = \mathbf{I}_2\mathbf{a},$$

where $\mathbf{I}_2$ is the unit matrix of order 2.

Substituting into (6.9.20b), we obtain

$$\begin{aligned} P_{\text{tot}} &= \mathbf{a}^+\,[\mathbf{I}_2\mathbf{a} - \mathbf{S}^+\mathbf{S}\mathbf{a}] \\ &= \mathbf{a}^+\,[\mathbf{I}_2 - \mathbf{S}^+\mathbf{S}]\mathbf{a}. \end{aligned} \qquad (6.9.21)$$

Thus, for a general *lossless* two-port, the total power absorbed is equal to zero and consequently

$$P_{\text{tot}} = 0$$

which in turn implies that

$$\mathbf{I}_2 - \mathbf{S}^+\mathbf{S} = \mathbf{0}_2, \qquad (6.9.22)$$

where $\mathbf{0}_2$ is the zero (null) matrix of order 2×2.

Let us now examine the matrix relationships to determine the constraints (6.9.22) placed on the two-port scattering parameters.

Equation (6.9.22) may be expressed as follows:

$$\mathbf{I}_2 = \mathbf{S}^+\,\mathbf{S}$$

which becomes

$$\begin{aligned} \begin{bmatrix} 1 & 0 \\ 0 & 1 \end{bmatrix} &= \begin{bmatrix} s_{11} & s_{12} \\ s_{21} & s_{22} \end{bmatrix}^+ \begin{bmatrix} s_{11} & s_{12} \\ s_{21} & s_{22} \end{bmatrix} \\ &= \begin{bmatrix} s_{11}^* & s_{21}^* \\ s_{12}^* & s_{22}^* \end{bmatrix}\begin{bmatrix} s_{11} & s_{12} \\ s_{21} & s_{22} \end{bmatrix} \\ &= \begin{bmatrix} s_{11}^*s_{11} + s_{21}^*s_{21} & s_{11}^*s_{12} + s_{21}^*s_{22} \\ s_{12}^*s_{11} + s_{22}^*s_{21} & s_{12}^*s_{12} + s_{22}^*s_{22} \end{bmatrix} \end{aligned}$$

This expression yields the four equations

$$|s_{11}|^2 + |s_{21}|^2 = 1, \tag{6.9.23a}$$

$$s_{11}^* s_{12} + s_{21}^* s_{22} = 0, \tag{6.9.23b}$$

$$s_{12}^* s_{11} + s_{22}^* s_{21} = 0, \tag{6.9.23c}$$

$$|s_{12}|^2 + |s_{22}|^2 = 1. \tag{6.9.23d}$$

Note that (6.9.23b) and (6.9.23c) are conjugates and, therefore, we have only three independent equations.

For a reciprocal two-port, we found in Chapter 5 that $z_{12} = z_{21}$. Utilizing this result, we will later show (in Example 6.9.5) that reciprocity consequently implies that $s_{12} = s_{21}$.

Let us assume for the time being that reciprocity does imply that s_{12} equals s_{21}, and make use of this property. Thus, for a reciprocal network, (6.9.23) can be reduced to the following three equations

$$|s_{11}|^2 + |s_{12}|^2 = 1, \tag{6.9.24a}$$

$$s_{11} s_{12}^* + s_{12} s_{22}^* = 0, \tag{6.9.24b}$$

$$|s_{22}|^2 + |s_{12}|^2 = 1. \tag{6.9.24c}$$

Employing (6.9.24), we will now show that a *lossless reciprocal* network can be completely specified by a knowledge of only its scattering parameters at ports 1 and 2.

Let s_{11}, s_{12}, and s_{22} be written in the form

$$s_{11} = |s_{11}|e^{j\phi_{11}}, \qquad s_{22} = |s_{22}|e^{j\phi_{22}}, \qquad s_{12} = |s_{12}|e^{j\phi_{12}}.$$

Solving (6.9.24a) and (6.9.24c) simultaneously yields the very important result

$$|s_{11}| = |s_{22}| \tag{6.9.25}$$

for a *lossless reciprocal* two-port.

Next, (6.9.24a) can be solved for $|s_{12}|$ to obtain

$$|s_{12}| = [1 - |s_{11}|^2]^{1/2}. \tag{6.9.26}$$

Substituting $s_{12} = |s_{12}|e^{j\phi_{12}}$ into (6.9.24b) gives

$$s_{11}s_{12}^* + s_{12}s_{22}^* = |s_{22}|e^{j\phi_{11}}|s_{12}|e^{-j\phi_{12}} + |s_{12}|e^{j\phi_{12}}|s_{22}|e^{-j\phi_{22}},$$

$$= 0$$

where the identity $|s_{11}| = |s_{22}|$ has been employed. Thus we obtain

$$e^{j\phi_{11} - j\phi_{12}} + e^{j\phi_{12} - j\phi_{22}} = 0.$$

Hence

$$e^{2j\phi_{12}} = -e^{j\phi_{11} + j\phi_{22}} = e^{j\phi_{11} + j\phi_{22} + j\pi}$$

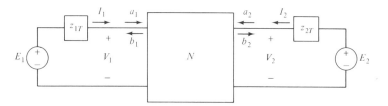

Fig. 6.9.7 A two-port network N with voltage generators.

and

$$\phi_{12} = \frac{\phi_{11} + \phi_{22} + \pi}{2};$$

or more generally

$$\phi_{12} = \frac{\phi_{11} + \phi_{22} + \pi + 2k\pi}{2}, \tag{6.9.27}$$

where $k = 0, \pm 1, \pm 2, \ldots$.

From (6.9.25), (6.9.26), and (6.9.27), we conclude that given the s_{11} and s_{22} parameters of a lossless reciprocal two-port, we may immediately determine the complete set of scattering parameters.†

For a general passive two-port, the total power absorbed by the network is greater than or equal to zero; thus $P_{\text{tot}} \geq 0$ implies that $\mathbf{I}_2 - \mathbf{S}^+\mathbf{S} \geq \mathbf{0}_2$.

The interrelationships between the open-circuit impedance parameters and the scattering parameters are derived in the following discussion.

Consider the general two-port shown in Fig. 6.9.7. We have relettered the generator impedances z_{1t} and z_{2t} for the sake of clarity in the ensuing discussion. Suppose that the passive network possesses an open-circuit impedance matrix

$$\mathbf{Z} = \begin{bmatrix} z_{11} & z_{12} \\ z_{21} & z_{22} \end{bmatrix},$$

where we may write

$$\begin{bmatrix} V_1 \\ V_2 \end{bmatrix} = \begin{bmatrix} z_{11} & z_{12} \\ z_{21} & z_{22} \end{bmatrix} \begin{bmatrix} I_1 \\ I_2 \end{bmatrix} \tag{6.9.28}$$

in the form $\mathbf{V} = \mathbf{ZI}$.

---

† See Problem 6.32.

From (6.9.16), we get

$$2\sqrt{r_{1t}}a_1 = V_1 + z_{1t}I_1$$
$$2\sqrt{r_{2t}}a_2 = V_2 + z_{2t}I_2,$$

where we have defined $z_{1t} = z_1$ and $z_{2t} = z_2$ to denote terminating impedances and to avoid confusion with previous notation. Of course, $r_{1t} = \text{Re}(z_{1t})$ and $r_{2t} = \text{Re}(z_{2t})$.

The above equation can be written in matrix form as follows:

$$\begin{bmatrix} 2\sqrt{r_{1t}}a_1 \\ 2\sqrt{r_{2t}}a_2 \end{bmatrix} = \begin{bmatrix} V_1 \\ V_2 \end{bmatrix} + \begin{bmatrix} z_{1t}I_1 \\ z_{2t}I_2 \end{bmatrix}$$

$$= \begin{bmatrix} z_{11} & z_{12} \\ z_{21} & z_{22} \end{bmatrix}\begin{bmatrix} I_1 \\ I_2 \end{bmatrix} + \begin{bmatrix} z_{1t}I_1 \\ z_{2t}I_2 \end{bmatrix},$$

where we have employed (6.9.28). This expression may be rewritten as

$$\begin{bmatrix} 2\sqrt{r_{1t}} & 0 \\ 0 & 2\sqrt{r_{2t}} \end{bmatrix}\begin{bmatrix} a_1 \\ a_2 \end{bmatrix} = \begin{bmatrix} z_{11} & z_{12} \\ z_{21} & z_{22} \end{bmatrix}\begin{bmatrix} I_1 \\ I_2 \end{bmatrix} + \begin{bmatrix} z_{1t} & 0 \\ 0 & z_{2t} \end{bmatrix}\begin{bmatrix} I_1 \\ I_2 \end{bmatrix}$$

or

$$\begin{bmatrix} 2 & 0 \\ 0 & 2 \end{bmatrix}\begin{bmatrix} \sqrt{r_{1t}} & 0 \\ 0 & \sqrt{r_{2t}} \end{bmatrix}\begin{bmatrix} a_1 \\ a_2 \end{bmatrix} = \left\{ \begin{bmatrix} z_{11} & z_{12} \\ z_{21} & z_{22} \end{bmatrix} + \begin{bmatrix} z_{1t} & 0 \\ 0 & z_{2t} \end{bmatrix} \right\}\begin{bmatrix} I_1 \\ I_2 \end{bmatrix}$$

$$(6.9.29)$$

denoting

$$\mathbf{M}_2 = \begin{bmatrix} 2 & 0 \\ 0 & 2 \end{bmatrix}; \quad \mathbf{R} = \begin{bmatrix} \sqrt{r_{1t}} & 0 \\ 0 & \sqrt{r_{2t}} \end{bmatrix};$$

$$\mathbf{Z}_t = \begin{bmatrix} z_{1t} & 0 \\ 0 & z_{2t} \end{bmatrix}. \qquad (6.9.30)$$

We are now in a position to develop the interrelationship between matrices $\mathbf{S}$ and $\mathbf{Z}$; i.e.,

$$\mathbf{S} = \mathbf{R}^{-1}(\mathbf{Z} - \mathbf{Z}_t^*)(\mathbf{Z} + \mathbf{Z}_t)^{-1}\mathbf{R}.$$

To proceed, (6.9.29) may be written in a more convenient form utilizing (6.9.30):

$$\mathbf{M}_2\mathbf{R}\mathbf{a} = (\mathbf{Z} + \mathbf{Z}_t)\mathbf{I}. \qquad (6.9.31a)$$

Similarly, (6.9.16b) and (6.9.16d) may be written in the form

$$\mathbf{M}_2 \mathbf{R} \mathbf{b} = (\mathbf{Z} - \mathbf{Z}_t^*)\mathbf{I}. \qquad (6.9.31b)\dagger$$

Recalling the identity $(\mathbf{AB})^{-1} = \mathbf{B}^{-1}\mathbf{A}^{-1}$ from matrix theory, we premultiply (6.9.31a) by $(\mathbf{M}_2 \mathbf{R})^{-1}$ and obtain

$$(\mathbf{M}_2 \mathbf{R})^{-1}(\mathbf{M}_2 \mathbf{R})\mathbf{a} = (\mathbf{M}_2 \mathbf{R})^{-1}(\mathbf{Z} + \mathbf{Z}_t)\mathbf{I}.$$

Therefore

$$\mathbf{a} = \mathbf{R}^{-1}\mathbf{M}_2^{-1}(\mathbf{Z} + \mathbf{Z}_t)\mathbf{I}$$

which, when premultiplied by $\mathbf{M}_2 \mathbf{R} \mathbf{S}$, gives

$$\mathbf{M}_2 \mathbf{R} \mathbf{S} \mathbf{a} = \mathbf{M}_2 \mathbf{R} \mathbf{S} \mathbf{R}^{-1}\mathbf{M}_2^{-1}(\mathbf{Z} + \mathbf{Z}_t)\mathbf{I}.$$

Since $\mathbf{b} = \mathbf{S}\mathbf{a}$, we may write this expression as

$$\mathbf{M}_2 \mathbf{R} \mathbf{b} = \mathbf{M}_2 \mathbf{R} \mathbf{S} \mathbf{R}^{-1}\mathbf{M}_2^{-1}(\mathbf{Z} + \mathbf{Z}_t)\mathbf{I}. \qquad (6.9.31c)$$

Comparing (6.9.31c) with (6.9.31b), we have

$$(\mathbf{Z} - \mathbf{Z}_t^*)\mathbf{I} = \mathbf{M}_2 \mathbf{R} \mathbf{S} \mathbf{R}^{-1}\mathbf{M}_2^{-1}(\mathbf{Z} + \mathbf{Z}_t)\mathbf{I}$$

or

$$\mathbf{M}_2 \mathbf{R} \mathbf{S} \mathbf{R}^{-1}\mathbf{M}_2^{-1}(\mathbf{Z} + \mathbf{Z}_t) = (\mathbf{Z} - \mathbf{Z}_t^*).$$

Postmultiplying both sides of this equation by $(\mathbf{Z} + \mathbf{Z}_t)^{-1}$ gives

$$\mathbf{M}_2 \mathbf{R} \mathbf{S} \mathbf{R}^{-1}\mathbf{M}_2^{-1}(\mathbf{Z} + \mathbf{Z}_t)(\mathbf{Z} + \mathbf{Z}_t)^{-1} = (\mathbf{Z} - \mathbf{Z}_t^*)(\mathbf{Z} + \mathbf{Z}_t)^{-1},$$

or

$$\mathbf{M}_2 \mathbf{R} \mathbf{S} \mathbf{R}^{-1}\mathbf{M}_2^{-1} = (\mathbf{Z} - \mathbf{Z}_t^*)(\mathbf{Z} + \mathbf{Z}_t)^{-1}. \qquad (6.9.32)$$

Next, since $\mathbf{M}_2$ and $\mathbf{R}$ are diagonal matrices, they commute for multiplication, that is, $\mathbf{M}_2 \mathbf{R} = \mathbf{R} \mathbf{M}_2$. Utilizing the matrix identity $(\mathbf{AB})^{-1} = \mathbf{B}^{-1}\mathbf{A}^{-1}$, we have $\mathbf{R}^{-1}\mathbf{M}_2^{-1} = \mathbf{M}_2^{-1}\mathbf{R}^{-1}$. For a general matrix $\mathbf{S}$, it is easily shown that $\mathbf{M}_2 \mathbf{S} = \mathbf{S} \mathbf{M}_2$. Using these identities, the left-hand side of (6.9.32) becomes

$$\mathbf{M}_2 \mathbf{R} \mathbf{S} \mathbf{R}^{-1}\mathbf{M}_2^{-1} = \mathbf{R} \mathbf{M}_2 \mathbf{S} \mathbf{M}_2^{-1}\mathbf{R}^{-1}$$

$$= \mathbf{R} \mathbf{S} \mathbf{M}_2 \mathbf{M}_2^{-1}\mathbf{R}^{-1} \qquad (6.9.33)$$

$$= \mathbf{R} \mathbf{S} \mathbf{R}^{-1}.$$

---

† See Problem 6.33.

Combining (6.9.32) and (6.9.33), we have

$$\mathbf{RSR}^{-1} = (\mathbf{Z} - \mathbf{Z}_t^*)(\mathbf{Z} + \mathbf{Z}_t)^{-1}.$$

Premultiplying by $\mathbf{R}^{-1}$, we obtain

$$\mathbf{SR}^{-1} = \mathbf{R}^{-1}(\mathbf{Z} - \mathbf{Z}_t^*)(\mathbf{Z} + \mathbf{Z}_t)^{-1}.$$

Postmultiplying by R, we have the desired relationship

$$\mathbf{S} = \mathbf{R}^{-1}(\mathbf{Z} - \mathbf{Z}_t^*)(\mathbf{Z} + \mathbf{Z}_t)^{-1}\mathbf{R}. \tag{6.9.34}$$

Similarly, it can be shown that

$$\mathbf{S} = \mathbf{R}^{-1}(\mathbf{I}_2 - \mathbf{Z}_t^*\mathbf{Y})(\mathbf{I}_2 + \mathbf{Z}_t\mathbf{Y})^{-1}\mathbf{R}. \tag{6.9.35}†$$

where $\mathbf{Y}$ is the short-circuit admittance matrix. Finally, the impedance matrix may be expressed in terms of the scattering matrix as

$$\mathbf{Z} = \mathbf{R}(\mathbf{I}_2 - \mathbf{S})^{-1}(\mathbf{Z}_t^* + \mathbf{SZ}_t)\mathbf{R}^{-1}. \tag{6.9.36}$$

Example 6.9.5 We are now in a position to easily prove the reciprocity condition for the two-port scattering matrix mentioned earlier (i.e., $s_{12} = s_{21}$).

Suppose, for the purposes of this example, that

$$\mathbf{R} = \begin{bmatrix} 1 & 0 \\ 0 & 1 \end{bmatrix} = \mathbf{R}^{-1}$$

and

$$\mathbf{Z}_t = \begin{bmatrix} 1 & 0 \\ 0 & 1 \end{bmatrix}.$$

Therefore, (6.9.34) becomes $\mathbf{S} = (\mathbf{Z} - \mathbf{I}_2)(\mathbf{Z} + \mathbf{I}_2)^{-1}$ which may be written in the form

$$\begin{bmatrix} s_{11} & s_{12} \\ s_{21} & s_{22} \end{bmatrix} = \begin{bmatrix} z_{11} - 1 & z_{12} \\ z_{21} & z_{22} - 1 \end{bmatrix}\begin{bmatrix} z_{11} + 1 & z_{12} \\ z_{21} & z_{22} + 1 \end{bmatrix}^{-1}$$

$$= \begin{bmatrix} z_{11} - 1 & z_{12} \\ \\ z_{21} & z_{22} - 1 \end{bmatrix}\begin{bmatrix} \dfrac{z_{22} + 1}{\Delta} & \dfrac{-z_{12}}{\Delta} \\ \\ \dfrac{-z_{21}}{\Delta} & \dfrac{z_{11} + 1}{\Delta} \end{bmatrix},$$

where

$$\Delta = (z_{11} + 1)(z_{22} + 1) - z_{12}z_{21} = \Delta_z + z_{11} + z_{22} + 1$$

---

† See Problem 6.34.

and

$$\Delta_z = z_{11}z_{22} - z_{12}z_{21}.$$

Thus

$$\begin{bmatrix} s_{11} & s_{12} \\ s_{21} & s_{22} \end{bmatrix}$$

$$= \begin{bmatrix} \dfrac{(z_{11} - 1)(z_{22} + 1) - z_{12}z_{21}}{\Delta} & \dfrac{(z_{11} - 1)(-z_{12}) + z_{12}(z_{11} + 1)}{\Delta} \\ \dfrac{z_{21}(z_{22} + 1) + (-z_{21})(z_{22} - 1)}{\Delta} & \dfrac{-z_{12}z_{21} + (z_{22} - 1)(z_{11} + 1)}{\Delta} \end{bmatrix}$$

and

$$s_{11} = \frac{(z_{11} - z_{22} - 1) + \Delta_z}{\Delta}, \qquad s_{12} = \frac{2z_{12}}{\Delta},$$

$$s_{21} = \frac{2z_{21}}{\Delta}, \qquad s_{22} = \frac{z_{22} - z_{11} - 1 + \Delta_z}{\Delta}.$$

From our previous results (Chapter 5), reciprocity implies that $z_{12} = z_{21}$; hence

$$s_{12} = 2z_{12}/\Delta \qquad \text{and} \qquad s_{21} = 2z_{12}/\Delta.$$

Consequently

$$s_{12} = s_{21}.$$

6.10 SUMMARY AND REMARKS

In the preceding sections, we have developed the theory of two-ports. The type of two-ports under consideration have been the linear two-ports which may be passive or active. In Section 6.3, we derived six sets of two-port parameters: namely, the immittance parameters, including the z_{ij}'s and the y_{ij}'s; the transmission parameters, which are the a_{ij}'s and the b_{ij}'s; and the hybrid parameters consisting of the h_{ij}'s and the g_{ij}'s. In Section 6.4, we derived from these parameters several equivalent networks or models which are very useful in the development of the theory of both network analysis and synthesis.

In addition to the interrelationships among the six sets of parameters as summarized in Table 6.3.1, further relationships among these parameters can be stated about reciprocal and electrically symmetrical two-ports. They are now listed in Table 6.10.1. The proofs of these statements are left as exercises.*

---

* See Problems 6.23 and 6.24.

Table 6.10.1 Relationships among Parameters for a
Reciprocal and Electrically Symmetrical Two-Port N

Parameters	Reciprocal N	Symmetrical N
z_{ij}	$z_{12} = z_{21}$	$z_{11} = z_{22}$
y_{ij}	$y_{12} = y_{21}$	$y_{11} = y_{12}$
a_{ij}	$\Delta_a = 1$	$a_{11} = a_{22}$
b_{ij}	$\Delta_b = 1$	$b_{11} = b_{22}$
h_{ij}	$h_{12} = -h_{21}$	$\Delta_h = 1$
g_{ij}	$g_{12} = -g_{21}$	$\Delta_g = 1$

In Section 6.5, we showed how two two-ports may be interconnected in various ways. The purpose of such a discussion was to illustrate how the network parameters of a complex two-port N may be considered as two simpler two-ports N_a and N_b interconnected in a certain way so that the two-port parameters of N can be calculated in terms of the corresponding parameters of N_a and N_b. This point was demonstrated in Examples 6.5.1 and 6.5.2. In the case where the Brune test fails for a certain connection of two given two-ports, it indicates that the terminal constraint will not be satisfied for that particular connection. Then the calculations of the parameters of the overall two-port should be done directly from the overall two-port. Alternatively, an ideal transformer with a unity turns ratio ($n = 1$) may be connected in cascade with one of the two two-ports to guarantee the satisfaction of the terminal constraints pertinent to that connection. This idea is depicted in Fig. 6.10.1. Such applications of the ideal transformer is an illustration of its *isolating property* which we have mentioned earlier.

Besides the ideal transformer, two other ideal two-port devices were studied in Section 6.6, namely, the gyrator, and the negative impedance converter (NIC). Although some practical applications have been developed in realizing the gyrator as a physical network element using operational amplifiers,* the prime importance of these three ideal devices lies in the application in the development of network theory both in analysis and synthesis.

The discussions on the general *n*-ports and the *n*-terminal networks presented in Sections 6.7 and 6.8 serve primarily as an introduction to the analysis of active multi-terminal networks. The interested reader is referred to literature of a more advanced level for more detailed discussion of these topics.†

In Section 6.9 we introduced the scattering parameters. The concept of incident and reflected waves was developed and utilized to define the scattering parameter for a one-port. Furthermore, we discussed the significance of these waves through their relationship with the power absorbed by the one-port. We

---

* See S. K. Mitra [MI 1], Chapter 11, and A. S. Morse and L. P. Huelsman [MO 1], for example.
† See S. K. Mitra [MI 1], K. L. Su [SU 1], P. M. Chirlian [CH 4], and L. P. Huelsman [HU 2], to name a few.

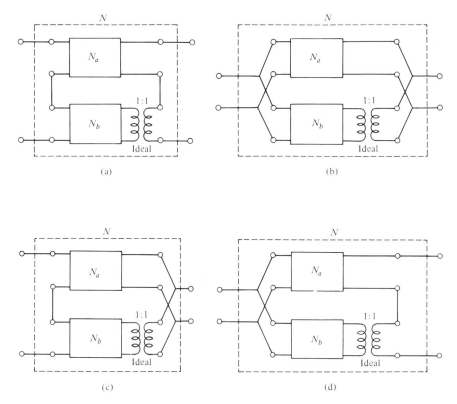

Fig. 6.10.1 Use of an ideal transformer in guaranteeing the satisfaction of the terminal constraints of (a) the series connection, (b) the parallel connection, (c) the series-parallel connection, and (d) the parallel-series connection of two-ports.

also mentioned that when the generator impedance is real, the s-parameter is commonly termed the *reflection coefficient* of the network. We defined the two-port scattering matrix, and derived the fundamental relations between scattering and impedance parameters. Finally, we developed the constraints imposed upon the scattering matrix by passivity, reciprocity, and losslessness.

It must be re-emphasized that the scattering matrix characterization becomes extremely important for design at high frequencies (especially above 100 MHz), since here, for certain devices, conventional parameter sets are not measurable.

It is worth noting that although we only developed the *two-port* scattering matrix in Section 6.9, extension to n-port networks is trivial when the incident and reflected waves are defined as in (6.9.16). The interested reader should consult literature elsewhere* for a complete discussion.

---

* See, for example, D. C. Youla [YO 1].

REFERENCES

1. H. Ruston and J. Bordogna, *Electric Networks: functions, filters, analysis*, McGraw-Hill, 1966, Chapter 4.

2. L. P. Huelsman, *Circuits, Matrices, and Linear Vector Spaces*, McGraw-Hill, 1963, Chapter 3.

3. S. Karni, *Network Theory: Analysis and Synthesis*, Allyn and Bacon, 1966, Chapter 1.

4. M. E. Van Valkenburg, *Network Analysis*, Second Edition, Prentice-Hall, 1964, Chapter 11.

5. S. K. Mitra, *Analysis and Synthesis of Linear Active Networks*, Wiley, 1969, Chapters 1 and 2.

6. K. L. Su, *Active Network Synthesis*, McGraw-Hill, 1965, Chapter 2.

7. P. M. Chirlian, *Integrated and Active Network Analysis and Synthesis*, Prentice-Hall, 1967, Chapter 3.

8. D. C. Youla, "On Scattering Matrices Normalized to Complex Port Numbers," *Proc. IRE*, **49**, 7, p. 1221, July 1961.

9. K. Kurokawa, "Power Waves and The Scattering Matrix," *IEEE Transactions on Microwave Theory and Techniques*, **MTT-13**, p. 194, March 1965.

10. H. J. Carlin and A. B. Giordano, *Network Theory—An Introduction to Reciprocal and Nonreciprocal Circuits*, Prentice-Hall, Englewood Cliffs, N.J., 1964.

11. T. Yanagisawa and Y. Kawashima, "Active Gyrator," *Electronics Letters*, **3**, pp. 105–107, March 1967.

12. R. W. Newcomb, *Active Integrated Circuit Synthesis*, Prentice-Hall, 1968, Chapter 2.

PROBLEMS

6.1 For each of the two-ports shown in Fig. P.6.1, determine:

    a) the open-circuit impedances or the z_{ij}-parameters, and
    b) the short-circuit admittances or the y_{ij}-parameters.

    Express the parameters in matrix form in each case.

6.2 Repeat Problem 6.1 for

    a) the transmission a_{ij}-parameters, and
    b) the transmission b_{ij}-parameters.

6.3 Repeat Problem 6.1 for

    a) the hybrid h_{ij}-parameters, and
    b) the hybrid g_{ij}-parameters.

6.4 Derive the single-voltage-source equivalent of a grounded two-port shown in Fig. 6.4.3 from its two-voltage-source equivalent depicted in Fig. 6.4.2.

6.5 Following a development parallel to that for deriving the two-voltage-source equivalent (shown in Fig. 6.4.1) for a linear two-port *N*, obtain the two-current-source equivalent of *N* shown in Fig. 6.4.4, based on the y_{ij}'s.

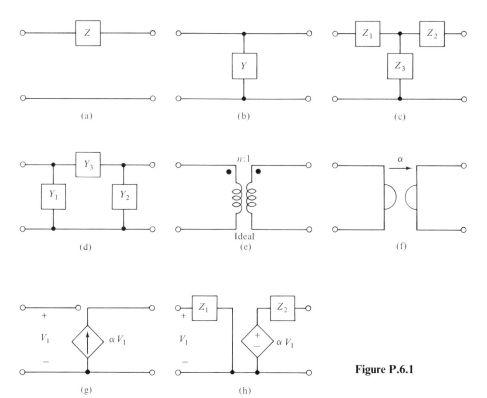

Figure P.6.1

6.6 Derive the single-current-source model depicted in Fig. 6.4.5 for a grounded two-port, following a development parallel to that used in Problem 6.4.

6.7 a) Derive the equivalent T and the equivalent Π models (shown in Figs. 6.4.6 and 6.4.7, respectively) of the linear two-port N of Fig. 6.2.1. Show all your work in detail.
   b) Repeat part (a) for the models shown in Figs. 6.4.8 and 6.4.9.

6.8 Consider each of the two-ports shown in Fig. P.6.8 as consisting of two subnetworks N_a and N_b (which are also two-ports) interconnected in such a way that port constraints are satisfied for the overall two-port N (as shown in the figure) as well as its subnetworks N_a and N_b. Obtain the appropriate set of two-port parameters for N in terms of the parameters of N_a and N_b by making use of the formulas developed in Section 6.5.

6.9 Show that for the series-parallel connection shown in Fig. 6.5.4, the h_{ij}-matrix of the overall two-port N is related to the h_{ij}-matrices of its subnetworks N_a and N_b by $[h_{ij}] = [h_{ija}] + [h_{ijb}]$.

6.10 Show that for the parallel-series connection depicted in Fig. 6.5.5, the g_{ij}-matrix of the overall two-port N and the g_{ij}-matrices of its subnetworks N_a and N_b are related by $[g_{ij}] = [g_{ija}] + [g_{ijb}]$.

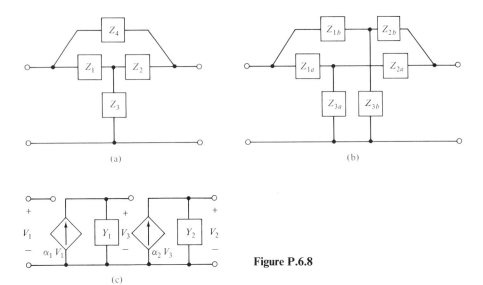

(a)

(b)

(c)

Figure P.6.8

6.11 Show that if a given parallel interconnection N of two two-ports N_a and N_b satisfies the Brune test described in Fig. 6.5.12, then the necessary (and sufficient) port constraints will also be satisfied for a valid parallel connection.

6.12 Derive the Brune test for making a valid series-parallel connection N of two two-ports N_a and N_b by first sketching the testing connections and the necessary measurements (similar to those shown in Fig. 6.5.12) and then providing the justification for making such connections and measurements.

6.13 Repeat Problem 6.12 for making a valid parallel-series connection.

6.14 Derive expressions (6.6.3) and (6.6.4) for the loaded perfect transformer (with load Z_L).

6.15 Characterize each of the two-ports shown in Fig. P.6.15 by an appropriate set of two-port parameters.

6.16 For each of the two-ports of Fig. P.6.15, assume that a load Z_L is connected across terminals 2–2′ at port 2 and the input impedance Z_{in} is to be determined with element values expressed in ohms, henrys, and farads as follows:

$$\text{For Fig. P.6.15(a): } Z_L = 1000, n = 2, \alpha = 1$$
$$\text{For Fig. P.6.15(b): } Z_L = 1, n = 1, \alpha = 2$$
$$\text{For Fig. P.6.15(c): } Z_L = 1 + S, \alpha = 2, Z = 1$$
$$\text{For Fig. P.6.15(d): } Z_L = 1, n = 2, \alpha = 1$$
$$\text{For Fig. P.6.15(e): } Z_L = 1 + 1/S, n = 1, \alpha = 2, C = 1$$

6.17 Repeat Problem 6.16 for the determination of the voltage ratio V_2/V_1, where V_2 is the voltage across terminals 2–2′ at port 2 and V_1 is the voltage across terminals 1–1′ at port 1 with the positive voltage polarities at the unprimed terminals in both cases.

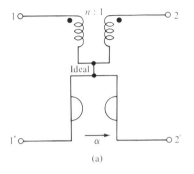

(a)

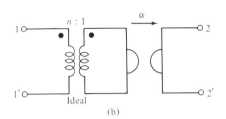

(b)

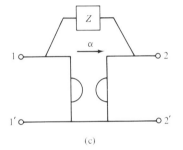

(c)

(With conversion ratio k)

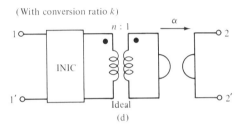

(d)

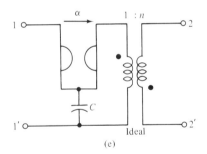

(e)

Figure P.6.15

6.18   For each of the n-ports shown in Fig. P.6.18, determine:
a) the open-circuit impedance matrix $\mathbf{Z} = [z_{ij}]$, and
b) the short-circuit admittance matrix $\mathbf{Y} = [y_{ij}]$.

6.19   Consider each of the networks shown in Fig. P.6.18 as an n-terminal network (with the terminals identified by the encircled numerals). Determine the indefinite admittance matrix $[y_{ij}]$ for each network with respect to a datum node 0 (not shown).

6.20   Making use of the properties of the indefinite admittance matrix $[y_{ij}]$, obtain an equivalent circuit for a grounded-base transistor with the help of (6.8.7) and (6.8.8). Show both the schematic diagram and the matrix equation relating the currents and voltages.

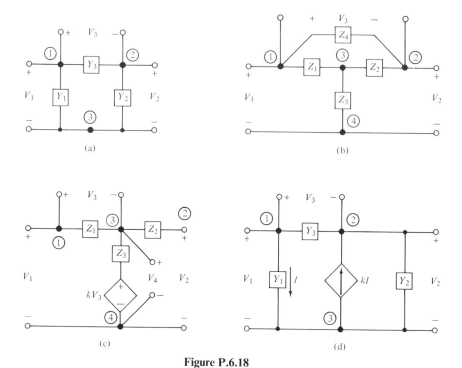

Figure P.6.18

6.21 a) Show that the y_{ij} matrix in (6.7.8) is the inverse of the z_{ij} matrix in (6.7.7) for the three-port shown in Fig. 6.7.4.
b) Derive the expressions for the two sets of hybrid parameters for a general linear n-port.
c) Repeat (b) for the transmission parameters.

6.22 Derive expression (6.8.6).

6.23 Derive the conditions for the six sets of parameters for a linear, *electrically symmetrical* two-port as listed in Table 6.10.1.

6.24 a) Repeat Problem 6.23 for a linear, *reciprocal* two-port as indicated in Table 6.10.1.
b) Show that the negative impedance converter (NIC) is an *active* device.

6.25 Show that the expressions $\cos(\omega t - kx)$ and $\cos(\omega t + kx)$ represent waves traveling to the right and left, respectively. [*Hint*: Plot $\cos(\omega t + kx)$ vs. kx for increasing time increments and note the progress of a fixed point on the wave.
For example, consider

$$A = \cos(\omega t + kx).$$

Then choose

$$\omega = 2\pi f = 2\pi;$$

thus

$$A = \cos(2\pi t + kx).$$

And we see that

for $t = 0$, $A = \cos(kx)$

for $t = \frac{1}{4}$, $A = \cos\left(\dfrac{\pi}{2} + kx\right)$

for $t = \frac{1}{2}$, $A = \cos(\pi + kx)$

for $t = \frac{3}{4}$, $A = \cos(\frac{3}{2}\pi + kx)$

for $t = 1$, $A = \cos(2\pi + kx)$

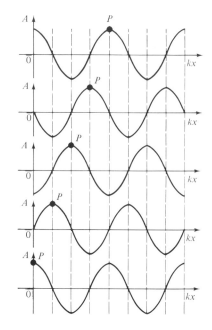

Figure P.6.25

Observe that a point P (for which the argument $(\omega t + kx)$ of cosine equals 2π), moves to the left as time t increases.] Repeat this process for the remaining cosine wave.

6.26 Show that $I = (k_2)^{-1}(a - b)$. (See 6.9.4b).

6.27 In the development of the expression for power (6.9.9), explain why it was necessary to assume that z was real.

6.28 From (6.9.10), derive the relations

$$V = \frac{z^*a}{\sqrt{r}} + \frac{zb}{\sqrt{r}}$$

$$I = \frac{a}{\sqrt{r}} - \frac{b}{\sqrt{r}}.$$

[*Hint*: First solve (6.9.10) for current (I).]

6.29 Using the values for V and I given in (6.9.10a), compute the power $\mathrm{Re}[VI^*]$ and show the result is the same as given in (6.9.9). [*Hint*: Use $I = (1/\sqrt{r})(a - b)$ and

$$V = \sqrt{r}(a + b) - \frac{\left(\dfrac{z - z^*}{2}\right)(a - b)}{\sqrt{r}}$$

from the previous problem.]

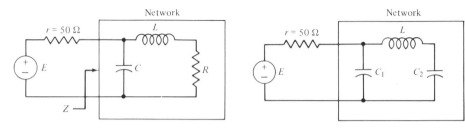

Figure P.6.30   **Figure P.6.31**

6.30 Compute the reflection coefficient ("*s*"-parameter) for the network of Fig. P.6.30.

6.31 Compute the reflection coefficient ("*s*"-parameter) for the lossless network shown in Fig. P.6.31, and demonstrate that for real frequencies ($j\omega$), the condition of (6.9.14) is satisfied.

6.32 A certain lossless reciprocal network possesses input and output scattering parameters $s_{11} = 0.4 + j0.1$ and $s_{22} = 0.1 + j0.4$, respectively. Compute the remaining scattering parameters for the network.

6.33 Derive equation (6.9.31b).

6.34 Verify the result of equation (6.9.35).

CHAPTER 7

STATE EQUATIONS FOR LINEAR SYSTEMS
AND METHODS OF SOLUTION

7.1 INTRODUCTION

In Chapter 3 we introduced the concept of state and outlined the use of state equations to describe the behavior of a physical system. When the *state model** for a system involves a large number of simultaneous differential equations, the solution is best obtained by using either the digital or analog computer. However, to attack the problem effectively, the programmer must be familiar with various techniques so that he can decide which one is most suitable for a given problem. In this chapter we present some basic techniques for obtaining the solution of a problem which is represented by a state model.

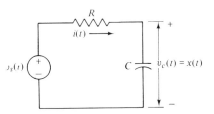

Fig. 7.2.1 The RC network N.

7.2 THE LINEAR TIME-INVARIANT SYSTEM

Consider the RC network N shown in Fig. 7.2.1. The problem is to determine the current $i(t)$ for $t \geq t_0$.

Since there is only one energy storage element, the state equation can be obtained by writing the KVL-equation around the loop with $x(t)$ taken to be the instantaneous voltage $v_c(t)$ across the capacitor C as indicated in N; thus,

$$v_s(t) = Ri(t) + x(t) \qquad (7.2.1)$$

$$x(t) = \frac{1}{C} \int i(t)\, dt. \qquad (7.2.2)$$

Differentiating (7.2.2) and combining with (7.2.1), we obtain the state equation

$$\dot{x}(t) = \frac{1}{C} i(t) = -\frac{1}{RC} x(t) + \frac{1}{RC} v_s(t). \qquad (7.2.3)$$

---

* When the system is represented by a set of first-order differential equations of the form $\dot{\mathbf{x}} = \mathbf{A}\mathbf{x} + \mathbf{B}\mathbf{u}$, then the representation is referred to as the state model.

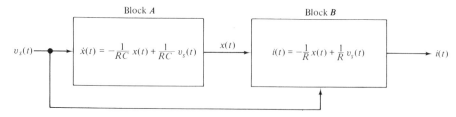

Fig. 7.2.2 Block diagram representation of the RC network N.

From (7.2.1) we can solve for $i(t)$ as the response of N:

$$i(t) = -\frac{1}{R} x(t) + \frac{1}{R} v_s(t). \tag{7.2.4}$$

The above system can be represented by a block diagram with a single input $v_s(t$ and a single output $i(t)$ as shown in Fig. 7.2.2, with its behavior completely described by equations (7.2.3) and (7.2.4). In Fig. 7.2.2, block A signifies the relationship between the state variable $x(t)$ and the input $v_s(t)$ of the system, namely, the RC network N. Similarly, block B represents the relationship between $x(t)$ and the output $i(t)$. It is obvious from (7.2.4) that $i(t)$ is uniquely determined once the state variable $x(t)$ is known. From (7.2.3), we write

$$\dot{x}(t) + \frac{1}{RC} x(t) = \frac{1}{RC} v_s(t), \tag{7.2.5}$$

which is in the same form as (A.1.5.47) in Appendix A.1.* An inspection of (A.1.5.48) gives the solution of (7.2.5); that is,

$$x(t) = e^{-(1/RC)(t-t_0)} x(t_0) + \int_{t_0}^{t} e^{-(1/RC)(t-\tau)} \left(\frac{1}{RC}\right) v_s(\tau)\, d\tau, \tag{7.2.6}$$

which is valid for $t \geq t_0$. Substituting (7.2.6) into (7.2.4), we obtain the current response

$$i(t) = -\frac{1}{R} e^{-(1/RC)(t-t_0)} x(t_0) - \frac{1}{R^2 C} \int_{t_0}^{t} e^{-(1/RC)(t-\tau)} v_s(\tau)\, d\tau + \frac{1}{R} v_s(t), \qquad t \geq t_0.$$

$$\tag{7.2.7}$$

If, for convenience, we define

$$a = -\frac{1}{RC}, \qquad b = \frac{1}{RC}, \qquad c = -\frac{1}{R}, \qquad d = \frac{1}{R},$$

and

$$u = v_s(t), \qquad y = i(t),$$

---

* Equations identified by numbers with an "A.1" prefix can be found in Appendix A.1.

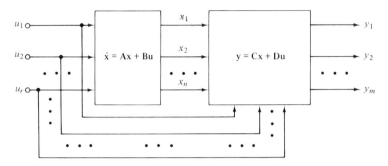

Fig. 7.2.3 A general linear time-invariant system.

then (7.2.3) and (7.2.4) can be written as

$$\dot{x} = ax + bu$$
$$y = cx + du. \tag{7.2.8}$$

The corresponding solution of the state equation is the same as (A.1.5.48) and is repeated here for convenience

$$x(t) = e^{a(t-t_0)}x(t_0) + \int_{t_0}^{t} e^{a(t-\tau)}bu(\tau)d\tau, \qquad t \geq t_0. \tag{7.2.9}$$

The above example is simply a special case of a general time-invariant system with r inputs and m outputs as depicted in Fig. 7.2.3.

The behavior of this general linear time-invariant system is described by the matrix equations

$$\dot{\mathbf{x}} = \mathbf{Ax} + \mathbf{Bu}$$

and

$$\mathbf{y} = \mathbf{Cx} + \mathbf{Du}, \tag{7.2.10}$$

where $\mathbf{x}$, $\mathbf{y}$, and $\mathbf{u}$ are vectors defined by

$$\mathbf{x} = \begin{bmatrix} x_1 \\ x_2 \\ \vdots \\ x_n \end{bmatrix}, \qquad \mathbf{y} = \begin{bmatrix} y_1 \\ y_2 \\ \vdots \\ y_m \end{bmatrix}, \qquad \mathbf{u} = \begin{bmatrix} u_1 \\ u_2 \\ \vdots \\ u_r \end{bmatrix}$$

and $\mathbf{A}$, $\mathbf{B}$, $\mathbf{C}$, and $\mathbf{D}$ are constant matrices of orders $n \times n$, $n \times r$, $m \times n$, and $m \times r$, respectively. Here the vectors $\mathbf{x}$, $\mathbf{y}$, and $\mathbf{u}$ are called, respectively, the state vector, output vector, and the input or control vector.*

---

* See Chapter 3 for the introduction of these quantities.

The solution of the vector differential equation is given by (A.1.5.51), which is written here for convenience:

$$\mathbf{x}(t) = e^{\mathbf{A}(t-t_0)}\mathbf{x}(t_0) + \int_{t_0}^{t} e^{\mathbf{A}(t-\tau)}\mathbf{B}\mathbf{u}(\tau)\,d\tau, \qquad t \geq t_0, \qquad (7.2.11)$$

where $e^{\mathbf{A}(t-t_0)}$, the state transition matrix of the system, is a function of the $\mathbf{A}$ matrix and is itself a matrix. Note that the output can be determined immediately by substituting (7.2.11) into the second equation of (7.2.10).

7.3 DETERMINATION OF $e^{\mathbf{A}t}$ BY LAPLACE TRANSFORMATION

We have seen in the previous section that, for a linear time-invariant system characterized by (7.2.10), we can determine the solution from (7.2.11) once we calculate the matrix function $e^{\mathbf{A}t}$. In this section we shall find the explicit expression of $e^{\mathbf{A}t}$ by means of the Laplace transform method.* We begin by considering the solution of the linear force-free system represented by

$$\dot{\mathbf{x}} = \mathbf{A}\mathbf{x}$$
$$\mathbf{x}(0) = \mathbf{x}_0, \qquad (7.3.1)$$

where $\mathbf{A}$ is again a constant matrix. The solution can be written down immediately from (7.2.11) by setting the input $\mathbf{u}$ equal to zero with $t_0 = 0$; thus

$$\mathbf{x}(t) = e^{\mathbf{A}t}\mathbf{x}_0. \qquad (7.3.2)$$

The next step is to determine $e^{\mathbf{A}t}$. Taking the Laplace transform of both sides of (7.3.1), we obtain

$$s\mathbf{X}(s) - \mathbf{x}_0 = \mathbf{A}\mathbf{X}(s)$$

or

$$\mathbf{X}(s) = [s\mathbf{I} - \mathbf{A}]^{-1}\mathbf{x}_0. \qquad (7.3.3)$$

The inverse Laplace transform of (7.3.3) is given by

$$\mathbf{x}(t) = \mathscr{L}^{-1}\{[s\mathbf{I} - \mathbf{A}]^{-1}\}\mathbf{x}_0 = \mathscr{L}^{-1}\left\{\frac{\text{Adjoint}\,[s\mathbf{I} - \mathbf{A}]}{\det\,[s\mathbf{I} - \mathbf{A}]}\right\}\mathbf{x}_0. \qquad (7.3.4)$$

Comparing (7.3.4) with (7.3.2), we arrive at the following equation

$$e^{\mathbf{A}t} = \mathscr{L}^{-1}\{[s\mathbf{I} - \mathbf{A}]^{-1}\}. \qquad (7.3.5)$$

---

* Several methods for finding $e^{\mathbf{A}t}$ are available. However, in this section, we discuss only the Laplace transform method.

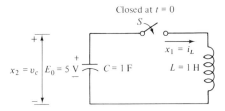

Fig. 7.3.1 An LC network.

A simple example will demonstrate the usefulness of this important formula.

Example 7.3.1 Consider the LC network shown in Fig. 7.3.1. The problem is to determine the expression for the current in the network as a function of time after the switch S is closed at $t = 0$. The initial voltage across the capacitor is assumed to be E_0.

Let us choose the state variables as $x_1 = i_L$, $x_2 = v_C$ and write the KVL-equation around the loop;

$$L\frac{dx_1}{dt} - x_2 = 0. \tag{7.3.6}$$

Differentiating the voltage $v_C \, (= x_2)$ across the capacitor, we obtain

$$\frac{dx_2}{dt} = -\frac{1}{C} x_1. \tag{7.3.7}$$

Rewriting (7.3.6) and (7.3.7) in normal form, we have

$$\begin{bmatrix} \dot{x}_1 \\ \dot{x}_2 \end{bmatrix} = \begin{bmatrix} 0 & \dfrac{1}{L} \\ -\dfrac{1}{C} & 0 \end{bmatrix} \begin{bmatrix} x_1 \\ x_2 \end{bmatrix}$$

$$\begin{bmatrix} x_1(0) \\ x_2(0) \end{bmatrix} - \begin{bmatrix} 0 \\ E_0 \end{bmatrix}. \tag{7.3.8}$$

Putting this expression in vector form with the parameter values properly substituted, we have

$$\dot{\mathbf{x}} = \mathbf{A}\mathbf{x}, \qquad \mathbf{x}(0) = \mathbf{x}_0, \tag{7.3.9}$$

where

$$\mathbf{A} = \begin{bmatrix} 0 & 1 \\ -1 & 0 \end{bmatrix}, \qquad \mathbf{x}_0 = \begin{bmatrix} 0 \\ 5 \end{bmatrix}.$$

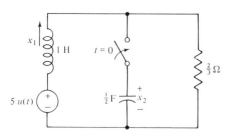

Fig. 7.3.2 The *RLC* network.

Thus, by virtue of (7.3.5), we obtain

$$[s\mathbf{I} - \mathbf{A}]^{-1} = \begin{bmatrix} s & -1 \\ 1 & s \end{bmatrix}^{-1} = \begin{bmatrix} \dfrac{s}{s^2 + 1} & \dfrac{1}{s^2 + 1} \\[2ex] \dfrac{-1}{s^2 + 1} & \dfrac{s}{s^2 + 1} \end{bmatrix}$$

and

$$e^{\mathbf{A}t} = \mathcal{L}^{-1} \begin{bmatrix} \dfrac{s}{s^2 + 1} & \dfrac{1}{s^2 + 1} \\[2ex] \dfrac{-1}{s^2 + 1} & \dfrac{s}{s^2 + 1} \end{bmatrix} = \begin{bmatrix} \cos t & \sin t \\ -\sin t & \cos t \end{bmatrix}.$$

Therefore, the solution is given by

$$\mathbf{x}(t) = e^{\mathbf{A}t}\mathbf{x}_0$$

$$= \begin{bmatrix} \cos t & \sin t \\ -\sin t & \cos t \end{bmatrix} \begin{bmatrix} 0 \\ 5 \end{bmatrix} \qquad (7.3.10)$$

$$= \begin{bmatrix} 5 \sin t \\ 5 \cos t \end{bmatrix}$$

and the current in the network is the first component of the vector $\mathbf{x}(t)$ in (7.3.10), namely, $i_L(t) = x_1(t) = 5 \sin t$.

It should be pointed out that the example we just discussed is an exceptionally simple problem. In general, even for systems of moderate complexity, partial fraction expansion must be used to calculate $[s\mathbf{I} - \mathbf{A}]^{-1}$. This can be best illustrated by means of another example which follows.

Example 7.3.2 Consider the *RLC* network shown in Fig. 7.3.2. Find the expressions for the current x_1 in the inductor and the voltage x_2 across the capacitor as functions of t after the switch is closed at $t = 0$. The initial voltage across the capacitor is 10 volts. The state equations are found to be:

$$\begin{bmatrix} \dot{x}_1 \\ \dot{x}_2 \end{bmatrix} = \begin{bmatrix} 0 & -1 \\ 2 & -3 \end{bmatrix} \begin{bmatrix} x_1 \\ x_2 \end{bmatrix} + \begin{bmatrix} 5 \\ 0 \end{bmatrix} u(t) \qquad (7.3.11)$$

with initial conditions given by

$$\begin{bmatrix} x_1(0) \\ x_2(0) \end{bmatrix} = \begin{bmatrix} 0 \\ 10 \end{bmatrix}. \tag{7.3.12}$$

The **A**-matrix in this case is

$$\mathbf{A} = \begin{bmatrix} 0 & -1 \\ 2 & -3 \end{bmatrix}.$$

To determine e^{At}, we first form

$$[s\mathbf{I} - \mathbf{A}] = \begin{bmatrix} s & 1 \\ -2 & s + 3 \end{bmatrix},$$

which gives the following inverse:

$$[s\mathbf{I} - \mathbf{A}]^{-1} = \frac{1}{s^2 + 3s + 2} \begin{bmatrix} s + 3 & -1 \\ 2 & s \end{bmatrix}$$

$$= \frac{1}{(s + 2)(s + 1)} \begin{bmatrix} s + 3 & -1 \\ 2 & s \end{bmatrix}$$

$$= \begin{bmatrix} \dfrac{s + 3}{(s + 2)(s + 1)} & \dfrac{-1}{(s + 2)(s + 1)} \\[3mm] \dfrac{2}{(s + 2)(s + 1)} & \dfrac{s}{(s + 2)(s + 1)} \end{bmatrix}.$$

If each of the elements of $[s\mathbf{I} - \mathbf{A}]^{-1}$ is decomposed by partial fraction expansion, we find that

$$[s\mathbf{I} - \mathbf{A}]^{-1} = \begin{bmatrix} \dfrac{2}{s + 1} - \dfrac{1}{s + 2} & -\dfrac{1}{s + 1} + \dfrac{1}{s + 2} \\[3mm] \dfrac{2}{s + 1} - \dfrac{2}{s + 2} & -\dfrac{1}{s + 1} + \dfrac{2}{s + 2} \end{bmatrix}.$$

Taking the inverse Laplace transform of the above equation yields

$$\mathscr{L}^{-1}\{[s\mathbf{I} - \mathbf{A}]^{-1}\} = \begin{bmatrix} 2e^{-t} - e^{-2t} & -e^{-t} + e^{-2t} \\ 2e^{-t} - 2e^{-2t} & -e^{-t} + 2e^{-2t} \end{bmatrix}$$

which is valid for $t \geq 0$. Finally, using (7.2.11) and (7.3.5), we obtain the solution of the problem:

$$
\begin{bmatrix} x_1(t) \\ x_2(t) \end{bmatrix} = \begin{bmatrix} 2e^{-t} - e^{-2t} & -e^{-t} + e^{-2t} \\ 2e^{-t} - 2e^{-2t} & -e^{-t} + 2e^{-2t} \end{bmatrix} \begin{bmatrix} 0 \\ 10 \end{bmatrix}
$$

$$
+ \int_0^t \begin{bmatrix} 2e^{-(t-\tau)} - e^{-2(t-\tau)} & -e^{-(t-\tau)} + e^{-2(t-\tau)} \\ 2e^{-(t-\tau)} - 2e^{-2(t-\tau)} & -e^{-(t-\tau)} + 2e^{-2(t-\tau)} \end{bmatrix} \begin{bmatrix} 5 \\ 0 \end{bmatrix} u(\tau) d\tau
$$

$$
= \begin{bmatrix} -10e^{-t} + 10e^{-2t} \\ -10e^{-t} + 20e^{-2t} \end{bmatrix} + \begin{bmatrix} 10(1 - e^{-t}) - 2.5(1 - e^{-2t}) \\ 10(1 - e^{-t}) - 5(1 - e^{-2t}) \end{bmatrix}
$$

$$
= \begin{bmatrix} 7.5 - 20e^{-t} + 12.5e^{-2t} \\ 5 - 20e^{-t} + 25e^{-2t} \end{bmatrix} \qquad t \geq 0.
$$

We have now demonstrated that Laplace transforms can be used to solve linear systems which are represented by state equations. However, for more complex systems, the matrix $\mathbf{A}$ will not be restricted to the second order and the computation of the inverse of $[s\mathbf{I} - \mathbf{A}]$ alone may become extremely complicated, not to mention the subsequent decomposition of each element by partial fraction expansion. This points out the need for some alternative techniques which can be used to simplify the computations involved in obtaining the solution of a given problem. In the following sections we shall discuss other methods of solution.

7.4 SOLUTION OF STATE EQUATIONS THROUGH DIAGONALIZATION

In the preceding section we have illustrated that, for a vector differential equation of the form

$$
\dot{\mathbf{x}} = \mathbf{A}\mathbf{x},
$$

$$
\mathbf{x}(0) = \mathbf{x}_0, \tag{7.4.1}
$$

the solution is given by

$$
\mathbf{x} = e^{\mathbf{A}t}\mathbf{x}_0. \tag{7.4.2}
$$

In this section we shall traverse another road to obtain the explicit solution of (7.4.1). This will lead us to the determination of eigenvalues and eigenvectors. We shall begin by looking for special solutions of the form $\mathbf{x} = e^{\lambda t}\mathbf{u}$, where λ is a scalar and $\mathbf{u}$ is a nonzero constant vector to be determined later. Substituting $\mathbf{x} = e^{\lambda t}\mathbf{u}$ into (7.4.1), we find

$$
\lambda e^{\lambda t}\mathbf{u} = \mathbf{A}e^{\lambda t}\mathbf{u}
$$

or

$$
\lambda\mathbf{u} = \mathbf{A}\mathbf{u}. \tag{7.4.3}
$$

Equation (7.4.3) can be written in the following form

$$[\mathbf{A} - \lambda\mathbf{I}]\mathbf{u} = 0 \qquad (7.4.4)$$

which indicates clearly that once the eigenvalues and the corresponding eigenvectors of $\mathbf{A}$ are determined, the special solution of (7.4.1) can be written down immediately.

In this section we shall discuss only the case in which the eigenvalues $\lambda_1, \dots, \lambda_n$ of $\mathbf{A}$ are *distinct* so that a set of linearly independent eigenvectors $\mathbf{u}_1, \dots, \mathbf{u}_n$ can be found by applying (7.4.4) directly. The general case will be taken up in the following section.

Let us change the state vector $\mathbf{x}$ to $\mathbf{z}$ by the linear nonsingular transformation*

$$\mathbf{x} = \mathbf{S}\mathbf{z} \qquad (7.4.5)$$

where the matrix $\mathbf{S}$ is formed by taking the eigenvectors $\mathbf{u}_1, \mathbf{u}_2, \dots, \mathbf{u}_n$ as its columns;

$$\mathbf{S} \triangleq [\mathbf{u}_1 \ \mathbf{u}_2 \cdots \mathbf{u}_n]. \qquad (7.4.6)$$

Substituting (7.4.5) into (7.4.1), we find that

$$\mathbf{S}\dot{\mathbf{z}} = \mathbf{A}\mathbf{S}\mathbf{z}. \qquad (7.4.7)$$

Premultiplying both sides of (7.4.7) by $\mathbf{S}^{-1}$ yields

$$\dot{\mathbf{z}} = \mathbf{S}^{-1}\mathbf{A}\mathbf{S}\mathbf{z} \qquad (7.4.8)$$

which, by virtue of (A.1.5.9), becomes

$$\dot{\mathbf{z}} = \Lambda\mathbf{z} \qquad (7.4.9)$$

with

$$\Lambda = \mathbf{S}^{-1}\mathbf{A}\mathbf{S}.$$

The matrix Λ is a diagonal matrix with the eigenvalues of $\mathbf{A}$ as its diagonal elements. In terms of components, (7.4.9) can be written as

$$\dot{z}_1 = \lambda_1 z_1$$

$$\dot{z}_2 = \lambda_2 z_2$$

$$\vdots \qquad \vdots \qquad\qquad (7.4.10)$$

$$\dot{z}_n = \lambda_n z_n$$

It becomes evident that the linear transformation given by (7.4.5) reduces the original system (7.4.1) with state variables $x_1, \dots, x_n$ to a new system (7.4.10) in which the state variables $z_1, \dots, z_n$ are completely decoupled. Consequently, the

---

* Refer to Appendix A.1.

solutions of (7.4.10) can be immediately obtained; i.e.,

$$z_1(t) = z_1(0)e^{\lambda_1 t}$$
$$z_2(t) = z_2(0)e^{\lambda_2 t}$$
$$\vdots$$
$$z_n(t) = z_n(0)e^{\lambda_n t}$$

$$(7.4.11)$$

or, in vector notation,

$$\mathbf{z}(t) = \begin{bmatrix} e^{\lambda_1 t} & 0 & 0 & \ldots & 0 \\ 0 & e^{\lambda_2 t} & 0 & \ldots & 0 \\ & & \cdot & \cdot & \\ 0 & 0 & 0 & \ldots & e^{\lambda_n t} \end{bmatrix} \mathbf{z}(0), \qquad (7.4.12)$$

where $\mathbf{z}(0)$ can be obtained from $\mathbf{x}(0)$ through (7.4.5) and is given by

$$\mathbf{z}(0) = \mathbf{S}^{-1}\mathbf{x}(0). \qquad (7.4.13)$$

Finally, substituting (7.4.12) into (7.4.5), we find the solution of the original system (7.4.1).

The same procedure can be used for determining the solution of the state equation (7.2.10) which is repeated here for convenience:

$$\dot{\mathbf{x}} = \mathbf{A}\mathbf{x} + \mathbf{B}\mathbf{u}. \qquad (7.4.14)$$

Manipulating between (7.4.5) and (7.4.14), we find

$$\dot{\mathbf{z}} = \mathbf{S}^{-1}\mathbf{A}\mathbf{S}\mathbf{z} + \mathbf{S}^{-1}\mathbf{B}\mathbf{u}$$
$$= \mathbf{\Lambda}\mathbf{z} + \mathbf{E}\mathbf{u}, \qquad (7.4.15)$$

where $\mathbf{E}$ is taken to be the product of $\mathbf{S}^{-1}$ and $\mathbf{B}$. In terms of components, (7.4.15) becomes

$$\dot{z}_1 = \lambda_1 z_1 + \sum_{j=1}^{r} e_{1j} u_j$$

$$\dot{z}_2 = \lambda_2 z_2 + \sum_{j=1}^{r} e_{2j} u_j \qquad (7.4.16)$$

$$\vdots$$

$$\dot{z}_n = \lambda_n z_n + \sum_{j=1}^{r} e_{nj} u_j,$$

where e_{ij} is the element located in the ith row and the jth column of the matrix $\mathbf{E}$ and r is the number of inputs in the given system. Again the variables z_i are completely decoupled (that is, z_i does not depend on z_j for $i \neq j$), and consequently the solution of (7.4.16), as well as that of original system (7.4.14), can be determined easily. A simple example will demonstrate the usefulness of the method.

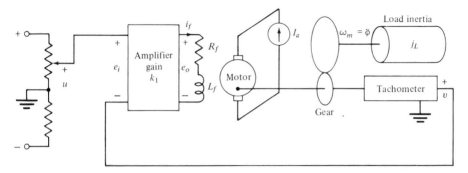

Fig. 7.4.1   Elementary closed-loop speed control system.

Example 7.4.1   As an example of practical interest, consider the elementary closed-loop speed control system* sketched in Fig. 7.4.1. The speed of the dc motor is controlled by varying the field current i_f which is in turn governed by the difference between the reference voltage u and the output voltage v of the tachometer. The tachometer voltage v is directly proportional to the output speed ω_m. In addition, it is assumed that the armature current I_a is kept constant externally at all times so that the motor torque† is directly proportional to the field current i_f ; that is,

$$T_m = Ki_f. \tag{7.4.17}$$

Neglecting the mechanical damping torque, as well as friction and windage, we can express the motor torque as

$$T_m = J_L \frac{d^2\phi}{dt^2} = J_L\ddot{\phi}, \tag{7.4.18}$$

where ϕ and J_L are, respectively, the angular position and the moment of inertia of the output load.

The equations governing the behavior of the system shown in Fig. 7.4.1 are given by

$$e_i = u - v, \tag{7.4.19}$$

$$v = K_2\omega_m = K_2\dot{\phi}, \tag{7.4.20}$$

$$e_o = K_1e_i, \tag{7.4.21}$$

$$e_o = R_fi_f + L_f\dot{i}_f, \tag{7.4.22}$$

---

* A closed loop control system is one in which the output has an influence on the input in such a way as to maintain the desired output.

† In general, the torque T_m developed in a dc shunt motor is taken to be directly proportional to the armature current I_a and the flux which is considered to vary directly with the field current i_f.

where K_1 and K_2 are, respectively, the gain of the amplifier and the tachometer constant. If i_f, ϕ, and $\dot{\phi}$ are taken to be the components of the state vector $\mathbf{x}$, i.e.,

$$\mathbf{x} = \begin{bmatrix} x_1 \\ x_2 \\ x_3 \end{bmatrix} = \begin{bmatrix} i_f \\ \phi \\ \dot{\phi} \end{bmatrix}, \tag{7.4.23}$$

and u and v are, respectively, the input and output of the system, i.e.,

$$\mathbf{u} = [u] \tag{7.4.24}$$

and

$$\mathbf{y} = [v], \tag{7.4.25}$$

then the state equation can be obtained by combining (7.4.17) through (7.4.22). We shall first combine (7.4.17), (7.4.18), (7.4.22), and (7.4.23) to form the following state equation with e_o as input. Thus

$$\begin{bmatrix} \dot{x}_1 \\ \dot{x}_2 \\ \dot{x}_3 \end{bmatrix} = \begin{bmatrix} -\dfrac{R_f}{L_f} & 0 & 0 \\ 0 & 0 & 1 \\ K_0 & 0 & 0 \end{bmatrix} \begin{bmatrix} x_1 \\ x_2 \\ x_3 \end{bmatrix} + \begin{bmatrix} \dfrac{1}{L_f} \\ 0 \\ 0 \end{bmatrix} e_o, \tag{7.4.26}$$

where $K_0 \triangleq K/J_L$. Incidentally, (7.4.26) is called the open-loop state model since, without feedback, the above equation describes the instantaneous behavior of an open-loop system. Substituting (7.4.19) and (7.4.20) into (7.4.21), we obtain

$$\begin{aligned} e_o &= K_1(u - v) = -K_1 K_2 x_3 + K_1 u \\ &= -K_3 x_3 + K_1 u, \end{aligned} \tag{7.4.27}$$

where $K_3 \triangleq K_1 K_2$ and $x_3 \triangleq \dot{\phi}$ [defined in (7.4.23)]. Finally, a combination of (7.4.26) and (7.4.27) yields the state equation for the closed-loop system of Fig. 7.4.1:

$$\begin{bmatrix} \dot{x}_1 \\ \dot{x}_2 \\ \dot{x}_3 \end{bmatrix} = \begin{bmatrix} -\dfrac{R_f}{L_f} & 0 & -\dfrac{K_3}{L_f} \\ 0 & 0 & 1 \\ K_0 & 0 & 0 \end{bmatrix} \begin{bmatrix} x_1 \\ x_2 \\ x_3 \end{bmatrix} + \begin{bmatrix} \dfrac{K_1}{L_f} \\ 0 \\ 0 \end{bmatrix} u. \tag{7.4.28}$$

The output is given by

$$\mathbf{y} = [v] = \begin{bmatrix} 0 & 0 & K_2 \end{bmatrix} \begin{bmatrix} x_1 \\ x_2 \\ x_3 \end{bmatrix}. \tag{7.4.29}$$

For the purpose of demonstrating the process of diagonalization, let us assume that the parameters of the control system of Fig. 7.4.1 are given by

$$R_f = 3, \qquad L_f = 1, \qquad K_0 = 2, \qquad \text{and} \qquad K_1 = K_2 = K_3 = 1.$$

Then (7.4.28) and (7.4.29) become, respectively,

$$\begin{bmatrix} \dot{x}_1 \\ \dot{x}_2 \\ \dot{x}_3 \end{bmatrix} = \begin{bmatrix} -3 & 0 & -1 \\ 0 & 0 & 1 \\ 2 & 0 & 0 \end{bmatrix} \begin{bmatrix} x_1 \\ x_2 \\ x_3 \end{bmatrix} + \begin{bmatrix} 1 \\ 0 \\ 0 \end{bmatrix} u \qquad (7.4.30)$$

and

$$\mathbf{y} = \begin{bmatrix} 0 & 0 & 1 \end{bmatrix} \begin{bmatrix} x_1 \\ x_2 \\ x_3 \end{bmatrix}. \qquad (7.4.31)$$

In terms of matrix constants defined in (7.2.10), we find, for this system,

$$\mathbf{A} = \begin{bmatrix} -3 & 0 & -1 \\ 0 & 0 & 1 \\ 2 & 0 & 0 \end{bmatrix}, \qquad \mathbf{B} = \begin{bmatrix} 1 \\ 0 \\ 0 \end{bmatrix}, \qquad (7.4.32)$$

$$\mathbf{C} = \begin{bmatrix} 0 & 0 & 1 \end{bmatrix}, \qquad \mathbf{D} = [0].$$

To determine the eigenvalues and the corresponding eigenvectors of $\mathbf{A}$, we write the characteristic equation

$$g(\lambda) = \det[\mathbf{A} - \lambda \mathbf{I}]$$

$$= \begin{vmatrix} -3 - \lambda & 0 & -1 \\ 0 & -\lambda & 1 \\ 2 & 0 & -\lambda \end{vmatrix}$$

$$= -\lambda(\lambda + 1)(\lambda + 2) = 0$$

which yields the distinct eigenvalues $\lambda_1 = 0$, $\lambda_2 = -1$, and $\lambda_3 = -2$. To determine the corresponding eigenvectors $\mathbf{u}_1$, $\mathbf{u}_2$, and $\mathbf{u}_3$, we let

$$[\mathbf{A} - \lambda_1 \mathbf{I}]\mathbf{u}_1 = 0$$

or

$$\begin{bmatrix} -3 & 0 & -1 \\ 0 & 0 & 1 \\ 2 & 0 & 0 \end{bmatrix} \begin{bmatrix} u_{11} \\ u_{12} \\ u_{13} \end{bmatrix} = 0$$

which yields $u_{11} = 0$ and $u_{13} = 0$. Hence:

$$\mathbf{u}_1 = u_{12} \begin{bmatrix} 0 \\ 1 \\ 0 \end{bmatrix} = \begin{bmatrix} 0 \\ 1 \\ 0 \end{bmatrix},$$

where the arbitrary constant u_{12} is taken to be unity for convenience. In exactly the same manner, we have, for $\lambda_2 = -1$, $[\mathbf{A} - \lambda_2 \mathbf{I}]\mathbf{u}_2 = 0$ yields

$$\begin{bmatrix} -2 & 0 & -1 \\ 0 & 1 & 1 \\ 2 & 0 & 1 \end{bmatrix} \begin{bmatrix} u_{21} \\ u_{22} \\ u_{23} \end{bmatrix} = 0,$$

or

$$-2u_{21} - u_{23} = 0$$
$$u_{22} + u_{23} = 0,$$

which give

$$\mathbf{u}_2 = u_{23} \begin{bmatrix} -\frac{1}{2} \\ -1 \\ 1 \end{bmatrix} = \begin{bmatrix} -1 \\ -2 \\ 2 \end{bmatrix}$$

with $u_{23} = 2$ for convenience. Similarly, for $\lambda_3 = -2$, $[\mathbf{A} - \lambda_3 \mathbf{I}]\mathbf{u}_3 = 0$ yields

$$\begin{bmatrix} -1 & 0 & -1 \\ 0 & 2 & 1 \\ 2 & 0 & 2 \end{bmatrix} \begin{bmatrix} u_{31} \\ u_{32} \\ u_{33} \end{bmatrix} = 0$$

or

$$-u_{31} - u_{33} = 0$$
$$2u_{32} + u_{33} = 0,$$

which give

$$\mathbf{u}_3 = u_{33} \begin{bmatrix} -1 \\ -\frac{1}{2} \\ 1 \end{bmatrix} = \begin{bmatrix} -2 \\ -1 \\ 2 \end{bmatrix}$$

with $u_{33} = 2$. Now, using the eigenvectors $\mathbf{u}_1$, $\mathbf{u}_2$, and $\mathbf{u}_3$ as columns of the matrix

S, we find

$$S = [\mathbf{u}_1 \quad \mathbf{u}_2 \quad \mathbf{u}_3] = \begin{bmatrix} 0 & -1 & -2 \\ 1 & -2 & -1 \\ 0 & 2 & 2 \end{bmatrix}$$

and

$$S^{-1} = \begin{bmatrix} 1 & 1 & \frac{3}{2} \\ 1 & 0 & 1 \\ -1 & 0 & -\frac{1}{2} \end{bmatrix}.$$

As a check, we form the matrix product

$$S^{-1}AS = \begin{bmatrix} 1 & 1 & \frac{3}{2} \\ 1 & 0 & 1 \\ -1 & 0 & -\frac{1}{2} \end{bmatrix} \begin{bmatrix} -3 & 0 & -1 \\ 0 & 0 & 1 \\ 2 & 0 & 0 \end{bmatrix} \begin{bmatrix} 0 & -1 & -2 \\ 1 & -2 & -1 \\ 0 & 2 & 2 \end{bmatrix}$$

$$= \begin{bmatrix} 1 & 1 & \frac{3}{2} \\ -1 & 0 & 1 \\ -1 & 0 & -\frac{1}{2} \end{bmatrix} \begin{bmatrix} 0 & 1 & 4 \\ 0 & 2 & 2 \\ 0 & -2 & -4 \end{bmatrix}$$

$$= \begin{bmatrix} 0 & 0 & 0 \\ 0 & -1 & 0 \\ 0 & 0 & -2 \end{bmatrix}$$

which is, as expected, a diagonal matrix with eigenvalues $\lambda_1 = 0$, $\lambda_2 = -1$, and $\lambda_3 = -2$ as elements on its main diagonal. Hence (7.4.15) becomes

$$\dot{\mathbf{z}} = \begin{bmatrix} 0 & 0 & 0 \\ 0 & -1 & 0 \\ 0 & 0 & -2 \end{bmatrix} \mathbf{z} + \begin{bmatrix} 1 \\ 1 \\ -1 \end{bmatrix} u, \qquad (7.4.33)$$

where $\mathbf{z}$ is related to $\mathbf{x}$ by (7.4.5) or

$$\mathbf{x} = S\mathbf{z}.$$

In terms of components, (7.4.33) can be expressed as

$$\dot{z}_1 = u,$$
$$\dot{z}_2 = -z_2 + u, \qquad (7.4.34)$$
$$\dot{z}_3 = -2z_3 - u,$$

and the corresponding solutions are readily found to be

$$z_1(t) = z_1(0) + \int_0^t u(\tau)\,d\tau,$$

$$z_2(t) = e^{-t}z_2(0) + \int_0^t u(\tau)e^{-(t-\tau)}\,d\tau, \qquad (7.4.35)$$

$$z_3(t) = e^{-2t}z_3(0) + \int_0^t u(\tau)e^{-2(t-\tau)}\,d\tau,$$

where $z_1(0)$, $z_2(0)$, and $z_3(0)$ can be expressed in terms of $x_1(0)$, $x_2(0)$, and $x_3(0)$ using (7.4.13). Hence

$$\mathbf{z}(0) = \mathbf{S}^{-1}\mathbf{x}(0) = \begin{bmatrix} 1 & 1 & \frac{3}{2} \\ 1 & 0 & 1 \\ -1 & 0 & -\frac{1}{2} \end{bmatrix} \begin{bmatrix} x_1(0) \\ x_2(0) \\ x_3(0) \end{bmatrix}$$

or

$$z_1(0) = x_1(0) + x_2(0) + \tfrac{3}{2}x_3(0),$$

$$z_2(0) = x_1(0) + x_3(0), \qquad (7.4.36)$$

$$z_3(0) = -x_1(0) - \tfrac{1}{2}x_3(0).$$

Finally, the solution of the system can be obtained by combining (7.4.5), (7.4.35), and (7.4.36):

$$\mathbf{x} = \mathbf{S}\mathbf{z} = \begin{bmatrix} 0 & -1 & -2 \\ 1 & -2 & 1 \\ 0 & 2 & 2 \end{bmatrix} \begin{bmatrix} z_1 \\ z_2 \\ z_3 \end{bmatrix} = \begin{bmatrix} -z_2 - 2z_3 \\ z_1 - 2z_2 + z_3 \\ 2z_2 + 2z_3 \end{bmatrix}$$

or

$$\begin{bmatrix} x_1(t) \\ x_2(t) \\ x_3(t) \end{bmatrix} = \begin{bmatrix} -e^{-t}\{x_1(0) + x_3(0)\} + 2e^{-2t}\{x_1(0) + \tfrac{1}{2}x_3(0)\} \\ \quad - \int_0^t u(\tau)\{e^{-(t-\tau)} - 2e^{-2(t-\tau)}\}\,d\tau \\ \{x_1(0) + x_2(0) + \tfrac{3}{2}x_3(0)\} - 2e^{-t}\{x_1(0) + x_3(0)\} - e^{-2t} \\ \quad \times\,\{x_1(0) + \tfrac{1}{2}x_3(0)\} + \int_0^t u(\tau)\{1 - 2e^{-(t-\tau)} + e^{-2(t-\tau)}\}\,d\tau \\ 2e^{-t}\{x_1(0) + x_3(0)\} - 2e^{-2t}\{x_1(0) + \tfrac{1}{2}x_3(0)\} + 2\int_0^t u(\tau) \\ \quad \{e^{-(t-\tau)} + e^{-2(t-\tau)}\}\,d\tau \end{bmatrix} \qquad (7.4.37)$$

7.5 SOLUTION OF STATE EQUATIONS THROUGH REDUCTION TO JORDAN FORM

As a generalization of the method used in the preceding section for obtaining the solution of a linear system described by

$$\dot{\mathbf{x}} = \mathbf{A}\mathbf{x} + \mathbf{B}\mathbf{u}$$
$$\mathbf{y} = \mathbf{C}\mathbf{x} + \mathbf{D}\mathbf{u},$$

(7.5.1)

the assumption on the *distinctness* of the eigenvalues of $\mathbf{A}$ is now removed. If the number of eigenvectors is equal to the order of $\mathbf{A}$, the method underlined in the preceding section can still be used to diagonalize the $\mathbf{A}$ matrix. However, if the number of eigenvectors is equal to the number of distinct eigenvalues of $\mathbf{A}$, the set of linearly independent vectors can be obtained by applying (A.1.5.19) in Appendix A.1. In other words, if the distinct eigenvalues of an $n \times n$-matrix $\mathbf{A}$ are denoted by $\lambda_1, \lambda_{p+1}, \lambda_{p+2}, \ldots, \lambda_n$ such that λ_1 is of multiplicity p, then the vectors $\mathbf{u}_1, \mathbf{u}_2, \ldots, \mathbf{u}_n$, which define the matrix $\mathbf{S}$

$$\mathbf{S} \triangleq [\mathbf{u}_1 \, \mathbf{u}_2 \cdots \mathbf{u}_n]$$

(7.5.2)

are given by

$$\mathbf{A}\mathbf{u}_1 = \lambda_1 \mathbf{u}_1$$
$$\mathbf{A}\mathbf{u}_2 = \lambda_1 \mathbf{u}_2 + \mathbf{u}_1$$
$$\mathbf{A}\mathbf{u}_3 = \lambda_1 \mathbf{u}_3 + \mathbf{u}_2$$
$$\vdots$$
$$\mathbf{A}\mathbf{u}_p = \lambda_1 \mathbf{u}_p + \mathbf{u}_{p-1}$$
$$\mathbf{A}\mathbf{u}_{p+1} = \lambda_{p+1} \mathbf{u}_{p+1}$$
$$\vdots$$
$$\mathbf{A}\mathbf{u}_n = \lambda_n \mathbf{u}_n.$$

(7.5.3)

Then, as before, substituting the linear transformation

$$\mathbf{x} = \mathbf{S}\mathbf{z}$$

(7.5.4)

into (7.5.1) and premultiplying the resultant equation by $\mathbf{S}^{-1}$, we find that

$$\dot{\mathbf{z}} = \mathbf{S}^{-1}\mathbf{A}\mathbf{S}\mathbf{z} + \mathbf{S}^{-1}\mathbf{B}\mathbf{u}$$
$$= \mathbf{J}\mathbf{z} + \mathbf{E}\mathbf{u},$$

(7.5.5)

where $\mathbf{J}$ is the Jordan canonical form (A.1.5.22) and $\mathbf{E}$ is taken to be the matrix product $\mathbf{S}^{-1}\mathbf{B}$. The variables z_i are partially decoupled (that is, some of the variables z_i can be determined only from the knowledge of other variables) in this case and once they are determined, the solution of the original system (7.5.1) can be calculated accordingly. Perhaps the above ideas can be best illustrated by the following example.

Example 7.5.1 In Example 7.4.1, the parameters of the feedback control system of Fig. 7.4.1 are given by $R_f = 2$, and $L_f = K_0 = K_1 = K_2 = K_3 = 1$ numerically. Then (7.4.28) and (7.4.29) become, respectively,

$$\begin{bmatrix} \dot{x}_1 \\ \dot{x}_2 \\ \dot{x}_3 \end{bmatrix} = \begin{bmatrix} -2 & 0 & -1 \\ 0 & 0 & 1 \\ 1 & 0 & 0 \end{bmatrix} \begin{bmatrix} x_1 \\ x_2 \\ x_3 \end{bmatrix} + \begin{bmatrix} 1 \\ 0 \\ 0 \end{bmatrix} u \tag{7.5.6}$$

and

$$\mathbf{y} = \begin{bmatrix} 0 & 0 & 1 \end{bmatrix} \begin{bmatrix} x_1 \\ x_2 \\ x_3 \end{bmatrix}. \tag{7.5.7}$$

The matrices $\mathbf{A}$, $\mathbf{B}$, $\mathbf{C}$, and $\mathbf{D}$ are given by

$$\mathbf{A} = \begin{bmatrix} -2 & 0 & -1 \\ 0 & 0 & 1 \\ 1 & 0 & 0 \end{bmatrix}, \quad \mathbf{B} = \begin{bmatrix} 1 \\ 0 \\ 0 \end{bmatrix}, \quad \mathbf{C} = \begin{bmatrix} 0 & 0 & 1 \end{bmatrix}, \quad \mathbf{D} = \begin{bmatrix} 0 \end{bmatrix} \tag{7.5.8}$$

and the characteristic equation of $\mathbf{A}$,

$$g(\lambda) = -\lambda(\lambda + 1)^2 = 0, \tag{7.5.9}$$

yields the distinct eigenvalues $\lambda_1 = -1$ (of multiplicity two) and $\lambda_3 = 0$ (of multiplicity one).

Applying (7.5.3), we find that

$$\mathbf{A}\mathbf{u}_1 = \lambda_1 \mathbf{u}_1$$

or

$$(\mathbf{A} - \lambda_1 \mathbf{I})\mathbf{u}_1 = 0:$$

$$\begin{bmatrix} -1 & 0 & -1 \\ 0 & 1 & 1 \\ 1 & 0 & 1 \end{bmatrix} \begin{bmatrix} u_{11} \\ u_{12} \\ u_{13} \end{bmatrix} = 0$$

which yields

$$\mathbf{u}_1 = u_{13} \begin{bmatrix} -1 \\ -1 \\ 1 \end{bmatrix} = \begin{bmatrix} -1 \\ -1 \\ 1 \end{bmatrix}$$

with $u_{13} = 1$. To determine $\mathbf{u}_2$, we write

$$\mathbf{A}\mathbf{u}_2 = \lambda_1 \mathbf{u}_2 + \mathbf{u}_1$$

or

$$(A - \lambda_1 I)u_2 = u_1:$$

$$\begin{bmatrix} -1 & 0 & -1 \\ 0 & 1 & 1 \\ 1 & 0 & 1 \end{bmatrix} \begin{bmatrix} u_{21} \\ u_{22} \\ u_{23} \end{bmatrix} = \begin{bmatrix} -1 \\ -1 \\ 1 \end{bmatrix},$$

which gives the following linearly independent equations

$$u_{21} + u_{23} = 1$$

$$u_{22} + u_{23} = -1.$$

If, for convenience, we set $u_{23} = 0$, then the vector u_2 becomes

$$u_2 = \begin{bmatrix} 1 \\ -1 \\ 0 \end{bmatrix}.$$

In exactly the same manner, the eigenvector u_3 can be found by solving $Au_3 = \lambda_3 u_3$ or $[A - \lambda_3 I]u_3 = 0$ (with $\lambda_3 = 0$):

$$\begin{bmatrix} -2 & 0 & -1 \\ 0 & 0 & 1 \\ 1 & 0 & 0 \end{bmatrix} \begin{bmatrix} u_{31} \\ u_{32} \\ u_{33} \end{bmatrix} = 0,$$

which yields

$$u_3 = \begin{bmatrix} 0 \\ 1 \\ 0 \end{bmatrix}$$

by setting $u_{32} = 1$. Consequently, the matrix S and its inverse are given by

$$S = [u_1 \quad u_2 \quad u_3] = \begin{bmatrix} -1 & 1 & 0 \\ -1 & -1 & 1 \\ 1 & 0 & 0 \end{bmatrix} \quad \text{and} \quad S^{-1} = \begin{bmatrix} 0 & 0 & 1 \\ 1 & 0 & 1 \\ 1 & 1 & 2 \end{bmatrix}$$

and (7.5.5) becomes

$$\dot{z} = S^{-1}ASz + S^{-1}Bu = Jz + Eu$$

$$= \begin{bmatrix} -1 & 1 & 0 \\ 0 & -1 & 0 \\ 0 & 0 & 0 \end{bmatrix} \begin{bmatrix} z_1 \\ z_2 \\ z_3 \end{bmatrix} + \begin{bmatrix} 0 \\ 1 \\ 1 \end{bmatrix} u. \qquad (7.5.10)$$

In terms of components, (7.5.10) can be expressed as

$$\dot{z}_1 = -z_1 + z_2,$$
$$\dot{z}_2 = -z_2 + u, \qquad (7.5.11)$$
$$\dot{z}_3 = u,$$

with the corresponding solutions given by

$$z_1(t) = e^{-t}z_1(0) + \int_0^t e^{-\tau}z_2(0)\, d\tau + \int_0^t \int_0^\alpha u(\alpha)e^{-(\alpha-\tau)}\, d\tau\, d\alpha$$

$$z_2(t) = e^{-t}z_2(0) + \int_0^t u(\tau)e^{-(t-\tau)}\, d\tau \qquad (7.5.12)$$

$$z_3(t) = z_3(0) + \int_0^t u(\tau)\, d\tau.$$

Remark. In obtaining the solutions in (7.5.12), z_2 and z_3 can be found directly by integrating the corresponding differential equations in (7.5.11). However, to obtain z_1, the solution of z_2 must be substituted in the first equation of (7.5.11) before it can be integrated. This is the usual procedure for solving a set of differential equations that are linked by a Jordan canonical form; that is, the solution of one variable depends on the solution of another variable.

With the variables z_1, z_2, and z_3 determined, the solution of the original system (7.5.6) can be easily obtained by simple algebraic manipulations and, hence, is left as an exercise.

7.6 SOLUTION OF STATE EQUATIONS BY ANALOG COMPUTER SIMULATION

Another method of obtaining the solution of a matrix state equation is by means of simulating the vector differential equation on an analog computer. The number of integrators used in the simulation is equal to the order of the system under consideration. Let us consider the following example.

Example 7.6.1 A third-order system can be described by a vector differential equation of the form

$$\begin{bmatrix} \dot{x}_1 \\ \dot{x}_2 \\ \dot{x}_3 \end{bmatrix} = \begin{bmatrix} a_{11} & a_{12} & a_{13} \\ a_{21} & a_{22} & a_{23} \\ a_{31} & a_{32} & a_{33} \end{bmatrix} \begin{bmatrix} x_1 \\ x_2 \\ x_3 \end{bmatrix} + \begin{bmatrix} b_{11} \\ b_{21} \\ b_{31} \end{bmatrix} u, \qquad (7.6.1)$$

where, for convenience, the elements a_{ij} and b_{i1} are assumed to satisfy the following

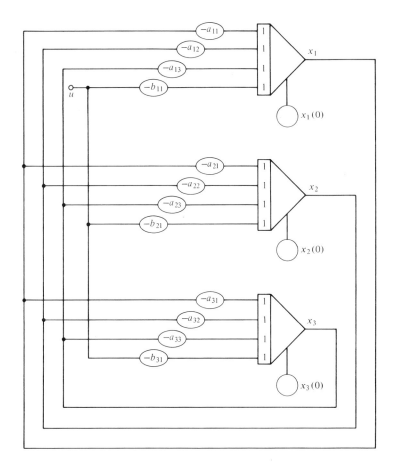

Fig. 7.6.1 Analog computer simulation of system (7.6.1).

conditions* for $i, j = 1, 2, 3$:

a) both a_{ij} and b_{i1} are negative,

b) $|a_{ij}| \leq 1$,
    $|b_{i1}| \leq 1$.

Figure 7.6.1 shows an analog computer simulation for the solution of (7.6.1).

It should be pointed out that with slight modifications, such as the use of summing devices and the adjustment of the gains of amplifiers, any nth-order

_____

* The conditions are not necessary. However, they are introduced here to simplify the simulation diagram so that better understanding can be achieved.

system of the form

$$\dot{\mathbf{x}} = \mathbf{A}\mathbf{x} + \mathbf{B}\mathbf{u}$$

$$\mathbf{y} = \mathbf{C}\mathbf{x} + \mathbf{D}\mathbf{u},$$

(7.6.2)

can be similarly simulated on an analog computer. Of course, the solution obtained by this method will be in graphical form.

7.7 SIGNAL FLOW GRAPHS

Several approaches have been introduced for determining the solution of the state equation. The analytical method is very efficient in the sense that the calculations of both the eigenvalues and the transformation matrix can be carried out by a digital computer. However, this method fails to provide a physical picture of the structure of a given system since it is difficult to visualize from its matrix representation how the state variables are affected by different signals applied to the system. If the output of each integrator in the analog computer simulation of a given system is taken as a state variable, then one can acquire a "feeling" about the relationships between the state variables inside the system. Unfortunately, however, no direct method can be utilized to determine the solution of the state equations other than through the use of an analog computer.

As will be discussed in detail in the following section, a technique using a so-called state-transition flow graph not only provides a clear picture on the structure of the system, but also gives a powerful means for solving the state equation. This method is based on the technique of signal flow graphs first introduced by Mason* as a topological method of representing and solving a set of simultaneous linear algebraic equations. In this section we shall give a brief discussion on this subject. The modification of signal flow graphs to state-transition flow graphs will be taken up in the next section.

Consider the simple linear equation

$$x_i = t_{ij}x_j,$$

(7.7.1)

where x_i and x_j are two variables and t_{ij} is a (constant) coefficient relating x_i and x_j. Topologically, we may represent this equation by a graph consisting of two nodes x_i and x_j and a branch t_{ij} as shown in Fig. 7.7.1. The *signal* x_j travels along branch t_{ij} (with a *transmittance* (or *gain*) also denoted by t_{ij}), from (the source) node x_j to (the sink) node x_i. Thus the graph of Fig. 7.7.1 is called a *signal flow graph* or a *Mason flow graph* which is a topological representation of (7.7.1).

Before we proceed to underline the necessary procedure for obtaining the signal flow graph of a general linear system of equations, the following simple example will clearly demonstrate the step-by-step approach to derive the signal flow graph from a given system of equations.

---

* Refer to S. J. Mason, [MA 2].

x_j t_{ij} x_i **Fig. 7.7.1** The signal flow graph of (7.7.1).

Example 7.7.1 Let the behavior of a physical system be described by the following set of simultaneous equations

$$a_{11}x_1 + a_{12}x_2 + a_{13}x_3 = b_{11}u,$$
$$a_{21}x_1 + a_{22}x_2 + a_{23}x_3 = b_{21}u, \qquad (7.7.2a)$$
$$a_{31}x_1 + a_{32}x_2 + a_{33}x_3 = b_{31}u,$$

or, in matrix notation,

$$\mathbf{Ax = Bu} \qquad (7.7.2b)$$

Our objective is to solve (7.7.2) for the variables x_1, x_2, and x_3 in terms of the forcing function u. Instead of computing $\mathbf{A}^{-1}$ and solving for $\mathbf{x}$ in the usual manner, we shall rewrite (7.7.2) in the form

$$\mathbf{0 = Ax - Bu}$$

and then add $\mathbf{x}$ to both sides of the equations yielding

$$\mathbf{x = Ax - Bu + x = (A + I)x - Bu}. \qquad (7.7.3)$$

In terms of components, we find

$$x_1 = (a_{11} + 1)x_1 + a_{12}x_2 + a_{13}x_3 - b_{11}u \triangleq t_{11}x_1 + t_{12}x_2 + t_{13}x_3 + t_{14}x_4$$
$$x_2 = a_{21}x_1 + (a_{22} + 1)x_2 + a_{23}x_3 - b_{21}u \triangleq t_{21}x_1 + t_{22}x_2 + t_{23}x_3 + t_{24}x_4$$
$$x_3 = a_{31}x_1 + a_{32}x_2 + (a_{33} + 1)x_3 - b_{31}u \triangleq t_{31}x_1 + t_{32}x_2 + t_{33}x_3 + t_{34}x_4$$

$$(7.7.4)$$

which can be written in matrix form as

$$\begin{bmatrix} x_1 \\ x_2 \\ x_3 \end{bmatrix} = \begin{bmatrix} t_{11} & t_{12} & t_{13} & t_{14} \\ t_{21} & t_{22} & t_{23} & t_{24} \\ t_{31} & t_{32} & t_{33} & t_{34} \end{bmatrix} \begin{bmatrix} x_1 \\ x_2 \\ x_3 \\ x_4 \end{bmatrix}, \qquad (7.7.5)$$

where $x_4 \triangleq u$ and t_{ij} is the *transmittance* of branch t_{ij} in the signal flow graph denoting the contribution of x_j to x_i. Comparing the coefficients of (7.7.4), we find that

$$t_{11} = a_{11} + 1, \qquad t_{12} = a_{12}, \qquad t_{13} = a_{13}, \qquad t_{14} = -b_{11},$$
$$t_{21} = a_{21}, \qquad t_{22} = a_{22} + 1, \qquad t_{23} = a_{23}, \qquad t_{24} = -b_{21}, \qquad (7.7.6)$$
$$t_{31} = a_{31}, \qquad t_{32} = a_{32}, \qquad t_{33} = a_{33} + 1, \qquad t_{34} = -b_{31}.$$

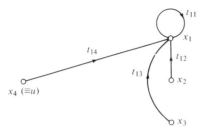

Fig. 7.7.2 Signal flow graph for
$x_1 = t_{11}x_1 + t_{12}x_2 + t_{13}x_3 + t_{14}x_4.$

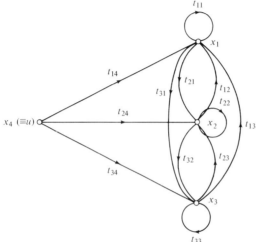

Fig. 7.7.3 Signal flow graph for (7.7.4).

Let us consider the first equation of (7.7.4). Mathematically, the variable x_1 is simply the sum of the variables x_1, x_2, x_3, and x_4, multiplied by the corresponding transmittances t_{11}, t_{12}, t_{13}, and t_{14}. Thus, this equation can be expressed topologically by the signal flow graph as shown in Fig. 7.7.2. In exactly the same manner, the three equations of (7.7.4) can be represented topologically in the form of a flow graph as shown in Fig. 7.7.3.

An inspection of Fig. 7.7.3 indicates that a signal flow graph of a given linear system is simply a *weighted* (with transmittances) *directed graph* satisfying the following rules:

1. The nodes of the signal flow graph represent the variables of the system.

2. A signal traveling along a branch is multiplied by the transmittance of that branch.

3. Signals can travel along branches only in the direction of arrows.

4. The value of the variable represented by any node is the sum of all signals directed toward that node.

5. The value of the variable represented by any node is transmitted to all branches directed away from that node.

It should be emphasized that *the rules stated above provide the basis for the construction of any signal flow graph.*

Now, let us demonstrate that the same procedure can be applied for the construction of the signal flow graph for a linear system whose behavior is described by

$$a_{11}x_1 + a_{12}x_2 + \cdots + a_{1n}x_n = b_{11}u$$

$$a_{21}x_1 + a_{22}x_2 + \cdots + a_{2n}x_n = b_{21}u$$

$$\vdots \qquad\qquad\qquad\qquad\qquad\qquad (7.7.7)$$

$$a_{n1}x_1 + a_{n2}x_2 + \cdots + a_{nn}x_n = b_{n1}u$$

or, in matrix form,

$$\mathbf{Ax} = \mathbf{B}u. \qquad (7.7.8)$$

Next, in (7.7.8), $\mathbf{B}u$ is first transposed and then the vector $\mathbf{x}$ is added to both sides of the resulting equation yielding

$$\mathbf{x} = (\mathbf{A} + \mathbf{I})\mathbf{x} - \mathbf{B}u = [(\mathbf{A} + \mathbf{I}) \quad\quad -\mathbf{B}] \begin{bmatrix} \mathbf{x} \\ u \end{bmatrix}. \qquad (7.7.9)$$

If we define

$$x_{n+1} \triangleq u \qquad \text{and} \qquad \mathbf{T} \triangleq [(\mathbf{A} + \mathbf{I}) \quad\quad -\mathbf{B}], \qquad (7.7.10)$$

then (7.7.9) becomes

$$\mathbf{x} = \mathbf{T} \begin{bmatrix} \mathbf{x} \\ x_{n+1} \end{bmatrix}. \qquad (7.7.11)$$

In terms of components, (7.7.11) becomes

$$\begin{bmatrix} x_1 \\ x_2 \\ \vdots \\ x_n \end{bmatrix} = \begin{bmatrix} t_{11} & t_{12}\cdots t_{1n} & t_{1(n+1)} \\ t_{21} & t_{22}\cdots t_{2n} & t_{2(n+1)} \\ \vdots & & \\ t_{n1} & t_{n2}\cdots t_{nn} & t_{n(n+1)} \end{bmatrix} \begin{bmatrix} x_1 \\ x_2 \\ \vdots \\ x_n \\ x_{n+1} \end{bmatrix}, \qquad (7.7.12)$$

from which the signal flow graph similar to Fig. 7.7.3 can be easily constructed.

In a signal flow graph, a node with *only outgoing branches* is referred to as an *input node* or a *source*, whereas a node with *only incoming branches* is called an *output node* or a *sink*. A loop from a node to itself, such as the three loops with transmittances t_{11}, t_{22}, and t_{33} in Fig. 7.7.3, is called a *self-loop*.

Before we proceed to determine the relationships between the input and the output variables, it is often desirable to reduce the given signal flow graph to a simpler form so that the subsequent algebraic manipulations can be greatly

Fig. 7.7.4 Combination of parallel branches.

Fig. 7.7.5 Combination of series branches.

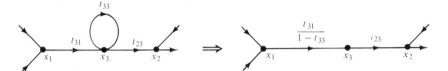

Fig. 7.7.6 Elimination of self-loop.

simplified. *Three basic reduction rules* for the simplification of signal flow graphs can be stated as follows:

Rule 1. Two *parallel paths* may be combined into a single path with a transmittance equal to the *sum* of the original transmittances. This rule is depicted in Fig. 7.7.4.

Rule 2. Two paths connected in *series* may be combined into a single path with a transmittance equal to the *product* of original transmittances as shown in Fig. 7.7.5.

Rule 3. A self-loop at x_j with transmittance $t_{jj} \neq 1$ can be removed by dividing the transmittances of *all incoming* branches at x_j by $(1 - t_{jj})$ as shown in Fig. 7.7.6.

Once the signal flow graph is determined, the final objective in the analysis is to obtain the *overall transmission* of the signal flow graph or simply the *graph gain G*, which is defined as the ratio of the output to the input variables.

The graph gain G can be obtained directly by means of the *Mason gain formula* which we shall state here without proof. The overall transmission G of a given signal flow graph is given by

$$G = \frac{1}{\Delta} \sum_k G_k \Delta_k, \tag{7.7.13}$$

where

$$\Delta = 1 - \sum_m P_{m1} + \sum_m P_{m2} - \sum_m P_{m3} + \cdots + (-1)^j \sum_m P_{mj}; \tag{7.7.14}$$

G_k = the gain of the kth *forward path* or *direct path* (i.e., the product of all the branch transmittances in the path) from the input to the output;

P_{m1} = the loop gain of the mth *feedback loop* (or *closed path*), i.e., the product of all the transmittances of the mth feedback loop;

P_{mi} = the product of the loop gains of the mth set of i *nontouching* feedback loops (i.e., feedback loops without any node or branch in common); or the product of the loop gains of the mth set of loops of *order i*;

Δ_k = value of Δ for that part of the graph not touching the kth forward path;

j = the highest order among all sets of loops in the given signal flow graph.

Let us demonstrate the ideas developed above by means of the examples that follow.

Example 7.7.2 Consider again the closed-loop speed control system of Example 7.4.1. The equations governing the performance of the system are repeated here for convenience

$$e_i = u - v, \qquad e_o = K_1 e_i, \qquad e_o = R_f i_f + L_f \dot{i}_f$$

$$T_m = K i_f, \qquad T_m = J_L \dot{\omega}_m, \qquad v = K_2 \omega_m.$$

Our problem is to determine the transfer function $H(s)$ relating the output speed $\Omega_m(s)$ to the input voltage $U(s)$ by the signal flow graph technique. Taking the Laplace transforms of the above equations with initial conditions set to zero, we obtain the corresponding set of transformed equations

$$E_i(s) = U(s) - V(s), \qquad E_o(s) = K_1 E_i(s),$$

$$I_f(s) = \frac{E_o(s)}{sL_f + R_f}, \qquad T_m(s) = K I_f(s),$$

$$\Omega_m(s) = \frac{T_m(s)}{J_L s}, \qquad V(s) = K_2 \Omega_m(s),$$

which, in matrix notation, is

$$
\begin{bmatrix} E_i(s) \\ E_o(s) \\ I_f(s) \\ T_m(s) \\ \Omega_m(s) \\ V(s) \end{bmatrix}
=
\begin{bmatrix}
0 & 0 & 0 & 0 & 0 & -1 & 1 \\
K_1 & 0 & 0 & 0 & 0 & 0 & 0 \\
0 & \dfrac{1}{sL_f + R_f} & 0 & 0 & 0 & 0 & 0 \\
0 & 0 & K & 0 & 0 & 0 & 0 \\
0 & 0 & 0 & \dfrac{1}{J_L s} & 0 & 0 & 0 \\
0 & 0 & 0 & 0 & K_2 & 0 & 0
\end{bmatrix}
\begin{bmatrix} E_i(s) \\ E_o(s) \\ I_f(s) \\ T_m(s) \\ \Omega_m(s) \\ V(s) \\ U(s) \end{bmatrix}
$$

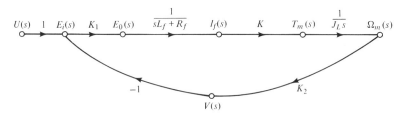

Fig. 7.7.7 Signal flow graph for the closed-loop control system of Example 7.4.1.

The corresponding signal flow graph is constructed in Fig. 7.7.7.

In Fig. 7.7.7, there is only one forward path from $U(s)$ to $\Omega_m(s)$ with the gain

$$G_1 = \frac{K_1 K}{(sL_f + R_f)sJ_L},$$

and only one feedback loop with the loop gain P_{11} given by

$$P_{11} = -K_2 G_1 = \frac{-K_2 K_1 K}{(sL_f + R_f)sJ_L}.$$

Hence, Δ and Δ_1 become, respectively,

$$\Delta = 1 - P_{11} = 1 + \frac{K_2 K_1 K}{(sL_f + R_f)sJ_L}$$

and

$$\Delta_1 = 1.$$

Finally, the transfer function $H(s)$ is found to be

$$H(s) = \frac{\Omega(s)}{U(s)} = \frac{1}{\Delta}(G_1\Delta_1) = \frac{\dfrac{K_1 K}{(sL_f + R_f)sJ_L}}{1 + \dfrac{K_2 K_1 K}{(sL_f + R_f)sJ_L}}$$

or

$$H(s) = \frac{K_1 K}{s^2 L_f J_L + sR_f J_L + K_2 K_1 K}.$$

Example 7.7.3 A motor-generator control is depicted in Fig. 7.7.8, in which a motor with a constant shunt field excitation is to drive a load with moment of inertia J_L and damping coefficient B. The generator, driven at constant speed by a prime mover, supplies power to the motor which, in turn, establishes a torque that drives the load.

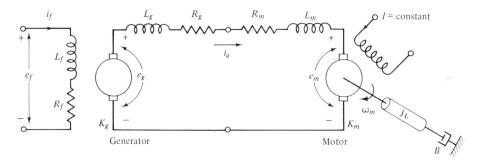

Fig. 7.7.8 A motor-generator control system.

The equations governing the performance of this system are

$$e_f = L_f i_f + R_f i_f, \qquad e_g = K_g i_f,$$

$$e_g = (L_g + L_m)\dot{i}_a + (R_g + R_m)i_a + e_m, \qquad (7.7.15)$$

$$e_m = K_m \omega_m, \qquad T_m = K i_a, \qquad T_m = J_L \dot{\omega}_m + B\omega_m.$$

Let us apply the Mason formula to determine the transfer function $H(s)$ defined by

$$H(s) = \frac{\Omega_m(s)}{E_f(s)}. \qquad (7.7.16)$$

As before, after taking the Laplace transform, the above set of equations can be rearranged as follows:

$$I_f(s) = \frac{E_f(s)}{sL_f + R_f}, \qquad E_g(s) = K_g I_f(s),$$

$$I_a(s) = \frac{E_g(s)}{(L_g + L_m)s + (R_g + R_m)} - \frac{E_m(s)}{(L_g + L_m)s + (R_g + R_m)} \qquad (7.7.17)$$

$$T_m(s) = K I_a(s), \qquad \Omega_m(s) = \frac{T_m(s)}{sJ_L + B}, \qquad E_m(s) = K_m \Omega_m(s).$$

The signal flow graph which corresponds to the above equations is illustrated in Fig. 7.7.9. From Fig. 7.7.9, we find that

$$G_1 = \frac{K_g K}{(sL_f + R_f)[(L_g + L_m)s + (R_g + R_m)](sJ_L + B)},$$

$$P_{11} = \frac{-KK_m}{(sJ_L + B)[(L_g + L_m)s + (R_g + R_m)]},$$

$$\Delta = 1 - P_{11} = 1 + \frac{KK_m}{(sJ_L + B)[(L_g + L_m)s + (R_g + R_m)]},$$

$$\Delta_1 = 1.$$

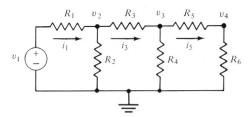

Fig. 7.7.9 Signal flow graph of the motor-generator system in Fig. 7.7.8.

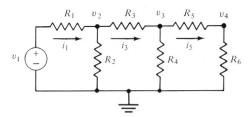

Fig. 7.7.10 A ladder network.

Therefore, by means of Mason's formula, we obtain

$$H(s) = \frac{1}{\Delta}(G_{11}\Delta_1) = \frac{K_g K}{(sL_f + R_f)\{(sJ_L + B)[(L_g + L_m)s + (R_g + R_m)] + KK_m\}}.$$

Example 7.7.4 Consider the ladder network shown in Fig. 7.7.10. The equations corresponding to this network are given by

$$i_1 = \frac{1}{R_1}(v_1 - v_2), \qquad v_2 = R_2(i_1 - i_3),$$

$$i_3 = \frac{1}{R_3}(v_2 - v_3), \qquad v_3 = R_4(i_3 - i_5), \qquad (7.7.18)$$

$$i_5 = \frac{1}{R_5}(v_3 - v_4), \qquad v_4 = R_6 i_5.$$

The signal flow graph for (7.7.18) is shown in Fig. 7.7.11. The voltage ratio v_4/v_1 can be determined by first finding the following:

$$G_1 = \frac{R_2 R_4 R_6}{R_1 R_3 R_5};$$

$$P_{11} = -\frac{R_2}{R_1}, \qquad P_{21} = -\frac{R_2}{R_3}, \qquad P_{31} = -\frac{R_4}{R_3},$$

$$P_{41} = -\frac{R_4}{R_5}, \qquad P_{51} = -\frac{R_6}{R_5};$$

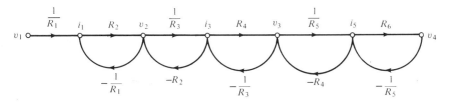

Fig. 7.7.11 Signal flow graph of the network in Fig. 7.7.10.

$$P_{12} = \frac{R_2 R_4}{R_1 R_3}, \qquad P_{22} = \frac{R_2 R_4}{R_1 R_5}, \qquad P_{32} = \frac{R_2 R_6}{R_1 R_5},$$

$$P_{42} = \frac{R_2 R_4}{R_3 R_5}, \qquad P_{52} = \frac{R_2 R_6}{R_3 R_5}, \qquad P_{62} = \frac{R_4 R_6}{R_3 R_5};$$

$$P_{13} = \frac{-R_2 R_4 R_6}{R_1 R_3 R_5}.$$

Therefore, we obtain

$$\Delta = 1 - \sum_{m=1}^{5} P_{m1} + \sum_{m=1}^{6} P_{m2} - P_{13}$$

$$= 1 + \frac{R_2}{R_1} + \frac{R_2}{R_3} + \frac{R_4}{R_3} + \frac{R_4}{R_5} + \frac{R_6}{R_5} + \frac{R_2 R_4}{R_1 R_3} + \frac{R_2 R_4}{R_1 R_5} + \frac{R_2 R_6}{R_1 R_5} + \frac{R_2 R_4}{R_3 R_5}$$

$$+ \frac{R_2 R_6}{R_3 R_5} + \frac{R_4 R_6}{R_3 R_5} + \frac{R_2 R_4 R_6}{R_1 R_3 R_5}$$

and $\Delta_1 = 1$. Finally, the voltage ratio is given by

$$\frac{v_4}{v_1} = \frac{1}{\Delta}(G_1 \Delta_1)$$

$$= \frac{\dfrac{R_2 R_4 R_6}{R_1 R_3 R_5}}{1 + \dfrac{R_2}{R_1} + \dfrac{R_2}{R_3} + \dfrac{R_4}{R_3} + \dfrac{R_4}{R_5} + \dfrac{R_6}{R_5} + \dfrac{R_2 R_4}{R_1 R_3} + \dfrac{R_2 R_4}{R_1 R_5} + \dfrac{R_2 R_6}{R_1 R_5} + \dfrac{R_2 R_4}{R_3 R_5} + \dfrac{R_2 R_6}{R_3 R_5} + \dfrac{R_4 R_6}{R_3 R_5} + \dfrac{R_2 R_4 R_6}{R_1 R_3 R_5}}$$

7.8 STATE-TRANSITION FLOW GRAPHS

In this section we shall discuss another approach for obtaining the solution of the state equation. This method of solution makes use of the properties of signal flow graphs (the Mason graphs).

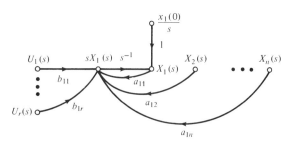

Fig. 7.8.1 Analog signal flow graph representing (7.8.3).

Suppose that we draw a signal flow graph corresponding to the system

$$\dot{x}_1 = a_{11}x_1 + a_{12}x_2 + \cdots + a_{1n}x_n + b_{11}u_1 + b_{12}u_2 + \cdots + b_{1r}u_r,$$

$$\dot{x}_2 = a_{21}x_1 + a_{22}x_2 + \cdots + a_{2n}x_n + b_{21}u_1 + b_{22}u_2 + \cdots + b_{2r}u_r, \quad (7.8.1)$$

$$\vdots$$

$$\dot{x}_n = a_{n1}x_1 + a_{n2}x_2 + \cdots + a_{nn}x_n + b_{n1}u_1 + b_{n2}u_2 + \cdots + b_{nr}u_r,$$

which, in matrix form, becomes

$$\dot{\mathbf{x}} = \mathbf{A}\mathbf{x} + \mathbf{B}\mathbf{u}, \quad (7.8.2)$$

such that the initial conditions $x_1(0)$, $x_2(0)$, ..., $x_n(0)$ are taken as inputs applied at the corresponding state variables; then the signal flow graph is referred to as the *state-transition flow graph*. Thus, for example, the subgraph of the state-transition flow graph corresponding to the first equation of (7.8.1), or

$$\dot{x}_1 = a_{11}x_1 + a_{12}x_2 + \cdots + a_{1n}x_n + b_{11}u_1 + \cdots + b_{1r}u_r, \quad (7.8.3)$$

is illustrated in Fig. 7.8.1, since the transformed equation of (7.8.3) can be arranged as

$$X_1(s) = \frac{x_1(0)}{s} + \frac{1}{s}[a_{11}X_1(s) + \cdots + a_{1n}X_n(s) + b_{11}U_1(s) + \cdots + b_{1r}U_r(s)].$$

Combining all the subgraphs for the n equations (similar to that of Fig. 7.8.1) into one graph, the state-transition flow graph is constructed.

After the state-transition flow graph for a given system is constructed, Mason's formula can be applied to determine the transformed variables $X_1(s)$, $X_2(s)$, ..., $X_n(s)$. Taking the inverse transform of these variables yields the solution of the problem. The following example will illustrate the steps involved in this approach.

Example 7.8.1 For the purpose of comparison, let us consider again the system in Example 7.4.1 with the equations of motion repeated here for convenience:

$$\begin{bmatrix} \dot{x}_1 \\ \dot{x}_2 \\ \dot{x}_3 \end{bmatrix} = \begin{bmatrix} -3 & 0 & -1 \\ 0 & 0 & 1 \\ 2 & 0 & 0 \end{bmatrix} \begin{bmatrix} x_1 \\ x_2 \\ x_3 \end{bmatrix} + \begin{bmatrix} 1 \\ 0 \\ 0 \end{bmatrix} u \quad (7.8.4)$$

7.8

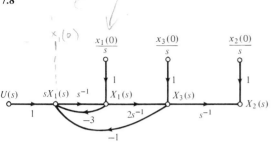

Fig. 7.8.2 State-transition flow graph for system (7.8.4) or (7.8.6).

and

$$y = x_3. \tag{7.8.5}$$

In terms of components, (7.8.4) becomes

$$\dot{x}_1 = -3x_1 - x_3 + u, \qquad \dot{x}_2 = x_3, \qquad \dot{x}_3 = 2x_1. \tag{7.8.6}$$

The state-transition flow graph corresponding to (7.8.6) is constructed in Fig. 7.8.2.

The next step is to write the expressions for the transformed state variables $X_1(s)$, $X_2(s)$, and $X_3(s)$. This step can be simplified considerably with the aid of Mason's formula. Thus, the determinant of the graph in Fig. 7.8.2 is

$$\Delta = 1 - (-3)(s^{-1}) - (s^{-1})(2s^{-1})(-1) = 1 + 3s^{-1} + 2s^{-2}. \tag{7.8.7}$$

To determine the expression for $X_1(s)$, we can consider the quantities $U(s)$, $[x_1(0)]/s$, and $[x_3(0)]/s$ as separate inputs to the system and the variable $X_1(s)$ as the output. The corresponding gains can be calculated separately by invoking the principle of superposition. Hence, we find that

$$X_1(s) = \frac{s^{-1}}{1 + 3s^{-1} + 2s^{-2}} U(s) + \frac{1}{1 + 3s^{-1} + {}^2s^{-2}} \left(\frac{x_1(0)}{s}\right)$$

$$+ \frac{-s^{-1}}{1 + 3s^{-1} + 2s^{-2}} \left(\frac{x_3(0)}{s}\right)$$

$$= \frac{sU(s)}{(s+1)(s+2)} + \frac{sx_1(0)}{(s+1)(s+2)} - \frac{x_3(0)}{(s+1)(s+2)}. \tag{7.8.8}$$

Similarly, the expressions for $X_2(s)$ and $X_3(s)$ are readily found to be

$$X_2(s) = \frac{2s^{-3}}{1 + 3s^{-1} + 2s^{-2}} U(s) + \frac{2s^{-2}}{1 + 3s^{-1} + 2s^{-2}} \left(\frac{x_1(0)}{s}\right)$$

$$+ \frac{1 + 3s^{-1} + 2s^{-2}}{1 + 3s^{-1} + 2s^{-2}} \left(\frac{x_2(0)}{s}\right) + \frac{s^{-1}(1 + 3s^{-1})}{1 + 3s^{-1} + 2s^{-2}} \left(\frac{x_3(0)}{s}\right) \tag{7.8.9}$$

$$= \frac{2U(s)}{s(s+1)(s+2)} + \frac{2x_1(0)}{s(s+1)(s+2)} + \frac{x_2(0)}{s} + \frac{(s+3)x_3(0)}{s(s+1)(s+2)}$$

and

$$X_3(s) = \frac{2s^{-2}}{1 + 3s^{-1} + 2s^{-2}} U(s) + \frac{2s^{-1}}{1 + 3s^{-1} + 2s^{-2}} \left(\frac{x_1(0)}{s}\right)$$

$$+ \frac{(1 + 3s^{-1})}{1 + 3s^{-1} + s^{-2}} \left(\frac{x_3(0)}{s}\right)$$

$$= \frac{2U(s)}{(s+1)(s+2)} + \frac{2x_1(0)}{(s+1)(s+2)} + \frac{(s+3)x_3(0)}{(s+1)(s+2)}. \qquad (7.8.10)$$

Expressed in partial fractions, (7.8.8), (7.8.9), and (7.8.10) become, respectively,

$$X_1(s) = -\frac{\{x_1(0) + x_3(0)\}}{s+1} + \frac{2\{x_1(0) + \frac{1}{2}x_3(0)\}}{s+2} - \left(\frac{1}{s+1} - \frac{2}{s+2}\right)U(s), \quad (7.8.11)$$

$$X_2(s) = \frac{1}{s}\{x_1(0) + x_2(0)\} - 2\frac{\{x_1(0) + x_3(0)\}}{s+1} + \frac{\{x_1(0) + \frac{1}{2}x_3(0)\}}{s+2}$$

$$+ \left(\frac{1}{s} - \frac{2}{s+1} + \frac{1}{s+2}\right)U(s), \qquad (7.8.12)$$

$$X_3(s) = \frac{2\{x_1(0) + x_3(0)\}}{s+1} - \frac{2}{s+2}\{x_1(0) + \frac{1}{2}x(0)\} + 2\left(\frac{1}{s+1} - \frac{1}{s+2}\right)U(s). \qquad (7.8.13)$$

Finally, taking the inverse Laplace transform of (7.8.11), (7.8.12), and (7.8.13), we obtain the solution of the problem, which, in this case, is identical to (7.4.37).

We are now in a position to summarize the steps in the procedure for solving the matrix state equation through a state-transition flow graph.

Step 1. Construct the state-transition flow graph.

Step 2. Apply Mason's formula to determine the transformed state variables.

Step 3. Simplify the expressions obtained in Step 2 by performing partial fraction expansions.

Step 4. Take the inverse Laplace transform of the expressions in Step 3 to obtain the solution of the problem.

7.9 STATE SPACE REPRESENTATION OF SYSTEMS

Consider the mass-spring-damper system of Fig. 7.9.1. If we denote the mass, the elastance (stiffness), and the damping (viscous friction) coefficient, respectively, by M, K, and B, then the equation of motion is described by

$$M\frac{d^2x}{dt^2} + B\frac{dx}{dt} + Kx = f(t), \qquad (7.9.1)$$

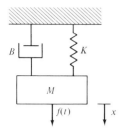

Fig. 7.9.1 Simple mass-spring-damper system.

where x denotes the displacement of the mass from the point of equilibrium and $f(t)$ is the applied force.

To determine the exact behavior of this system at any time t, we need to know both the displacement and the velocity of the mass; that is, $x(t)$ and $[dx(t)]/dt$. Thus, it is quite natural to define, for convenience,

$$x_1 = x, \qquad x_2 = \frac{dx}{dt}, \tag{7.9.2}$$

and to plot the solution curve, viewed in the x_1–x_2 plane with t as a parameter. This curve is called the *trajectory* of the system and the x_1–x_2 plane is referred to as the *state plane*. Thus, combining (7.9.1) with (7.9.2), the system of Fig. 7.9.1 can be described by a set of first-order differential equations:

$$\dot{x}_1 = x_2, \qquad \dot{x}_2 = -\frac{K}{M}x_1 - \frac{B}{M}x_2 + \frac{1}{M}f(t),$$

the solution of which, when plotted on the x_1–x_2 plane, describes the exact behavior of the system.

In practice, the engineer is usually faced with systems which are described either by differential equations of orders higher than two or by transfer functions of various complexity. Let us consider the following cases.

Case 1. The nth-order differential equation with only one forcing term

$$\frac{d^n x}{dt^n} + a_{n-1}\frac{d^{n-1}x}{dt^{n-1}} \mid \quad \mid a_1 \frac{dx}{dt} + a_0 x = bu. \tag{7.9.3}$$

If a system is described by (7.9.3), its state-space representation can be obtained by defining

$$x_1 = x$$

$$x_2 = \dot{x}_1$$

$$x_3 = \dot{x}_2 \tag{7.9.4}$$

$$\vdots$$

$$x_n = \dot{x}_{n-1}.$$

Then (7.9.3) is equivalent to

$$\dot{x}_1 = x_2$$
$$\dot{x}_2 = x_3$$
$$\vdots$$
$$\dot{x}_n = -a_0 x_1 - \cdots - a_{n-1} x_n + bu$$

(7.9.5)

which can be written in matrix form as

$$\dot{\mathbf{x}} = \mathbf{A}\mathbf{x} + \mathbf{B}\mathbf{u},$$

(7.9.6)

where

$$\mathbf{x} = \begin{bmatrix} x_1 \\ x_2 \\ \vdots \\ x_n \end{bmatrix}, \quad \mathbf{A} = \begin{bmatrix} 0 & 1 & 0 & \cdots & 0 \\ 0 & 0 & 1 & \cdots & 0 \\ \vdots & & & & \\ 0 & 0 & 0 & \cdots & 1 \\ -a_0 & -a_1 & -a_2 & \cdots & -a_{n-1} \end{bmatrix}, \quad \mathbf{B} = \begin{bmatrix} 0 \\ 0 \\ \vdots \\ 0 \\ b \end{bmatrix}, \quad \mathbf{u} = [u].$$

Example 7.9.1 The third-order differential equation

$$\frac{d^3 x}{dt^3} + 3\frac{d^2 x}{dt^2} + \frac{dx}{dt} + 2x = 3u$$

(7.9.7)

with the initial conditions

$$x(0) = 1, \quad \frac{dx(0)}{dt} = 2, \quad \text{and} \quad \frac{d^2 x(0)}{dt^2} = 3$$

can be made equivalent to a state-space representation as follows:
Let

$$x_1 = x, \quad x_2 = \frac{dx}{dt}, \quad x_3 = \frac{d^2 x}{dt^2},$$

then (7.9.7) can be represented by

$$\begin{bmatrix} \dot{x}_1 \\ \dot{x}_2 \\ \dot{x}_3 \end{bmatrix} = \begin{bmatrix} 0 & 1 & 0 \\ 0 & 0 & 1 \\ -2 & -1 & -3 \end{bmatrix} \begin{bmatrix} x_1 \\ x_2 \\ x_3 \end{bmatrix} + \begin{bmatrix} 0 \\ 0 \\ 3 \end{bmatrix} u, \quad \begin{bmatrix} x_1(0) \\ x_2(0) \\ x_3(0) \end{bmatrix} = \begin{bmatrix} 1 \\ 2 \\ 3 \end{bmatrix}.$$

(7.9.8)

Case 2. The nth-order differential equation involving derivatives of the forcing function u

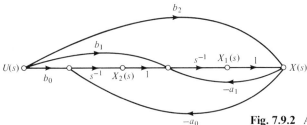

Fig. 7.9.2 A signal flow graph of (7.9.13).

Before we proceed to underline the necessary steps for the determination of a state-space representation of a system whose behavior is described by

$$\frac{d^n x}{dt^n} + a_{n-1}\frac{d^{n-1}x}{dt^{n-1}} + a_{n-2}\frac{d^{n-2}x}{dt^{n-2}} + \cdots + a_1\frac{dx}{dt} + a_0 x$$

$$= b_n\frac{d^n u}{dt^n} + b_{n-1}\frac{d^{n-1}u}{dt^{n-1}} + b_{n-2}\frac{d^{n-2}u}{dt^{n-2}} + \cdots + b_1\frac{du}{dt} + b_0 u, \qquad (7.9.9)$$

we shall first demonstrate the method by applying it to a simple special case of (7.9.9) using the following example.

Example 7.9.2 Consider the following system:

$$\frac{d^2 x}{dt^2} + a_1\frac{dx}{dt} + a_0 x = b_2\frac{d^2 u}{dt^2} + b_1\frac{du}{dt} + b_0 u. \qquad (7.9.10)$$

Taking the Laplace transform of (7.9.10) with initial conditions deleted (i.e., being set to zero), we find that

$$(s^2 + a_1 s + a_0)X(s) = (b_2 s^2 + b_1 s + b_0)U(s) \qquad (7.9.11)$$

which can be rearranged by collecting the terms with like powers of s so that the term with the highest power of s appears on the left-hand side:

$$s^2[X(s) - b_2 U(s)] = s[b_1 U(s) - a_1 X(s)] + [b_0 U(s) - a_0 X(s)]. \qquad (7.9.12)$$

Multiplying (7.9.12) by s^{-2} and solving for $X(s)$, we find

$$X(s) = b_2 U(s) + \frac{1}{s}[b_1 U(s) - a_1 X(s)] + \frac{1}{s^2}[b_0 U(s) - a_0 X(s)]. \qquad (7.9.13)$$

A signal flow graph of (7.9.13) is constructed in Fig. 7.9.2. If the sink nodes of all the branches with transmittance s^{-1} are designated as $X_1(s)$ and $X_2(s)$ (the state variables) as indicated in Fig. 7.9.2, the set of relations can be obtained by inspection:

$$X(s) = X_1(s) + b_2 U(s)$$
$$X_1(s) = s^{-1}[X_2(s) + b_1 U(s) - a_1 X(s)] \qquad (7.9.14)$$
$$X_2(s) = s^{-1}[b_0 U(s) - a_0 X(s)].$$

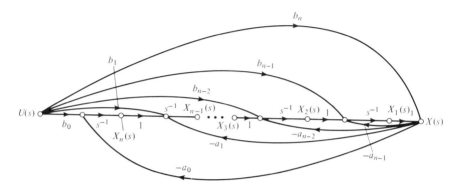

Fig. 7.9.3 A signal flow graph for (7.9.20).

When we rearrange (7.9.14), we obtain

$$X_1(s) = X(s) - b_2 U(s)$$
$$X_2(s) = sX_1(s) + a_1 X(s) - b_1 U(s) \qquad (7.9.15)$$
$$sX_2(s) = -a_0 X(s) + b_0 U(s).$$

Note that the first two equations of (7.9.15) provide the definitions of x_1 and x_2, respectively, and, of course, $sX(s)$ corresponds to dx/dt in the time domain. Eliminating $X(s)$ between the equations in (7.9.15), we obtain the state equations of the system

$$\dot{x}_1 = -a_1 x_1 + x_2 + (b_1 - a_1 b_2)u$$
$$\dot{x}_2 = -a_0 x_1 + (b_0 - a_0 b_2)u \qquad (7.9.16)$$

or, in matrix form,

$$\begin{bmatrix} \dot{x}_1 \\ \dot{x}_2 \end{bmatrix} = \begin{bmatrix} -a_1 & 1 \\ -a_0 & 1 \end{bmatrix} \begin{bmatrix} x_1 \\ x_2 \end{bmatrix} + \begin{bmatrix} b_1 - a_1 b_2 \\ b_0 - a_0 b_2 \end{bmatrix} u. \qquad (7.9.17)$$

We are now ready to derive a set of state equations for the general system expressed by (7.9.9). Just as in the above example, we begin by taking the Laplace transform of (7.9.9) with all the initial conditions deleted (set to zero) yielding:

$$(s^n + a_{n-1}s^{n-1} + a_{n-2}s^{n-2} + \cdots + a_1 s + a_0)X(s)$$
$$= (b_n s^n + b_{n-1}s^{n-1} + b_{n-2}s^{n-2} + \cdots + b_1 s + b_0)U(s) \qquad (7.9.18)$$

or, after rearranging terms,

$$s^n[X(s) - b_n U(s)] = s^{n-1}[b_{n-1}U(s) - a_{n-1}X(s)] + s^{n-2}[b_{n-2}U(s) - a_{n-2}X(s)]$$
$$+ \cdots + s[b_1 U(s) - a_1 X(s)] + [b_0 U(s) - a_0 X(s)]. \qquad (7.9.19)$$

After dividing through by s^n, we can write (7.9.19) as

$$X(s) = b_n U(s) + \frac{1}{s}[b_{n-1}U(s) - a_{n-1}X(s)] + \frac{1}{s^2}[b_{n-2}U(s) - a_{n-2}X(s)]$$

$$+ \cdots + \frac{1}{s^{n-1}}[b_1 U(s) - a_1 X(s)] + \frac{1}{s^n}[b_0 U(s) - a_0 X(s)] \qquad (7.9.20)$$

from which a signal flow graph can be constructed as shown in Fig. 7.9.3. An inspection of the signal flow graph indicates that the following relations can be readily established

$$
\begin{aligned}
X(s) &= X_1(s) + b_n U(s) \\
X_1(s) &= s^{-1}[X_2(s) + b_{n-1}U(s) - a_{n-1}X(s) \\
X_2(s) &= s^{-1}[X_3(s) + b_{n-2}U(s) - a_{n-2}X(s)] \\
&\vdots \\
X_{n-1}(s) &= s^{-1}[X_n(s) + b_1 U(s) - a_1 X(s)] \\
X_n(s) &= s^{-1}[b_0 U(s) - a_0 X(s)].
\end{aligned}
\qquad (7.9.21)
$$

Just as in Example 7.9.2, (7.9.21) can be rearranged to give

$$
\begin{aligned}
X_1(s) &= X(s) - b_n U(s) \\
X_2(s) &= sX_1(s) + a_{n-1}X(s) - b_{n-1}U(s) \\
X_3(s) &= sX_2(s) + a_{n-2}X(s) - b_{n-2}U(s) \\
&\vdots \\
X_n(s) &= sX_{n-1}(s) + a_1 X(s) - b_1 U(s) \\
sX_n(s) &= a_0 X(s) + b_0 U(s).
\end{aligned}
\qquad (7.9.22)
$$

Eliminating $X(s)$ in (7.9.22), we obtain the following set of state equations for the system (7.9.9):

$$
\begin{aligned}
\dot{x}_1 &= -a_{n-1}x_1 + x_2 + (b_{n-1} - a_{n-1}b_n)u \\
\dot{x}_2 &= -a_{n-2}x_1 + x_3 + (b_{n-2} - a_{n-2}b_n)u \\
&\vdots \\
\dot{x}_{n-1} &= -a_1 x_1 + x_n + (b_1 - a_1 b_n)u \\
\dot{x}_n &= -a_0 x_1 + (b_0 - a_0 b_n)u
\end{aligned}
\qquad (7.9.23)
$$

or

$$
\begin{bmatrix} \dot{x}_1 \\ \dot{x}_2 \\ \vdots \\ \dot{x}_{n-1} \\ \dot{x}_n \end{bmatrix} = \begin{bmatrix} -a_{n-1} & 1 & 0 \cdots 0 \\ -a_{n-2} & 0 & 1 \cdots 0 \\ \vdots & & \\ -a_1 & 0 & 0 \cdots 1 \\ -a_0 & 0 & 0 \cdots 0 \end{bmatrix} \begin{bmatrix} x_1 \\ x_2 \\ \vdots \\ x_{n-1} \\ x_n \end{bmatrix} + \begin{bmatrix} b_{n-1} - a_{n-1}b_n \\ b_{n-2} - a_{n-2}b_n \\ \vdots \\ b_1 - a_1 b_n \\ b_0 - a_0 b_n \end{bmatrix} u.
$$
(7.9.24)

The initial conditions for the state variables $x_1, \ldots, x_n$ can be obtained in terms of the original variables through (7.9.22). If, in (7.9.22), the state variables on the right-hand side of the equations are eliminated, a set of equations expressing state variables $x_1, \ldots, x_n$ in terms of x, u, and their derivatives results and is readily found to be (after substituting $sX(s)$ by dx/dt, $s^2 X(s)$ by d^2x/dt^2, and so on):

$$
\begin{bmatrix} x_1 \\ x_2 \\ x_3 \\ \vdots \\ x_n \end{bmatrix} = \begin{bmatrix} 1 & 0 & 0 \cdots 0 & 0 \\ a_{n-1} & 1 & 0 & 0 & 0 \\ a_{n-2} & a_{n-1} & 1 & 0 & 0 \\ \vdots & & & \\ a_1 & a_2 & a_3 & a_{n-1} & 1 \end{bmatrix} \begin{bmatrix} x \\ \dfrac{dx}{dt} \\ \dfrac{d^2x}{dt^2} \\ \vdots \\ \dfrac{d^{n-1}x}{dt^{n-1}} \end{bmatrix}
$$

$$
+ \begin{bmatrix} -b_n & 0 & 0 \cdots 0 \\ -b_{n-1} & -b_n & 0 & 0 \\ -b_{n-2} & -b_{n-1} & -b_n & 0 \\ \vdots & & \\ -b_1 & -b_2 & -b_3 & -b_n \end{bmatrix} \begin{bmatrix} u \\ \dfrac{du}{dt} \\ \dfrac{d^2u}{dt^2} \\ \vdots \\ \dfrac{d^{n-1}u}{dt^{n-1}} \end{bmatrix}
$$
(7.9.25)

From (7.9.25) the initial conditions for the state variables can be expressed in terms of the initial conditions of the original variables. A simple example will illustrate the ideas involved.

Example 7.9.3 Determine a set of state equations for the third-order system

$$
\frac{d^3x}{dt^3} + 3\frac{d^2x}{dt^2} + \frac{dx}{dt} + 2x = 2\frac{d^2u}{dt^2} + \frac{du}{dt} + 3u
$$
(7.9.26)

Fig. 7.9.4 A linear time-invariant system.

with the initial conditions

$$\frac{d^2u(0)}{dt^2} = \frac{du(0)}{dt} = u(0) = \frac{d^2x(0)}{dt^2} = \frac{dx(0)}{dt} = x(0) = 1 \qquad (7.9.27)$$

(where the values are equal only numerically).

Using (7.9.24), we find the set of state equations corresponding to the third-order system ($n = 3$) to be

$$\begin{bmatrix} \dot{x}_1 \\ \dot{x}_2 \\ \dot{x}_3 \end{bmatrix} = \begin{bmatrix} -3 & 1 & 0 \\ -1 & 0 & 1 \\ -2 & 0 & 0 \end{bmatrix} \begin{bmatrix} x_1 \\ x_2 \\ x_3 \end{bmatrix} + \begin{bmatrix} 2 \\ 1 \\ 3 \end{bmatrix} u. \qquad (7.9.28)$$

The initial conditions of the state variables can be obtained by substituting (7.9.27) into (7.9.25), with $a_3 = 1$, $a_2 = 3$, $a_1 = 1$, $a_0 = 2$, and $b_3 = 0$, $b_2 = 2$, $b_1 = 1$, $b_0 = 3$ obtained from (7.9.26):

$$\begin{bmatrix} x_1(0) \\ x_2(0) \\ x_3(0) \end{bmatrix} = \begin{bmatrix} 1 & 0 & 0 \\ 3 & 1 & 0 \\ 1 & 3 & 1 \end{bmatrix} \begin{bmatrix} x(0) \\ \dfrac{dx(0)}{dt} \\ \dfrac{d^2x(0)}{dt^2} \end{bmatrix} + \begin{bmatrix} 0 & 0 & 0 \\ -2 & 0 & 0 \\ -1 & -2 & 0 \end{bmatrix} \begin{bmatrix} u(0) \\ \dfrac{du(0)}{dt} \\ \dfrac{d^2u(0)}{dt^2} \end{bmatrix} = \begin{bmatrix} 1 \\ 2 \\ 2 \end{bmatrix}.$$

Case 3. Transfer functions—partial fraction expansion

For a single-input–single-output, time-invariant, linear system as shown in Fig. 7.9.4, the direct relationship between the output $y(t)$ and the input $u(t)$ is usually described by the transfer function of the system, $H(s)$, which is defined as the ratio of the Laplace transformed output $Y(s)$ to the Laplace transformed input $U(s)$ with initial conditions set to zero, or

$$H(s) = \frac{Y(s)}{U(s)}. \qquad (7.9.29)$$

Of course, the transfer function $H(s)$ describes the system behavior correctly only if there is no initial energy in the system. Otherwise, additional terms must be used to account for the portion of the output caused by the initial stored energy.

We shall assume that the initial stored energy of the system of Fig. 7.9.4 is negligible and that the transfer function $H(s)$ is given in the form of a rational function (i.e., a ratio of two polynomials) in s

$$H(s) = \frac{N(s)}{D(s)}. \qquad (7.9.30)$$

Furthermore, the order of $N(s)$ is assumed to be less than or equal to that of $D(s)$ which is usually the case for most physical systems. We shall consider the following two possibilities.

a) *The poles of $H(s)$ (which are the zeros of $D(s)$) are distinct.* Then for an nth-order polynomial $D(s)$ with distinct zeros $\lambda_1, \lambda_2, \ldots, \lambda_n$, the transfer function $H(s)$ can be expanded into partial fractions of the following form

$$H(s) = K_0 + \frac{K_1}{s - \lambda_1} + \frac{K_2}{s - \lambda_2} + \cdots + \frac{K_n}{s - \lambda_n} = K_0 + \sum_{i=1}^{n} \frac{K_i}{s - \lambda_i}, \qquad (7.9.31)$$

where the constants are given by

$$K_0 = \lim_{s \to \infty} H(s),$$

$$K_i = \{(s - \lambda_i)H(s)\}_{s = \lambda_i}, \qquad i = 1, 2, \ldots, n. \qquad (7.9.32)$$

Combining (7.9.29) and (7.9.31), we can write the transformed output $Y(s)$ as

$$Y(s) = H(s)U(s) = K_0 U(s) + \sum_{i=1}^{n} \frac{K_i}{s - \lambda_i} U(s). \qquad (7.9.33)$$

If we define the state variables $X_i(s)$ by the relation

$$X_i(s) \triangleq \frac{U(s)}{s - \lambda_i}, \qquad i = 1, 2, \ldots, n \qquad (7.9.34)$$

which can be rearranged as

$$sX_i(s) = \lambda_i X_i(s) + U(s), \qquad i = 1, 2, \ldots, n, \qquad (7.9.35)$$

the time-domain equivalent of (7.9.35) is simply

$$\dot{x}_i(t) = \lambda_i x_i(t) + u(t), \qquad i = 1, 2, \ldots, n. \qquad (7.9.36)$$

Substituting (7.9.34) into (7.9.33), we find that

$$Y(s) = K_0 U(s) + \sum_{i=1}^{n} K_i X_i(s) \qquad (7.9.37)$$

or, its time-domain equivalent,

$$y(t) = K_0 u(t) + \sum_{i=1}^{n} K_i x_i(t). \qquad (7.9.38)$$

Expressing (7.9.36) and (7.9.38) in matrix form, we obtain the state equations

$$\begin{bmatrix} \dot{x}_1 \\ \dot{x}_2 \\ \vdots \\ \dot{x}_n \end{bmatrix} = \begin{bmatrix} \lambda_1 & 0 & \cdots & 0 \\ 0 & \lambda_2 & \cdots & 0 \\ \vdots & & \ddots & \\ 0 & 0 & \cdots & \lambda_n \end{bmatrix} \begin{bmatrix} x_1 \\ x_2 \\ \vdots \\ x_n \end{bmatrix} + \begin{bmatrix} 1 \\ 1 \\ \vdots \\ 1 \end{bmatrix} u \qquad (7.9.39)$$

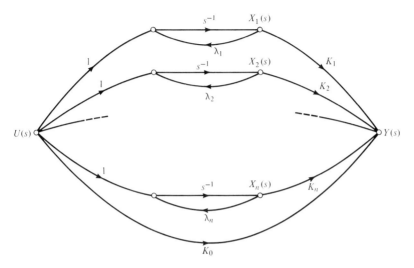

Fig. 7.9.5 Signal flow graph corresponding to the system of (7.9.39) and (7.9.40).

and

$$y = [K_1 \; K_2 \ldots K_n] \begin{bmatrix} x_1 \\ x_2 \\ \cdot \\ \cdot \\ \cdot \\ x_n \end{bmatrix} + [K_0]u. \tag{7.9.40}$$

The above development can be best summarized by means of the signal flow graph (Fig. 7.9.5) which shows how the state variables x_j, the input $u(t)$, and the output $y(t)$ are interrelated.

Example 7.9.4 Let us use the above technique to determine the state equations for a system whose transfer function is given by

$$H(s) = \frac{Y(s)}{U(s)} = \frac{3}{(s+1)(s+2)}.$$

Performing the partial fraction expansion of $H(s)$, we find that

$$H(s) = \frac{3}{s+1} - \frac{3}{s+2}.$$

Hence

$$Y(s) = \frac{3U(s)}{s+1} - \frac{3U(s)}{s+2} = 3X_1(s) - 3X_2(s),$$

where

$$X_1(s) = \frac{U(s)}{s + 1},$$

$$X_2(s) = \frac{U(s)}{s + 2}.$$

Therefore, the state equation becomes

$$\begin{bmatrix} \dot{x}_1 \\ \dot{x}_2 \end{bmatrix} = \begin{bmatrix} -1 & 0 \\ 0 & -2 \end{bmatrix} \begin{bmatrix} x_1 \\ x_2 \end{bmatrix} + \begin{bmatrix} 1 \\ 1 \end{bmatrix} u,$$

and the output is

$$y(t) = 3x_1 - 3x_2.$$

b) *The poles of H(s) are not distinct.* Let us assume that the transfer function $H(s)$ in this case is given by

$$H(s) = \frac{N(s)}{(s - \lambda_1)^r (s - \lambda_{r+1}) \cdots (s - \lambda_n)}. \tag{7.9.41}$$

Note that no generality is lost by this assumption since the same technique can be applied equally well for other cases. Again, taking the partial fraction expansion of (7.9.41), we obtain

$$H(s) = K_0 + \frac{K_{11}}{(s - \lambda_1)^r} + \frac{K_{12}}{(s - \lambda_1)^{r-1}} + \cdots + \frac{K_{1r}}{s - \lambda_1} + \frac{K_{r+1}}{s - \lambda_{r+1}} + \cdots + \frac{K_n}{s - \lambda_n}$$

$$\tag{7.9.42}$$

$$= K_0 + \sum_{j=1}^{r} \frac{K_{1j}}{(s - \lambda_1)^{r-j+1}} + \sum_{i=r+1}^{n} \frac{K_i}{(s - \lambda_i)},$$

where

$$K_0 = \lim_{s \to \infty} H(s),$$

$$K_{1j} = \frac{1}{(j - 1)!} \frac{d^{j-1}}{ds^{j-1}} \{(s - \lambda_1)^r H(s)\}_{s=\lambda_1}, \qquad j = 1, 2, \ldots, r, \tag{7.9.43}$$

and

$$K_i = \{(s - \lambda_i) H(s)\}_{s=\lambda_i}, \qquad i = r + 1, \ldots, n.$$

Thus, the output $Y(s) = H(s)U(s)$ can be expressed as

$$Y(s) = H(s)U(s) = K_0 U(s) + \sum_{j=1}^{r} \frac{K_{1j} U(s)}{(s - \lambda_1)^{r-j+1}} + \sum_{i=r+1}^{n} \frac{K_i U(s)}{(s - \lambda_i)}. \tag{7.9.44}$$

Let us define the state variables for the system of (7.9.41) as

$$X_j(s) = \frac{U(s)}{(s - \lambda_1)^{r-j+1}}, \qquad j = 1, 2, \ldots, r \tag{7.9.45}$$

and

$$X_i(s) = \frac{U(s)}{s - \lambda_i}, \qquad i = r + 1, r + 2, \ldots, n. \tag{7.9.46}$$

Then (7.9.44) can be expressed in terms of the state variables as

$$Y(s) = K_0 U(s) + \sum_{j=1}^{r} K_{1j} X_j(s) + \sum_{i=r+1}^{n} K_i X_i(s). \tag{7.9.47}$$

Our next step is to write the state equations by finding the corresponding expressions of (7.9.45) and (7.9.46) in the time domain. This can be done by first writing out the expressions of (7.9.45) for different values of the index j. Thus, for $j = 1, 2, \ldots, r$, we find, respectively:

$$X_1(s) = \frac{U(s)}{(s - \lambda_1)^r}$$

$$X_2(s) = \frac{U(s)}{(s - \lambda_1)^{r-1}}$$

$$\vdots$$

$$X_{r-1}(s) = \frac{U(s)}{(s - \lambda_1)^2}$$

$$X_r(s) = \frac{U(s)}{s - \lambda_1}.$$

The above set of state variables is linked by the relations

$$X_{k+1}(s) = (s - \lambda_1) X_k(s), \qquad k = 1, 2, \ldots, r - 1$$

and

$$X_r(s) = \frac{U(s)}{s - \lambda_1} \tag{7.9.48}$$

which, as in the case of distinct roots of $D(s)$, can be expressed in the time domain as

$$\dot{x}_k = \lambda_1 x_k + x_{k+1}, \qquad k = 1, 2, \ldots, r - 1$$

and

$$\dot{x}_r = \lambda_1 x_r + u(t). \tag{7.9.49}$$

In like manner, (7.9.46) can also be expressed in the time domain as

$$\dot{x}_i = \lambda_i x_i + u(t), \qquad i = r + 1, r + 2, \ldots, n. \tag{7.9.50}$$

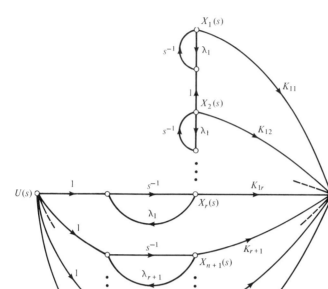

Fig. 7.9.6 Signal flow graph corresponding to the system of (7.9.51) and (7.9.52).

Finally, expressing (7.8.49) and (7.8.50) in matrix form, we obtain the required state equation

$$
\begin{bmatrix} \dot{x}_1 \\ \dot{x}_2 \\ \vdots \\ \dot{x}_r \\ \dot{x}_{r+1} \\ \vdots \\ \dot{x}_n \end{bmatrix} = \begin{bmatrix} \lambda_1 & 1 & 0 & \cdots & 0 & 0 & \cdots & 0 \\ 0 & \lambda_1 & 1 & \cdots & 0 & 0 & \cdots & 0 \\ \vdots & & & & & & & \\ 0 & 0 & 0 & \cdots & \lambda_1 & 0 & \cdots & 0 \\ 0 & 0 & 0 & \cdots & 0 & \lambda_{r+1} & \cdots & 0 \\ \vdots & & & & & & & \\ 0 & 0 & 0 & \cdots & 0 & 0 & \cdots & \lambda_n \end{bmatrix} \begin{bmatrix} x_1 \\ x_2 \\ \vdots \\ x_r \\ x_{r+1} \\ \vdots \\ x_n \end{bmatrix} + \begin{bmatrix} 0 \\ 0 \\ \vdots \\ 1 \\ 1 \\ \vdots \\ 1 \end{bmatrix} u. \quad (7.9.51)
$$

The output $y(t)$ can be obtained from (7.9.47) as

$$y(t) = [K_{11} \quad K_{12} \quad \cdots \quad K_{1r} \quad K_{r+1} \quad \cdots \quad K_n] \begin{bmatrix} x_1 \\ x_2 \\ \vdots \\ x_r \\ x_{r+1} \\ \vdots \\ x_n \end{bmatrix} + [K_0]u(t). \qquad (7.9.52)$$

The signal flow graph corresponding to the system represented by (7.9.51) and (7.9.52) is shown in Fig. 7.9.6.

Example 7.9.5   Consider a system with the transfer function

$$H(s) = \frac{Y(s)}{U(s)} = \frac{s^3}{(s + 1)^2(s + 2)}.$$

Expanding $H(s)$ into partial fractions, we obtain

$$H(s) = K_0 + \frac{K_{11}}{(s + 1)^2} + \frac{K_{12}}{s + 1} + \frac{K_3}{s + 2}$$

$$= 1 - \frac{1}{(s + 1)^2} + \frac{4}{s + 1} - \frac{8}{s + 2}.$$

Hence

$$Y(s) = U(s) - \frac{U(s)}{(s + 1)^2} + \frac{4U(s)}{s + 1} - \frac{8U(s)}{s + 2}$$

$$= U(s) - X_1(s) + 4X_2(s) - 8X_3(s),$$

where

$$X_1(s) = \frac{U(s)}{(s + 1)^2},$$

$$X_2(s) = \frac{U(s)}{s + 1},$$

and

$$X_3(s) = \frac{U(s)}{s + 2}.$$

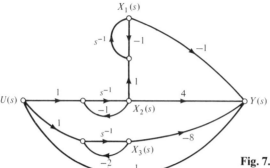

Fig. 7.9.7 Signal flow graph corresponding to the system of Example 7.9.5.

By virtue of (7.9.49) and (7.9.50) with $\lambda_1 = -1$, and $\lambda_3 = -2$, we find that

$$\dot{x}_1 = -x_1 + x_2, \qquad \dot{x}_2 = -x_2 + u, \qquad \dot{x}_3 = -2x_3 + u,$$

which, when expressed in matrix form (7.9.51) become

$$\begin{bmatrix} \dot{x}_1 \\ \dot{x}_2 \\ \dot{x}_3 \end{bmatrix} = \begin{bmatrix} -1 & 1 & 0 \\ 0 & -1 & 0 \\ 0 & 0 & -2 \end{bmatrix} \begin{bmatrix} x_1 \\ x_2 \\ x_3 \end{bmatrix} + \begin{bmatrix} 0 \\ 1 \\ 1 \end{bmatrix} u.$$

The output $y(t)$ is simply

$$y(t) = \begin{bmatrix} -1 & 4 & -8 \end{bmatrix} \begin{bmatrix} x_1 \\ x_2 \\ x_3 \end{bmatrix} + [1]u(t).$$

The signal flow graph corresponding to this system is shown in Fig. 7.9.7.

It should be pointed out that the method developed in this section always leads to a Jordan canonical form (of which the diagonal form for distinct poles of $H(s)$ is simply a special case). However, in a system with multiple inputs and outputs, if a Jordan form for the **A**-matrix of $\dot{\mathbf{x}} = \mathbf{Ax} + \mathbf{Bu}$ is desired, it can be obtained directly by means of a similarity transformation.

In the next section we shall present a brief discussion on linear time-varying systems.

7.10 LINEAR TIME-VARYING SYSTEMS
As discussed previously, a linear time-invariant system can always be described by the equations of the form

$$\dot{\mathbf{x}} = \mathbf{Ax} + \mathbf{Bu}$$

and

$$\mathbf{y} = \mathbf{Cx} + \mathbf{Du} \tag{7.10.1}$$

with $\mathbf{A}$, $\mathbf{B}$, $\mathbf{C}$, and $\mathbf{D}$ being constant matrices. If any of the above matrices is not fixed, but rather varies with time, the system is said to be a *time-varying* system. In general, a linear time-varying system with multiple inputs and outputs can be described by the following equations:

$$\dot{\mathbf{x}} = \mathbf{A}(t)\mathbf{x} + \mathbf{B}(t)\mathbf{u}$$
$$\mathbf{y} = \mathbf{C}(t)\mathbf{x} + \mathbf{D}(t)\mathbf{u}. \tag{7.10.2}$$

As indicated in Appendix A.1, the solution of the state equation is given by

$$\mathbf{x}(t) = \Phi(t, t_0)\mathbf{x}(t_0) + \int_{t_0}^{t} \Phi(t, \tau)\mathbf{B}(\tau)\mathbf{u}(\tau)\, d\tau. \tag{7.10.3}$$

Substituting (7.10.3) into the second equation of (7.10.2), we obtain the expression of $\mathbf{y}(t)$:

$$\mathbf{y}(t) = \mathbf{C}(t)\Phi(t, t_0)\mathbf{x}(t_0) + \int_{t_0}^{t} \mathbf{C}(t)\Phi(t, \tau)\mathbf{B}(\tau)\mathbf{u}(\tau)\, d\tau + \mathbf{D}(t)\mathbf{u}(t), \tag{7.10.4}$$

where $\Phi(t, \tau)$, the *state-transition matrix* of the state equation of (7.10.2), is the solution of the matrix differential equation

$$\dot{\Phi}(t, t_0) = \mathbf{A}(t)\Phi(t, t_0)$$

with

$$\Phi(t_0, t_0) = \mathbf{I}. \tag{7.10.5}$$

An inspection of (7.10.3) and (7.10.4) indicates that once the state-transition matrix $\Phi(t, \tau)$ is found, the solution of the system (7.10.2) can be readily determined. In other words, the exact behavior of the linear time-varying system depends directly on the properties of the state-transition matrix. It can be shown* that $\Phi(t, t_0)$ satisfies the following important properties:

a) If the matrices $\displaystyle\int_{t_0}^{t} \mathbf{A}(\tau)\, d\tau$ and $\mathbf{A}(t)$ commute for all t with respect to matrix

multiplication, then

$$\Phi(t, t_0) = e^{\int_{t_0}^{t} \mathbf{A}(\tau)\, d\tau}. \tag{7.10.6}$$

b) $\Phi(t_1, t_2)\Phi(t_2, t_3) = \Phi(t_1, t_3)$ for any t_1, t_2, and t_3.   (7.10.7)

c) If $\mathbf{A}(t)$ is continuous for all finite intervals of t, then $\Phi(t, t_0)$ is nonsingular for all finite t.

d) $\Phi^{-1}(t_1, t_2) = \Phi(t_2, t_1)$.   (7.10.8)

In contrast to time-invariant systems, it is quite difficult to determine the state-transition matrix of a time-varying system except in some specific cases. Therefore, the only effective means for determining the response of a time-varying system lies in the use of a digital computer.

---

* For example, refer to L. A. Zadeh and C. A. Desoer, [ZA 1], Chapter 6.

7.11 SUMMARY

In the preceding sections of this chapter, various techniques were presented for determining the solution of the (matrix) state equation of a linear time-invariant system. Several methods for deriving the state equations of different systems were also discussed. A brief discussion of time-varying systems was also given. However, due to the broad spectrum of available topics, the methods presented in this chapter are by no means exhaustive.

REFERENCES

1. K. Ogata, *State Space Analysis of Control Systems*, Prentice-Hall, 1967.
2. P. M. DeRusso, R. J. Roy, and C. M. Close, *State Variables for Engineers*, John Wiley & Sons, 1967, Chapter 5.
3. L. A. Zadeh and C. A. Desoer, *Linear System Theory*, McGraw-Hill, 1963, Chapters 4, 5, and 6.
4. B. Friedman, *Principles and Techniques of Applied Mathematics*, John Wiley & Sons, 1965, Chapter 2.

PROBLEMS

7.1 The transfer function of a single-input–single-output system in Fig. P.7.1 is given by

$$H(s) = \frac{s^2 + 14s + 36}{s^2 + 7s + 12}.$$

    a) Determine the state equations for this system such that the matrix **A** is either in diagonal or Jordan form.

    b) Find the state-transition matrix of the system.

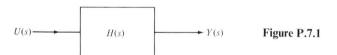

$U(s)$ $H(s)$ $Y(s)$ **Figure P.7.1**

7.2 Reduce the matrix

$$\mathbf{A} = \begin{bmatrix} 0 & 1 & 0 \\ 0 & 0 & 1 \\ 1 & -3 & 3 \end{bmatrix}$$

to Jordan canonical form by finding a matrix **S**. Verify your result.

7.3 The behavior of a second-order system is described by the state equations

$$\begin{bmatrix} x_1 \\ \dot{x}_2 \end{bmatrix} = \begin{bmatrix} -3 & 0 \\ 0 & -4 \end{bmatrix} \begin{bmatrix} x_1 \\ x_2 \end{bmatrix} + \begin{bmatrix} 1 \\ 1 \end{bmatrix} u(t); \qquad \begin{bmatrix} x_1(0) \\ x_2(0) \end{bmatrix} = \begin{bmatrix} 1 \\ 1 \end{bmatrix}$$

and

$$y(t) = 3x_1(t) + 4x_2(t) + u(t).$$

If the input to the system is a unit-step function, determine the output $y(t)$ for $t > 0$ by first finding the state-transition matrix.

7.4 Determine a set of state equations for the third-order system

$$2\frac{d^3x}{dt^3} + 6\frac{d^2x}{dt^2} + \frac{dx}{dt} + 2x = 4\frac{du}{dt} + u$$

with the initial conditions

$$u(0) = 2 \quad \text{and} \quad \frac{d^2x}{dt^2}(0) = \frac{dx}{dt}(0) = x(0) = 1.$$

7.5 Let $\lambda_1, \lambda_2, \ldots, \lambda_n$ be the distinct eigenvalues of a matrix $\mathbf{A}$. Show that the eigenvalues of the matrix $\mathbf{B} \triangleq \mathbf{A}^m$ are given by $\lambda_1^m, \lambda_2^m, \ldots, \lambda_n^m$, where m is a positive integer.

7.6 Let $\mathbf{A}$ be a triangular matrix, i.e., all the elements either above or below the principal diagonal are zero. Show that the elements of the principal diagonal of $\mathbf{A}$ are the eigenvalues of $\mathbf{A}$.

7.7 Draw the signal flow graph for the network of Fig. P.7.7 and then apply Mason's method to determine the transfer function

$$H(s) = \frac{E_2(s)}{E_1(s)}.$$

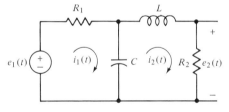

Figure P.7.7

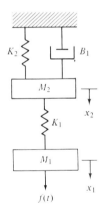

Figure P.7.8

7.8 Construct the signal flow graph for the mechanical system of Fig. P.7.8, and then determine the transfer function

$$H(s) = \frac{X_2(s)}{F(s)}.$$

Note that x_1 and x_2 represent the displacements of the masses M_1 and M_2 from the point of equilibrium, respectively.

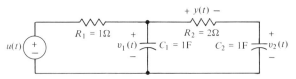

Figure P.7.9

7.9 Consider the RC network of Fig. P.7.9. If the output $y(t)$ is defined as the voltage across the resistor R_2 as indicated, obtain the state equations for the network by letting $x_1(t) = v_1(t)$ and $x_2(t) = v_2(t)$, and then draw the state-transition flow graph.

7.10 Construct the state-transition flow graph for the system of Problem 7.3 and then apply Mason's formula to determine the output $y(t)$.

7.11 Repeat Problem 7.10 for the system

$$\begin{bmatrix} \dot{x}_1 \\ \dot{x}_2 \end{bmatrix} = \begin{bmatrix} 2 & 2 \\ 1 & 3 \end{bmatrix} \begin{bmatrix} x_1 \\ x_2 \end{bmatrix}; \qquad \begin{bmatrix} x_1(0) \\ x_2(0) \end{bmatrix} = \begin{bmatrix} 1 \\ 2 \end{bmatrix}.$$

7.12 Determine the solution of the differential system

$$\begin{bmatrix} \dot{x}_1 \\ \dot{x}_2 \end{bmatrix} = \begin{bmatrix} 2 & 1 \\ 3 & 4 \end{bmatrix} \begin{bmatrix} x_1 \\ x_2 \end{bmatrix}; \qquad \begin{bmatrix} x_1(0) \\ x_2(0) \end{bmatrix} = \begin{bmatrix} 1 \\ 2 \end{bmatrix}$$

by the method of the Laplace transformation.

7.13 Repeat Problem 7.12 for the system of Problem 7.11.

7.14 Construct a signal flow graph corresponding to a system whose transfer function is given by

$$H(s) = \frac{s^2 + s + 1}{(s + 1)(s + 2)^2}.$$

7.15 A third-order system is described by the state equations

$$\begin{bmatrix} \dot{x}_1 \\ \dot{x}_2 \\ \dot{x}_3 \end{bmatrix} = \begin{bmatrix} -\frac{1}{3} & -\frac{1}{2} & 0 \\ -\frac{1}{2} & 0 & -1 \\ -1 & \frac{1}{2} & 0 \end{bmatrix} \begin{bmatrix} x_1 \\ x_2 \\ x_3 \end{bmatrix} + \begin{bmatrix} -1 \\ -\frac{1}{2} \\ -\frac{1}{3} \end{bmatrix} [u]; \qquad \begin{bmatrix} x_1(0) \\ x_2(0) \\ x_3(0) \end{bmatrix} = \begin{bmatrix} 1 \\ 0 \\ 0 \end{bmatrix}$$

Construct an analog computer diagram for solving the equations.

CHAPTER 8

DISCRETE-TIME SYSTEMS

8.1 INTRODUCTION

Up to now we have tacitly assumed that the systems under investigation have variables which are continuous with respect to time. A system in which all variables are known or specified at all instants of time is called a *continuous* or *analog* system. A system in which at least one signal is sampled intermittently in the form of pulses or numbers is referred to as a *sampled-data* or a *discrete-time* system. A typical example of a signal sampled at equal intervals of time T is depicted in Fig. 8.1.1 in which the sampled signal $f^*(t)$ is described by the sequence of numbers

$$f(0), f(T), f(2T), f(3T), \ldots, f(nT), \ldots .$$

The device which converts a continuous signal $f(t)$ into a sequence of numbers is known as the ideal *sampling switch* or *sampler*. A schematic representation is shown in Fig. 8.1.2 where the ideal sampler is assumed to close every T seconds for an infinitesimally short period of time so that the output of the sampler $f^*(t)$ contains a train of pulses with infinitesimally small pulse width (Fig. 8.1.1). The constant T is referred to as the *sampling period*.

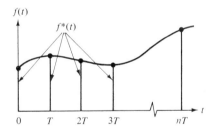

Fig. 8.1.1 A typical signal sampled at equal intervals of time T.

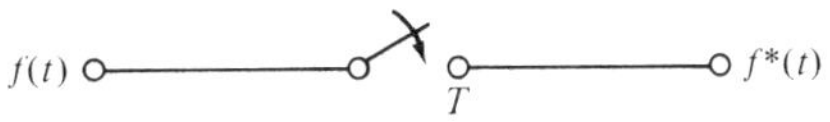

Fig. 8.1.2 An ideal sampler with sampling period T.

Discrete-time systems may arise in various ways. In some systems a sampling operation may be purposely introduced to obtain transient performance whose quality cannot be achieved by fully continuous systems. Another class of discrete-time systems is one in which signals are available only in sampled form. Some important examples of systems with inherent sampling are radar tracking systems in which input signals are in the form of a sequence of pulses, or a *pulse train*, and feedback control systems using digital computers as system components.

The present chapter is devoted entirely to discrete-time systems whose dynamic behaviors are governed by difference equations. Due to space limitations, only state-space techniques will be studied, since, with the availability of digital computers, these techniques have become very effective tools in both analysis and design of discrete-time systems. The more traditional z-transform approach will be discussed only to the extent* of providing the necessary background for the introduction of the topic on digital filters which will also be presented very briefly in this chapter.

8.2 REPRESENTATION OF DISCRETE SYSTEMS BY STATE EQUATIONS

In this section we shall present a method, based on the same approach used in Section 7.9 for the continuous case, for determining the state-space representation of a discrete system whose performance is governed by an nth-order difference equation. In the continuous case, the state variables are taken to be the sink nodes of all the branches with transmittance s^{-1}. Physically, s^{-1} corresponds to the transfer function of an integrator. In a discrete-time system the element analogous to the integrator is referred to as the *unit delay* or the *shift operator* denoted by Z_τ.†

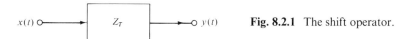

$x(t)$ ○────→ | Z_τ | ────→○ $y(t)$ **Fig. 8.2.1** The shift operator.

The schematic representation is shown in Fig. 8.2.1. The input-output relationship of Z_τ is given by

$$y(t) \triangleq Z_\tau[x(t)]$$
$$= x(t - \tau) \qquad \text{for all } t, \tag{8.2.1}$$

where the time delay τ is assumed to be a positive real constant. If we define the inverse of Z_τ by the relation

$$Z_\tau^{-1}[x(t)] = x(t + \tau), \tag{8.2.2}$$

then it can be readily shown that

$$Z_\tau^{-1}\{Z_\tau[x(t)]\} = Z_\tau^{-1}[x(t - \tau)] = x(t) \tag{8.2.3}$$

---

* See, for example, J. R. Ragazzini and G. F. Franklin [RA 1] for a detailed discussion.
† Z_τ is sometimes called a translation operator.

and that

$$Z_\tau Z_\tau^{-1}[x(t)] = Z_\tau[x(t + \tau)]$$

$$= x(t). \tag{8.2.4}$$

In addition, it can be easily verified that both Z_τ and its inverse are linear operators.*
    With the definition of the shift operator and its inverse we are ready to derive the state equations for a discrete-time system. Let us consider the following cases.

Case 1. The nth-order difference equation with only one forcing term
Consider a discrete-time system described by an nth-order difference equation in the form

$$x(k + n) + a_{n-1}x(k + n - 1) + \cdots + a_1x(k + 1) + a_0x(k) = bu(k),$$

$$k = 0, 1, 2, \ldots, \tag{8.2.5}\dagger$$

where k denotes the kth sampling instant and the coefficients a_i $(i = 0, \ldots, n - 1)$ and b are assumed to be real constants.
    The state-space representation of (8.2.5) can be obtained by defining

$$x_1(k) = x(k)$$
$$x_2(k) = x_1(k + 1)$$
$$x_3(k) = x_2(k + 1) \tag{8.2.6}$$
$$\vdots$$
$$x_n(k) = x_{n-1}(k + 1).$$

Then, (8.2.5) is equivalent to

$$x_1(k + 1) = x_2(k)$$
$$x_2(k + 1) = x_3(k)$$
$$\vdots \tag{8.2.7}$$
$$x_{n-1}(k + 1) = x_n(k)$$
$$x_n(k + 1) = -a_0x_1(k) - a_1x_2(k) - \cdots - a_{n-1}x_n(k) + bu(k)$$

which can be written in matrix form as

$$\mathbf{x}(k + 1) = \mathbf{A}\mathbf{x}(k) + \mathbf{B}u(k), \qquad k = 0, 1, 2, \ldots,$$

---

* See Problem 8.7 (a).
† For simplicity, the term $x(k)$ instead of $x(kT)$ is used.

where

$$\mathbf{x}(k) = \begin{bmatrix} x_1(k) \\ x_2(k) \\ \vdots \\ x_n(k) \end{bmatrix}, \qquad \mathbf{A} = \begin{bmatrix} 0 & 1 & 0 & \cdots & 0 \\ 0 & 0 & 1 & \cdots & 0 \\ \vdots & & & \cdots & \\ 0 & 0 & 0 & \cdots & 1 \\ -a_0 & -a_1 & -a_2 & \cdots & -a_{n-1} \end{bmatrix},$$

$$(8.2.8)$$

$$\mathbf{B} = \begin{bmatrix} 0 \\ 0 \\ \vdots \\ 0 \\ b \end{bmatrix}, \qquad \mathbf{u}(k) = [u(k)].$$

Example 8.2.1   The third-order difference equation

$$x(k + 3) + 3x(k + 2) + x(k + 1) + 2x(k) = 3u(k) \qquad (8.2.9)$$

with the initial conditions $x(0) = 1$, $x(1) = 2$, and $x(2) = 3$ can be made equivalent to a state-space representation as follows. Let

$$\begin{aligned} x_1(k) &= x(k), \\ x_2(k) &= x_1(k + 1), \\ x_3(k) &= x_2(k + 1). \end{aligned}$$

Then, (8.2.9) can be represented by

$$\begin{bmatrix} x_1(k + 1) \\ x_2(k + 1) \\ x_3(k + 1) \end{bmatrix} = \begin{bmatrix} 0 & 1 & 0 \\ 0 & 0 & 1 \\ -2 & -1 & -3 \end{bmatrix} \begin{bmatrix} x_1(k) \\ x_2(k) \\ x_3(k) \end{bmatrix} + \begin{bmatrix} 0 \\ 0 \\ 3 \end{bmatrix} u(k), \qquad \begin{bmatrix} x_1(0) \\ x_2(0) \\ x_3(0) \end{bmatrix} = \begin{bmatrix} 1 \\ 2 \\ 3 \end{bmatrix} \quad (8.2.10)$$

Remark.   Upon comparison between the two equations (7.9.3) and (8.2.5), together with their corresponding state-space representations (7.9.6) and (8.2.8), respectively, it is evident that the method used for obtaining the state-space representation for a continuous system with only one forcing term applies equally well to a discrete-time system once the equivalence of $d^i x(t)/dt^i$ and $x(k + i)$ for $i = 0, 1, \ldots$ between the two classes of systems is established. The reader should compare the results of Examples 7.9.1 and 8.2.1 to appreciate this fact.

Case 2.   The nth-order difference equation with more than one forcing term

When the difference equation describing the behavior of a discrete-time system has more than one forcing term, a convenient method for obtaining the state-space representation of the system starts with the construction of a signal flow graph in a manner parallel to that discussed in Case 2 of Section 7.9. Then we determine the state-space equations after state variables are assigned to various nodes of the graph.

Before underlining the necessary steps for obtaining a state-space representation of a general discrete-time system, we use the following example to illustrate the ideas involved in the approach.

Example 8.2.2 Consider a discrete-time system described by the difference equation

$$x(k + 2) + a_1 x(k + 1) + a_0 x(k) = b_2 u(k + 2) + b_1 u(k + 1) + b_0 u(k). \qquad (8.2.11)$$

To obtain a state-space representation, (8.2.11) can be written in the following form

$$Z_T^{-2}[x(k)] + a_1 Z_T^{-1}[x(k)] + a_0 x(k) = b_2 Z_T^{-2}[u(k)] + b_1 Z_T^{-1}[u(k)] + b_0 u(k), \qquad (8.2.12)$$

where the shift operator Z_τ with $\tau = T$ is applied, and where, by definition:

$$Z_T^{-n}[x(k)] \triangleq Z_T^{-(n-1)}\{Z_T^{-1}[x(k)]\}, \qquad \text{for } n = 2, 3, \dots .$$

Rearranging the terms, we can write (8.2.12) as

$$Z_T^{-2}[x(k) - b_2 u(k)] = Z_T^{-1}[-a_1 x(k) + b_1 u(k)] + [-a_0 x(k) + b_0 u(k)] \qquad (8.2.13)$$

which can be simplified to yield

$$x(k) = b_2 u(k) + Z_T[-a_1 x(k) + b_1 u(k)] + Z_T^2[-a_0 x(k) + b_0 u(k)] \qquad (8.2.14)$$

with

$$Z_T^n[x(k)] \triangleq Z_T^{n-1}\{Z_T[x(k)]\} \qquad \text{for } n = 2, 3, \dots .$$

If the shift operator is regarded as a branch with transmittance Z_T, the signal flow graph corresponding to (8.2.14) can be drawn as shown in Fig. 8.2.2.

If the sink nodes of all the branches with transmittance Z_T are identified as state variables as shown in Fig. 8.2.2, the following set of relations can be readily arrived at by inspection:

$$x(k) = x_1(k) + b_2 u(k),$$

$$x_1(k) = Z_T[x_2(k) + b_1 u(k) - a_1 x(k)], \qquad (8.2.15)$$

$$x_2(k) = Z_T[b_0 u(k) - a_0 x(k)].$$

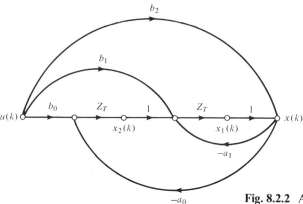

Fig. 8.2.2 A signal flow graph of (8.2.14).

Applying the inverse operator Z_T^{-1} to both sides of the last two equations of (8.2.15) and then eliminating the variable $x(k)$, we obtain the required state equations

$$x_1(k + 1) = -a_1 x_1(k) + x_2(k) + (b_1 - a_1 b_2)u(k)$$
$$x_2(k + 1) = -a_0 x_1(k) + (b_0 - a_0 b_2)u(k) \tag{8.2.16}$$

or, in matrix form,

$$\begin{bmatrix} x_1(k + 1) \\ x_2(k + 1) \end{bmatrix} = \begin{bmatrix} -a_1 & 1 \\ -a_0 & 0 \end{bmatrix} \begin{bmatrix} x_1(k) \\ x_2(k) \end{bmatrix} + \begin{bmatrix} b_1 - a_1 b_2 \\ b_0 - a_0 b_2 \end{bmatrix} u(k) \tag{8.2.17}$$

which is the desired state-space representation of the system.

Remarks. In obtaining the state equations (8.2.15), we must exercise caution for we should always keep in mind that Z_T, being a shift operator, must precede the variable to be operated on. In other words, a term like $x(k)Z_T$ does not have any meaning while $Z_T[x(k)]$ is equal to $x(k-1)$.

In deriving the state equations (8.2.15), the linearity property of the shift operator Z_T and its inverse, e.g.,

$$Z_T[a_1 y_1(k) + a_2 y_2(k)] = a_1 Z_T[y_1(k)] + a_2 Z_T[y_2(k)]$$

for arbitrary constants a_1 and a_2, has been applied. This property can be easily verified by direct substitution into the definition given in (8.2.2).

With the above development, we are ready to determine a set of state equations for a discrete-time system described by the general difference equation:

$$x(k + n) + a_{n-1}x(k + n - 1) + \cdots + a_1 x(k + 1) + a_0 x(k)$$
$$= b_n u(k + n) + b_{n-1}u(k + n - 1) + \cdots + b_1 u(k + 1) + b_0 u(k). \tag{8.2.18}$$

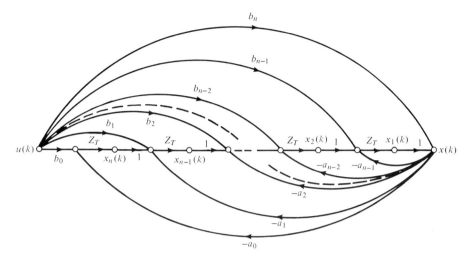

Fig. 8.2.3 A signal flow graph of (8.2.19).

As in the continuous case, many physical linear systems are indeed governed by difference equations of the form shown in (8.2.18). A typical example is a sampled-data feedback control system using a linear digital controller; that is, the relation between the input number sequence and the output number sequence of the controller is linear and hence describable by a linear difference equation.

Proceeding in exactly the same manner as in Example 8.2.2, we write (8.2.18) in the form

$$x(k) = b_n u(k) + Z_T[-a_{n-1}x(k) + b_{n-1}u(k)]$$
$$+ \cdots + Z_T^{n-1}[-a_1 x(k) + b_1 u(k)] + Z_T^n[-a_0 x(k) + b_0 u(k)] \quad (8.2.19)$$

from which the corresponding signal flow graph can be constructed as depicted in Fig. 8.2.3.

If the sink nodes of all the branches with transmittance Z_T are again identified as state variables as indicated in the figure, we can obtain the following set of state equations by inspection:

$$x(k) = x_1(k) + b_n u(k)$$
$$x_1(k) = Z_T[x_2(k) + b_{n-1}u(k) - a_{n-1}x(k)]$$
$$\vdots \qquad\qquad\qquad\qquad\qquad (8.2.20)$$
$$x_{n-1}(k) = Z_T[x_n(k) + b_1 u(k) - a_1 x(k)]$$
$$x_n(k) = Z_T[b_0 u(k) - a_0 x(k)].$$

The expression (8.2.20) can be simplified to yield the set of state equations:

$$
\begin{bmatrix} x_1(k+1) \\ x_2(k+1) \\ \vdots \\ x_{n-1}(k+1) \\ x_n(k+1) \end{bmatrix} = \begin{bmatrix} -a_{n-1} & 1 & 0 & \cdots & 0 \\ -a_{n-2} & 0 & 1 & \cdots & 0 \\ \vdots & & & & \\ -a_1 & 0 & 0 & \cdots & 1 \\ -a_0 & 0 & 0 & \cdots & 0 \end{bmatrix} \begin{bmatrix} x_1(k) \\ x_2(k) \\ \vdots \\ x_{n-1}(k) \\ x_n(k) \end{bmatrix} + \begin{bmatrix} b_{n-1} - a_{n-1}b_n \\ b_{n-2} - a_{n-2}b_n \\ \vdots \\ b_1 - a_1 b_n \\ b_0 - a_0 b_n \end{bmatrix} u(k)
$$

(8.2.21)

The initial conditions of the state equations (8.2.21) can be determined as follows: Applying the relation $Z_T^{-1}[x(k)] = x(k+1)$, we can write the first n equations of (8.2.20) as

$$x_1(k) = x(k) - b_n u(k)$$

$$x_2(k) = x_1(k+1) - b_{n-1}u(k) + a_{n-1}x(k)$$

$$\vdots \tag{8.2.22}$$

$$x_n(k) = x_{n-1}(k+1) - b_1 u(k) + a_1 x(k)$$

The first equation of (8.2.22) defines the initial condition of the state variable $x_1(k)$ in terms of $x(k)$ and $u(k)$ with $k = k_0$ corresponding to the initial time $k_0 T$. Next, the first two equations of (8.2.22) are combined so that the variable $x_1(k+1)$ is eliminated, resulting in

$$x_2(k) = a_{n-1}x(k) + x(k+1) - b_{n-1}u(k) - b_n u(k+1). \tag{8.2.23}$$

Setting $k = k_0$ in (8.2.23), we obtain the initial condition of $x_2(k)$. Thus, if in (8.2.22) all the state variables on the right-hand side of the equations are eliminated, the following set of equations expressing the state variables $x_1(k), x_2(k), \ldots, x_n(k)$ in terms of x and u at various time instants results:

$$
\begin{bmatrix} x_1(k) \\ x_2(k) \\ x_3(k) \\ \vdots \\ x_n(k) \end{bmatrix} = \begin{bmatrix} 1 & 0 & 0 & \cdots & 0 \\ a_{n-1} & 1 & 0 & \cdots & 0 \\ a_{n-2} & a_{n-1} & 1 & \cdots & 0 \\ \vdots & & & & \\ a_1 & a_2 & a_3 & \cdots & 1 \end{bmatrix} \begin{bmatrix} x(k) \\ x(k+1) \\ x(k+2) \\ \vdots \\ x(k+n-1) \end{bmatrix}
$$

$$
+ \begin{bmatrix} -b_n & 0 & 0 & \cdots & 0 \\ -b_{n-1} & -b_n & 0 & \cdots & 0 \\ -b_{n-2} & -b_{n-1} & -b_n & \cdots & 0 \\ \vdots & & & & \\ -b_1 & -b_2 & -b_3 & \cdots & -b_n \end{bmatrix} \begin{bmatrix} u(k) \\ u(k+1) \\ u(k+2) \\ \vdots \\ u(k+1-n) \end{bmatrix}
$$

(8.2.24)

From (8.2.24), by setting $k = k_0$, the initial conditions for the state variables can be expressed in terms of x and u evaluated at the time instants $k_0 T, (k_0 + 1)T, \ldots, (k_0 + n - 1)T$.

Example 8.2.3 Determine a set of state equations for the third-order system

$$x(k + 3) + 3x(k + 2) + x(k + 1) + 2x(k) = 2u(k + 2) + u(k + 1) + 3u(k),$$
(8.2.25)

with initial conditions

$$u(2) = u(1) = u(0) = 1; \qquad x(2) = x(1) = x(0) = 1.$$
(8.2.26)

Using (8.2.21) with $n = 3$, the set of state equations corresponding to the third-order system is given by

$$\begin{bmatrix} x_1(k + 1) \\ x_2(k + 1) \\ x_3(k + 1) \end{bmatrix} = \begin{bmatrix} -3 & 1 & 0 \\ -1 & 0 & 1 \\ -2 & 0 & 0 \end{bmatrix} \begin{bmatrix} x_1(k) \\ x_2(k) \\ x_3(k) \end{bmatrix} + \begin{bmatrix} 2 \\ 1 \\ 3 \end{bmatrix} u(k).$$
(8.2.27)

The initial conditions of the state variables can be obtained by substituting (8.2.26) into (8.2.24) with $a_2 = 3$, $a_1 = 1$, $a_0 = 2$, $b_3 = 0$, $b_2 = 2$, $b_1 = 1$, and $b_0 = 3$ (where the a_j and b_j values are determined by comparing the corresponding coefficients in (8.2.18) and (8.2.25)):

$$\begin{bmatrix} x_1(0) \\ x_2(0) \\ x_3(0) \end{bmatrix} = \begin{bmatrix} 1 & 0 & 0 \\ 3 & 1 & 0 \\ 1 & 3 & 1 \end{bmatrix} \begin{bmatrix} x(0) \\ x(1) \\ x(2) \end{bmatrix} + \begin{bmatrix} 0 & 0 & 0 \\ -2 & 0 & 0 \\ -1 & -2 & 0 \end{bmatrix} \begin{bmatrix} u(0) \\ u(1) \\ u(2) \end{bmatrix} = \begin{bmatrix} 1 \\ 2 \\ 2 \end{bmatrix}.$$

The reader should observe the similarity in results between this example and Example 7.9.3 for the continuous case.

8.3 THE STATE-TRANSITION MATRIX AND ITS PROPERTIES

In the preceding section, we showed that a single-input–single-output linear discrete-time system described by a set of difference equations with constant coefficients can always be represented by a vector difference equation of the form

$$\mathbf{x}(k + 1) = \mathbf{A}\mathbf{x}(k) + \mathbf{B}\mathbf{u}(k),$$
(8.3.1)

where $\mathbf{A}$ and $\mathbf{B}$ are constant matrices. For a linear discrete-time system with multiple inputs and outputs, the behavior of the system can be described by the following set of matrix equations:

$$\mathbf{x}(k + 1) = \mathbf{A}\mathbf{x}(k) + \mathbf{B}\mathbf{u}(k)$$
$$\mathbf{y}(k) = \mathbf{C}\mathbf{x}(k) + \mathbf{D}\mathbf{u}(k),$$
(8.3.2)

where $\mathbf{x}(k)$, $\mathbf{y}(k)$, and $\mathbf{u}(k)$ are, respectively, the state, the output, and the input vectors at the time instant kT, and $\mathbf{A}$, $\mathbf{B}$, $\mathbf{C}$, $\mathbf{D}$ are constant matrices.

In the case of a linear time-varying system, the equations of motion take the following form:

$$\mathbf{x}(k + 1) = \mathbf{A}(k)\mathbf{x}(k) + \mathbf{B}(k)\mathbf{u}(k)$$
$$\mathbf{y}(k) = \mathbf{C}(k)\mathbf{x}(k) + \mathbf{D}(k)\mathbf{u}(k)$$

(8.3.3)

with $\mathbf{A}(k)$, $\mathbf{B}(k)$, $\mathbf{C}(k)$, and $\mathbf{D}(k)$ being time-varying matrices.

It is evident that the output vector of both systems (8.3.2) and (8.3.3) can be easily calculated once the state vector $\mathbf{x}(k)$ is determined. In the following discussion, we shall proceed to find the solution of the force-free system of (8.3.3), which can be expressed in terms of the state-transition matrix. Then in Section 8.4 we shall determine the solution of (8.3.3) taking the solution of the time-invariant system as a special case.

Consider first the force-free system described by

$$\mathbf{x}(k + 1) = \mathbf{A}(k)\mathbf{x}(k).$$

(8.3.4)

In terms of the initial conditions $\mathbf{x}(k_0)$, we find

$$\mathbf{x}(k_0 + 1) = \mathbf{A}(k_0)\mathbf{x}(k_0).$$

(8.3.5)

Similarly,

$$\mathbf{x}(k_0 + 2) = \mathbf{A}(k_0 + 1)\mathbf{x}(k_0 + 1)$$
$$= \mathbf{A}(k_0 + 1)\mathbf{A}(k_0)\mathbf{x}(k_0).$$

(8.3.6)

Continuing this process, it can be readily shown that $\mathbf{x}(k)$, for $k > k_0$, is given by

$$\mathbf{x}(k) = \mathbf{A}(k - 1)\mathbf{A}(k - 2) \cdots \mathbf{A}(k_0 + 1)\mathbf{A}(k_0)\mathbf{x}(k_0).$$

(8.3.7)

Let us define, for $k > k_0$

$$\mathbf{\Phi}(k, k_0) \triangleq \prod_{i=k_0}^{k-1} \mathbf{A}(i) = \mathbf{A}(k - 1)\mathbf{A}(k - 2) \cdots \mathbf{A}(k_0 + 1)\mathbf{A}(k_0)$$

(8.3.8)

and

$$\mathbf{\Phi}(k_0, k_0) = \mathbf{I} \quad \text{(identity matrix)}.$$

(8.3.9)

Then (8.3.7) can be written as

$$\mathbf{x}(k) = \mathbf{\Phi}(k, k_0)\mathbf{x}(k_0)$$

(8.3.10)

for $k > k_0$. The matrix $\mathbf{\Phi}(k, k_0)$, as for the continuous case, is referred to as the *state-transition matrix* for the discrete system (8.3.4) and is readily seen to satisfy the equation

$$\mathbf{\Phi}(k + 1, k_0) = \mathbf{A}(k)\mathbf{\Phi}(k, k_0), \quad \text{for } k > k_0.$$

(8.3.11)

From (8.3.8), for $k \geq j \geq k_0$, $\mathbf{\Phi}(k, k_0)$ can be written as

$$\mathbf{\Phi}(k, k_0) = \mathbf{A}(k - 1)\mathbf{A}(k - 2) \cdots \mathbf{A}(j)\mathbf{A}(j - 1) \cdots \mathbf{A}(k_0 + 1)\mathbf{A}(k_0)$$
$$= \prod_{r=j}^{k-1} \mathbf{A}(r) \prod_{s=k_0}^{j-1} \mathbf{A}(s),$$

and hence

$$\mathbf{\Phi}(k, k_0) = \mathbf{\Phi}(k, j)\mathbf{\Phi}(j, k_0). \qquad (8.3.12)$$

If, in addition, $\mathbf{A}(k)$ is nonsingular for all k, then we can define, for $k < k_0$,

$$\mathbf{\Phi}(k, k_0) \triangleq \mathbf{A}^{-1}(k)\mathbf{A}^{-1}(k + 1)\cdots\mathbf{A}^{-1}(k_0 - 1); \qquad (8.3.13)$$

and setting $k_0 = k$ in (8.3.12) leads to

$$\mathbf{\Phi}(k, j) = \mathbf{\Phi}^{-1}(j, k) \qquad (8.3.14)$$

for any k and j.

In the case of a time-invariant system, i.e., $\mathbf{A}(k) = \mathbf{A}$, a constant matrix for all k, the state-transition matrix $\mathbf{\Phi}(k, k_0)$, for $k > k_0$, reduces to,

$$\begin{aligned} \mathbf{\Phi}(k, k_0) &= \mathbf{\Phi}(k - k_0) \\ &= \mathbf{A}^{(k-k_0)} \end{aligned} \qquad (8.3.15)$$

from which the following interesting property

$$\mathbf{\Phi}(k_1 + k_2) = \mathbf{\Phi}(k_1)\mathbf{\Phi}(k_2) \qquad (8.3.16)$$

can be readily derived for any k_1 and k_2. By setting $k_1 = -k_2 = k$ in (8.3.16), we find that

$$\mathbf{\Phi}(k)\mathbf{\Phi}(-k) = \mathbf{I} \qquad (8.3.17)$$

which reduces to

$$\mathbf{\Phi}(k) = \mathbf{\Phi}^{-1}(-k)$$

provided, of course, that the matrix $\mathbf{A}$ is nonsingular.

The useful properties of the state-transition matrix for discrete systems can now be summarized as follows:

a) For $k > k_0$,

$$\mathbf{\Phi}(k, k_0) = \prod_{i=k_0}^{k-1} \mathbf{A}(i) = \mathbf{A}(k - 1)\mathbf{A}(k - 2)\cdots\mathbf{A}(k_0);$$

and for $k = k_0$,

$$\mathbf{\Phi}(k_0, k_0) = \mathbf{I}.$$

b) For $k \geq j \geq k_0$,

$$\mathbf{\Phi}(k, k_0) = \mathbf{\Phi}(k, j)\mathbf{\Phi}(j, k_0).$$

c) If $\mathbf{A}(k)$ is nonsingular for all k, then

$$\mathbf{\Phi}(k, j) = \mathbf{\Phi}^{-1}(j, k)$$

for any k and j.

Furthermore, for time-invariant systems, we have

d) $\mathbf{\Phi}(k, k_0) = \mathbf{\Phi}(k - k_0) = \mathbf{A}^{(k-k_0)}$, and

e) $\mathbf{\Phi}(k) = \mathbf{\Phi}^{-1}(-k)$.

8.4 SOLUTION OF STATE EQUATIONS

Let us return to the determination of the solutions of the difference equations (8.3.3) describing the behavior of a linear time-varying multiple input–output discrete system. For convenience, these equations are repeated here:

$$\mathbf{x}(k + 1) = \mathbf{A}(k)\mathbf{x}(k) + \mathbf{B}(k)\mathbf{u}(k)$$
$$\mathbf{y}(k) = \mathbf{C}(k)\mathbf{x}(k) + \mathbf{D}(k)\mathbf{u}(k). \tag{8.4.1}$$

Again, in terms of the initial conditions $\mathbf{x}(k_0)$ and $\mathbf{u}(k_0)$, we write

$$\mathbf{x}(k_0 + 1) = \mathbf{A}(k_0)\mathbf{x}(k_0) + \mathbf{B}(k_0)\mathbf{u}(k_0). \tag{8.4.2}$$

Similarly,

$$\mathbf{x}(k_0 + 2) = \mathbf{A}(k_0 + 1)\mathbf{x}(k_0 + 1) + \mathbf{B}(k_0 + 1)\mathbf{u}(k_0 + 1)$$
$$= \mathbf{A}(k_0 + 1)\mathbf{A}(k_0)\mathbf{x}(k_0) + \mathbf{A}(k_0 + 1)\mathbf{B}(k_0)\mathbf{u}(k_0) + \mathbf{B}(k_0 + 1)\mathbf{u}(k_0 + 1) \tag{8.4.3}$$

and

$$\mathbf{x}(k_0 + 3) = \mathbf{A}(k_0 + 2)\mathbf{x}(k_0 + 2) + \mathbf{B}(k_0 + 2)\mathbf{u}(k_0 + 2)$$
$$= \mathbf{A}(k_0 + 2)\mathbf{A}(k_0 + 1)\mathbf{A}(k_0)\mathbf{x}(k_0) + \mathbf{A}(k_0 + 2)\mathbf{A}(k_0 + 1)\mathbf{B}(k_0)\mathbf{u}(k_0)$$
$$+ \mathbf{A}(k_0 + 2)\mathbf{B}(k_0 + 1)\mathbf{u}(k_0 + 1) + \mathbf{B}(k_0 + 2)\mathbf{u}(k_0 + 2). \tag{8.4.4}$$

Thus, in light of (8.4.4), it is readily seen that for $k > k_0$:

$$\mathbf{x}(k) = \mathbf{A}(k - 1)\mathbf{A}(k - 2)\cdots\mathbf{A}(k_0)\mathbf{x}(k_0) + \mathbf{A}(k - 1)\mathbf{A}(k - 2)\cdots\mathbf{A}(k_0 + 1)\mathbf{B}(k_0)\mathbf{u}(k_0)$$
$$+ \mathbf{A}(k - 1)\mathbf{A}(k - 2)\cdots\mathbf{A}(k_0 + 2)\mathbf{B}(k_0 + 1)\mathbf{u}(k_0 + 1) + \cdots$$
$$+ \mathbf{A}(k - 1)\mathbf{B}(k - 2)\mathbf{u}(k - 2) + \mathbf{B}(k - 1)\mathbf{u}(k - 1) \tag{8.4.5}$$

which, in terms of the state-transition matrix $\mathbf{\Phi}(k, i)$, can be written as

$$\mathbf{x}(k) = \mathbf{\Phi}(k, k_0)\mathbf{x}(k_0) + \sum_{i=k_0}^{k-1} \mathbf{\Phi}(k, i + 1)\mathbf{B}(i)\mathbf{u}(i). \tag{8.4.6}$$

As in the case of a continuous system, the first term on the right-hand side of (8.4.6) is the zero-input component of the state vector $\mathbf{x}(k)$ while the second term represents the zero-state component of $\mathbf{x}(k)$. Substituting (8.4.6) into the second equation of (8.4.1) yields the output $\mathbf{y}(k)$ of the system.

For a linear time-invariant discrete system, i.e., for the case when the matrices $\mathbf{A}(k) = \mathbf{A}$, $\mathbf{B}(k) = \mathbf{B}$, $\mathbf{C}(k) = \mathbf{C}$, and $\mathbf{D}(k) = \mathbf{D}$ are constant matrices, the solution (8.4.6) reduces to

$$\mathbf{x}(k) = \mathbf{\Phi}(k - k_0)\mathbf{x}(k_0) + \sum_{i=k_0}^{k-1} \mathbf{\Phi}(k - i - 1)\mathbf{B}\mathbf{u}(i). \tag{8.4.7}$$

With the foregoing development, it should be evident that systems described by difference equations can be treated in much the same manner as their continuous analogs. Much similarity can be drawn by comparing (8.4.6) with (7.10.3). Both equations have practically the same appearance except that the zero-state component of the state vector is represented by a convolution integral in the continuous system, while the discrete analog is expressed by a convolution summation. It should be pointed out that the techniques developed in Chapter 7 are, in general, applicable to discrete systems as the following example will illustrate.

Example 8.4.1   Consider the nth-order time-invariant discrete system

$$\mathbf{x}(k + 1) = \mathbf{A}\mathbf{x}(k) \tag{8.4.8}$$

such that the eigenvalues $\lambda_1, \lambda_2, \ldots, \lambda_n$ of $\mathbf{A}$ are distinct.

As discussed in Section 7.4, a set of linearly independent eigenvectors $\mathbf{u}_1, \mathbf{u}_2, \ldots, \mathbf{u}_n$ can be found. Let us change the state vector $\mathbf{x}$ to $\mathbf{z}$ by the linear nonsingular transformation

$$\mathbf{x}(k) = \mathbf{S}\mathbf{z}(k), \tag{8.4.9}$$

where the matrix $\mathbf{S}$ is formed by taking the eigenvectors $\mathbf{u}_1, \mathbf{u}_2, \ldots, \mathbf{u}_n$ as its columns:

$$\mathbf{S} \triangleq [\mathbf{u}_1 \quad \mathbf{u}_2 \quad \cdots \quad \mathbf{u}_n]. \tag{8.4.10}$$

Substituting (8.4.9) into (8.4.8), we find that

$$\mathbf{S}\mathbf{z}(k + 1) = \mathbf{A}\mathbf{S}\mathbf{z}(k), \tag{8.4.11}$$

which reduces to

$$\mathbf{z}(k + 1) = \mathbf{\Lambda}\mathbf{z}(k) \tag{8.4.12}$$

with

$$\mathbf{\Lambda} = \mathbf{S}^{-1}\mathbf{A}\mathbf{S}. \tag{8.4.13}$$

The matrix $\mathbf{\Lambda}$ is again a diagonal matrix with the eigenvalues of $\mathbf{A}$ as its diagonal elements. Denoting the state-transition matrix of the system (8.4.12) by $\mathbf{\Phi}_z(k, k_0)$, then, in light of (8.3.15) with k_0 set to zero for simplicity, we find

$$\mathbf{\Phi}_z(k) = \mathbf{\Lambda}^k = \begin{bmatrix} \lambda_1^k & 0 & 0 \cdots 0 \\ 0 & \lambda_2^k & 0 \cdots 0 \\ 0 & 0 & \lambda_3^k \cdots 0 \\ \vdots & & \ddots \\ 0 & 0 & 0 \cdots \lambda_n^k \end{bmatrix} \tag{8.4.14}$$

which, together with (8.3.15) and (8.4.13), yields an explicit expression for the state-transition matrix of the original system (8.4.8); that is

$$\mathbf{\Phi}(k) = \mathbf{S}\mathbf{\Lambda}^k\mathbf{S}^{-1}. \tag{8.4.15}$$

8.5 DISCRETIZATION OF CONTINUOUS-TIME SYSTEMS

As discussed in Section 8.1, discrete-time systems may arise in various ways. A specific example is that of calculating the response of a continuous system by means of a digital computer. Under this situation, a continuous system must first be converted into an equivalent discrete system, which is the main topic of this section.

In the process of discretizing a continuous system described by a vector differential equation of the form

$$\dot{\mathbf{x}} = \mathbf{A}\mathbf{x} + \mathbf{B}\mathbf{u}, \tag{8.5.1}$$

where $\mathbf{A}$ and $\mathbf{B}$ are assumed to be constant matrices, we shall impose the following conditions on the system:

a) The sampling period T is constant at all times, i.e., signals are sampled at equally spaced time instants of T seconds.
b) The input vector $\mathbf{u}$ can change only at the sampling instants; that is, at the kth sampling instant

$$\mathbf{u}(kT + \tau) = \mathbf{u}(kT), \qquad 0 < \tau < T \tag{8.5.2}$$

for $k = 0, 1, 2, \ldots$

With the above assumptions it can be shown that the equivalent linear discrete-time system of (8.5.1) can be expressed as

$$\mathbf{x}[(k + 1)T] = \mathbf{G}\mathbf{x}(kT) + \mathbf{H}\mathbf{u}(kT). \tag{8.5.3}$$

In the following development, we shall proceed to determine the constant matrices $\mathbf{G}$ and $\mathbf{H}$ in (8.5.3). The solution of (8.5.1) is given by (7.2.11), which is repeated here for convenience:

$$\mathbf{x}(t) = e^{\mathbf{A}(t - t_0)}\mathbf{x}(t_0) + \int_{t_0}^{t} e^{\mathbf{A}(t - \tau)}\mathbf{B}\mathbf{u}(\tau)\, d\tau, \qquad t \geq t_0. \tag{8.5.4}$$

Substituting $t = (k + 1)T$ and $t_0 = k_0 T$ into (8.5.4), we obtain

$$\mathbf{x}[(k + 1)T] = e^{\mathbf{A}(k + 1 - k_0)T}\mathbf{x}(k_0 T)$$

$$+ \int_{k_0 T}^{(k + 1)T} e^{\mathbf{A}[(k + 1)T - \tau]}\mathbf{B}\mathbf{u}(\tau)\, d\tau. \tag{8.5.5}$$

Expression (8.5.5) can be written in terms of $\mathbf{x}(kT)$ as

$$\mathbf{x}[(k + 1)T] = e^{\mathbf{A}T}\mathbf{x}(kT) + \int_{kT}^{(k + 1)T} e^{\mathbf{A}[(k + 1)T - \tau]}\mathbf{B}\mathbf{u}(kT)\, d\tau \tag{8.5.6}$$

in which the assumption (8.5.2) has been applied.

Consider the second term on the right-hand side of (8.5.6). Introducing a new variable $\sigma \triangleq \tau - kT$, we find that

$$\int_{kT}^{(k+1)T} e^{\mathbf{A}[(k+1)T-\tau]}\mathbf{B}u(k\tau)\, d\tau = \int_0^T e^{\mathbf{A}(T-\sigma)}\mathbf{B}u(kT)\, d\sigma$$

$$= \left[\int_0^T e^{\mathbf{A}\sigma}\mathbf{B}\, d\sigma\right]u(kT). \qquad (8.5.7)^*$$

In light of (8.5.7), (8.5.6) can be written as

$$\mathbf{x}[(k+1)T] = e^{\mathbf{A}T}\mathbf{x}(kT) + \left[\int_0^T e^{\mathbf{A}\sigma}\mathbf{B}\, d\sigma\right]u(kT) \qquad (8.5.8)$$

or

$$\mathbf{x}[(k+1)T] = \mathbf{G}\mathbf{x}(kT) + \mathbf{H}u(kT)$$

with **G** and **H** given by

$$\mathbf{G} = e^{\mathbf{A}T} \qquad (8.5.9)$$

and

$$\mathbf{H} = \int_0^T e^{\mathbf{A}\sigma}\mathbf{B}\, d\sigma. \qquad (8.5.10)$$

For a linear time-varying continuous system of the form

$$\dot{\mathbf{x}} = \mathbf{A}(t)\mathbf{x} + \mathbf{B}(t)\mathbf{u}, \qquad (8.5.11)$$

it can be readily shown that the equivalent discrete-time system is given by

$$\mathbf{x}[(k+1)T] = \mathbf{G}(kT)\mathbf{x}(kT) + \mathbf{H}(kT)\mathbf{u}(kT), \qquad (8.5.12)$$

where

$$\mathbf{G}(kT) = \mathbf{\Phi}[(k+1)T, kT] \qquad (8.5.13)$$

and

$$\mathbf{H}(kT) = \int_{kT}^{(k+1)T} \mathbf{\Phi}[(k+1)T, \tau]\mathbf{B}(\tau)\, d\tau \qquad (8.5.14)$$

with $\mathbf{\Phi}(t, \tau)$ being the state-transition matrix of the continuous system defined in (7.10.5). The reader is urged to verify the results expressed in (8.5.13) and (8.5.14).

8.6 STATE-TRANSITION FLOW GRAPHS

In Section 8.2, the state-space representation of a discrete-time system was discussed in detail when the system behavior was governed by an nth-order difference

---

* See Problem 8.7 (b).

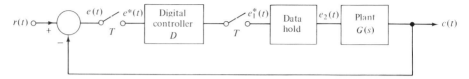

Fig. 8.6.1 Closed-loop computer control system.

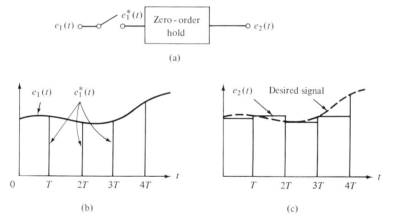

Fig. 8.6.2 Impulse response of zero-order hold.

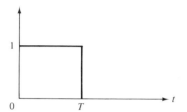

Fig. 8.6.3 Operation of a zero-order hold in t-domain. (a) Block diagram of zero-order hold.
(b) Input signal $e_1(t)$ and sampled signal $e_1^*(t)$. (c) Actual output signal $e_2(t)$ and desired
output signal.

equation. With the aid of the shift operator, we first constructed a discrete signal
flow graph and then identified the sink nodes of all the branches with transmittance
Z_T as state variables. In the present section, we shall show by means of constructing
a *hybrid state-transition flow graph** how the state equations describing the
behavior of a system can be obtained in discrete form readily programmable for
computer solution. Consider the block diagram for a typical closed-loop computer
control system, shown in Fig. 8.6.1, which contains both continuous and discrete

_____

* The term "hybrid" refers to the fact that both continuous and discrete signals are present
in the signal flow graph.

signals as usually found in most sampled-data systems. The *plant*, i.e., the object to be controlled, is in general a continuous system so that its input-output relationship can be described by a transfer function $G(s)$ in the s-domain as shown. The *data hold* is a device used to reconstruct continuous data from a sequence of samples or numbers. For convenience we shall assume that the data hold in Fig. 8.6.1 is of zero-order,* i.e., its impulse response has the waveform as indicated in Fig. 8.6.2. Figure 8.6.3 shows the input and output waveforms of a zero-order hold circuit. An inspection of Fig. 8.6.3 indicates that the input and the output of the zero-order hold are related by

$$e_2(kT + \tau) = e_1(kT), \qquad 0 < \tau \le T. \tag{8.6.1}$$

A difference equation which represents the input-output characteristic of a digital controller may be written as

$$e_1[(k + n)T] + a_{n-1}e_1[(k + n - 1)T] + \cdots + a_1e_1[(k + 1)T] + a_0e_1(kT)$$

$$= b_n e[(k + n)T] + b_{n-1}e[(k + n - 1)T] + \cdots + b_1 e[(k + 1)T] + b_0 e(kT). \tag{8.6.2}$$

To obtain the discrete difference equations describing the behavior of the closed-loop system, we first construct a hybrid state-transition flow graph. This hybrid flow graph consists of three parts: (a) the state-transition flow graph corresponding to the continuous elements of the system, (b) the flow graph corresponding to the digital controller, and (c) the link, connecting the continuous and the discrete elements which together represent the sampled and hold device. We discussed the state-transition flow graph for a continuous system in detail in Sections 7.6 and 7.7, whereas we studied the representation of the discrete system in Section 8.2. Once we have drawn the hybrid transition flow graph for the closed-loop system, we can apply Mason's formula to determine the expressions for the transformed state variables representing the continuous signals. Then we perform the inverse Laplace transformation on these transformed equations. These equations, evaluated at $t = kT$, together with those representing the discrete signals provide the required set of state equations in discrete form. To illustrate the step-by-step approach to the determination of the set of discrete-time state equations for a closed-loop control system, we shall consider the following example.

Example 8.6.1 Figure 8.6.4 depicts a closed-loop sampled data control system. Let the input-output characteristic of the digital controller D be described by

$$e_1[(k + 1)T] + e_1(kT) = 4e[(k + 1)T] + 2e(kT). \tag{8.6.3}$$

Determine the discrete-time state equations of the system.

The differential equation corresponding to

$$G(s) = \frac{C(s)}{E_2(s)} = \frac{2}{(s + 1)(s + 2)}$$

---

* Data holds of various orders may be used for data reconstruction. The effectiveness of the zero-order hold as an extrapolating device depends on the sampling rate as well as other factors. For further details, see, for example, [KU 2], Chapter 3.

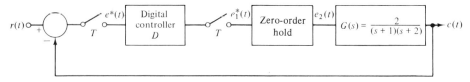

Fig. 8.6.4 Closed-loop digital control system of Example 8.6.1.

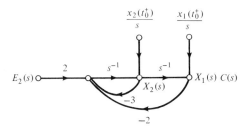

Fig. 8.6.5 State transition flow graph of the plant $G(s) = \dfrac{2}{(s+1)(s+2)}$.

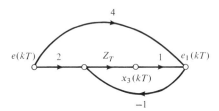

Fig. 8.6.6 Signal flow graph of (8.6.3).

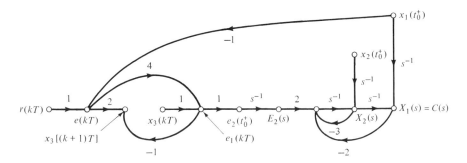

Fig. 8.6.7 Hybrid state transition flow graph of the control system in Example 8.6.1.

is given by

$$\frac{d^2 c}{dt^2} + 3\frac{dc}{dt} + 2c = 2e_2(t) \tag{8.6.4}$$

which, in light of (7.9.3) and (7.9.6), has the following state-space representation:

$$\begin{bmatrix} \dot{x}_1 \\ \dot{x}_2 \end{bmatrix} = \begin{bmatrix} 0 & 1 \\ -2 & -3 \end{bmatrix} \begin{bmatrix} x_1 \\ x_2 \end{bmatrix} + \begin{bmatrix} 0 \\ 2 \end{bmatrix} e_2 \tag{8.6.5}$$

with $c \triangleq x_1$. The state-transition flow graph representing (8.6.5) is shown in Fig. 8.6.5.

Next, we construct the flow graph for the discrete elements of the system. Using the technique developed in Section 8.2, we can obtain the flow graph corresponding to (8.6.3) as depicted in Fig. 8.6.6 where the discrete state variable has been designated as $x_3(k)$. From Fig. 8.6.4, we find, for the zero-order hold,

$$e_2(t_0^+) = e_2(t_0 + \tau) = e_1(t_0^+), \qquad 0 < \tau < T \tag{8.6.6}$$

and, for the input to the digital controller,

$$e(kT) = r(kT) - c(kT) \tag{8.6.7}$$

which, for $k = k_0$, becomes

$$e(t_0) = r(t_0) - c(t_0). \tag{8.6.8}$$

In light of (8.6.6) through (8.6.8), together with Figs. 8.6.5 and 8.6.6, the hybrid state-transition flow graph for the closed-loop system can be constructed as depicted in Fig. 8.6.7 with Z_T omitted for convenience.

Note that in the figure a branch of transmittance s^{-1} was connected between $e_2(t_0^+)$ and $E_2(s)$ representing the transition from the t-domain to the s-domain. For the discrete portion of the system, we find that

$$x_3[(k + 1)T] = 2e(kT) - e_1(kT) = 2r(kT) - 2x_1(t_0^+) - e_1(kT) \tag{8.6.9}$$

and

$$e_1(kT) = x_3(kT) + 4e(kT) = x_3(kT) + 4r(kT) - 4x_1(t_0^+). \tag{8.6.10}$$

Eliminating $e_1(kT)$ between (8.6.9) and (8.6.10) yields

$$x_3[(k + 1)T] = 2x_1(t_0^+) - x_3(kT) - 2r(kT). \tag{8.6.11}$$

Utilizing Mason's gain formula, we obtain the state equations for $X_1(s)$ and $X_2(s)$:

$$\begin{aligned}
X_1(s) &= \frac{8s^{-3}}{1 + 3s^{-1} + 2s^{-2}} [r(kT) - x_1(t_0^+)] + \frac{s^{-1}(1 + 3s^{-1})}{1 + 3s^{-1} + 2s^{-2}} x_1(t_0^+) \\
&\quad + \frac{s^{-2} x_2(t_0^+)}{1 + 3s^{-1} + 2s^{-2}} + \frac{2s^{-3}}{1 + 3s^{-1} + 2s^{-2}} x_3(kT) \\
&= \frac{s^2 + 3s - 8}{s(s + 1)(s + 2)} x_1(t_0^+) + \frac{1}{(s + 1)(s + 2)} x_2(t_0^+) + \frac{2}{s(s + 1)(s + 2)} x_3(kT) \\
&\quad + \frac{8}{s(s + 1)(s + 2)} r(kT)
\end{aligned} \tag{8.6.12}$$

and

$$X_2(s) = \frac{8s^{-2}}{1 + 3s^{-1} + 2s^{-2}}[r(kT) - x_1(t_0^+)] + \frac{s^{-1}}{1 + 3s^{-1} + 2s^{-2}}x_2(t_0^+)$$

$$+ \frac{2s^{-2}}{1 + 3s^{-1} + 2s^{-2}}x_3(kT)$$

$$= -\frac{8}{(s+1)(s+2)}x_1(t_0^+) + \frac{s}{(s+1)(s+2)}x_2(t_0^+) + \frac{2}{(s+1)(s+2)}x_3(kT)$$

$$+ \frac{8}{(s+1)(s+2)}r(kT). \tag{8.6.13}$$

Taking the inverse Laplace transform of (8.6.12) and (8.6.13) and substituting $t_0^+ = kT$ and $t = (k+1)T$ in (8.6.11) through (8.6.13), we obtain the discrete-time state equations

$$x_1[(k+1)T] = (-4 + 10e^{-kT} - 5e^{-2kT})x_1(kT) + (e^{-kT} - e^{-2kT})x_2(kT)$$

$$+ (1 - 2e^{-kT} + e^{-2kT})x_3(kT) + 4(1 - 2e^{-kT} + e^{-2kT})r(kT) \quad (8.6.14)$$

$$x_2[(k+1)T] = 8(-e^{-kT} + e^{-2kT})x_1(kT) + (-e^{-kT} + 2e^{-2kT})x_2(kT)$$

$$+ 2(e^{-kT} - e^{-2kT})x_3(kT) + 8(-e^{-kT} + e^{-2kT})r(kT) \quad (8.6.15)$$

$$x_3[(k+1)T] = 2x_1(kT) - x_3(kT) - 2r(kT). \tag{8.6.16}$$

The reader is invited to verify the above results.

8.7 THE Z TRANSFORMATION

Laplace transformation is a very powerful tool for solving linear differential equations with constant coefficients: Differential equations are first transformed into an equivalent set of algebraic equations, partial fraction expansions are then carried out, and the solution is obtained with the aid of a table of Laplace transform-pairs. In the case of discrete-time systems, application of the Laplace transformation method will usually lead to nonalgebraic equations from which both partial fraction expansions and inverse Laplace transforms are difficult to obtain. In this section, we introduce the Z transformation primarily for two purposes: (1) to provide an analogy between continuous and discrete-time systems, and (2) to establish the mathematical background necessary for the discussion of the interesting topic of digital filter in the next section.

The output of an ideal sampler to continuous input such as that shown in Fig. 8.1.1 can be regarded as the product of a continuous function $f(t)$ and a train of impulses of unit area as depicted in Fig. 8.7.1.

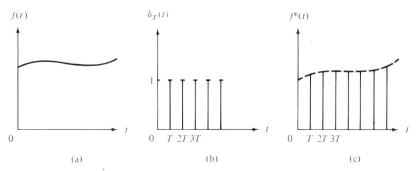

Fig. 8.7.1 Waveforms illustrating the operation of an ideal sampler. (a) Input signal to sampler. (b) Train of impulses of unit area. (c) Output of sampler.

In terms of the train of impulses, the sampled function $f^*(t)$ can be expressed as

$$f^*(t) = f(t)\delta_T(t)$$

$$= f(t) \sum_{k=0}^{\infty} \delta(t - kT)$$

$$= \sum_{k=0}^{\infty} f(kT)\delta(t - kT), \qquad (8.7.1)$$

where the impulse sequence $\delta_T(t)$, defined by

$$\delta_T(t) \triangleq \sum_{k=0}^{\infty} \delta(t - kT), \qquad (8.7.2)$$

is interpreted as being a sequence of impulses whose areas are equal to unity. Thus, in (8.7.1), the sampled function $f^*(t)$ can be regarded as a sequence of delta functions whose respective areas are equal to $f(t)$ evaluated at the corresponding time instants. In other words, no ambiguity should arise if (8.7.1) is properly interpreted.

Taking the Laplace transform of both sides of (8.7.1), we obtain

$$F^*(s) = \mathscr{L}\{f^*(t)\}$$

$$= \sum_{k=0}^{\infty} f(kT)e^{-kTs}, \qquad (8.7.3)$$

where the relation $\mathscr{L}\{f(t - \tau)u(t - \tau)\} = e^{-s\tau}F(s)$ has been applied. An inspection of (8.7.3) reveals the interesting property that the Laplace transform of a sampled function is in general a transcendental function whose inverse Laplace transform is usually difficult to obtain. To overcome this difficulty, the Z transformation is introduced by defining the variable z as

$$z \triangleq e^{Ts}. \qquad (8.7.4)$$

Substituting (8.7.4) into (8.7.3), we obtain the (one-sided) Z transform of the function $f(t)$; viz.,

$$F(z) \triangleq Z[f(t)] \triangleq F^*(s)\Big|_{s=(1/T)\ln z} = \sum_{k=0}^{\infty} f(kT)z^{-k}. \qquad (8.7.5)$$

From the foregoing discussion, it is obvious that to obtain the Z transform of a function $f(t)$ the following three steps must be performed: (a) sample the function $f(t)$ with sampling period T to yield $f^*(t)$; (b) take the Laplace transform of $f^*(t)$, that is, $F^*(s) = \mathscr{L}\{f^*(t)\}$, and then (3) replace s by $(1/T)\ln z$ to obtain

$$F(z) = F^*(s)\big|_{s=(1/T)\ln z}$$

Example 8.7.1 Let us determine the Z transform of the simple function

$$f(t) = e^{-at},$$

where a is a real constant. Following the above steps, we have

$$f^*(t) = e^{-at}\delta_T(t) = \sum_{k=0}^{\infty} e^{-akT}\delta(t - kT)$$

whose Laplace transform is

$$F^*(s) = \sum_{k=0}^{\infty} e^{-akT}e^{-kTs} = \sum_{k=0}^{\infty} e^{-k(s+a)T}.$$

Assuming that $|e^{-(s+a)T}| < 1$, $F^*(s)$ can be expressed in compact form as

$$F^*(s) = \frac{1}{1 - e^{-(s+a)T}}.$$

Finally, applying (8.7.4) to the above expression, we obtain the required Z transform

$$F(z) = \frac{1}{1 - z^{-1}e^{-aT}} = \frac{z}{z - e^{-aT}}.$$

Often it is desired to determine the Z transform $F(z)$ directly from a prescribed Laplace transform $F(s)$ of some function $f(t)$. Using impulse approximation, we find that the Laplace and Z transforms of the sampled function $f(t)\delta_T(t)$ are, respectively, given by

$$F^*(s) = \sum_{\substack{\text{all poles} \\ \text{of } F(p)}} \text{residue}\left[F(p)\frac{1}{1 - e^{-T(s-p)}} \right] \qquad (8.7.6)$$

and

$$F(z) = \sum_{\substack{\text{all poles} \\ \text{of } F(p)}} \text{residue}\left[F(p)\frac{1}{1 - e^{pT}z^{-1}} \right]. \qquad (8.7.7)\dagger$$

---

† For detailed derivation, see, for example, Ragazzini and Franklin [RA 1].

Example 8.7.2 The *Z* transform corresponding to the Laplace transform

$$F(s) = \frac{1}{(s + a)(s + b)} \qquad (8.7.8)$$

can be determined as follows: The poles of $F(s)$ occur at $s = -a$ and $s = -b$. By means of (8.7.7), we find

$$\text{Residue}\left[\frac{F(p)}{1 - e^{pT}z^{-1}}\right]_{p = -a} = \frac{1}{(p + b)(1 - e^{pT}z^{-1})}\bigg|_{p = -a}$$

$$= \frac{1}{(b - a)(1 - e^{-aT}z^{-1})} \qquad (8.7.9)$$

and

$$\text{Residue}\left[\frac{F(p)}{1 - e^{pT}z^{-1}}\right]_{p = -b} = \frac{1}{(p + a)(1 - e^{pT}z^{-1})}\bigg|_{p = -b}$$

$$= \frac{1}{(a - b)(1 - e^{-bT}z^{-1})}. \qquad (8.7.10)$$

Combining the terms in (8.7.9) and (8.7.10), we obtain

$$F(z) = \frac{1}{(b - a)} \frac{1}{(1 - e^{-aT}z^{-1})} + \frac{1}{(a - b)} \frac{1}{(1 - e^{-bT}z^{-1})}$$

$$= \frac{1}{(a - b)} \frac{z^{-1}(e^{-bT} - e^{-aT})}{(1 - e^{-bT}z^{-1})(1 - e^{-aT}z^{-1})}$$

$$= \frac{1}{(a - b)} \frac{z(e^{-bT} - e^{-aT})}{(z - e^{-bT})(z - e^{-aT})}$$

A more direct technique than the formal application of the residue formula makes use of the expansion of the function $F(p)$ into partial fractions. Then, by referring to a table such as Table 8.7.1, listing functions in the *t*-domain, their Laplace transforms, as well as their *Z* transforms, we can obtain either the pulse sequence $f(kT)$ or the *Z* transform. A glance at Table 8.7.1 reveals a very important point; that is, all the basic *Z* transform expressions have *z* as a common factor in the numerator. Thus, if a function $F(z)$ is given in the form of a rational function; that is,

$$F(z) = \frac{a_m z^m + a_{m-1} z^{m-1} + \cdots + a_1 z + a_0}{z^n + b_{n-1} z^{n-1} + \cdots + b_1 z + b_0}$$

$$= \frac{a_m z^m + a_{m-1} z^{m-1} + \cdots + a_1 z + a_0}{(z - z_1)(z - z_2) \cdots (z - z_n)}, \qquad (8.7.11)$$

Table 8.7.1 Z transform pairs.

	Time function $f(t)$, $t \geq 0$	Laplace transform $F(s)$	Z transform $F(z)$
1	Unit impulse $\delta(t)$	1	1
2	Unit step $u(t)$	$\dfrac{1}{s}$	$\dfrac{z}{z-1}$
3	$\delta_T(t) = \displaystyle\sum_{k=0}^{\infty} \delta(t-kT)$	$\dfrac{1}{1-e^{-Ts}}$	$\dfrac{z}{z-1}$
4	t	$\dfrac{1}{s^2}$	$\dfrac{Tz}{(z-1)^2}$
5	t^2	$\dfrac{2}{s^3}$	$\dfrac{T^2 z(z+1)}{(z-1)^3}$
6	e^{-at}	$\dfrac{1}{s+a}$	$\dfrac{z}{z-e^{-aT}}$
7	a^t	$\dfrac{1}{s-\ln a}$	$\dfrac{z}{z-a}$
8	$1 - e^{-at}$	$\dfrac{a}{s(s+a)}$	$\dfrac{z(1-e^{-aT})}{(z-1)(z-e^{-aT})}$
9	te^{-at}	$\dfrac{1}{(s+a)^2}$	$\dfrac{Tze^{-aT}}{(z-e^{-aT})^2}$
10	$\sin \omega t$	$\dfrac{\omega}{s^2+\omega^2}$	$\dfrac{z \sin \omega T}{z^2 - 2z \cos \omega T + 1}$
11	$e^{-at} \sin \omega t$	$\dfrac{\omega}{(s+a)^2+\omega^2}$	$\dfrac{ze^{-aT} \sin \omega T}{z^2 - 2ze^{-aT} \cos \omega T + e^{-2aT}}$
12	$\cos \omega t$	$\dfrac{s}{s^2+\omega^2}$	$\dfrac{z(z - \cos \omega T)}{z^2 - 2z \cos \omega T + 1}$
13	$e^{-at} \cos \omega t$	$\dfrac{s+a}{(s+a)^2+\omega^2}$	$\dfrac{z^2 - ze^{-aT} \cos \omega T}{z^2 - 2ze^{-aT} \cos \omega T + e^{-2aT}}$

where $n \geq m$ for physically realizable systems, it can be expanded into the form

$$F(z) = \frac{k_1 z}{z - z_1} + \frac{k_2 z}{z - z_2} + \cdots + \frac{k_n z}{z - z_n} \tag{8.7.12}$$

which is obtained by first expanding $F(z)/z$ [not $F(z)$] into partial fractions and then multiplying the result by z. The following examples will illustrate some of the techniques employed in obtaining the Z transform corresponding to a given Laplace transform as well as the inverse Z transform, that is $f(kT) = Z^{-1}[F(z)]$.

Example 8.7.3 The Z transform corresponding to the Laplace transform given in (8.7.8) can be obtained by first expanding $F(s)$ into partial fractions

$$F(s) = \frac{1}{(s + a)(s + b)} = \frac{1}{b - a}\left[\frac{1}{s + a} - \frac{1}{s + b}\right].$$

By referring to the sixth Z transform pair in Table 8.7.1, we then find the corresponding Z transform to be

$$F(z) = \frac{1}{b - a}\left[\frac{z}{z - e^{-aT}} - \frac{z}{z - e^{-bT}}\right]$$

$$= \frac{z(e^{-bT} - e^{-aT})}{(a - b)(z - e^{-bT})(z - e^{-aT})}.$$

Example 8.7.4 Determine the inverse Z transform of

$$F(z) = \frac{Az}{[z - (1 - A)](z - 1)}.$$

Expanding $F(z)/z$ into partial fractions, we get

$$\frac{F(z)}{z} = \frac{1}{z - (1 - A)} - \frac{1}{z - 1}.$$

Therefore,

$$F(z) = \frac{z}{z - (1 - A)} - \frac{z}{z - 1}.$$

From Table 8.7.1, we obtain

$$f(kT) = (1 - A)^{kT} - 1.$$

Another way of determining the inverse Z transform is the power-series method which is based upon the technique of expanding $F(z)$ into powers of z^{-1}; that is,

$$F(z) = f_0 + f_1 z^{-1} + f_2 z^{-2} + \cdots + f_k z^{-k} + \cdots. \qquad (8.7.13)$$

Comparing (8.7.13) with the defining series for a Z transform in (8.7.5) which we repeat here for convenience:

$$F(z) = f(0) + f(T)z^{-1} + f(2T)z^{-2} + \cdots + f(kT)z^{-k} + \cdots, \qquad (8.7.5)$$

we find that the coefficient of the z^{-k} term in (8.7.13) corresponds to the value of $f(t)$ at $t = kT(k = 0, 1, 2, \ldots)$. We shall illustrate this method by means of an example that follows.

Example 8.7.5 Determine the inverse Z transform of

$$F(z) = \frac{z^2}{z^2 + z + 2} \qquad (8.7.14)$$

by the power-series method.

The expression (8.7.14) can be written as

$$F(z) = \frac{1}{1 + z^{-1} + 2z^{-2}}$$

which can be expanded into a power series in z^{-1} by long division,

$$
\begin{array}{r}
1 - z^{-1} - z^{-2} + 3z^{-3} - z^{-4} + \cdots \\
1 + z^{-1} + 2z^{-2} \,\big|\, \overline{1 \phantom{+ z^{-1} + 2z^{-2}}} \\
1 + z^{-1} + 2z^{-2} \\
\hline
- z^{-1} - 2z^{-2} \\
- z^{-1} - z^{-2} - 2z^{-3} \\
\hline
- z^{-2} + 2z^{-3} \\
- z^{-2} - z^{-3} - 2z^{-4} \\
\hline
3z^{-3} + 2z^{-4} \\
3z^{-3} + 3z^{-4} + 6z^{-5} \\
\hline
- z^{-4} - 6z^{-5} \\
- z^{-4} - z^{-5} - 2z^{-6} \\
\hline
- 5z^{-5} + \cdots
\end{array}
$$

yielding

$$F(z) = 1 - z^{-1} - z^{-2} + 3z^{-3} - z^{-4} + \cdots .$$

By inspection of the above infinite series, we obtain the sampled function

$$f^*(t) = 1\delta(t) - \delta(t - T) - \delta(t - 2T) + 3\delta(t - 3T) - \delta(t - 4T) + \cdots .$$

The reader is familiar with the fact that manipulation of Laplace transforms is greatly facilitated by applying a handful of special properties. Needless to say, the same is true in the case of Z transforms. These important properties are presented in the form of theorems, a majority of which can be proved simply by direct substitution into the definition of the Z transform.

Theorem 8.7.1 (*Linearity*) The Z transformation is a *linear* transformation; that is,

$$Z[af_1(t) + bf_2(t)] = aF_1(z) + bF_2(z). \tag{8.7.15}$$

Theorem 8.7.2 (*Real translation*) Let $F(z)$ be the Z transform of the function $f(t)u(t)$. Then the Z transform of the function $f(t - nT)u(t - nT)$ for any positive integer n is given by

$$Z[f(t - nT)u(t - nT)] = z^{-n}F(z). \tag{8.7.16}$$

Theorem 8.7.3 (*Time advance*) Let $F(z)$ be the Z transform of the function $f(t)u(t)$. Then the Z transform of $f(t + nT)u(t)$ for any positive integer n is given by

$$Z[f(t + nT)u(t)] = z^n\left[F(z) - \sum_{k=0}^{n-1} f(kT)z^{-k}\right]. \tag{8.7.17}$$

For $n = 1$, (8.7.9) reduces to

$$Z[f(t + T)u(t)] = zF(z) - zf(0) \qquad (8.7.18)$$

which performs the same role in the solution of difference equations as the well-known property

$$\mathscr{L}\left[\frac{d}{dt}f(t)\right] = sF(s) - f(0)$$

does in differential equations.

Theorem 8.7.4 (*Multiplication by e^{at}*) Let $F(z)$ be the Z transform of the function $f(t)$. Then the Z transform of the product $f(t)e^{at}$ for any real constant a is given by

$$Z[e^{at}f(t)] = F(ze^{aT}). \qquad (8.7.19)$$

As an application of the above theorem, let us determine the Z transform of e^{-at}. Suppose that $Z[u(t)] = z/(z - 1)$ is known. Then

$$Z[e^{-at}] = Z[e^{-at}u(t)] = \frac{ze^{-aT}}{ze^{-aT} - 1} = \frac{z}{z - e^{aT}}$$

which is identical to the result shown in Table 8.7.1.

Theorem 8.7.5 (*Initial value*) Let $F(z)$ be the Z transform of $f(t)$. Then

$$\lim_{t \to 0} f^*(t) = \lim_{k \to 0} f(kT) = \lim_{z \to \infty} F(z). \qquad (8.7.20)$$

Theorem 8.7.6 (*Final value*) With the same notation as in Theorem 8.7.5, if the function $(1 - z^{-1})F(z)$ does not have poles on or outside the unit circle in the z-plane, i.e., $|z| = 1$, then

$$\lim_{t \to \infty} f^*(t) = \lim_{k \to \infty} f(kT) = \lim_{z \to 1}[(1 - z^{-1})F(z)]. \qquad (8.7.21)$$

The final-value theorem (8.7.21) is of great value to the designer of discrete-time systems since in general the requirement of steady-state performance is included in the specification.

The following examples provide an illustration of the applications of some of the theorems discussed in this section.

Example 8.7.6 Determine the solution of the difference equation

$$8x[(k + 2)T] + 6x[(k + 1)T] + x(kT) = 8, \qquad (8.7.22)$$

with the initial conditions $x(0) = 0$ and $x(T) = 1$, by the Z-transform method. Let $X(z)$ be the Z transform of $x(t)$. Then multiplying (8.7.22) by z^{-k} and summing up each of the terms in the resulting equation with $k = 0, 1, \ldots, \infty$, we get

$$8\sum_{k=0}^{\infty} x[(k + 2)T]z^{-k} + 6\sum_{k=0}^{\infty} x[(k + 1)T]z^{-k} + \sum_{k=0}^{\infty} x(kT)z^{-k} - 8\sum_{k=0}^{\infty} z^{-k}. \qquad (8.7.23)$$

Next, note the fact that the third term on the left-hand side of (8.7.23) is, by definition (8.7.5), the Z transform of $x(t)$. Then if we make use of (8.7.17) and (8.7.18), with $Z[1] = Z[u(t)] = z/(z - 1)$, we can obtain from (8.7.23)

$$8z^2 \left[X(z) - \sum_{k=0}^{1} x(kT)z^{-k} \right] + 6z[X(z) - x(0)] + X(z) = \frac{8z}{z - 1}.$$

After substituting the initial conditions and simplifying, we obtain

$$(8z^2 + 6z + 1)X(z) = \frac{8z^2}{z - 1}$$

or

$$X(z) = \frac{8z^2}{(z - 1)(8z^2 + 6z + 1)} = \frac{z^2}{(z - 1)(z + \frac{1}{2})(z + \frac{1}{4})}. \qquad (8.7.24)$$

Next, expanding $X(z)/z$ into partial fractions, we find that

$$\frac{X(z)}{z} = \frac{\frac{8}{15}}{z - 1} - \frac{\frac{4}{3}}{z + \frac{1}{2}} + \frac{\frac{4}{5}}{z + \frac{1}{4}}$$

or

$$X(z) = \tfrac{8}{15}z/(z - 1) - \tfrac{4}{3}z/(z + \tfrac{1}{2}) + \tfrac{4}{5}z/(z + \tfrac{1}{4}).$$

Hence, the solution $x(kT)$ is given by

$$x(kT) = \tfrac{8}{15}u(kT) - \tfrac{4}{3}(-\tfrac{1}{2})^{kT} + \tfrac{4}{5}(-\tfrac{1}{4})^{kT}. \qquad (8.7.25)$$

Example 8.7.7   Determine the final value of $x(kT)$ of the difference equation (8.7.22) by means of the final-value theorem. Making use of (8.7.24), we form the product

$$(1 - z^{-1})X(z) = \frac{z}{(z + \frac{1}{2})(z + \frac{1}{4})}$$

which does not possess poles on or outside the unit circle $|z| = 1$. Therefore, the final-value theorem may be applied, yielding

$$x(\infty) = \lim_{z \to 1} [(1 - z^{-1})X(z)]$$

$$= \tfrac{8}{15}$$

which can be easily checked to be the correct answer by allowing k to approach infinity in (8.7.25).

In like manner, the Z-transform method can be applied to solve the matrix difference equations for a linear discrete-time system with multiple inputs and outputs as in the case of solving matrix differential equations discussed in the preceding sections. That is, consider the following matrix difference equations:

$$\mathbf{x}[(k+1)T] = \mathbf{A}\mathbf{x}(kT) + \mathbf{B}\mathbf{u}(kT) \tag{8.7.26}$$

$$\mathbf{y}(kT) = \mathbf{C}\mathbf{x}(kT) + \mathbf{D}\mathbf{u}(kT). \tag{8.7.27}$$

Taking the Z transform on both sides of (8.7.26) yields

$$z\mathbf{X}(z) - z\mathbf{x}(kT)|_{k=0} = \mathbf{A}\mathbf{X}(z) + \mathbf{B}\mathbf{U}(z)$$

which can be solved for $\mathbf{X}(z)$ as

$$\mathbf{X}(z) = (z\mathbf{I} - \mathbf{A})^{-1}z\mathbf{x}(kT)|_{k=0} + (z\mathbf{I} - \mathbf{A})^{-1}\mathbf{B}\mathbf{U}(z). \tag{8.7.28}$$

The inverse Z transform of (8.7.28) is

$$\mathbf{x}(kT) = Z^{-1}[(z\mathbf{I} - \mathbf{A})^{-1}z]\mathbf{x}(0) + Z^{-1}[(z\mathbf{I} - \mathbf{A})^{-1}\mathbf{B}\mathbf{U}(z)], \tag{8.7.29}$$

where

$$\mathbf{x}(0) = \mathbf{x}(kT)|_{k=0}$$

has been substituted.

By referring to (8.3.15) and (8.4.6), the following identities

$$Z^{-1}[(z\mathbf{I} - \mathbf{A})^{-1}z] = \mathbf{A}^k \tag{8.7.30}$$

and

$$Z^{-1}[(z\mathbf{I} - \mathbf{A})^{-1}\mathbf{B}\mathbf{U}(z)] = \sum_{i=0}^{k-1} \mathbf{A}^{k-i-1}\mathbf{B}\mathbf{u}(iT) \tag{8.7.31}$$

can be readily obtained. The details are left as exercises.

8.8 INTRODUCTION TO DIGITAL FILTERS

An electric filter is usually regarded as a device made up of resistors, capacitors, and inductors which performs the processing of signal information. The advances made in the past decade have tremendously broadened the scope of filter theory. Electric filters can be classified in many different ways such as in terms of pass band (low-pass, band-pass, or high-pass), network configuration (lattice, ladder, or other), or active or passive, etc. In addition, since the processing of signals can be performed either by filters made up of analog components such as inductors, capacitors, amplifiers or by digital or sampling techniques, filters can be classified as analog or digital filters according to the means of processing signal information.

   This section is intended to give the reader a brief introduction to the subject of digital filtering. The term *digital filter* can be regarded as meaning the computational process of transforming a sequence of numbers (the input) into another sequence of numbers (the output). The advantages of using digital filters are that they not only provide an excellent degree of reliability, accuracy, and stability, but

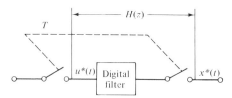

Fig. 8.8.1 A digital filter.

also yield, as a consequence of availability of low-cost digital integrated circuits, reductions in size, cost, and weight, especially in the low frequency range.

Let $u^*(t)$ and $x^*(t)$ be, respectively, the sampled input and output of a digital filter as depicted in Fig. 8.8.1. Then the input and output pulse sequences can be related by a difference equation of the form

$$x(nT) + b_1 x[(n-1)T] + b_2 x[(n-2)T] + \cdots + b_k c[(n-k)T]$$

$$= a_0 u(nT) + a_1 u[(n-1)T] + a_2 u[(n-2)T] + \cdots + a_m u[(n-m)T], \qquad (8.8.1)$$

where the a's and b's are constants. Equation (8.8.1) is a linear relation through which the present output number $x(nT)$ can be determined by taking the weighted sum of a finite number of input and output values. If $b_k \neq 0$, then the filter represented by (8.8.1) is referred to as a *kth-order digital filter*. In other words, a kth-order digital filter is described by a kth-order difference equation. If, in (8.8.1), all the b's are zero, then the filter is said to be a *nonrecursive* or *transversal digital filter*; otherwise, the filter is of *recursive* type.

Taking the Z transform on both sides of (8.8.1), we obtain

$$X(z) + b_1 z^{-1} X(z) + b_2 z^{-2} X(z) + \cdots + b_k z^{-1} X(z)$$

$$= a_0 U(z) + a_1 z^{-1} U(z) + a_2 z^{-2} U(z) + \cdots + a_m z^{-m} U(z) \qquad (8.8.2)$$

from which the system transfer function $H(z) \triangleq X(z)/U(z)$ can be written as

$$H(z) = \frac{X(z)}{U(z)} = \frac{a_0 + a_1 z^{-1} + a_2 z^{-2} + \cdots + a_m z^{-m}}{1 + b_1 z^{-1} + b_2 z^{-2} + \cdots + b_k z^{-k}}. \qquad (8.8.3)$$

Making a term-by-term comparison between (8.8.1) and (8.8.2), it can be readily seen that z^{-1} is the *unit delay operator*.

The implementation of the transfer function $H(z)$ can be done in several ways depending upon how we manipulate the expression (8.8.3). Thus, for example, we write

$$H(z) = H_1(z) H_2(z), \qquad (8.8.4)$$

where

$$H_1(z) \triangleq \frac{X_1(z)}{U(z)} \triangleq \frac{1}{1 + b_1 z^{-1} + b_2 z^{-2} + \cdots + b_k z^{-k}} \qquad (8.8.5)$$

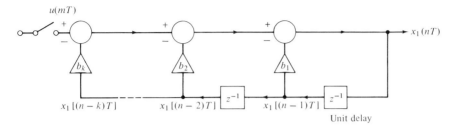

Fig. 8.8.2 Implementation of $H_1(z)$.

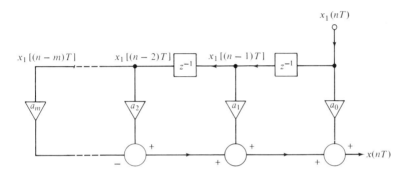

Fig. 8.8.3 Implementation of $H_2(z)$.

and

$$H_2(z) \triangleq \frac{X(z)}{X_1(z)} \triangleq a_0 + a_1 z^{-1} + \cdots + a_m z^{-m}. \qquad (8.8.6)$$

The block diagram showing the implementation of $H_1(z)$ is depicted in Fig. 8.8.2 from which the nth pulse of the output $x_1(nT)$ can be readily seen to be

$$x_1(nT) = u(nT) - b_1 x_1[(n-1)T] - b_2 x_1[(n-2)T] - \cdots - b_k x_1[(n-k)T]. \qquad (8.8.7)$$

Taking the Z transform on both sides of (8.8.7) yields

$$X_1(z) = U(z) - b_1 z^{-1} X(z) - b_2 z^{-2} X(z) - \cdots - b_k z^{-k} X(z) \qquad (8.8.8)$$

which is identical to (8.8.5) after simple algebraic manipulations.

In like manner, the block diagram for the implementation of $H_2(z)$ in (8.8.6) can be drawn as shown in Fig. 8.8.3 from which $x(nT)$ can be written as

$$x(nT) = a_0 x_1(nT) + a_1 x_1[(n-1)T] + a_2 x_1[(n-2)T] + \cdots + a_m x[(n-m)T]. \qquad (8.8.9)$$

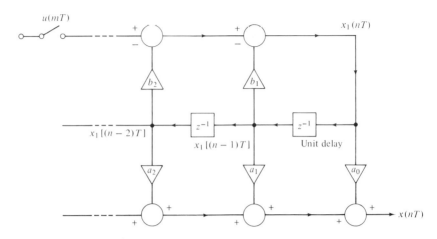

Fig. 8.8.4 Implementation of nth-order recursive filter with transfer function $H(z)$.

It should be noted that Fig. 8.8.3 is an example of a nonrecursive mth-order digital filter. Next, combining Figs. 8.8.2 and 8.8.3, we obtain the implementation of the transfer function $H(z)$ for a recursive kth-order digital filter as depicted in Fig. 8.8.4.

The implementation in Fig. 8.8.4 requires a minimum number of delay units [that is, the number of delay units is equal to the order of $H(z)$] and is referred to as the *canonical form* of realization. Other forms of realization such as the *cascade* and the *parallel* can also be accomplished. The cascade configuration can be obtained by first factoring the transfer function $H(z)$ into pole-zero pairs, i.e.,

$$H(z) = a_0 \frac{\displaystyle\prod_{i=1}^{m} (1 + \alpha_i z^{-1})}{\displaystyle\prod_{j=1}^{k} (1 + \beta_j z^{-1})}$$

(8.8.10)

$$= a_0 \left(\frac{1 + \alpha_1 z^{-1}}{1 + \beta_1 z^{-1}}\right)\left(\frac{1 + \alpha_2 z^{-1}}{1 + \beta_2 z^{-1}}\right) \cdots \left(\frac{1 + \alpha_m z^{-1}}{1 + \beta_m z^{-1}}\right)\left(\frac{1}{1 + \beta_{m+1} z^{-1}}\right) \cdots \left(\frac{1}{1 + \beta_k z^{-1}}\right).$$

The parallel configuration can be realized by expanding $H(z)$ into partial fractions, i.e.,

$$H(z) = \sum_{j=1}^{k} \frac{c_j}{1 + \beta_j z^{-1}} = \frac{c_1}{1 + \beta_1 z^{-1}} + \frac{c_2}{1 + \beta_2 z^{-1}} + \cdots + \frac{c_k}{1 + \beta_k z^{-1}}. \quad (8.8.11)$$

The cascade and parallel configuration are shown in Figs. 8.8.5 and 8.8.6, respectively.

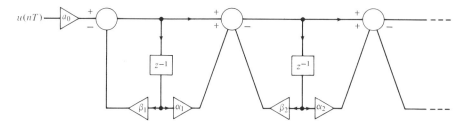

Fig. 8.8.5 Realization of $H(z)$ in cascade configuration.

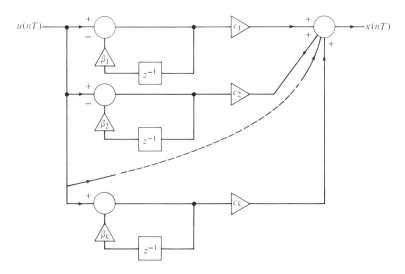

Fig. 8.8.6 Realization of $H(z)$ in parallel configuration.

Different methods are available for the design of digital filters. Due to space limitation, a detailed treatment on filter design methods is beyond the intent of our coverage. The interested reader is referred to the literature listed at the end of this chapter. However, to get a feel for this subject, we shall discuss briefly in the following simple example a procedure for designing a digital filter which will provide the same behavior at the sampling time instants as a continuous filter.

Example 8.8.1 Let the transfer function of a continuous filter be given by

$$G(s) = \frac{1}{(s + a)(s + b)},\tag{8.8.12}$$

where a and b are positive real numbers such that $a \neq b$. Expanding $G(s)$ into partial fractions, we find

$$G(s) = \frac{1}{b - a}\frac{1}{s + a} - \frac{1}{b - a}\frac{1}{s + b} \tag{8.8.13}$$

from which the impulse response $g(t)$ is

$$g(t) = \frac{1}{b - a}e^{-at} - \frac{1}{b - a}e^{-bt}. \tag{8.8.14}$$

At the sampling instants, we have

$$g(nT) = \frac{1}{b - a}(e^{-anT} - e^{-bnT}). \tag{8.8.15}$$

To ensure that the digital filter has the same behavior at sampling instants as the continuous filter, the transfer function of the digital filter

$$H(z) = \sum_{n=0}^{\infty} h(nT)z^{-n} \tag{8.8.16}$$

must satisfy the condition

$$h(nT) = g(nT) \qquad \text{for all} \quad n. \tag{8.8.17}$$

Thus, substituting (8.8.15) into (8.8.17) and then into (8.8.16), we get

$$
\begin{aligned}
H(z) &= \sum_{n=0}^{\infty} \frac{1}{b - a}(e^{-anT} - e^{-bnT})z^{-n} \\
&= \frac{1}{b - a}\left[\sum_{n=0}^{\infty} e^{-anT}z^{-n} - \sum_{n=0}^{\infty} e^{-bnT}z^{-n}\right] \\
&= \frac{1}{b - a}\left[\frac{1}{1 - e^{-aT}z^{-1}} - \frac{1}{1 - e^{-bT}z^{-1}}\right]
\end{aligned}
\tag{8.8.18}
$$

from which a parallel configuration can be realized immediately.

Note that the result in (8.8.18) can be obtained directly by determining the Z transform corresponding to the continuous transfer function $H(s)$. The general approach using the design method presented in this example is usually referred to as the *impulse response invariance technique*.

8.9 SUMMARY

In this chapter, a general linear discrete-time system has been discussed in some detail. In Section 8.2, by means of shift operator Z_τ, the state representation of a linear discrete-time system described by an nth-order difference equation can be easily obtained. We pointed out that Z_τ, sometimes referred to as the unit delay, is analogous to the integrator in a continuous system (with s^{-1} as its transfer function). In the discussion of linear digital filters using the Z transform, we indicated that z^{-1} is also a unit delay operator. In other words, the operators s^{-1} and z^{-1} are analogous operators in continuous and discrete-time systems, respectively. The analysis of a linear time-invariant continuous system by means of the Laplace transform method is equivalent to that of the discrete-time counterpart through the Z transform. Indeed, a very close analogy can be drawn between continuous and discrete-time systems. As an example, state-space representations of discrete-time systems discussed in Section 8.2 can be achieved by first taking the Z transform of the difference equations and then using z^{-1} as the unit delay operator.

In Sections 8.3 and 8.4, we discussed the state-transition matrix and its properties. Once the state-transition matrix is known, the solution of the state equation can be easily obtained. We investigated the conversion of continuous systems into their discrete-time equivalents in Section 8.5. A specific example on its application is that of calculating the response of a continuous system by means of a digital computer.

The topics of state-transition flow graphs and digital filters were briefly presented in Sections 8.6 and 8.8, respectively. The reader interested in exploring these subjects further is invited to consult the literature listed at the end of this chapter.

REFERENCES

1. H. Freeman, *Discrete-Time Systems*, John Wiley & Sons, 1965, Chapters 1 and 2.

2. B. C. Kuo, *Analysis and Synthesis of Sampled-Data Control Systems*, Prentice-Hall, 1963, Chapters 2, 3, and 5.

3. F. F. Kuo and J. F. Kaiser, *System Analysis by Digital Computers*, John Wiley & Sons, 1966, Chapter 7.

4. L. P. Huelsman, *Active Filters*, McGraw-Hill, 1970, Chapter 5.

PROBLEMS

8.1 A certain discrete-time system is described by the difference equation

$$x(k + 2) + 3x(k + 1) + 4x(k) = 2u(k).$$

Obtain two state-space representations for this system.

8.2 Determine a state-space representation for the system described by

$$x(k + 2) + 3x(k + 1) + 4x(k) = 2u(k + 2) + 3u(k + 1) + u(k).$$

8.3 Let the initial conditions of the system in Problem 8.2 be $x(1) = 1$, $x(0) = 3$, $u(1) = 2$, and $u(0) = 0$. Determine the initial conditions of the corresponding state-space representation.

8.4 Obtain an equivalent discrete-time system for the electrical network shown in Fig. P.8.4. Assume unity value for the sampling period T.

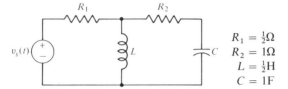

$R_1 = \frac{1}{2}\Omega$
$R_2 = 1\Omega$
$L = \frac{1}{2}H$
$C = 1F$

Figure P.8.4

8.5 Let the parameter values of the system of Fig. P.8.5 be given by $R_1 = 1\Omega$, $L = 1H$, $C = \frac{1}{2}F$, and $R_2 = 2\Omega$. If the system is initially in zero-state and if the voltage source is a unit-step, obtain:

a) an equivalent discrete-time representation for the system by using $i_L(t)$ and $v_C(t)$ as state variables;

b) the solution for the discrete-time representation found in (a) by assuming $T = 1$ sec.

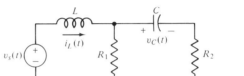

Figure P.8.5

8.6 Discretize the second-order continuous system described by

$$\ddot{x} = v_s(t),$$

where $v_s(t)$ is the input to the system.

8.7 a) Verify that both Z_t and its inverse (Section 8.2) are linear operators.
     b) Verify the expression given in (8.5.7).

8.8 Discretize the continuous system

$$\begin{bmatrix} \dot{x}_1 \\ \dot{x}_2 \end{bmatrix} = \begin{bmatrix} 1 & 0 \\ 1 & 1 \end{bmatrix} \begin{bmatrix} x_1 \\ x_2 \end{bmatrix}$$

and determine the corresponding transition matrix.

8.9 Determine an explicit expression for the state-transition matrix for the system described by

$$\mathbf{x}(k + 1) = \mathbf{A}\mathbf{x}(k) \qquad \text{where} \qquad \mathbf{A} = \begin{bmatrix} 1 & 4 \\ 1 & 1 \end{bmatrix}.$$

8.10 Determine the solution of the discrete system described by the equation

$$x(k + 2) + 3x(k + 1) + 2x(k) = 0, \qquad k = 1, 2, \ldots$$

subject to the initial conditions $x(0) = 1$, and $x(1) = -1$ by first finding an equivalent state-space representation.

8.11 For the closed-loop system depicted in Fig. P.8.11:

  a) draw the hybrid state-transition flow graph, and
  b) determine the output $c(t)$ for a unit-step input, i.e., $r(t) = u(t)$.

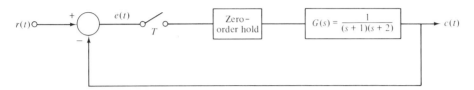

Figure P.8.11

8.12 Draw the state-transition flow graph for the difference equation

$$e_1[(k + 1)T] + 2e_1(kT) = 3e[(k + 1)T] + e(kT).$$

8.13 If the digital controller D of the sampled-data system shown in Fig. P.8.13 is described by the difference equation of Problem 8.12, determine the corresponding hybrid state-transition flow graph.

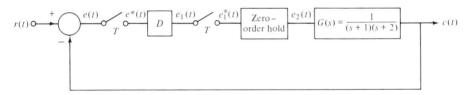

Figure P.8.13

8.14 Repeat Problem 8.10 by the Z-transform method.

8.15 Verify Theorem 8.7.3.

8.16 Verify Theorems 8.7.5 and 8.7.6.

8.17 The transfer function of a linear discrete-time system is given by

$$H(z) = \frac{X(z)}{U(z)} = \frac{64z^3 + 128z^2}{64z^3 + 56z^2 + 14z + 1}.$$

  a) Determine the discrete state equations for the system.
  b) Determine the difference equation relating $x(k)$ and $u(k)$.

8.18 Determine the canonical, the cascade, and the parallel forms of realization for the digital system in Problem 8.17.

CHAPTER 9

COMPUTER-AIDED NETWORK ANALYSIS I

9.1 INTRODUCTION

In recent years the digital computer has become a very powerful tool in engineering design, and consequently computer-aided network analysis and design have become important disciplines in the electrical engineering curriculum both at the undergraduate and the graduate levels. FORTRAN programming and numerical analysis techniques are now taught as lower division courses in many colleges and universities.

This chapter is intended to serve as an introduction to some of the basic methods which are used frequently in network analysis and design with the digital computer as an aid. The programs and subroutines to be presented in the subsequent sections will be kept relatively simple so that emphasis can be placed on theory rather than on the details of sophisticated programming techniques.* Examples will be used whenever appropriate to illustrate important ideas in connection with a theory or to show steps involved in an algorithm or method. Although an appendix has been included to review some of the basic numerical methods that are used in this chapter, the reader is assumed to have taken an introductory course in numerical analysis and to have some working knowledge in FORTRAN programming.

9.2 ANALYSIS OF SIMPLE *RL* AND *RC* NETWORKS

In this section, we shall discuss computer applications of certain basic numerical techniques in solving simple *RL* and *RC* networks and, in the meantime, introduce two subroutines, namely ISIMS (*Integration by SIMpSon's rule*) and RKAMC (*Runge-Kutta-Adams-Moulton predictor-Corrector method*). The former subroutine is used to perform the numerical integration required in the solution of the voltage across a capacitor or the current through an inductor, while the latter is for solving a single first-order ordinary differential equation in connection with the analysis of a simple network with its order of complexity equal to one.†

---

* The authors indeed realize that a programmer or even a user of a program should consider such factors as the speed, accuracy, rate of convergence, etc., when he writes or uses a program. However, a detailed discussion of such topics is clearly beyond the scope of the intended coverage here in this chapter.

† Refer to Chapter 3, Section 3.5, for the definition of the order of complexity of a network. Analysis of networks of higher order will be discussed in a later section of this chapter.

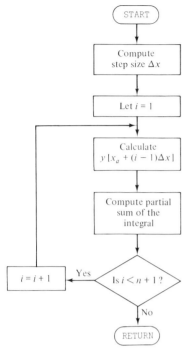

Fig. 9.2.1 A simplified flow chart for subroutine ISIMS.

```
SUBROUTINE ISIMS(XA,XB,N,Z)
DIMENSION A(100)
DX=(XB-XA)/N
M=N+1
DO 10  I=1,M
10 A(I)=Y(XA+(I-1)*DX)
YEVEN=0.
DO 20   J=2,N,2
20 YEVEN = YEVEN + A(J)
YODD = 0.
JI = N - 1
IF (N .LT. 4) GO TO 40
DO 30 K= 3,JI,2
30 YODD = YODD + A(K)
40 Z = (A(1)+2.*YODD +4.*YEVEN + A(M))* DX/3.
RETURN
END
```

Fig. 9.2.2 FORTRAN listing for subroutine ISIMS.

    Also, for the purpose of displaying computer results, we shall present a plotting subroutine, VPLOT, which provides the plotting of two-dimensional arrays in discrete form.

    Subroutine ISIMS is a simple subroutine for evaluating integrals of the form

$$\int_{x_a}^{x_b} y(x)\,dx$$

by the application of Simpson's rule, which is one of the most widely known methods in numerical integration and is given by

$$\int_{x_a}^{x_b} y(x)\,dx \cong \frac{\Delta x}{3}(y_1 + 4y_2 + 2y_3 + 4y_4 + \cdots + 2y_{n-1} + 4y_n + y_{n+1}), \quad (9.2.1)^*$$

where n is an even integer denoting the number of subintervals between x_a and x_b, and where

$$\Delta x = \frac{x_b - x_a}{n}, \quad \text{the length of each subinterval or the step size,}$$

$$y_i = y(x_a + (i - 1)\Delta x) \quad i = 1, 2, \ldots, n + 1.$$

    Based on (9.2.1), a simple FORTRAN program is written, namely subroutine ISIMS, of which the flowchart and the FORTRAN listing are given in Figs. 9.2.1 and 9.2.2, respectively.

---

* See Appendix A.3, Section A.3.2.

It is evident from Fig. 9.2.2 that subroutine ISIMS requires a function sub-program denoted by FUNCTION Y(X) which defines the function $y(x)$ to be integrated. This subprogram must be supplied by the user of the subroutine.

For ease of future reference, the characteristics of this subroutine are summarized as follows:

Subroutine ISIMS

1. *Purpose*: To evaluate

$$\int_{x_a}^{x_b} y(x)\, dx.$$

2. *Calling statement*: CALL ISIMS (XA, XB, N, Z).

3. *Input arguments*: XA = lower limit of the integral; XB = upper limit of the integral; N = number of iterations used (an even number must be used).

4. *Output argument*: Z = result of the integral.

5. *User-supplied subprogram*: FUNCTION Y(X).

The application of subroutine ISIMS is illustrated by the following examples.

Example 9.2.1   For the network shown in Fig. 9.2.3, determine the voltage across the capacitor for $t = 4$ using subroutine ISIMS, with $v_c(0) = 0$ and

$$i(t) = (1 + t + t^2). \tag{9.2.2}$$

Since

$$v_c(t) = \frac{1}{C}\int_0^t i(t)\, dt + v_c(0), \tag{9.2.3}$$

we obtain, by substituting (9.2.2) into (9.2.3) and noting that $v_c(0) = 0$,

$$v_c(4) = \int_0^4 (1 + t + t^2)\, dt. \tag{9.2.4}$$

Thus we write a function subprogram for $i(t)$ and a main program, which will call subroutine ISIMS to evaluate (9.2.4) with the value of n set to $2, 4, 6, \ldots, 30$, as shown in (a) and (b) of Fig. 9.2.4.

The computer output is shown in Fig. 9.2.5. It is interesting to note that in this simple example the choice of N is not critical for the accuracy of the computed result.

Next, we shall discuss the method of obtaining the digital solution of a first-order ordinary differential equation of the form

$$\frac{dy}{dx} = f(x, y) \tag{9.2.5}$$

which is the typical equation describing the behavior of a network with its order of complexity equal to one. For this purpose we introduce the subroutine RKAMC

```
REAL FUNCTION Y(X)
Y = 1.+X+X*X
RETURN
END
```
(a)

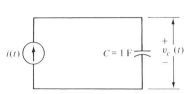

Fig. 9.2.3 Network for Example 9.2.1.

```
C     MAIN PROGRAM FOR EXAMPLE 9-2-1
      DO 5 N=2,30,2
      CALL ISIMS (0.,4.,N,Z)
      WRITE (6,7) N,Z
7     FORMAT (I5,1PE20.8)
5     CONTINUE
      STOP
      END
```
(b)

Fig. 9.2.4 FORTRAN listings for
(a) the function subprogram and
(b) the main program for
Example 9.2.1.

N	$V_c(4)$
2	3.33333282E 01
4	3.33333282E 01
6	3.33332977E 01
8	3.33333282E 01
10	3.33332977E 01
12	3.33332672E 01
14	3.33332825E 01
16	3.33333282E 01
18	3.33332672E 01
20	3.33332520E 01
22	3.33332367E 01
24	3.33332520E 01
26	3.33332672E 01
28	3.33332825E 01
30	3.33332672E 01

Fig. 9.2.5 Computer output for Example 9.2.1.

which begins with the fourth-order Runge-Kutta method* to generate the first five solution values and then completes the solution of the differential equation by applying the Adams-Moulton method.*

The fourth-order Runge-Kutta method for generating an approximate value y_{n+1} for the differential equation (9.2.5) with a given initial point (x_0, y_0) is characterized by the following iterative expressions

$$y_{n+1} = y_n + \tfrac{1}{6}(k_1 + 2k_2 + 2k_3 + k_4), \qquad (9.2.6)$$

where

$$k_1 = hf(x_n, y_n),$$
$$k_2 = hf(x_n + \tfrac{1}{2}h, y_n + \tfrac{1}{2}k_1),$$
$$k_3 = hf(x_n + \tfrac{1}{2}h, y_n + \tfrac{1}{2}k_2), \qquad (9.2.7)$$
$$k_4 = hf(x_n + h, y_n + k_3)$$

with

$$x_{n+1} = x_n + h, \qquad h = \text{a fixed step size}, \qquad n = 0, 1, 2, \ldots$$

---

* See Appendix A.3, Section A.3.3.

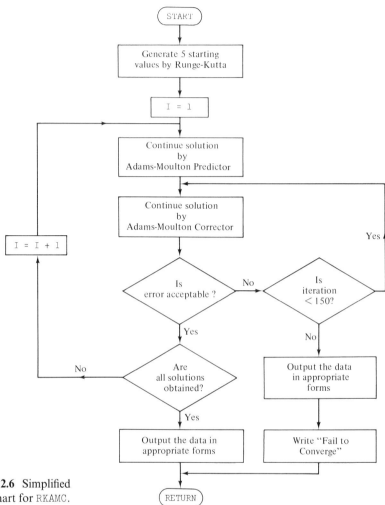

Fig. 9.2.6 Simplified flowchart for RKAMC.

With five known values of $y(x)$, the Adams-Moulton predictor and corrector formulas can be applied, which are given by, respectively,

$$y_{n+1}^{(0)} = y_n + \frac{h}{720}(1901f_n - 2984f_{n-1} + 2616f_{n-2} - 1274f_{n-3} + 251f_{n-4})$$

(9.2.8)

and

$$y_{n+1}^{(k)} = y_n + \frac{h}{720}\{251f(x_{n+1}, y_{n+1}^{(k-1)}) + 646f_n - 264f_{n-1} + 106f_{n-2} - 19f_{n-3}\}.$$

(9.2.9)

```
C       SUBROUTINE RKAMC (XI,YI,XSTOP,ITER,Y,IN)
C
        DIMENSION X(501),Y(501),F(501),D(501,2),JXY(2)
        READ(5,17) IP1,IP2
17      FORMAT (2I5)
        INC=0
        EPS=.8E-05
        X(1)=XI
        Y(1)=YI
C
C       GENERATE STARTING VALUES
C
        QUIT=ITER-5
        DX=(XSTOP-XI)/ITER
        DO 1 I=1,5
        C1=FCT(X(I),Y(I))
        C2=FCT(X(I)+DX/2.,Y(I)+DX*C1/2.)
        C3=FCT(X(I)+DX/2.,Y(I)+DX*C2/2.)
        C4=FCT(X(I)+DX,Y(I)+DX*C3)
        X(I+1)=X(I)+DX
        Y(I+1)=Y(I)+(C1+2.*C2+2.*C3+C4)*DX/6.
1       F(I+1)=FCT(X(I+1),Y(I+1))
C
C       USE ADAMS MOULTON PREDICTOR-CORRECTOR METHOD TO CONTINUE
C
        I=1
2       F(5+I)=FCT(X(5+I),Y(5+I))
        YP=Y(I+5)+(DX/720.)*(1901.*F(I+5)-2984.*F(I+4)+2616.*F(I+3)-
       &1274.*F(I+2)+251.*F(I+1))
        M=0
        X(I+6)=X(I+5)+DX
3       F(I+6)=FCT(X(I+6),YP)
        YC=Y(I+5)+(DX/720.)*(251.*F(I+6)+646.*F(I+5)-264.*F(I+4)+
       &106.*F(I+3)-19.*F(I+2))
        DY=ABS(YC-YP)
        IF(DY-EPS) 6,6,4
4       IF(M-150) 5,5,11
5       YP=YC
        M=M+1
        GO TO 3
6       Y(I+6)=YC
        IF(I-QUIT) 7,7,8
7       I=I+1
        GO TO 2
8       IF (IN.LE.0) GO TO 9
        IF(IP1) 14,14,18
18      WRITE (6,12) (X(I),Y(I),I=1,ITER,IN)
14      IF(IP2) 19,19,20
20      DO 15 I=1,ITER,IN
        INC=INC+1
        D(INC,1)=X(I)
15      D(INC,2)=Y(I)
        DO 16 J=1,2
16      JXY(J)=J
        CALL VPLOT (D,JXY,INC,501,1,0,0.,0.,0.,0.)
19      CONTINUE
10      RETURN
11      ILAST=I
        IF(IP1) 21,21,22
22      WRITE (6,12) (X(I),Y(I),I=1,ILAST,IN)
        WRITE(6,13)
21      IF(IP2) 23,23,24
24      DO 25 I=1,ILAST,IN
        INC=INC+1
        D(INC,1)=X(I)
25      D(INC,2)=Y(2)
        DO 26 J=1,2
26      JXY(J)=J
        CALL VPLOT (D,JXY,INC,501,1,0,0.,0.,0.,0.)
23      GOTO 10
12      FORMAT (E25.3,1PF15.4)
13      FORMAT(2X,'FAILS TO CONVERGE. CHANGE STEP SIZE')
9       WRITE (6,12) X(ITER+1),Y(ITER+1)
        RETURN
        END
```

Fig. 9.2.7 FORTRAN listing of RKAMC.

Subroutine RKAMC implements the equations (9.2.6) through (9.2.9). A simplified flowchart and the **FORTRAN** listing of RKAMC are shown, respectively, in Figs. 9.2.6 and 9.2.7.

The characteristics of RKAMC are described as follows:

Description of Subroutine RKAMC

1. *Purpose*: To solve a first-order differential equation.
2. *Calling statement*: CALL RKAMC (XI, YI, XSTOP, ITER, Y, IN).

```
      SUBROUTINE VPLOT(XY,JXY,N,NDIM,NCUR,ISCALE,XL,XU,YL,YU)
      DIMENSION IGRID(101),XS(11),YS(13),ICHAR(7),XY(1),JXY(1)
      DATA ICHAR/1H+,1H*,1H¬,1H$,1H=,1H.,1H /
      XS(1)=XL
      XMAX=XU
      YMIN=YL
      YS(1)=YU
      IF(ISCALE.NE.0)GO TO 32
      XMAX=-1.0E 20
      XS(1)=-XMAX
      YS(1)=XMAX
      YMIN=XS(1)
      J2=0
      DO 31 J=1,NCUR
      J2=J2+2
      JIX=(JXY(J2-1)-1)*NDIM
      JIY=(JXY(J2)-1)*NDIM
      DO 31 I=1,N
      IJX=JIX+I
      IJY=JIY+I
      IF(XY(IJX).GT.XMAX)XMAX=XY(IJX)
      IF(XY(IJX).LT.XS(1))XS(1)=XY(IJX)
      IF(XY(IJY).GT.YS(1))YS(1)=XY(IJY)
      IF(XY(IJY).LT.YMIN)YMIN=XY(IJY)
   31 CONTINUE
   32 XR=XMAX-XS(1)
      IF(XR.EQ.0.0)XR=1.0E-20
      YR=YS(1)-YMIN
      IF(YR.EQ.0.0)YR=1.0E-20
      XT=XMAX*XS(1)
      YT=YMIN*YS(1)
      IF(XT.LT.0.0)IYAX=100.0*(-XS(1))/XR+1.5
      IF(YT.LE.0.0)IXAX=48.0*YS(1)/YR+1.5
      XMAX=XR/10.
      DO 46 I=2,11
   46 XS(I)=XS(I-1)+XMAX
      XMAX=YR/12.
      DO 47 I=2,13
   47 YS(I)=YS(I-1)-XMAX
      WRITE(6,10)(XS(I),I=1,11)
      II=1
      KK=0
      DO 146 LINE=1,49
      DO 101 J=1,101
  101 IGRID(J)=ICHAR(7)
      IF(YT.GT.0.0)GO TO 109
      IF(LINE.NE.IXAX)GO TO 109
      DO 105 J=1,101
  105 IGRID(J)=ICHAR(6)
  109 IF(XT.LT.0.0)IGRID(IYAX)=ICHAR(6)
      J2=0
      DO 125 J=1,NCUR
      J2=J2+2
      JIX=(JXY(J2-1)-1)*NDIM
      JIY=(JXY(J2)-1)*NDIM
      JC=MOD(J,5)
      DO 125 I=1,N
      IJX=JIX+I
      IJY=JIY+I
      IPTY=48.0*(YS(1)-XY(IJY))/YR+1.5
      IF(IPTY.GT.49)IPTY=49
      IF(IPTY.LT.1)IPTY=1
      IF(IPTY.NE.LINE)GO TO 125
      IPTX=100.0*(XY(IJX)-XS(1))/XR+1.5
      IF(IPTX.LT.1)IPTX=1
      IF(IPTX.GT.101)IPTX=101
      IF(JC.NE.0)GO TO 119
      IGRID(IPTX)=ICHAR(5)
      GO TO 125
  119 IGRID(IPTX)=ICHAR(JC)
  125 CONTINUE
      IF(KK.GT.0)GO TO 134
      WRITE(6,20)YS(II),(IGRID(I),I=1,101),YS(II)
      II=II+1
      GO TO 135
  134 WRITE(6,30)(IGRID(I),I=1,101)
  135 KK=KK+1
      IF(KK.NE.4)GO TO 146
      KK=0
  146 CONTINUE
      WRITE(6,40)(XS(I),I=1,11)
   10 FORMAT(1H1,1PE15.2,10E10.2/10X,1H*,20(5H****),2H**)
   20 FORMAT(1PE10.2,1H+,101A1,1H+,E9.2)
   30 FORMAT(10X,1H*,101A1,1H*)
   40 FORMAT(10X,1H*,20(5H****),2H**/1PE16.2,10E10.2)
      RETURN
      END
```

Fig. 9.2.8 FORTRAN listing for VPLOT.

3. *Input arguments*: XI = initial value of the independent variable x; YI = initial value of the dependent variable y; XSTOP = final value of x for which the solution $y(x)$ is desired; ITER = number of iteration steps from XI to XSTOP; IN = output increment desired. If IN = 0, only the final value, Y(XSTOP), will be printed.

4. *Output argument*: Y = dependent variable y and thus the solution value.

5. *User-supplied subprograms*:

   a) FUNCTION FCT(X, Y), which specifies the differential equation
   $$\frac{dy}{dx} = f(x, y).$$

   b) SUBROUTINE VPLOT for plotting results.

6. *Data Card*:

   Card 1: IP1, IP2
   Format: 2I5

   Note: If IP1 > 0, output will be printed; otherwise they will be suppressed. If IP2 > 0, output will be plotted; otherwise they will not be plotted.

As indicated in the description of the subroutine RKAMC, the plotting subroutine VPLOT is required for displaying the results graphically. The solution of a typical network problem is usually desired over either a finite interval of time or a finite range of frequency and the analysis of the data obtained in tabulated form becomes more and more difficult as the number of solution points increases. Consequently a graphical display of the computed values will in general be very helpful or sometimes even necessary in interpreting the results. The subroutine VPLOT will serve this purpose. The summary of the characteristics of this subroutine is given next with the FORTRAN listing shown in Fig. 9.2.8.

Description of Subroutine VPLOT

1. *Purpose*: To plot one or more curves on one page by a line printer.

2. *Calling statement*: CALL VPLOT (XY, JXY, N, NDIM, NCUR, ISCALE, XL, XU,
                                    YL, YU)

3. *Input arguments*:

   XY(I, J) = a two-dimensional array in which each column represents the values of a variable (dependent or independent).

   JXY(J) = a one-dimensional vector used to specify which of the columns of XY are considered to be independent variables and which dependent ones. The odd rows of this vector are used to specify the independent variables, while the even rows are for the corresponding dependent variables. Thus, for example, if the first, second, third columns of the matrix XY represent, respectively, the values of the variables x_1, x_2, x_3 and so forth, then the specification of the vector such as

```
JXY (1) =5
JXY (2) =3
JXY (3) =3
JXY (4) =4
JXY (5) =1
JXY (6) =2
```

will result in the plotting of the following three curves: x_3 vs. x_5, x_4 vs. x_3, and x_2 vs. x_1. Since a curve must be specified by a pair of variables, the dimension of JXY must be exactly twice the total number of curves to be plotted.

N   = the number of points to be plotted per curve.

NDIM = the storage allocation number equal to the number of rows in XY specified in the program which calls VPLOT.

NCUR = the number of curves to be plotted on a single plot.

XL   = the minimum value specified for the independent variables.

XU   = the maximum value specified for the independent variables.

YL   = the minimum value specified for the dependent variables.

YU   = the maximum value specified for the dependent variables.

ISCALE = a parameter indicating whether the boundary limits of the plots are to be determined by VPLOT or to be preset by XL, XU, YL, and YU. If ISCALE = 0, the data will be automatically scaled by VPLOT (that is, the largest and the smallest values of all the curves to be plotted are used as, respectively, the upper and lower limits in the plot). If ISCALE = 1, the bounds are set by XL, XU, YL, and YU. Points outside these limits are plotted on the edge of the plot with no error indication.

4. *Output argument*: The output will be plotted on a line printer.

5. *User-supplied subprogram*: None required.

6. *Remarks*: The following remarks may be helpful in using the subroutine:

   a) NDIM must be equal to the number of rows in XY;

   b) N is less than or equal to NDIM;

   c) The dimension of JXY is equal to 2 × NCUR.

It should be noted that VPLOT is primarily written as a supporting subroutine to be used in conjunction with a main program so that the results obtained from this main program will be simultaneously plotted. This subroutine can also be incorporated into another subroutine by including a calling statement in that subroutine, as was done in the case of RKAMC, which, for example, contains the following FORTRAN statements:

```
DO 26 J =1, 2
JXY(J) =J
CALL VPLOT (D, JXY, INC, 501, 1, 0, 0., 0., 0., 0.)
```

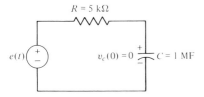

Fig. 9.2.9 A series *RC* network.

```
C        FUNCTION FCT(T,VCAP) FOR   EXAMPLE 9-2-2
         FUNCTION FCT(T,VCAP)
         IF(T-.02) 10,10,20
   10    FCT=(10.-VCAP)/5.E-3
         RETURN
   20    FCT=(-VCAP)/5.E-3
         RETURN
         END
```

Fig. 9.2.10 FORTRAN listing for the function *FCT* for Example 9.2.2.

```
C        MAIN PROGRAM FOR EXAMPLE 9-2-2
         DIMENSION T(501), VCAP(501)
         CALL RKAMC(0.,0.,.05,100,VCAP,1)
         STOP
         END
```

Fig. 9.2.11 FORTRAN listing of the main program for Example 9.2.2.

A comparison of the third statement above with the calling statement of VPLOT reveals the fact that due to the value 0 of ISCALE, the automatic scaling routine of VPLOT will be used so that the values of XL, XU, YL, and YU will be ignored. (Refer to subroutine description of VPLOT under ISCALE.)

In the next example, we shall show how VPLOT is utilized with RKAMC in addition to the illustration of the application of RKAMC in solving simple network problems described by a single first-order differential equation.

Example 9.2.2 We shall use subroutine RKAMC to determine the voltage $v_c(t)$ across the capacitor C of the network shown in Fig. 9.2.9. With the element values as indicated in the figure and an input voltage $e(t)$ given by

$$e(t) = 10[u(t) - u(t - 0.02)] \quad \text{V},$$

the differential equation governing the behavior of $v_c(t)$ can be readily shown to be

$$\frac{dv_c}{dt} = \frac{1}{5 \times 10^{-3}} [10 - v_c(t)] \qquad \text{for} \quad 0 \le t \le 0.02 \qquad (9.2.10)$$

and

$$\frac{dv_c}{dt} = \frac{1}{5 \times 10^{-3}} [-v_c(t)] \qquad \text{for} \quad t > 0.02. \qquad (9.2.11)$$

The differential equations in (9.2.10) and (9.2.11) are used for writing the function subprogram FCT(X, Y), as shown in Fig. 9.2.10, in which the symbols T and VCAP denote the time variable t and the voltage $v_c(t)$, respectively.

Since the input voltage $e(t)$ drops to zero at $t = 0.02$ second and the time constant in this circuit is $RC = 5 \times 10^{-3}$ second, it is sufficient to obtain the solution for the interval of 0 to 0.05 seconds with 100 iterations and a plot including all the solution points to display the transient behavior. The FORTRAN listing of the necessary main program for calling RKAMC is given in Fig. 9.2.11.

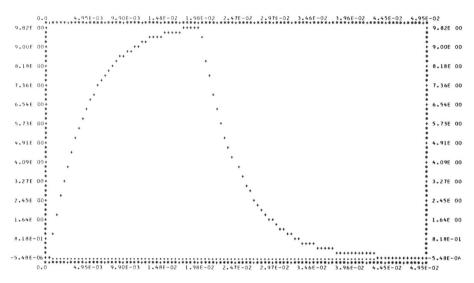

Fig. 9.2.12 Output for Example 9.2.2.

As indicated in the description of RKAMC, the user has to prepare a data card to be used with the function subprogram and the main program. The purpose of this data card is to indicate the option of printing and/or plotting the output. Since, in this example, only a plot is called for, we shall use 0 for IP1 and 1 for IP2 in column 5 and column 10, respectively, in the data card.

The output is plotted in Fig. 9.2.12 in which the ordinate represents $v_c(t)$ in volts, and the abscissa t in seconds.

In some cases, a nonlinear network can also be represented by an ordinary differential equation, as the following example will illustrate.

Example 9.2.3 Consider the network containing a half-wave rectifier and an RC filter as shown in Fig. 9.2.13. Writing the KCL-equation at node A and the VCR-equation for the capacitor C, we have respectively

(a)
$$i_c = i_1 - i_2$$

(b)
$$i_c = C \frac{dv_o}{dt}.$$

$$(9.2.12)$$

Solving (9.2.12) for dv_o/dt, we obtain

$$\frac{dv_o}{dt} = \frac{1}{C}(i_1 - i_2). \tag{9.2.13}$$

Since the diode is assumed to be ideal, we write

$$i_1 = \begin{cases} 0, & \text{if } (v_i - v_o) \leq 0; \\ \dfrac{1}{R_1}(v_i - v_o), & \text{if } (v_i - v_o) > 0. \end{cases} \tag{9.2.14}$$

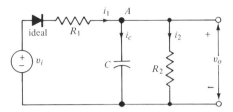

Fig. 9.2.13 The network for Example 9.2.3.

In addition, we have the VCR-equation for resistor R_2:

$$i_2 = \frac{1}{R_2} v_o. \tag{9.2.15}$$

It is evident that RKAMC can now be used to solve (9.2.13) for the filter output voltage $v_o(t)$ for a given input voltage $v_i(t)$ with the aid of (9.2.14) and (9.2.15). The actual solution is left as an exercise.*

9.3 RESISTIVE NETWORKS

In general, the solution of a multi-loop (node) resistive network problem amounts to the task of inverting a (nonsingular) matrix of constant entries. In the following paragraphs, we shall introduce a subroutine, called GAUSEI (based on the *GAU*ss-*SEI*del method), which will enable us to obtain computer solutions of multi-loop (node) resistive network problems.

Consider a set of n independent linear algebraic equations in n unknowns written in the form

$$a_{11}x_1 + a_{12}x_2 + \cdots + a_{1n}x_n = b_1$$
$$a_{21}x_1 + a_{22}x_2 + \cdots + a_{2n}x_n = b_2 \tag{9.3.1}$$
$$\vdots$$
$$a_{n1}x_1 + a_{n2}x_2 + \cdots + a_{nn}x_n = b_n.$$

The Gauss-Seidel iterative formulas for solving (9.3.1) are given by

$$x_1^{k+1} = \frac{1}{a_{11}} (b_1 - a_{12}x_2^k - a_{13}x_3^k - \cdots - a_{1n}x_n^k)$$

$$x_2^{k+1} = \frac{1}{a_{22}} (b_2 - a_{21}x_1^{k+1} - a_{23}x_3^k - \cdots - a_{2n}x_n^k) \tag{9.3.2}†$$

$$\vdots$$

$$x_n^{k+1} = \frac{1}{a_{nn}} (b_n - a_{n1}x_1^{k+1} - a_{n2}x_2^{k+1} - \cdots - a_{n,n-1}x_{n-1}^{k+1}),$$

where x_i^k denotes the value of the ith variable after the kth iteration.

---

* See Problem 9.3.

† See Appendix A.3, Section A.3.6.

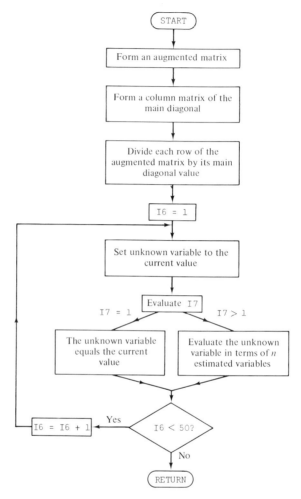

Fig. 9.3.1 A simplified flowchart for GAUSEI.

Based on (9.3.2), subroutine GAUSEI is written for $n \le 6$. A simplified flowchart and the listing of the subroutine are given, respectively, in Figs. 9.3.1 and 9.3.2. The program is described as follows.

Description of Subroutine GAUSEI

1. *Purpose*: To solve a set of N independent linear algebraic equations given in matrix form as $[a_{ij}][b_i] = [c_i]$.

2. *Calling statement*: CALL GAUSEI(N, A, B, C).

3. *Input arguments*: A = the (two-dimensional) entries of the matrix $[a_{ij}]$; C = the (one-dimensional) entries of the (column) matrix $[c_i]$; N = number of equations.

```
      SUBROUTINE GAUSEI (N,A,B,C)
      DIMENSION A(6,7),B(10),C(10),E(10)
      NA=N+1
      DO  100   I=1,N
100   A(I,NA)=C(I)
      DO  200   I1=1,N
200   E(I1)=A(I1,I1)
      DO  300   I2=1,N
      DO  300   I3=1,NA
300   A(I2,I3)=A(I2,I3)/E(I2)
      DO  500   I6=1,50
      DO  400   I4=1,N
      B(I4)=A(I4,NA)
      IF (I6 ,EQ. 1)  GO  TO  400
      DO  410   I5=1,N
      I7=I4-I5
      IF (I7 ) 10,11,10
10    AN=1.0
      GO  TO  12
11    AN=0.0
      GO  TO  12
12    CONTINUE
      B(I4)=B(I4)-AN*A(I4,I5)*B(I5)
410   CONTINUE
400   CONTINUE
500   CONTINUE
      RETURN
      END
```

Fig. 9.3.2 FORTRAN listing for GAUSEI.

4. *Output argument*: B = the (one-dimensional) entries of the (column) matrix $[b_i]$ of the variables.

5. *User-supplied subprogram*: None required.

6. *Data cards*: As required by the main program to read in matrices.

7. *Remarks*:

   a) The number N must be ≤6. Dimensions may be changed to increase N if it is necessary.

   b) A total of 50 iterations are used in the subroutine.

The following example will illustrate the use of the subroutine.

Example 9.3.1 We shall use GAUSEI to determine the loop currents in the network shown in Fig. 9.3.3. From Fig. 9.3.3, we write the loop equations in matrix form as follows:

$$
\begin{bmatrix}
2 & -1 & 0 & 1 & 0 & 0 \\
-1 & 2 & -1 & 0 & 0 & 0 \\
0 & -1 & 4 & -1 & 0 & -1 \\
-1 & 0 & -1 & 2 & 0 & 0 \\
0 & 0 & 0 & 0 & 2 & -1 \\
0 & 0 & -1 & 0 & -1 & 4
\end{bmatrix}
\begin{bmatrix}
i_1 \\ i_2 \\ i_3 \\ i_4 \\ i_5 \\ i_6
\end{bmatrix}
=
\begin{bmatrix}
-4 \\ 0 \\ 0 \\ 4 \\ -4 \\ 20
\end{bmatrix}.
\tag{9.3.3}
$$

The main program which calls GAUSEI to solve (9.3.3) is given in Fig. 9.3.4. It should be noted that, in Fig. 9.3.4, the FORMAT for reading the entries of the coefficient

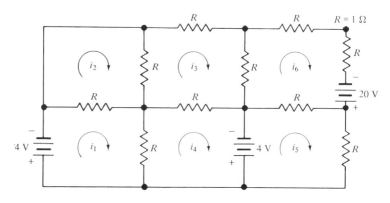

Fig. 9.3.3 Network for Example 9.3.1.

```
C       MAIN PROGRAM FOR EXAMPLE 9-3-1
        DIMENSION AI(6,7),B(10),C(10)
        READ (5,10) N
10      FORMAT (I5)
        DO 20 I=1,N
20      READ (5,30) (AI(I,J), J=1,N)
        READ (5,30) (C(I), I=1,N)
30      FORMAT (6F10.0)
        CALL GAUSEI (N,AI,B,C)
        WRITE (6,40) (B(I),I=1,N)
40      FORMAT (1H0, E15.7)
        STOP
        END
```

Fig. 9.3.4 Main program for Example 9.3.1.

```
0.9999790E 00

0.1999982E 01

0.2999988E 01

0.3999983E 01

0.9999971E 00

0.5999995E 01
```

Fig. 9.3.5 Computer output for
Example 9.3.1.

matrix $[a_{ij}]$ is 6F10.0, which is also used for reading the column matrix of the known voltages. Thus, we supply the cards containing the data:

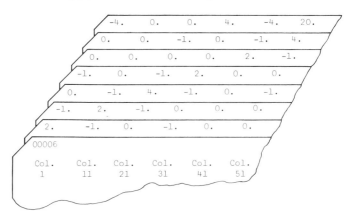

The computer output is shown in Fig. 9.3.5. Since the FORMAT for the output is E15.7, we obtain from Fig. 9.3.5 approximately

$$i_1 = 1, \quad i_2 = 2, \quad i_3 = 3, \quad i_4 = 4, \quad i_5 = 1, \quad \text{and} \quad i_6 = 6$$

all in amperes.

9.4 NETWORK ANALYSIS BY LAPLACE TRANSFORM TECHNIQUE

As discussed in Chapter 5, network equations can be expressed either in the t-domain or in the s-domain. If the network under consideration is linear and time-invariant, the solution of the network equations given in the t-domain can be obtained by the powerful method of Laplace transformation which consists of the following three steps: (1) Take the Laplace transforms of these equations; (2) solve these transform equations for the desired quantities; and (3) take the inverse Laplace transforms of the quantities to determine the solutions in the t-domain. In the case of simple network problems, solutions can be easily obtained by hand cal-

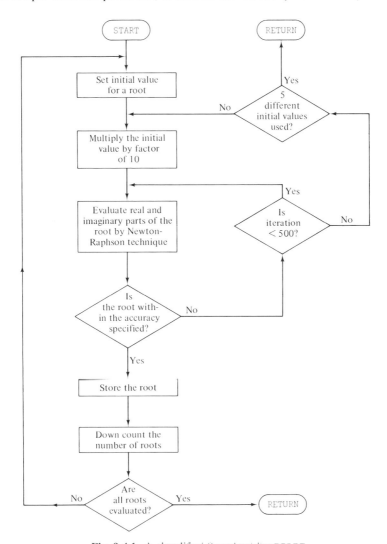

Fig. 9.4.1 A simplified flowchart for DPLRT.

culations.   However, for more complicated networks, digital computers may be needed for determining the solutions.

Let the transform quantity, say $F(s)$, found in Step 2 above be expressed in the form of a rational function

$$F(s) = \frac{a_n s^n + a_{n-1} s^{n-1} + \cdots + a_1 s + a_0}{b_m s^m + b_{m-1} s^{m-1} + \cdots + b_1 s + b_0}. \tag{9.4.1}$$

The inverse Laplace transform $f(t)$ of $F(s)$ can be found as required in Step 3, by first rewriting (9.4.1) as

$$F(s) = \frac{N(s)}{(s - p_1)(s - p_2) \cdots (s - p_m)} \tag{9.4.2}$$

and then expanding it into partial fractions, from which the inverse transform of $F(s)$ can be obtained.

In this section we shall present two subroutines, namely DPLRT (a root-finding routine with double precision) and PFEX (to evaluate the coefficients associated with the terms in the partial fraction expansion).

Subroutine DPLRT* is designed to compute the real and/or complex roots of the equation with real coefficients $a_j, j = 0, 1, \ldots, n$:

$$P(z) = a_n z^n + a_{n-1} z^{n-1} + \cdots + a_1 z + a_0 = 0 \tag{9.4.3}$$

based on the Newton-Raphson iterative formula

$$z_{n+1} = z_n - \frac{f(z_n)}{f'(z_n)} \tag{9.4.4}\dagger$$

where $n = 0, 1, 2, \ldots$.

A simplified flowchart and the listing of subroutine DPLRT are given in Figs. 9.4.1 and 9.4.2, respectively.

Description of Subroutine DPLRT

1. *Purpose*: To evaluate the real and complex roots of the equation

$$P(z) = a_n z^n + a_{n-1} z^{n-1} + \cdots + a_1 z + a_0 = 0.$$

2. *Calling statement*: CALL DPLRT (XCOF, COF, M, ROOTR, ROOTI, IER).

3. *Input arguments*: XCOF = one-dimensional array of M+1 coefficients of the polynomial in ascending order of power; M = order of the polynomial.

---

* This subroutine is a modified version of subroutine POLRT of IBM Scientific Subroutine Library.

† See Appendix A.3, Section A.3.4.

```
      SUBROUTINE DPLRT(XCOF,COF,M,ROOTR,ROOTI,IER)
      DIMENSION XCOF(1),COF(1),ROOTR(1),ROOTI(1)
      DOUBLE PRECISION XO,YO,X,Y,XPR,YPR,UX,UY,V,YT,XT,U,XT2,YT2,SUMSQ, P
     1 DX,DY,TEMP,ALPHA
      DOUBLE PRECISION XCOF,COF,ROOTR,ROOTI
      IFIT=0
      N=M
      IER=0
      IF(XCOF(N+1))10,25,10
   10 IF(N) 15,15,32
C
C         SET ERROR CODE TO 1
C
   15 IER=1
   20 RETURN
C
C         SET ERROR CODE TO 4
C
   25 IER=4
      GO TO 20
C
C         SET ERROR CODE TO 2
C
   30 IER=2
      GO TO 20
   32 IF(N-36) 35,35,30
   35 NX=N
      NXX=N+1
      N2=1
      KJ1 = N+1
      DO 40 L=1,KJ1
      MT=KJ1-L+1
   40 COF(MT)=XCOF(L)
C
C         SET INITIAL VALUES
C
   45 XO=.00500101
      YO=0.01000101
C
C         ZERO INITIAL VALUE COUNTER
C
      IN=0
   50 X=XO
C
C         INCREMENT INITIAL VALUES AND COUNTER
C
      XO=-10.0*YO
      YO=-10.0*X
C
C         SET X AND Y TO CURRENT VALUE
C
      X=XO
      Y=YO
      IN=IN+1
      GO TO 59
   55 IFIT=1
      XPR=X
      YPR=Y
C
C         EVALUATE POLYNOMIAL AND DERIVATIVES
C
   59 ICT=0
   60 UX=0.0
      UY=0.0
      V =0.0
      YT=0.0
      XT=1.0
      U=COF(N+1)
      IF(U) 65,130,65
   65 DO 70 I=1,N
      L =N-I+1
      TEMP=COF(L)
      XT2=X*XT-Y*YT
      YT2=X*YT+Y*XT
      U=U+TEMP*XT2
      V=V+TEMP*YT2
      FI=I
      UX=UX+FI*XT*TEMP
      UY=UY-FI*YT*TEMP
      XT=XT2
   70 YT=YT2
      SUMSQ=UX*UX+UY*UY
      IF(SUMSQ) 75,110,75
   75 DX=(V*UY-U*UX)/SUMSQ
      X=X+DX
      DY=-(U*UY+V*UX)/SUMSQ
      Y=Y+DY
   78 IF(DABS(DY)+DABS(DX)-1.0D-12) 100,80,80
C
C         STEP ITERATION COUNTER
C
   80 ICT=ICT+1
      IF(ICT-500) 60,85,85
   85 IF(IFIT)100,90,100
   90 IF(IN-5) 50,95,95
```

Fig. 9.4.2 FORTRAN listing for DPLRT (*continued*).

```
C
C            SET ERROR CODE TO 3
C
    95 IER=3
       GO TO 20
   100 DO 105 L=1,NXX
       MT=KJ1-L+1
       TEMP=XCOF(MT)
       XCOF(MT)=COF(L)
   105 COF(L)=TEMP
       ITEMP=N
       N=NX
       NX=ITEMP
       IF(IFIT) 120,55,120
   110 IF(IFIT) 115,50,115
   115 X=XPR
       Y=YPR
   120 IFIT=C
   122 IF(DABS(Y/X)-1.0D-10) 135,125,125
   125 IF (DABS(X).LT.1.0D-20) X=0
       ALPHA=X+X
       SUMSQ = X*X + Y*Y
       N=N-2
       GO TO 140
   130 X=0.0
       NX=NX-1
       NXX=NXX-1
   135 Y=0.0
       SUMSQ=0.0
       ALPHA=X
       N=N-1
   140 COF(2)=COF(2)+ALPHA*COF(1)
   145 DO 150 L=2,N
   150 COF(L+1)=COF(L+1)+ALPHA*COF(L)-SUMSQ*COF(L-1)
   155 ROOTI(N2)=Y
       ROOTR(N2)=X
       N2=N2+1
       IF(SUMSQ) 160,165,160
   160 Y=-Y
       SUMSQ=0.0
       GO TO 155
   165 IF(N) 20,20,45
       END
```

Fig. 9.4.2 FORTRAN listing for DPLRT (*concluded*).

4. *Output arguments*: ROOTR = resultant array of dimension M containing the real parts of the zeros of the polynomial; ROOTI = resultant array of dimension M containing the imaginary parts of the zeros of the polynomial; IER = a single-valued error code defined as follows:

IER	Message
0	No error
1	M less than 1
2	M greater than 36
3	Unable to determine root after 500 iterations on 5 different initial values
4	Highest-order coefficient equals zero.

5. *User-supplied subprogram*: None required.

6. *Data cards*: None required.

7. *Remarks*:

a) COF = one-dimensional temporary storage array of dimension M + 1.

b) M must be equal to or less than 36.

c) The calling program must contain the following statements:

```
DOUBLE PRECISION XCOF, CQF, ROOTR, ROOTI
DIMENSION XCOF(M+1), COF(M+1), ROOTR(M), ROOTI(M)
```

```
C          MAIN PROGRAM FOR EXAMPLE 9-4-1
           DOUBLE PRECISION XCOF,COF,ROOTR,ROOTI
           DIMENSION XCOF(5),COF(5),ROOTR(4),ROOTI(4)
           M=4
           XCOF(1) = 6.
           XCOF(2) = 5.
           XCOF(3) = 7.
           XCOF(4) = 5.
           XCOF(5) = 1.
           CALL DPLRT (XCOF,COF,M,ROOTR,ROOTI,IER)
           WRITE (6,10) IER, M, (ROOTR(I),ROOTI(I), I=1,M)
     10    FORMAT (2I10,/(2F10.3))
           STOP
           END
```

Fig. 9.4.3 FORTRAN listing for the main program in Example 9.4.1.

```
        0              4
-2.000         0.0
-3.000         0.0
-0.000        -1.000
-0.000         1.000
```

Fig. 9.4.4 Computer output for Example 9.4.1.

d) To retrieve the roots found, the user must include a WRITE and a FORMAT statement for IER, M, ROOTR and ROOTI.

Example 9.4.1 As an illustration for the use of subroutine DPLRT, we shall factor the following polynomial:

$$P(s) = 6 + 5s + 7s^2 + 5s^3 + s^4. \qquad (9.4.5)$$

The main program for factoring (9.4.5) is shown in Fig. 9.4.3. The computer output is given in Fig. 9.4.4, from which we obtain

$$P(s) = (s + 2)(s + 3)(s + j)(s - j)$$

$$= (s + 2)(s + 3)(s^2 + 1)$$

which is the desired result.

We shall next turn our attention to the problem of determining the coefficients associated with the partial-fraction expansion of a given rational function. This problem is one of the laborious steps involved in the process of taking the inverse Laplace transform.

Consider the function

$$F(s) = \frac{N(s)}{(s + p_0)^n(s + p_1)D_1(s)}, \qquad (9.4.6)$$

where $s = -p_0$ is a pole of $F(s)$ of multiplicity (or order) n, $s = -p_1$ is a simple pole of $F(s)$, $D_1(s)$ is the factor in the denominator of $F(s)$ contributed by all other poles (if any) of $F(s)$, and $N(s)$ is the numerator of $F(s)$. Without loss of generality, we assume that $D_1(s)$ in (9.4.6) is equal to 1. Under this assumption, the partial-fraction expansion of $F(s)$ is given by

$$F(s) = \frac{K_1}{(s + p_0)^n} + \frac{K_2}{(s + p_0)^{n-1}} + \cdots + \frac{K_n}{(s + p_0)} + \frac{K_{p_1}}{(s + p_1)} \qquad (9.4.7)^*$$

---

* See Appendix A.2, Section A.2.4.

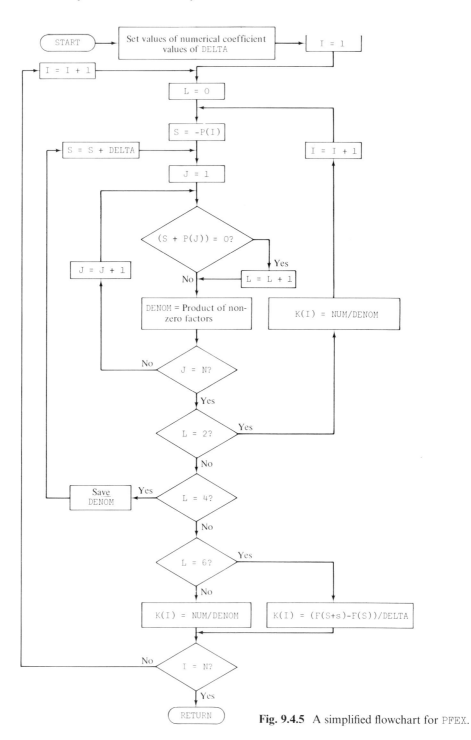

Fig. 9.4.5 A simplified flowchart for PFEX.

```
      SUBROUTINE PFEX(N,A,P,K)
      COMPLEX K,P,S,Z,NUM,DENOM,FUNCS,FUNCDS,DELTA
      DIMENSION A(6),P(6),K(6)
      NUM(Z)=A1+A2*Z+A3*Z**2+A4*Z**3+A5*Z**4+A6*Z**5
      A1=A(1)
      A2=A(2)
      A3=A(3)
      A4=A(4)
      A5=A(5)
      A6=A(6)
      DELTA=(.001,0.)
      I=1
    5 L=0
    6 S=P(I)
    7 DENOM=(1.,0.)
      DO 50 J=1,N
      IF(CABS(S-P(J))-.05)9,9,8
    8 DENOM=DENOM*(S-P(J))
      GO TO 50
    9 L=L+1
   50 CONTINUE
      GO TO(11,12,13,13,14,14),L
   11 K(I)=NUM(S)/DENOM
      GO TO 10
   12 K(I)=NUM(S)/DENOM
      I=I+1
      GO TO 6
   13 FUNCS=NUM(S)/DENOM
      S=S+DELTA
      GO TO 7
   14 FUNCDS=NUM(S)/DENOM
      K(I)=(FUNCDS-FUNCS)/DELTA
   10 IF(I.EQ.N) GO TO 90
      I=I+1
      GO TO 5
   90 RETURN
      END
```

Fig. 9.4.6 FORTRAN listing for PFEX.

from which we can obtain K_{p_1}, the residue of the single pole $s = -p_1$, using the formula

$$K_{p_1} = (s + p_1)F(s)\Big|_{s=-p_1} = \frac{N(s)}{(s + p_0)^n D_1(s)}\Big|_{s=-p_1}, \qquad (9.4.8)$$

and also the coefficients K_j $(j = 1, 2, \ldots, n)$ associated with the multiple pole $s = -p_0$, by means of the general expression

$$K_j = \frac{1}{(j-1)!}\left\{\frac{d^{(j-1)}}{ds^{(j-1)}}[(s + p_0)^n F(s)]\right\}_{s=-p_0}, \qquad j = 1, 2, \ldots, n. \quad (9.4.9)$$

Based on (9.4.7) through (9.4.9), the subroutine PFEX is developed to evaluate the coefficients associated with poles of *order 2 or less*. A simplified flowchart and the FORTRAN listing of PFEX are given in Fig. 9.4.5 and Fig. 9.4.6, respectively. The characteristics of this program are given in the following subroutine description.

Description of Subroutine PFEX

1. *Purpose*: To evaluate the coefficients associated with the simple and double poles of a rational function which is a *proper fraction*.*

2. *Calling statement*: CALL PFEX(N, A, P, K).

3. *Input arguments*: N = the degree of the denominator polynomial; A = one-dimensional array of six real coefficients of the numerator polynomial in the order of increasing power; P = one-dimensional array of N complex roots of the denominator polynomial. (Double roots must be listed adjacent to each other.)

---

* For the definition of a proper fraction, refer to Appendix A.2, Section A.2.4.

4. *Output argument*: K = one-dimensional array of the values of the N complex coefficients.

5. *User-supplied subprogram*: None required.

6. *Data cards*:

Card 1: N
Format: I1

Card 2: A(1), A(2), A(3), A(4), A(5), A(6)
Format: 6F10.0

Card 3   P(1), P(2), P(3),...
Format: 2F10.0, 2F10.0, 2F10.0,...

7. *Remarks*:

a) Multiplicity of the poles must be ≤2.
b) All six coefficients of the numerator polynomial must be specified.

The next example will illustrate the use of subroutine PFEX.

Example 9.4.2 With the aid of subroutine PFEX, we shall obtain the partial-fraction expansion of the function

$$F(s) = \frac{s^2 + s + 1}{(s^2 + 1)(s + 2)(s + 3)}. \tag{9.4.10}$$

From (9.4.10) we obtain the following information for the preparation of the three data cards:

N =4, A(1) =A(2) =A(3) =1, A(4) =A(5) =A(6) =0

P(1) = (0, -1), P(2) = (0, +1), P(3) = (-2, 0), P(4) = (-3, 0),

P(5) =P(6) = (0, 0)

The main program and the computer output are given respectively in Figs. 9.4.7 and 9.4.8.

With the coefficients found in Fig. 9.4.8, the partial-fraction expansion of $F(s)$ is found to be

$$F(s) = \frac{0.05 - j0.05}{(s - j)} + \frac{0.05 + j0.05}{(s + j)} + \frac{0.6}{(s + 2)} + \frac{-0.7}{(s + 3)}$$

which is the desired result.

In the case of a double pole (i.e., a pole of order 2) subroutine PFEX will compute the coefficients in the order K_1, K_2 as defined in (9.4.7) and computer output will be so printed as the next example will illustrate.

Example 9.4.3 Let us determine the partial-fraction expansion for

$$F(s) = \frac{2s + 2}{(s + 2)^2(s + 3)}. \tag{9.4.11}$$

```
C     MAIN PROGRAM FOR EXAMPLE 9-4-2
C
      COMPLEX K,P
      DIMENSION A(6),P(6),K(6)
    1 READ(5,100)N
  100 FORMAT(I1)
      IF(N.EQ.0) GO TO 99
      READ(5,101)(A(I),I=1,6)
  101 FORMAT(6F10.0)
      READ(5,102)(P(I),I=1,6)
  102 FORMAT(12F5.0)
      CALL PFEX(N,A,P,K)
      WRITE(6,200)N,(A(I),I=1,6)
  200 FORMAT(1H1,6X,'THE DEGREE OF THE DENOMINATOR POLYNOMIAL IS ',I1//6
     *X,' THE COEF OF THE NUMERATOR POLYNOMIAL ARE:'//(10X,6F10.3))
      WRITE(6,201)(P(I),K(I),I=1,N)
  201 FORMAT(1H0,16X,'POLE FACTOR',10X,'COEFFICIENTS'//15X,'REAL',6X,'IM
     *AG',6X,'REAL',6X,'IMAG'//(10X,4F10.3))
      GO TO 1
   99 STOP
      END
```

Fig. 9.4.7 Main program for Example 9.4.2.

```
THE DEGREE OF THE DENOMINATOR POLYNOMIAL IS 4
THE COEF OF THE NUMERATOR POLYNOMIAL ARE:
        1.000      1.000      1.000      0.0         0.0        0.0
           POLE FACTOR              COEFFICIENTS

        REAL       IMAG       REAL       IMAG
        0.0       -1.000      0.050      0.050
        0.0        1.000      0.050     -0.050
       -2.000      0.0        0.600      0.0
       -3.000      0.0       -0.700      0.0
```

Fig. 9.4.8 Computer output for Example 9.4.2.

```
THE DEGREE OF THE DENOMINATOR POLYNOMIAL IS 3
THE COEF OF THE NUMERATOR POLYNOMIAL ARE:
        2.000      2.000      0.0        0.0         0.0        0.0
           POLE FACTOR              COEFFICIENTS

        REAL       IMAG       REAL       IMAG
       -2.000      0.0       -2.000      0.0
       -2.000      0.0        3.998      0.0
       -3.000      0.0       -4.000      0.0
```

Fig. 9.4.9 Computer output for Example 9.4.3.

Using PFEX and following the same procedure as in Example 9.4.2, we find that the computer output will take the form shown in Fig. 9.4.9, from which we obtain

$$F(s) = \frac{-2}{(s+2)^2} + \frac{3.999}{(s+2)} + \frac{-4}{(s+3)}. \tag{9.4.12}$$

By comparing (9.4.12) with the coefficient values in Fig. 9.4.9, it should be clear that the first two values, $-2.000 + j0.0$ and $3.999 + j0.0$ in the figure identify the coefficients K_1 and K_2 as indicated in (9.4.7).

In the next example, we shall discuss how the subroutines PFEX and DPLRT can be applied together for the analysis of a relatively simple network using the Laplace transform technique.

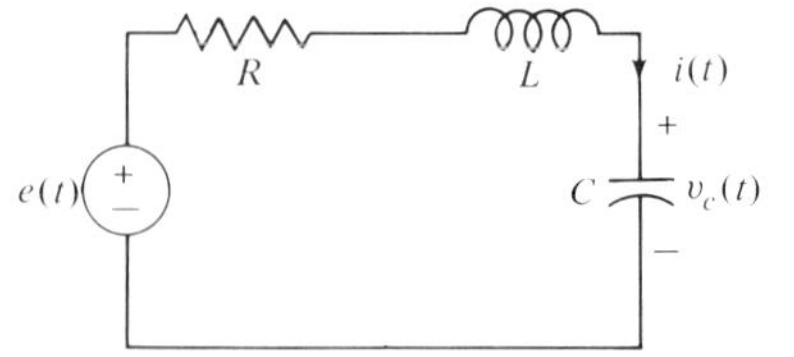

Fig. 9.4.10 Network for Example 9.4.4.

Example 9.4.4 We shall determine the voltage $v_c(t)$ across the capacitor C in the network shown in Fig. 9.4.10 with the aid of subroutines DPLRT and PFEX.

First, the transform loop equation for the network can be written as

$$\left(R + Ls + \frac{1}{Cs}\right) I(s) = E(s) \tag{9.4.13}$$

and the voltage transform $V_c(s)$ across the capacitor C is given by

$$V_c(s) = \frac{1}{Cs} I(s). \tag{9.4.14}$$

Next, substituting element values and $E(s) = 10/s$ into (9.4.13) and (9.4.14) and solving for $V_c(s)$, we obtain

$$V_c(s) = \frac{250}{s^3 + 6s^2 + 25s}. \tag{9.4.15}$$

The denominator of $V_c(s)$ in (9.4.15) can be factored by applying subroutine DPLRT as

$$s^3 + 6s^2 + 25s = s(s + 3 + j4)(s + 3 - j4).$$

With the poles of $V_c(s)$ found, subroutine PFEX can be used to compute the residues of the poles, yielding

$$V_c(s) = \frac{10}{s} + \frac{(-5 - j3.75)}{(s + 3 + j4)} + \frac{(-5 + j3.75)}{(s + 3 - j4)}. \tag{9.4.16}$$

Finally, taking the inverse Laplace transform of (9.4.16) and rearranging the terms, we obtain the desired result

$$v_c(t) = 10 + 12.5e^{-3t} \cos (4t + 143.2^0), \qquad t \geq 0.$$

Although subroutines DPLRT and PFEX are written separately, they can easily be incorporated into a single program* so that the user may obtain the coefficients in the partial-fraction expansion of a given rational function by providing the information on its numerator and denominator polynomials as input data. They can also be used in some other manner, such as to facilitate the frequency response analysis of networks, which will be discussed in a later section of this chapter.

_____

* See Problem 9.7.

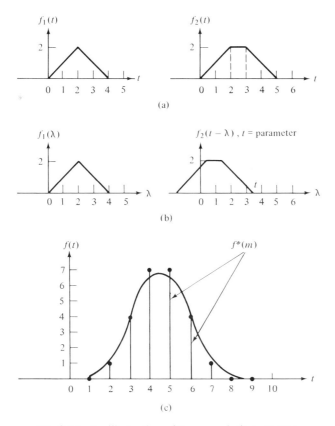

Fig. 9.5.1 An illustration of the convolution process.

9.5 NETWORK RESPONSES BY CONVOLUTION

In Chapter 5, we pointed out that the response of a given network to an arbitrary input can be obtained by applying the convolution integral once the impulse response of the network is known. Furthermore, we emphasized (in Section 5.5) that the convolution integral is especially useful when both the impulse response and the input exist only in graphical form—a form most conveniently handled by means of a digital computer.

Subroutine CONVOL (*CONVOL*ution) is designed to compute the convolution of two functions, $f_1(t)$ and $f_2(t)$, defined by

$$f(t) = f_1(t) * f_2(t) = \int_0^t f_1(\lambda) f_2(t - \lambda)\, d\lambda. \qquad (9.5.1)\dagger$$

To evaluate numerically the convolution integral (9.5.1), the continuous functions $f_1(t)$ and $f_2(t)$ must first be converted into two equivalent discrete functions. Let

---

† See Appendix A.2, Section A.2.3.

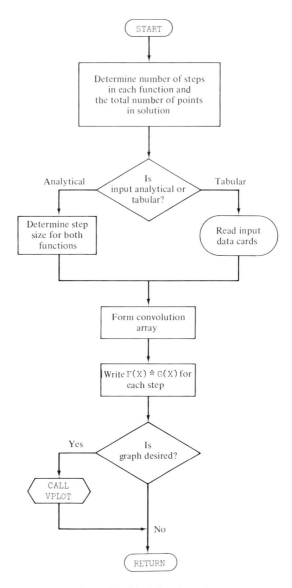

Fig. 9.5.2 A simplified flowchart for CONVOL.

the values of $f_1(\lambda)$ and $f_2(\lambda)$ evaluated at $\lambda = i\Delta\lambda$ ($i = 1, 2, \ldots$) be denoted by $f_1(i)$ and $f_2(i)$, respectively. Then (9.5.1) can be expressed approximately as

$$f^*(m) \cong \sum_{i=0}^{m} \{f_1(i)f_2(m-i)\}\,\Delta\lambda, \qquad m = 1, 2, \ldots, \tag{9.5.2}$$

where

$$f^*(m) \triangleq f(t)|_{t=m\Delta\lambda}, \qquad \text{with } \Delta\lambda \text{ being the step size.}$$

```
      SUBROUTINE CONVOL(H,TF1,TF2,ID)
      DIMENSION A(50),B(50),C(50,50),D(101,2),JXY(2)
      N=TF1/H
      M=TF2/H
      NM=N+M+1
      DO 1 I=1,NM
    1 D(I,2)=0.
      IF(ID)2,2,3
    2 READ(5,102) (A(I),I=1,N)
  102 FORMAT(8F10.0)
      READ(5,102) (B(I),I=1,M)
      GO TO 8
    3 DO 10 I=1,N
   10 A(I)=F(I*H-.5*H)
      DO 11 I=1,M
   11 B(I)=G(I*H-.5*H)
    8 DO 5 I=1,N
      DO 4 J=1,M
      C(I,J)=A(I)*B(J)
      K=I+J
    4 D(K,2)=D(K,2)+C(I,J)*H
    5 CONTINUE
      D(1,1)=.0
      DO 7 L=2,NM
      AA=L-1.
    7 D(L,1)=AA*H
      WRITE(6,100)
  100 FORMAT(1H0,7X,'TIME',10X,'F-STAR-G',/)
      DO 12 I=1,NM
   12 WRITE(6,101)D(I,1),D(I,2)
      IF(ID.EQ.1.OR.ID.EQ.-1)GO TO 6
      JXY(1)=1
      JXY(2)=2
      CALL VPLOT(D,JXY,NM,101,1,0,W,X,Y,Z)
  101 FORMAT(2X,2E15.7)
    6 RETURN
      END
```

Fig. 9.5.3 FORTRAN listing for CONVOL.

To develop a subroutine for performing the convolution of two functions, we shall assume, without loss of generality, that the functions $f_1(t)$ and $f_2(t)$ vanish outside of the (closed) intervals $[0, T_1]$ and $[0, T_2]$, respectively. Then, for a given step-size h, $[0, T_1]$ and $[0, T_2]$ can be divided into $M \triangleq [T_1/h]$† and $N \triangleq [T_2/h]$ equal subintervals. Under this assumption about $f_1(t)$ and $f_2(t)$, an inspection of (9.5.2) indicates that it is sufficient to determine $f^*(m)$ for $m = 1, 2, \ldots, M + N$, since $f^*(m)$ is identically equal to zero for all the values of $m \geq M + N$. A simple example on the convolution process is shown in Fig. 9.5.1 where (a) depicts the functions $f_1(t)$ and $f_2(t)$ with $T_1 = 2$ and $T_2 = 5$; (b) indicates the corresponding $f_1(\lambda)$ and $f_2(t - \lambda)$; and (c) shows the waveform of $f(t)$ and the spectrum of $f^*(m)$ as obtained by applying (9.5.1) and (9.5.2), respectively.

Based on the foregoing discussion, the subroutine CONVOL is developed with a simplified flowchart and the FORTRAN listing shown in Figs. 9.5.2 and 9.5.3, respectively. A summary of the characteristics of the subroutine CONVOL follows.

Description of Subroutine CONVOL

1. *Purpose* To perform convolution of two functions each of which is given either analytically as a function subprogram or in tabular form with data cards; that is,

$$\int_0^x f(\lambda)g(x - \lambda) \, d\lambda.$$

---

† The symbol $[x]$ known as the *greatest-integer function* may be defined as follows: If x is an integer, then $[x] = x$; and if x is not an integer, then $[x]$ assumes the value of the greatest integer less than x.

2. *Calling statement*: CALL CONVOL (H, TF1, TF2, IN).

3. *Input arguments*: H = step size of the independent variable x; TF1 = final value for $f(x)$; TF2 = final value for $g(x)$; IN = an integer code indicating to the subroutine whether the input is in tabular or analytical form as well as whether a plot for the output is desired:

   a) IN ≤ 0 (i.e., a zero or *any* negative integer for IN) indicates a tabular input.

   b) IN > 0 (i.e., *any* positive integer) indicates an analytical input.

   c) IN = ±1 indicates that a computer plot is not required; any other integer assigned for IN will result in the plotting of the output data.

4. *Output argument*: Result of the convolution will be given in the form as specified by the value of IN.

5. *User-supplied subprograms*:

   a) Subroutine VPLOT.

   b) FUNCTION F(X) for the first function.

   c) FUNCTION G(X) for the second function.

6. *Data cards*: Data cards are needed for graphical function input in the format of 8F10.0.

7. *Remarks*:

   a) If the input data are given in tabulated form, the number of data points must be equal to

   $$TF1/H \leq 50 \text{ for } F(X)$$

   and

   $$TF2/H \leq 50 \text{ for } G(X)$$

   If more data points are desired, the subroutine must be redimensioned.

   b) If the functions F(X) and G(X) are given in analytical form, one must first decide on the interval of interest [0, *T*] for the convolution integral and then assign values to the final times, TF1 and TF2 of the functions F(X) and G(X) respectively. A convenient way is to assign both TF1 and TF2 equal to *T*.

The use of subroutine CONVOL will be illustrated in the following examples.

Example 9.5.1 We shall use the subroutine CONVOL to evaluate the convolution integral of the following functions:

$$f_1(x) = \sin x\, u(x), \qquad f_2(x) = \cos x\, u(x)$$

and plot the result for the interval $0 \leq x \leq 4$.

Since the two functions are given in analytical form, we shall use analytical inputs with step size H = 0.1. To obtain a plot for the output, one must assign a positive integral value other than unity for IN since we have analytical inputs. Let us use, say, 5 for this problem. The main program and the function subprograms are shown in Fig. 9.5.4. The computer output is given in Fig. 9.5.5.

```
0001          FUNCTION F(X)
0002          F=SIN(X)
0003          RETURN
0004          END
```

```
0001          FUNCTION G(X)
0002          G=COS(X)
0003          RETURN
0004          END
```

```
              C   MAIN PROGRAM FOR EXAMPLE 9-5-1
0001              CALL CONVOL(.1,2.0,2.0,5)
0002              STOP
0003              END
```

Fig. 9.5.4 Function subprograms and main program for Example 9.5.1.

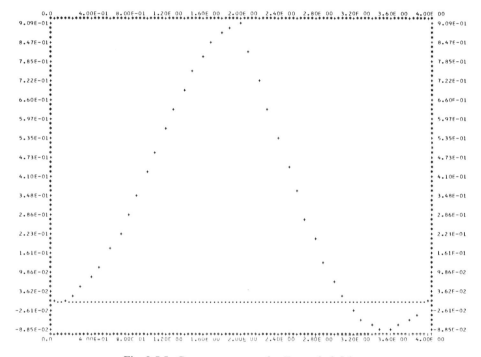

Fig. 9.5.5 Computer output for Example 9.5.1.

In the next example, we shall demonstrate the use of CONVOL with the given data input in tabular form.

Example 9.5.2   The impulse response $h(t)$ of a certain linear system is assumed to be approximately equal to the waveform shown in Fig. 9.5.6(b). Determine the response $r(t)$ of the system to an input $f(t)$ as given in Fig. 9.5.6(a).

To obtain $r(t)$, we shall evaluate the convolution $f(t) * h(t)$ by digital computation with the aid of CONVOL. However, to use CONVOL in this example with a tabular input, we need to prepare data cards to describe the two functions in the format of

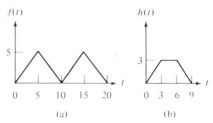

Fig. 9.5.6 (a) Input function $f(t)$ and (b) impulse response of a system.

(a)   (b)

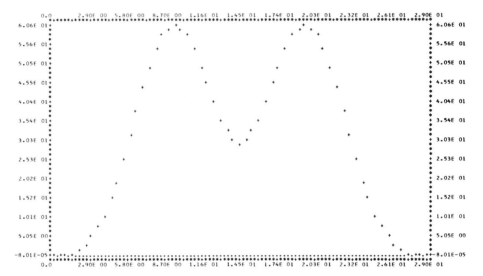

Fig. 9.5.7 Computer output for Example 9.5.2.

8F10.0. The number of data cards depends on how large the value of the step size H is chosen as well as the values of the final times, TF1 and TF2, of the given functions. An inspection of the waveforms in Fig. 9.5.6 reveals that a step size of H =0.5 will result in a fairly good approximation. Thus, using H =0.5, we have 40 steps for $f(t)$ and 18 steps for $h(t)$ so that a total of 8 data cards [5 for $f(t)$ and 3 for $h(t)$] are required. The calling statement is

```
CALL CONVOL (.5, 20., 9., -2)
```

in which we use IN = −2 (or any negative integer other than −1) to signify the fact that the input is in tabular form and that a plot is requested as the output. Figure 9.5.7 shows the computer output for this example.

It is perhaps a good exercise at this point to repeat the entire problem by choosing a different value for the step size H and then compare the results.†

---

† See Problem 9.11.

It should be noted also that although the functions given in Example 9.5.2 are in graphical form, we can still enter the data analytically. For instance, the function $f(t)$ can be described by means of the following statements in the function subprogram for F(X):

```
         FUNCTION F(X)
         F = X
         IF (X.LT.5.) GO TO 10
         IF (X.GT.10.) GO TO 20
         F = 10.0 - X
         RETURN
      20 IF (X.GT.15.) GO TO 30
         F = -10.0 + X
         RETURN
      30 F = 20.0 - X
      10 RETURN
         END
```

The reader may wish to compare the result in Example 9.5.2 with that obtained by using the analytical input.†

Regardless of the form used for the input, function subprograms must be supplied together with the main program. In the case that a tabular input is used, "dummy" function subprograms can be supplied to satisfy the subprograms requirement during the linkage step performed by the computer. Such a dummy program may take the following form:

```
         FUNCTION F(X)
         RETURN
         END
```

We have seen in the last two examples the application of the subroutine CONVOL, which accepts inputs either in analytical or in tabular form. Obviously, when the tabular form is used, a function is represented by a set of discrete values, which are entered as input data. We note that the subroutine requires equal step size for both functions to be convolved. Thus, if the two functions are given in discrete form but not in the same step size, we must convert the data of one of these two functions so that both functions form a compatible set of input data for CONVOL. Subroutine XYFUN, which is developed based on the method of linear interpolation and the application of Shannon's sampling theorem, is designed to serve this purpose. It is used to generate, at a specified sampling time interval, a set of discrete sampled values of a given function which is described by a sufficiently large number of randomly selected data points.

The simplified flowchart and the FORTRAN listing of subroutine XYFUN are given in Figs. 9.5.8 and 9.5.9, respectively. The characteristics of this program are described as follows.

---

† See Problem 9.12.

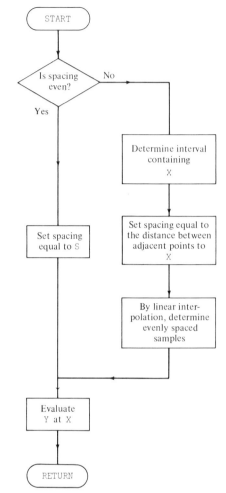

Fig. 9.5.8 A simplified flowchart for
XYFUN.

Description of Subroutine XYFUN

1. *Purpose*: To reconstruct a function from a given set of data points.

2. *Calling statement*: CALL XYFUN (YDATA, S, XDATA, NU, X, Y).

3. *Input arguments*: YDATA = the one-dimensional array of the input values of
   the dependent variable; S = a real number to be assigned as follows:

   a) If data points are evenly spaced, S equals the step size.
   b) If data points are not evenly spaced, S equals 0.

   XDATA = the one-dimensional array of the input values of the independent
   variable; NU = number of input data points ≤ 100.

4. *Output arguments*: X = the value of the independent variable at which the
   function value is to be generated (reconstructed); Y = the generated value of
   the function.

5. *User-supplied subprogram*: None required.

```
         SUBROUTINE XYFUN(YDATA,S,XDATA,NU,X,Y)
         DIMENSION YDATA(100), XDATA(100), XDAT(100) ,YDAT(100)
         XDAT(1)=XDATA(1)
         DX=S
         N=NU
         IF (DX.GT.0) GO TO 10
C
C        PROCEDURE FOR RANDOMLY SPACED SAMPLES
C        DETERMINE INTERVAL CONTAINING X
         DO 3 J=1,N
         IF(X-XDATA(J)) 4,11,3
     3   CONTINUE
C        DETERMINE  SAMPLING RANGE FOR N SAMPLES EVENLY SPACED
     4   DX=XDATA(J)-XDATA(J-1)
         N1=(XDATA(J-1)-XDATA(1))/DX
         IF(N1.EQ.0) GO TO 19
         IF(N1.GT.J-1) N1=J-1
         DO 5 K=1,N1
     5   XDAT(J-K  )=XDATA(J-1)-K*DX
    19   N2=(XDATA(N)-XDATA(J))/DX
         IF(N2.EQ.0) GO TO 20
         IF (N2.GT.(N-J))   N2=N-J
         DO 6 K=1,N2
     6   XDAT(J+K+1)=XDATA(J)+K*DX
    20   XDAT(J+1)=XDATA(J)
         XDAT(J  )=XDATA(J-1)
         L=J-N1
         N=N1+N2+2+L
         J=L
C        USE LINEAR INTERPOLATION TO DETERMINE SAMPLE POINTS
         J1= J
         DO 7 K=J1,N
    18   IF(XDAT(K)-XDATA(J)) 15, 17, 17
    17   J=J+1
         GO TO 18
    15   YDAT(K)=YDATA(J-1)+(YDATA(J)-YDATA(J-1))/(XDATA(J)-XDATA(J-1))
        **(XDAT(K)-XDATA(J-1)))
     7   CONTINUE
         GO TO 12
C        PROCEDURE FOR EVENLY SPACED INTERVALS
    10   DO 9 J1=1,N
     9   YDAT(J1)=YDATA(J1)
         L=1
    12   X=X-XDAT(L)
         K=1
         Y=0.
         B=3.1415927*X/DX
         DO 13J=L,N
         A=B-3.1415927*(K-1)
         IF(A.EQ.0.0) GO TO 14
         Y1=YDAT (J)*SIN(A)/A+Y
         K=K+1
    13   Y=Y1
         GO TO 16
    11   YDAT(J)=YDATA(J)
    14   Y=YDAT(J)
    16   RETURN
         END
```

Fig. 9.5.9 FORTRAN listing for XYFUN.

6. *Data cards.*: In format as called for by the main program.

7. *Remarks*: The main program must contain statements to read in XDATA and YDATA, and appropriate statements to call VPLOT if a display of the reconstructed function is desired.

The next example will illustrate the use of subroutine XYFUN.

Example 9.5.3 We shall use XYFUN to generate 101 sets of data points for reconstructing the function

$$f(t) = 70e^{-t} \sin\left(\frac{\pi}{1.25}t\right)$$

which is described by the following 17 values evaluated sequentially at steps of 0.3 seconds from $t = 0$ to $t = 4.8$ seconds:

f_i = 0, 35.50, 38.34, 21.92, 2.64, -9.18, -11.37, -7.24, -1.58, 2.26,
   3.31, 2.34, 0.71, -0.52, -0.99, -0.74, -0.28,

where $i = 1, 2, \ldots, 17$.

```
FUNCTION F(T)
F = 70. * EXP(-T) * SIN(3.1415*T/1.25)
RETURN
END


C        MAIN PROGRAM FOR EXAMPLE 9-5-3
         DIMENSION XY(101,4),YDATA(100),XDATA(100),JXY(4)
         S = 0.3
         NU = 17
         READ (5,1) (XDATA(I),YDATA(I),I=1,NU)
    1    FORMAT (2F10.0)
         T = 0.0
         DO 2 I = 1,101
         XY(I,1) = T
         XY(I,2) = F(T)
         CALL XYFUN (YDATA,S,XDATA,NU,T,Y)
         XY(I,3) = Y
    2    T = T + 0.05
         JXY(1) = 1
         JXY(2) = 2
         JXY(3) = 1
         JXY(4) = 3
         CALL VPLOT (XY,JXY,101,101,2,0,0,0,0,0)
         STOP
         END
```

Fig. 9.5.10 Function subprogram and main program for Example 9.5.3.

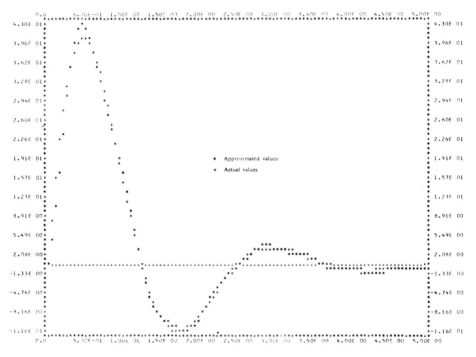

Fig. 9.5.11 Computer output for Example 9.5.3.

For comparison we shall plot both the original function and the reconstructed function using the subroutine VPLOT. The function subprogram and the main program are shown in Fig. 9.5.10. The computer output in Fig. 9.5.11 indicates that the reconstructed curve follows the original one very closely except at those points where the values of the function are nearly zero or approaching a maximum. This situation can be improved, of course, by having more sampled points.

9.6 SOLUTION OF STATE EQUATIONS
AND OF AN nth-ORDER DIFFERENTIAL EQUATION

In Chapter 3, we discussed how state equations can be formulated for a given linear network, and various examples were used to illustrate how the algorithm for formulating state equations may be applied to obtain, and then express, these equations in standard form $\dot{\mathbf{x}} = \mathbf{A}\mathbf{x} + \mathbf{B}\mathbf{u}$ (Sections 3.5 through 3.7).

In Chapter 7, we investigated the solution of state equations and presented several methods for solving the state equations, which are simply a set of simultaneous first-order ordinary differential equations (Sections 7.3 through 7.6). Furthermore, we showed (in Section 7.9) that an nth-order differential equation may be described by a set of n first-order differential equations (namely, as state equations) redefining the derivatives as new variables and hence may be expressed in the standard matrix form

$$\dot{\mathbf{x}} = \mathbf{A}\mathbf{x} + \mathbf{B}\mathbf{u}.$$

In this section, we shall show how computer techniques may be employed in obtaining the solution of a physical system, once the state model (namely, the matrix state equation) of the system is known. In particular, we shall introduce and describe the subroutine called RKAM (*R*unge-*K*utta *A*dams-*M*oulton), which is designed primarily to solve a set of n first-order ordinary differential equations based on the fourth-order Runge-Kutta method together with the Adams-Moulton predictor-corrector formulas.* (These formulas were also used for developing subroutine RKAMC, which, as discussed earlier, can solve a first-order differential equation only). Subroutine RKAM starts by calling a function subroutine, DERFUN, which defines the set of differential equations to be solved. The solution of an nth-order differential equation can also be obtained by using RKAM, but we must first convert the equation into a set of n simultaneous first-order differential equations, expressed in standard matrix form. The applications of RKAM will be illustrated with examples later.

Figure 9.6.1 shows a simplified flowchart for RKAM, indicating both the available choices of solution routines and error checking routines. The FORTRAN listing of RKAM is given in Fig. 9.6.2, and this program is now described as follows.

Description of Subroutine RKAM

1. *Purpose:* To solve a set of n simultaneous first-order differential equations, with arbitrary initial conditions.

2. *Calling statement:* CALL RKAM (Y, DY, YD, VF, UY).

3. *Input arguments:* YD = a working storage array with dimension $4n$; UY, VF = two working storage arrays with dimension $5n$ each.

4. *Output arguments:* Y = the array of dependent variables with a dimension equal to n; DY = the array of derivatives of the dependent variable with the same dimension as that of Y.

---

* See Appendix A.3, Section A.3.5.

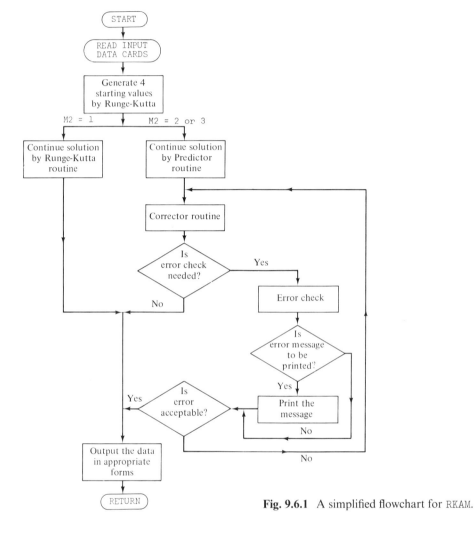

Fig. 9.6.1 A simplified flowchart for RKAM.

5. *User-supplied subprograms*:

   a) Subroutine DERFUN used to define the set of *n* simultaneous first-order differential equations to be solved.

   b) VPLOT for plotting.

6. *Data cards*: The following data cards are to be prepared by the user to inform the subroutine RKAM of the given initial conditions, the integration method to be used, the step size, allowable errors, the print interval desired, and the number and type of graphical outputs requested. These cards are read according to the READ and FORMAT statements in RKAM which are repeated here for convenience.

```
      SUBROUTINE RKAM(Y,DY,YD,VF,UY)
      DOUBLE PRECISION UY
      DIMENSION Y(1),UY(1),VF(1),YD(1),A(20),D(20),ID(20),B(20),DY(1)
     *,XV(5),BET(4),IOR(10),IOW(10),JXY(10),YX(100,6),MYY(6),YY(100,6),
     *IYX(4),IYY(4)
      EQUIVALENCE (M1,NN),(M2,MODE),(M3,KKA),(A2,E1MAX),(A3,E1MIN),(A4,E
     *2MAX),(A5,E2MIN),(A6,FACT)
      DATA BET/0.5,0.5,1.0,0.0/,MYY/1,2,3,4,5,6/,JXY/1,2,1,3,1,4,1,5,1,
     *6/,IYP/3H Y(/,MP/3H Y(/
C RUNGE-KUTTA ADAMS MOULTON SOLUTION OF N SIMULTANEOUS FIRST ORDER
C ORDINARY DIFFERENTIAL EQUATIONS USING PARTIAL DOUBLE PRECISION ARITH-
C METIC
C
C     A1=THE INITIAL VALUE OF THE STEP SIZE***************************
C     A2=THE MAXIMUM ABSOLUTE VALUE OF THE ALLOWABLE SINGLE STEP ERROR*****
C     A3= THE MINIMUM ABSOLUTE VALUE OF THE ALLOWABLE SINGLE STEP ERROR****
C     A4=THE MAXIMUM ABSOLUTE VALUE OF STEP SIZE*********************
C     A5=THE MINIMUMM ABSOLUTE VALUE OF STEP SIZE********************
C     A6=REDUCTION FACTOR FOR THE STEP SIZE*************************
C     M1=NUMBER OF FIRST ORDER EQUATIONS***************************
C     M2=MODE OF INTEGRATION, M2=1=RK,M2=2=RKAM,M2=3=RKAM WITH ERROR CHECK*
C     M3=TYPE OF ERROR CHECK,M3=1=REL. ERROR,M3=2=ABSOLUTE ERROR**********
C     M4=PRINT OR SUPPRESS ERROR MESSAGES,M4=1=PRINT MESSAGES,M4=2=DELETE
C     MESSAGES*****************************************************
C     M5=PRINTING INTERVAL,M5>0=PRINT ACCORDING TO SPECIFICATIONS,M5=0=
C     PRINT AT EACH INTEGRATION STEP*******************************
C
C IDENTIFICATION DATA-BLANK CARD WILL SUPPRESS*****************
C
      READ(5,30)(ID(I),I=1,20)
C
C IOR=INPUT READ FORMAT,IOW=OUTPUT WRITE FORMAT********************
C
      READ(5,30)(IOR(I),I=1,10),(IOW(J),J=1,10)
      READ(5,20)M1,M2,M3,M4,M5,IYVSX,IYVSY
      READ(5,IOR)A1,A2,A3,A4,A5,A6
C
C INITIAL AND FINAL VALUE OF THE INDEPENDENT VARIABLE****************
C
      READ(5,IOR)ALPHA,OMEGA
C
C INITIAL VALUES OF THE DEPENDENT VARIABLES*********************
      READ(5,IOR)(Y(I),I=1,M1)
C
      LT=2
      IF(M5)121,121,119
  119 READ(5,IOR)(A(I),D(I),B(I),I=1,M5)
      LT=1
  121 CONTINUE
      WRITE(6,220)
      WRITE(6,10)(ID(I),I=1,20)
      WRITE(6,40)(IOR(I),I=1,10),(IOW(J),J=1,10)
      WRITE(6,90)
      WRITE(6,120)A1,A2,A3,A4,A5,A6
      WRITE(6,70)
      WRITE(6,80)M1,M2,M3,M4,M5,IYVSX,IYVSY
      WRITE(6,60)ALPHA,OMEGA
      WRITE(6,110)
      WRITE(6,IOW)(I,Y(I),I=1,M1)
      IF(IYVSX)335,335,331
  331 READ(5,20)NYX,(IYX(I),I=1,NYX)
      WRITE(6,50)NYX,(IYP,IYX(I),I=1,NYX)
  335 IF(IYVSY)337,337,336
  336 READ(5,20)NYY,(IYY(2*I-1),IYY(2*I),I=1,NYY)
      WRITE(6,150)(MP,IYY(2*I-1),IYY(2*I),I=1,NYY)
  337 CONTINUE
      LL=1
      KM=1
      INP=0
      IND=0
      JNP=0
      JND=0
      T=ALPHA
      WRITE(6,220)
C
C SET CONSTANTS ON RUNGE-KUTTA AND (IF USED), ADAMS-MOULTON**************
C
      F16TH=1./6.
      F24=1./24.
      GO TO (7,88,88),M2
    7 MM=4
      J1=4
      GO TO 9
   88 MM=1
      J1=1
    9 EPM=2.0*ABS(OMEGA-ALPHA)
      E3=EPM
      ZOT=.5+2.0**(-30.0)
      N2=NN+2
      DT=A1
      R=19.0/270.0
      XV(MM)=T
      IF(E1MIN)2,2,1
    2 E1MIN=E1MAX/55.0
    1 IF(FACT)4,4,3
    4 FACT=.5
```

Fig. 9.6.2 FORTRAN listing of RKAM (continued).

```
    3 CALL DERFUN(Y,DY,T)
      MMM=(MM-1)*NN
      DO 320 I=1,NN
      IMM=MMM+I
      VF(IMM)=DY(I)
  320 UY(IMM)=Y(I)
      GO TO (501,502),LT
  501 E3=ABS(A(1)-B(1))
      IF(D(1))505,504,505
  504 E3=ABS(DT*0.5)
  505 ABM=E3
      WRITE(6,400)KM,A(KM),D(KM),B(KM)
      WRITE(6,130)
      GO TO 1001
  502 WRITE(6,100)
C
C RUNGE-KUTTA ROUTINE**********************************************************
C
 1001 MMM=(MM-1)*NN
      KI=-NN
      DO 1034 K=1,4
      KI=KI+NN
      DO 1350 I=1,NN
      IK=KI+I
      IMM=MMM+I
      YD(IK)=DT*VF(IMM)
 1350 Y(I)=UY(IMM)+BET(K)*YD(IK)
      T=BET(K)*DT+XV(MM)
      CALL DERFUN(Y,DY,T)
      DO 1100 I=1,NN
 1100 VF(I+MMM)=DY(I)
 1034 CONTINUE
      DO 1039 I=1,NN
      I2=I+NN
      I3=I2+NN
      I4=I3+NN
      IMM=MMM+I
      UY(IMM+NN)=UY(IMM)+(YD(I)+2.*(YD(I2)+YD(I3))+YD(I4))*F16TH
 1039 CONTINUE
      MM=MM+1
      XV(MM)=XV(MM-1)+DT
      MMM=(MM-1)*NN
      DO 1400 I=1,NN
 1400 Y(I)=UY(MMM+I)
      T=XV(MM)
      CALL DERFUN(Y,DY,T)
      GO TO (42,99,99),MODE
   99 MMM=(MM-1)*NN
      DO 151 I=1,NN
  151 VF(MMM+I)=DY(I)
      IF(MM-3)1001,1001,2000
C
C ADAMS-MOULTON ROUTINE*******************************************************
C
 2000 DO 2048 I=1,NN
      I2=I+NN
      I3=I2+NN
      I4=I3+NN
      I5=I4+NN
      Y(I)=UY(I4)+DT*(55.*VF(I4)-59.*VF(I3)+37.*VF(I2)-9.*VF(I))*F24
 2048 YD(I)=Y(I)
      T=XV(4)+DT
      CALL DERFUN(Y,DY,T)
      XV(5)=T
      DO 2051 I=1,NN
      I2=I+NN
      I3=I2+NN
      I4=I3+NN
      I5=I4+NN
      UY(I5)=UY(I4)+DT*(9.*DY(I)+19.*VF(I4)-5.*VF(I3)+VF(I2))*F24
 2051 Y(I)=UY(I5)
      CALL DERFUN(Y,DY,T)
      GO TO (42,42,3000),MODE
C
C ERROR CHECKING ROUTINE********************************-*********************Q
C
 3000 SSE=0.0
      DO 3033 I=1,NN
      EPSIL=R*ABS(Y(I)-YD(I))
      GO TO (3301,3307),KKA
 3301 IF(Y(I))3650,3307,3650
 3650 EPSIL=EPSIL/(ABS(Y(I)))
 3307 IF(SSE-EPSIL)3032,3033,3033
 3032 SSE=EPSIL
 3033 CONTINUE
      IF(E1MAX-SSE)3034,3034,3035
 3034 IF(ABS(DT)-E2MIN)42,42,4340
 3035 IF(SSE-E1MIN)3036,42,42
 3036 IF(E2MAX-ABS(DT))42,42,5360
 4340 LL=1
      MM=1
      GO TO (4560,4660),M4
 4560 WRITE(6,800)SSE,T,DT
      DT=DT*FACT
      WRITE(6,820)XV(1),DT
      GO TO 1001
```

Fig. 9.6.2 FORTRAN listing of RKAM (*continued*).

```
 4660 DT=DT*FACT
      GO TO 1001
 5360 GO TO (42,5361),LL
 5361 XV(2)=XV(3)
      XV(3)=XV(5)
      DO 5363 I=1,NN
      I2=I+NN
      I3=I2+NN
      I5=I3+I3-I
      VF(I2)=VF(I3)
      VF(I3)=DY(I)
      UY(I2)=UY(I3)
 5363 UY(I3)=UY(I5)
      LL=2
      MM=3
      GO TO (5270,5370),M4
 5270 WRITE(6,800)SSE,T,DT
      DT=DT*2.
      WRITE(6,840)DT
      WRITE(6,130)
      GO TO 1001
 5370 DT=2.*DT
      GO TO 1001
C
C  EXIT ROUTINE*************************************************************
C
   42 II5=4*NN
      DO 322 I=1,NN
  322 VF(II5+I)=DY(I)
      GO TO (701,602),LT
  142 KI=-NN
      DO 12 K=1,3
      KI=KI+NN
      XV(K)=XV(K+1)
      DO 12 I=1,NN
      IK=KI+I
      VF(IK)=VF(IK+NN)
   12 UY(IK)=UY(IK+NN)
      LL=2
      MM=4
      XV(4)=XV(5)
      II4=3*NN
      DO 52 I=1,NN
      I4=II4+I
      VF(I4)=VF(I4+NN)
   52 UY(I4)=UY(I4+NN)
      GO TO (1001,2000,2000),MODE
C
C  WRITE OUTPUT WITH SPECIFICATIONS**************************************
C
  701 TZ=ABS(DT)*ZOT
      GO TO (700,750,750),MODE
  700 KT=4
      GO TO 430
  750 KT=1
  430 SPACE=A(KM)-XV(KT)
      Z=ABS(SPACE)
      IF(Z-TZ)437,437,436
  437 IF(SPACE)413,403,413
  403 KTM=(KT-1)*NN
      DO 404 I=1,NN
  404 Y(I)=UY(KTM+I)
      GO TO 443
  436 KT=KT+1
      IF(KT-5)430,430,790
  413 KTM=(KT-1)*NN
      DO 438 I=1,NN
  438 DY(I)=VF(I+KTM)
      KI=-NN
      DO 439 K=1,4
      KI=KI+NN
      DO 440 I=1,NN
      IK=KI+I
      YD(IK)=SPACE*DY(I)
  440 Y(I)=UY(I+KTM)+BET(K)*YD(IK)
      T=BET(K)*SPACE+XV(KT)
  439 CALL DERFUN(Y,DY,T)
      DO 442 I=1,NN
      I2=I+NN
      I3=I2+NN
      I4=I3+NN
  442 Y(I)=UY(KTM+I)+(YD(I)+2.*(YD(I2)+YD(I3))+YD(I4))*F16TH
  443 WRITE(6,140)A(KM)
      WRITE(6,IOW)(I,Y(I),I=1,NN)
C
C  GRAPH ROUTINE *********************************************************
C
      IF(IYVSX)471,471,451
  451 INP=INP+1
      IF(INP-IYVSX)471,455,471
  455 INP=0
      IND=IND+1
      YX(IND,1)=A(KM)
      DO 457 I=1,NYX
  457 YX(IND,I+1)=Y(IYX(I))
      IF(IND-100)471,459,471
  459 IND=0
```

Fig. 9.6.2　FORTRAN listing of RKAM (*continued*).

```
          CALL VPLOT(YX,JXY,100,100,NYX,0,0.0,0.0,0.0,0.0)
          IF(JND+1-100)461,471,461
461   WRITE(6,220)
          WRITE(6,130)
471   IF(IYVSY)491,491,473
473   JNP=JNP+1
          IF(JNP-IYVSY)491,475,491
475   JNP=0
          JND=JND+1
          I2=0
          DO 477 I=1,NYY
          I2=I2+2
          YY(JND,I2-1)=Y(IYY(I2-1))
477   YY(JND,I2)=Y(IYY(I2))
          IF(JND-100)491,479,491
479   CALL VPLOT(YY,MYY,100,100,NYY,0,0.0,0.0,0.0,0.0)
          JND=0
          WRITE(6,220)
          WRITE(6,130)
491   CCNTINUE
733   IF(D(KM))715,746,715
715   A(KM)=A(KM)+D(KM)
          E=ABS(A(KM)-B(KM))
          IF(E-ABM)744,746,746
744   ABM=E
          GO TO 430
746   KM=KM+1
          IF(KM-M5)702,702,648
702   WRITE(6,220)
          WRITE(6,400)KM,A(KM),D(KM),B(KM)
          E3=ABS(A(KM)-B(KM))
          IF(D(KM))714,713,714
713   E3=ABS(DT*0.5)
714   ABM=E3
          GO TO (700,750,750),MODE
790   E=ABS(XV(J1)-OMEGA)
          IF(E-EPM)706,648,648
706   EPM=E
          GO TO 142
C
C  WRITE OUTPUT AT EACH MESHPOINT********************************************
C
602   E=ABS(OMEGA-XV(J1))
          IF(E-EPM)672,672,648
672   EPM=E
          GO TO (600,650,650),MODE
600   KT=4
          GO TO 630
650   KT=1
630   E=ABS(OMEGA-XV(KT))
          IF(E-E3)647,645,645
647   E3=E
          KTM=(K-1)*NN
          DO 666 I=1,NN
666   Y(I)=UY(KTM+I)
          CALL DERFUN(Y,DY,T)
          WRITE(6,140)XV(KT)
          WRITE(6,IOW)(I,Y(I),I=1,NN)
          IF(LT-1)733,733,645
645   KT=KT+1
          IF(KT-5)630,630,142
648   IF(IYVSX)651,651,649
649   CALL VPLOT(YX,JXY,IND,100,NYX,0,0.0,0.0,0.0,0.0)
651   IF(IYVSY)655,655,653
653   CALL VPLOT(YY,MYY,JND,100,NYY,0,0.0,0.0,0.0,0.0)
655   WRITE(6,220)
          RETURN
10    FORMAT(1X,20A4)
20    FORMAT(16I5)
30    FORMAT(20A4)
40    FORMAT(//21H INPUT-OUTPUT FORMATS/10A4,10A4)
50    FORMAT(17H PRINTER PLOT OF ,I2,38H VARIABLE(S) VS. INDEPENDENT VAR
     *IABLE./22H VARIABLES PLOTTED ARE,3(A3,I2,1H)))
60    FORMAT(/9H ALPHA = ,1PE12.5,10X,9H OMEGA = ,E12.5/)
70    FORMAT(/42H     M1     M2     M3     M4     M5 IYVSX IYVSY)
80    FORMAT(16I6)
90    FORMAT(71H    .INITIAL    MAX. ALLOW. MIN. ALLOW. MAX. ALLOW. MIN. A
     *LLOW. STEP SIZE/72H      STEP      SINGLE STEP SINGLE STEP    ABSOLUTE
     *      ABSOLUTE    REDUCTION/69H        SIZE              ERROR       ERROR
     *  STEP SIZE    STEP SIZE   FACTOR/)
100   FORMAT(40X,28H WRITING AT EACH MESH POINT//)
110   FORMAT(39H STARTING VALUES OF DEPENDENT VARIABLES//)
120   FORMAT(1P6E12.4)
130   FORMAT(12H INDEPENDENT,8X,32HDEPENDENT VARIABLE NO. AND VALUE/
     *10H  VARIABLE/)
140   FORMAT(1PE12.4)
150   FORMAT(23H PRINTER PHASE PLOT OF ,2(A3,I2,8H) VS. Y(,I2,1H)))
220   FORMAT(1H1)
400   FORMAT(30X,18HWRITEING INTERVAL,I3//10H START AT ,1PE14.6,5X, 17H
     +INCREMENTING BY ,1E14.6,5X,15H AND FINISH AT ,1E14.6//)
800   FORMAT(/18H MAXIMUM ERROR OF ,1PE12.5,4H AT ,1PE12.5,19H WITH ST
     *EP SIZE OF ,1PE12.5)
820   FORMAT(  27H ERROR TOO LARGE, BEGIN AT ,1PE12.5,19H WITH STEP SIZ
     *E OF ,1PE12.5/)
840   FORMAT( 32H ERROR TOO SMALL. NEW STEP SIZE ,1PE12.5/)
          END
```

Fig. 9.6.2 FORTRAN listing of RKAM (*concluded*).

Card 1. A user identification statement to be printed at the beginning of the output. The READ and FORMAT statements in RKAM are:

```
   READ (5, 30) (ID(I), I =1, 20)
30 FORMAT (20A4)
```

Card 2. The user specifies any input and output formats desired for his data. The READ and FORMAT statements in RKAM are

```
   READ(5, 30) (IØR(I), I =1, 10), (IØW(J), J =1, 10)
30 FORMAT (20A4)
```

Note: A convenient choice of formats is (8E10.3) and (1H +, 11 X 6 (I4, 1PE12.4)/(12X, 6 (I4, E12.4))); the former is for the input and is to be punched anywhere between columns 1 and 40 and the latter, to be punched anywhere between columns 41 and 80, is for the output.

Card 3. This card controls the integration method, error check, and form of output desired. The READ and FORMAT statements in RKAM are

```
   READ (5, 20) M1, M2, M3, M4, M5, IYVSX, IYVSY
20 FORMAT (16I5)
```

where M1 is set equal to n, representing the order of the system; M2 specifies the integration method as follows:

> If M2 =1, the Runge-Kutta method will be used.
>
> If M2 =2, the Runge-Kutta, Adams-Moulton method without error check will be performed.
>
> If M2 =3, the Runge-Kutta, Adams-Moulton method with error check will be performed.

M3 specifies the type of error check as follows:

> If M3 =1, relative error check will be performed.
>
> If M3 =2, absolute error check will be performed.*

M4 specifies whether error messages are to be printed:

> If M4 =1, the messages will be printed.
>
> If M4 =2, the messages will be suppressed.

M5 specifies the number of desired intervals for printing out solutions as follows:

> If M5 =0, every integration step will be printed out.

---

* The absolute error is based on the magnitude of the difference between the predicted and the corrected values of the dependent variable (at a specified value of the independent variable) obtained by applying the Adams-Moulton predictor-corrector formulas whereas the relative error is equal to the absolute error divided by the magnitude of the corrected value of the dependent variable.

If M5 =1, a single "print-out interval" starting at the value of the independent variable specified by A(1) and stopping at the value specified by B(1) with increments between print-out points specified by D(1) will result.

If M5 =2, solution points will be printed out at two separate intervals; that is, in addition to the print-out intervals specified by the two values of A(1) and B(1) discussed above, there will be another print-out interval starting at the initial value specified by A(2) and ending at the final value specified by B(2) with increments dictated by D(2). *Note*: In like manner, if M5 =3, 4, ..., I, there will be 3, 4, ..., I separate print-out intervals starting at initial values specified by A(1), A(2),..., A(1), respectively, and ending at the final values specified by B(1), B(2),..., B(I), respectively, with increments dictated by the values of D(1), D(2),..., D(I) respectively. A maximum number of print-out intervals is 20; that is, I =20. The contents of these three arrays A, B, and D will be entered in Data Card 7.

IYVSX specifies the desired "Point plot intervals" as follows:

If IYVSX =0, no plot will result.

If IYVSX =1, every point printed will also be plotted.

If IYVSX =2, every second point printed will also be plotted, and so on.

IYVSY specifies the desired "phase plot* intervals" the same way as in IYVSX for the point plot.

Card 4. This card specifies the step size and error controls by the following parameters:

A1 = the initial step size;

A2 = the maximum (absolute) value of the allowable single-step error;

A3 = the minimum (absolute) value of the allowable single-step error;

A4 = the maximum (absolute) allowable step size;

A5 = the minimum (absolute) allowable step size;

A6 = the multiplication reduction factor for the step size. If the error being checked exceeds A2, the step size will be reduced by A6 and then another attempt will be made to integrate the equations.

This data card is read by the following statement in RKAM:

READ (5, IØR) A1, A2, A3, A4, A5, A6

---

* When two functions $f_1(x)$ and $f_2(x)$ of the same independent variable x are plotted on the f_1, f_2-plane, the plot is referred to as a *phase* plot.

Note: The format IØR is supplied by the user in Card 2. The sample format of (8E10.3) will serve this purpose well.

Card 5. This card specifies the initial and final values of the independent variable over which the solution is desired. This card is read by the following statement:

```
READ (5, IØR) ALPHA, ØMEGA
```

where ALPHA and ØMEGA indicate the initial value and the final value, respectively.

Card 6. Y(I), the initial values of the dependent variables, is read by the following statement:

```
READ (5, IØR)(Y(I), I =1, M1)
```

Card 7. A(I),D(I),B(I) specify, respectively, the initial independent variable print point, the increment between print points, and the final print point of the Ith desired print-out interval as requested by M5 in Data Card 3. This card is read by the following statement:

```
READ (5, IØR)(A(I), D(I), B(I), I =1, M5)
```

Card 8. This card will be read only if IYVSX > 0, and it contains the following parameters:

      NYX = number of dependent variables to be plotted versus the independent variable. Also NYX ≤ 5.

  IYX(I) = the index of the dependent variable to be plotted. For example, if IYX(I) = 2, then Y(2) will be plotted.

The READ statement for this card is

```
READ (5, 20) NYX, (IYX(I), I =1, NYX)
```

and the format is (16I5).

Card 9. This card is read only if IYVSY > 0, and it contains the following parameters:

        NYY = number of phase plots to be made on a single plot. NYY ≤ 3.

  IYY (odd) = the abscissa of the phase plot (the independent variable of the plot).

 IYY (even) = the ordinate of the phase plot (the corresponding dependent variable of the plot).

This card is read by the following statement

```
READ (5, 20) NYY, (IYY(2*I −1), IYY(2*I), I =1, NYY)
```

7. *Remarks*:

  a) The main program which calls RKAM must be in the following form:

```
C MAIN PROGRAM MUST INCLUDE FOLLOWING STATEMENTS
  DOUBLE PRECISION UY
  DIMENSION Y(n), DY(n), YD(4n), VF(5n), UY(5n)
  CALL RKAM (Y, DY, YD, VF, UY)
  END
```

  b) Subroutine DERFUN must include the following statements:

```
SUBROUTINE DERFUN (Y, DY, T)
DIMENSION Y(1), DY(1)
  .

  .

  .

RETURN
END
```

  c) It may be necessary to supply more than one card in a particular data card section. For example, if there are 10 values to be read in a field of 8E10.3, then two cards are needed.

The following example will illustrate the use of subroutine RKAM for obtaining the solution of a given set of state equations.

Example 9.6.1 We shall obtain the unit-step response of the low-pass network, shown in Fig. 9.6.3, with zero initial conditions.

Following the method for formulating the state equations as discussed in Chapter 3, and noting that $v_1(t)$, $v_2(t)$, and $i_L(t)$ are the state variables of the network, we obtain

$$\frac{dv_1(t)}{dt} = -\frac{1}{R_1 C_1} v_1(t) - \frac{1}{C_1} i_L(t) + \frac{1}{R_1 C_1} e(t),$$

$$\frac{dv_2(t)}{dt} = -\frac{1}{R_2 C_2} v_2(t) + \frac{1}{C_2} i_L(t), \qquad (9.6.1)$$

$$\frac{di_L(t)}{dt} = \frac{1}{L} v_1(t) - \frac{1}{L} v_2(t).$$

Next, we define

$$v_1(t) \triangleq x_1(t),$$

$$v_2(t) \triangleq x_2(t), \qquad (9.6.2)$$

$$i_L(t) \triangleq x_3(t).$$

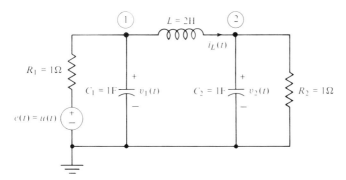

Fig. 9.6.3 Low-pass network for Example 9.6.1.

```
SUBROUTINE DERFUN(Y,DY,T)
DIMENSION Y(1),DY(1)
DY(1)=-Y(1)-Y(3)+1.
DY(2)=-Y(2)+Y(3)
DY(3)=.5*Y(1)-.5*Y(2)
RETURN
END
```

Fig. 9.6.4 Function subroutine and main program for Example 9.6.1.

```
C      MAIN PROGRAM FOR EXAMPLE 9-6-1
       DOUBLE PRECISION UY
       DIMENSION UY(15),VF(15),YD(12),Y(3),DY(3)
       CALL RKAM(Y,DY,YD,VF,UY)
       END
```

Substituting (9.6.2) and the element values into (9.6.1) gives

$$\dot{x}_1(t) = -x_1(t) - x_3(t) + 1,$$

$$\dot{x}_2(t) = -x_2(t) - x_3(t),$$ (9.6.3)

$$\dot{x}_3(t) = .5x_1(t) - .5x_2(t)$$

with $x_1(0) = x_2(0) = x_3(0) = 0$.

Using (9.6.3) the subroutine DERFUN and the main program can be written as shown in Fig. 9.6.4. The necessary data cards are given in Fig. 9.6.5. The computer output is shown in Figs. 9.6.6 and 9.6.7.

It was mentioned earlier that RKAM may be used to solve an nth-order differential equation. This fact will now be illustrated by the next example.

Example 9.6.2 Consider the differential equation

$$\frac{d^2y}{dt^2} + \frac{1}{t}\frac{dy}{dt} + y - 0.$$ (9.6.4)

The solution can be obtained by first transforming (9.6.4) into a set of first-order differential equations. First we can define the terms as:

$$y_1 \triangleq y, \qquad y_2 \triangleq \frac{dy}{dt},$$ (9.6.5)

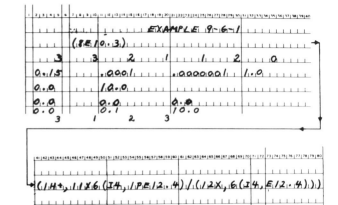

Fig. 9.6.5 Data cards.

```
        OUTPUT FOR EXAMPLE 9-6-1

INPUT-OUTPUT FORMATS
         (8E10.3)                        (1H+,11X6(I4,1PE12.4)/(12X,6(I4,E12.4)))
    INITIAL    MAX. ALLOW.  MIN. ALLOW.  MAX. ALLOW.  MIN. ALLOW.  STEP SIZE
    STEP       SINGLE STEP  SINGLE STEP  ABSOLUTE     ABSOLUTE     REDUCTION
    SIZE       ERROR        ERROR        STEP SIZE    STEP SIZE    FACTOR

  1.5000E-01   1.0000E-C4   1.0000E-07   1.0000E 00   1.0000E-06   5.0000E-01

    M1    M2    M3    M4    M5 IYVSX IYVSY
     3     3     2     1     1    1     2     0

ALPHA =  0.0                    OMEGA =  1.00000E 01
STARTING VALUES OF DEPENDENT VARIABLES
                1  0.0           2  0.0           3  0.0
PRINTER PLOT OF  3 VARIABLE(S) VS. INDEPENDENT VARIABLE.
VARIABLES PLOTTED ARE Y( 1) Y( 2) Y( 3)

INDEPENDENT              DEPENDENT VARIABLE NO. AND VALUE
VARIABLE

0.0           1  0.0          2  0.0          3  0.0
1.0000E-01    1  9.5083E-02   2  7.8611E-05   3  2.4164E-03
2.0000E-C1    1  1.8067E-01   2  6.0201E-04   3  9.3343E-03
3.0000E-01    1  2.5725E-01   2  1.9311E-03   3  2.0259E-02
4.0000E-01    1  3.2533E-01   2  4.3475E-03   3  3.4706E-02
5.0000E-01    1  3.8541E-01   2  8.0605E-03   3  5.2202E-02
6.0000E-01    1  4.3797E-01   2  1.3216E-02   3  7.2292E-02
7.0000E-01    1  4.8352E-01   2  1.9901E-02   3  9.4538E-02
8.0000E-01    1  5.2252E-01   2  2.8152E-02   3  1.1852E-01
9.0000E-01    '  5.5547E-01   2  3.7969E-02   3  1.4?^5E-01
 0000E 00       ^?81E-01      ?                3
  ^00E
                            -0L
              ..136E-01      2
4.800UE       ..5139E-01     2  5.     -U1    3  5.461.
4.9000E 00    4.5184E-01     2  5.4C72E-01    3  5.4172E-U.
5.0000E 00    1  4.5266E-01  2  5.4060E-01    3  5.3729E-01
5.1000E 00    1  4.5383E-01  2  5.4007E-01    3  5.3294E-01
MAXIMUM ERROR OF  9.22767E-08 AT  5.24999E 00 WITH STEP SIZE OF  1.50000E-01
ERROR TOO SMALL. NEW STEP SIZE  3.00000E-01
INDEPENDENT              DEPENDENT VARIABLE NO. AND VALUE
VARIABLE

5.20C0E 00    1  4.5530E-01  2  5.3919E-01   3  5.2868E-01
5.3000E 00    1  4.57C2E-01  2  5.3799E-01   3  5.2456E-01
5.40CCE 00    1  4.5857E-01  2  5.3651E-01   3  5.2059E-01
5.5000E 00    1  4.6110E-01  2  5.3482E-0^      5.1681E-01
5.60C0E ^^    1  4.6338F-^^  ?  5.3?^^        ^323E-^'
  7^             4.65^^
```

Fig. 9.6.6 Partial computer output for Example 9.6.1.

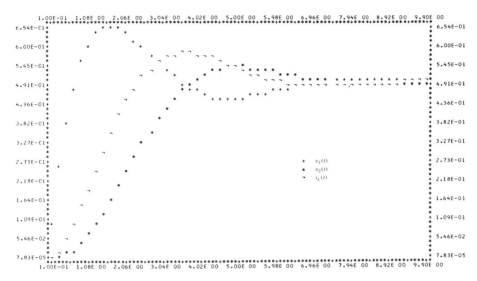

9.6.7 Computer output plot for Example 9.6.1.

```
      SUBROUTINE DERFUN (Y,DY,T)
      DIMENSION Y(1), DY(1)
      DY(1) = Y(2)
      IF(T) 2,1,2
    1 DY(2) =-Y(1)
      RETURN
    2 DY(2) =-((1./T)*Y(2) + Y(1))
      RETURN
      END
```

Fig. 9.6.8 Subroutine DERFUN for Example 9.6.2.

Then (9.6.4) may be rewritten in the following form:

$$\frac{dy_2}{dt} = -\frac{1}{t}y_2 - y_1 \qquad (9.6.6)$$

$$\frac{dy_1}{dt} = y_2. \qquad (9.6.7)$$

With the aid of expressions (9.6.6) and (9.6.7), the subroutine DERFUN for RKAM can be written as shown in Fig. 9.6.8. The completion of Example 9.6.2 is left as an exercise.*

9.7 NETWORK DESIGN THROUGH ANALYSIS TECHNIQUES

Analysis techniques play an important role in the problem of network design. For example, if an electrical engineer is asked to investigate whether a given network is best suited for a certain application, the first step he may take is to analyze the network and see if it satisfies the requirements of the intended application. This may mean the checking of the load characteristics of the network if the device is to be a power supply or the checking of the frequency responses of the network if an amplifier or a filter is under investigation.

---

* See Problem 9.15.

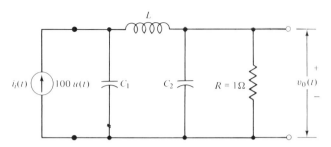

Fig. 9.7.1 Low-pass filter with a step current input.

In the preceding sections of this chapter we presented a number of subroutines which can be used as an aid in network analysis and design. In this section, we shall demonstrate how some of the analysis techniques may be applied, with the aid of a computer, to solve a problem in network design.

Example 9.7.1 Consider the filter network given in Fig. 9.7.1. Suppose that the elements in the network can take on one of the six sets of values as indicated in Table 9.7.1. We shall investigate which set of element values will result in the least settling time (as measured by the time beyond which the variation of the output voltage is within 10% of its final value) and with essentially no overshoot for the output voltage.

Table 9.7.1 Combination of Element Values for the Network in Figure 9.7.1.

Case	L	C_1	C_2
1	$\frac{3}{2}$	$\frac{4}{3}$	$\frac{1}{2}$
2	$\frac{3}{2}$	$\frac{4}{9}$	$\frac{3}{2}$
3	$\frac{3}{2}$	$\frac{1}{6}$	4
4	$\frac{1}{3}$	6	$\frac{1}{2}$
5	$\frac{1}{3}$	2	$\frac{3}{2}$
6	$\frac{1}{3}$	$\frac{3}{4}$	4

It can be readily shown that the transfer function of the system, $V_o(s)/I_{in}(s)$, is given by

$$\frac{V_o(s)}{I_{in}(s)} = \frac{\dfrac{1}{LC_1C_2}}{s^3 + \dfrac{1}{C_2}s^2 + \dfrac{(C_1 + C_2)}{LC_1C_2}s + \dfrac{1}{LC_1C_2}} \qquad (9.7.1)$$

Based on the expression (9.7.1), together with the aid of subroutines DPLRT, PFEX, and VPLOT, a computer program can be written for determining the response

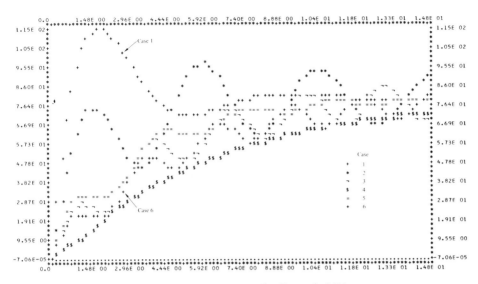

Fig. 9.7.2   Computer output for Example 9.7.1.

$v_o(t)$ for various sets of element values. For the input current transform $I_{in}(s)$ = 100/s, the output $v_o(t)$ obtained by the computer is shown in Fig. 9.7.2, from which we see that while Case 1 reveals the shortest settling time, its overshoot is too large and hence cannot be accepted. Case 5, on the other hand, gives a settling time slightly longer than that of Case 1 but it produces no overshoot. Thus, for the specification given in the problem, we shall select the following parameter values for the filter: $L = \frac{1}{3}$ henry, $C_1 = 2$ farads, and $C_2 = 1.5$ farads.

In the above discussion, a set of values for the network elements is chosen to suit a particular set of requirements in a network design problem. That is, we "optimize" the values of the network parameters to satisfy the given specifications. This process of network design is generally known as *network optimization*. In network optimization we may wish to optimize the network performance by changing one or more parameters with their variations usually limited to within certain ranges of values. This type of optimization is called the *optimization with "constraints."*

Computer-aided network optimization with or without constraints usually relies on one or more optimization programs, using certain iterative "algorithms" such that a specified function can be optimized (for example, minimization of an error function). However, a discussion of optimization is beyond the scope of this text. The interested reader is referred to literature elsewhere.*

The transfer function of a network is quite useful to a circuit designer since it contains the information he needs for the investigation of various network prob-

---

* See, for example, F. F. Kuo and W. G. Magnuson [KU 6] Chapter 5.

```
      SUBROUTINE FRQ(A,B,NMPTS,FMAX,FMIN,W,AMAG,PHASE)
C          THIS SUBROUTINE ACCEPTS COEFFICIENTS OF A TRANSFER
C     FUNCTION AND RETURNS VECTORS DESCRIBING THE MAGNITUDE
C     AND PHASE OF THE FUNCTION VERSUS RADIAN FREQUENCY.
C          A IS A VECTOR CONTAINING THE SIX COEFFICIENTS
C     OF THE NUMERATOR POLYNOMIAL OF THE TRANSFER FUNCTION IN ORDER OF
C     INCREASING  POWERS OF S.
C          B IS A VECTOR CONTAINING THE SEVEN COEFFICIENTS
C     OF THE DENOMINATOR POLYNOMIAL OF THE TRANSFER FUNCTION
C     IN ORDER OF INCREASING POWERS OF S.
C          NMPTS IS THE NUMBER OF FREQUENCY INTERVALS DESIRED
C     IN THE RETURN VECTORS CONTAINING THE SOLUTION.
C          FMAX IS THE MAXIMUM FREQUENCY TO BE CONSIDERED IN
C     THE SOLUTION.
C          FMIN IS THE MINIMUM FREQUENCY TO BE CONSIDERED.
C          W, AMAG, AND PHASE ARE THE VECTORS THAT WILL
C     CONTAIN THE SOLUTION RADIAN FREQ, THE FUNCTION MAGNITUDE
C     AND PHASE RESPECTIVELY UPON RETURN.
      COMPLEX Y,S,T
      DIMENSION A(6), B(7), W(101),AMAG(101), PHASE(101)
C
      WMAX=FMAX*6.2831853
      WMIN=FMIN*6.2831853
      DW=(WMAX-WMIN)/NMPTS
      W(1)=WMIN
      DO 10 I=1,NMPTS
      D=W(I)
      S=CMPLX(0.,D)
      Y=T(A,B,S)
      AMAG(I)=CABS(Y)
      U=AIMAG(Y)
      V=REAL(Y)
      PHASE(I)=ATAN2(U,V)*360./6.2831853
   10 W(I+1)=W(I)+DW
      RETURN
      END
```

Fig. 9.7.3 FORTRAN listing of subroutine FREQ for Example 9.7.2.

lems such as "sensitivity"* of the network due to parameter variations and "stability."* Earlier in this section we considered the response $v_o(t)$ of the network of Fig. 9.7.1 to a step input for various sets of parameter values. A more complex study in network design may, for example, call for an investigation of a network in which one or more parameters are varied continuously (approximated by very small steps) within the specified ranges of values as certain characteristics of a network function are being observed.

The determination of the frequency response of the network (as discussed in Chapter 5, Section 5.7) is, of course, another type of study. It is true that a circuit designer can construct approximately the magnitude and phase plots for a given set of parameter values without too much difficulty provided that the network function is relatively simple and the zeros and poles of the function can be found easily. However, in general, the task of determining and plotting the frequency response curves is tedious unless it is done with the aid of a computer.

In the following example we shall discuss how to determine the frequency response of a network with the aid of some of the subroutines already presented in this chapter.

Example 9.7.2 Write a program with necessary subroutines to obtain the frequency response of the following transfer function.

$$H(s) = \frac{0.2085s^2 + 1.075}{s^3 + 1.454s^2 + 1.812s + 1.075} \tag{9.7.2}$$

We shall use the following steps to achieve our goal: (1) use subroutine DPLRT to factor the numerator and denominator polynomials of $H(s)$; then (2) write and

---

* See Chapter 11 for more discussion of these subjects.

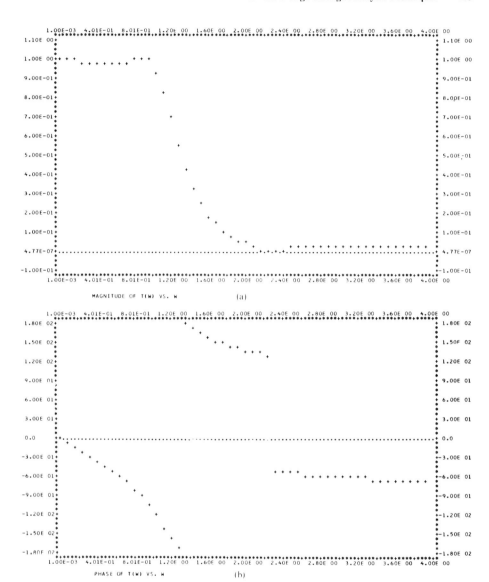

Fig. 9.7.4 (a) Magnitude of $H(\omega)$ vs. ω. (b) phase of $H(\omega)$ vs. ω.

call a subroutine designated by FREQ, which computes the magnitude and phase of $H(s)$; finally, (3) use subroutine VPLOT to display the results.

The subroutine FREQ, which will perform the computations in Step 2, is given in Fig. 9.7.3. Following the steps specified in this example, we find that the result will be as shown in Fig. 9.7.4. However, the writing of the main program to complete this example is left as an exercise.*

---

* See Problem 9.18.

9.8 SUMMARY

We have shown in this chapter how numerical analysis techniques may be utilized in designing various subroutines which are employed as tools in solving problems in computer-aided analysis and design. In particular, several subroutines have been introduced to perform: (a) analysis of simple RL and RC networks; (b) the study of multi-loop (node) resistive networks; (c) network analysis by means of Laplace transform technique; (d) convolution in obtaining network responses; (e) analysis of linear networks using state-variable methods, and finally; (f) network design making use of all the analysis techniques studied.

It should be emphasized that the methods as well as the subroutines presented in this chapter serve as an introduction on the subject of network analysis by means of a digital computer; they are by no means exhaustive. Furthermore, the dimensions in the subroutines used here are purposely kept small so that the required computer storage space will not be excessive. As a consequence, with the exception of RKAM, these subroutines are applicable only for solving relatively simple network problems. However, if these subroutines are to be used to handle larger systems, the user can easily increase the dimensions to suit his needs. Some of the subroutines in this chapter may be combined to form a more complex program for network analysis as suggested in some of the exercises included at the end of this chapter. Finally, it should be pointed out that a complete network analysis program must include, among other requirements, a routine which generates a set of network equations for a given network. Some of the more commonly used network analysis programs will be discussed in the next chapter.

REFERENCES

1. L. P. Huelsman, *Digital Computations in Basic Circuit Theory*, McGraw-Hill, 1968.

2. F. F. Kuo and J. F. Kaiser, *System Analysis by Digital Computer*, John Wiley & Sons, 1966, Chapters 1, 3, and 4.

PROBLEMS

9.1 Determine the voltage across the capacitor in the network shown in Fig. 9.2.3. for $t = 1$ sec using subroutine ISIMS with $v_c(0) = 0$ and $i(t) = 1 + t^2 + t^3 + t^4$. Use several different numbers of iterations (between 2 and 50) and determine the optimum value for accuracy.

9.2 Derive an expression for the capacitor voltage $v_c(t)$. Use subroutine RKAMC to solve for $v_c(t)$ if the input voltage is given by

$$v_i(t) = \sin\left(\frac{2\pi}{5}\right) t \text{ V}.$$

Obtain a plot to show the transient behavior of the solution. [*Hint*: Estimate the amount of time the response takes to reach the steady state.]

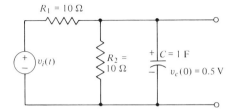

Figure P.9.2

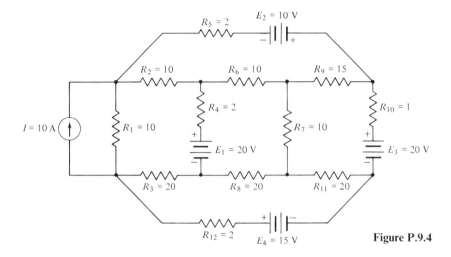

Figure P.9.4

9.3 Use RKAMC to determine the filter output voltage $v_o(t)$ in the network of Fig. 9.2.13 if it is given that

$$R_1 = 100\Omega, \quad R_2 = 1000\Omega, \quad C = 2 \times 10^{-4}\text{F}, \quad \text{and} \quad v_i(t) = \sin(337t) \text{ V}.$$

9.4 Use GAUSEI to determine the loop currents in the network shown in Fig. P.9.4.

9.5 Use DPLRT to factor the following polynomials:

a) $p(s) = s^7 + 3s^6 + 8s^5 + 15s^4 + 17s^3 + 12s^2 + 4s$,
b) $p(s) = s^6 + 5s^4 + 5s^3 + 2s^2 + 7s + 13$,
c) $p(s) = s^6 + 1$,
d) $p(s) = s^6 - 5s^5 + 4s^4 - 3s^3 + 2s^2 + s - 1$.

9.6 Use PFEX to obtain the partial-fraction expansion of the following function

$$F(s) = \frac{2s + 2}{(s + 2)^2 + (s + 3)}.$$

[*Note*: The answer is given in (9.4.12), Example 9.4.3.]

9.7 With the aid of subroutines DPLRT and PFEX, obtain the partial-fraction expansion of the function

$$H(s) = \frac{4.5s^3 + 65s^2 + 516s + 1208}{s^4 + 12s^3 + 152s^2 + 896s + 1664}.$$

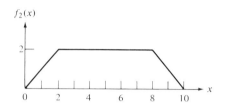

Figure P.9.10

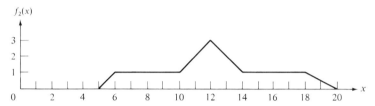

Figure P.9.13

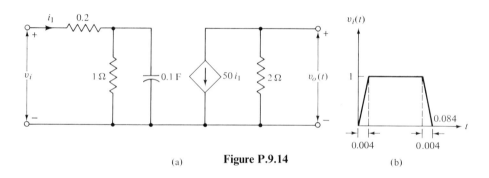

(a)   **Figure P.9.14**   (b)

9.8 Repeat Problem 9.7 for

$$H(s) = \frac{4s + 1}{s^3 - s^2 + s - 1}.$$

9.9 With the aid of subroutines DPLRT and PFEX, determine the inverse Laplace transform of

$$F(s) = \frac{s^3 + 3s^2 - 4s - 7}{s^5 + 2s^4 - 3s^2 - 4s - 8}.$$

9.10 Use CONVOL to determine the convolution of

$$f_1(x) = 10xe^{-2x}$$

and the function shown in Fig. P.9.10.

9.11 Use CONVOL to determine the convolution of the two functions given in Fig. 9.5.6, Example 9.5.2. Use a step size H = 1 and then H = 2 with the input in tabular form. Compare the results with that of Example 9.5.2.

9.12 Repeat Problem 9.11 with the input in analytical form.

9.13 Using Fig. P.9.13 repeat Problem 9.10 for

$$f_1(x) = 2 \sin (1.57x).$$

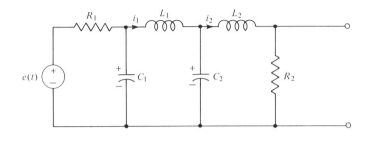

$e(t) = u(t),$
$R_1 = R_2 = 1,$
$L_1 = L_2 = 2,$
$C_1 = C_2 = 1.$
All initial conditions = 0

Figure P.9.16

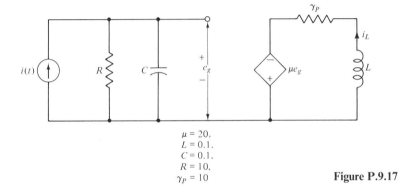

$\mu = 20,$
$L = 0.1,$
$C = 0.1,$
$R = 10,$
$\gamma_P = 10$

Figure P.9.17

9.14 The input to the approximate hybrid-π transistor model shown in Fig. P.9.14(a) is a pulse depicted in Fig. P.9.14(b). Determine the output $v_o(t)$ by first solving for the impulse response, $h(t)$, of the circuit and then use CONVOL to obtain $v_o(t) = h(t) * v_i(t)$.

9.15 Use RKAM to obtain the solution of the differential equation (9.6.4) in Example 9.6.2, with zero initial conditions except $y(o) = -70$.

9.16 a) For the network shown in Fig. P.9.16, derive the state equations using the voltages across the capacitors and the currents through the inductors as state variables.
      b) Use RKAM to solve the state equations obtained in (a).

9.17 a) Derive the state equations for the network shown in Fig. P.9.17, using i_L and e_g as state variables.
      b) Use RKAM to solve the state equations obtained in (a) for a unit-step input.

9.18 Project: Write a main program and the necessary subroutines to complete the problem described in Example 9.7.2.

CHAPTER 10

COMPUTER-AIDED NETWORK ANALYSIS II

10.1 INTRODUCTION

In Chapter 9 we presented a number of subroutines which can be used as an aid to solving network problems. Each of these subroutines performs a specific type of computation that may be required for analyzing a given network. To form a network analysis program, a number of subroutines are grouped together and linked by a "calling program," which acts as an interpreter between the user and the program. Such a network analysis program, in general, contains: (1) a subprogram which derives a set of (independent) system equations (for example, a set of nodal equations from the incidence matrix described in the computer input information); (2) a subprogram which solves the equations derived in (1); and (3) a subprogram that determines, from the input data, the kinds of calculations to be performed, the particular subroutines to be used, and the forms of output desired.

In general, the types of analyses performed in a network analysis program can be classified into (a) dc-analysis, (b) ac-analysis, and (c) transient analysis. The dc-analysis produces the steady-state solution of a network with constant inputs. The solution may include, for instance, the node voltages, branch currents, element power losses, etc. The ac-analysis gives the sinusoidal steady-state solutions for either a specified frequency or a range of frequencies. Thus the ac-analysis is to be used to obtain the frequency response of a network. The transient analysis produces the transient response of a network to a specified input such as a step input, impulse input, or any other arbitrarily described input.

Many computer programs are available today for network analysis. Classifications of these programs may be based, for example, on the types of networks the programs are designed to analyze, on the types of solutions obtainable from the programs, etc. If the types of system equations and the methods of solution are used as a basis for classification, the programs may be divided into the following four groups: (1) the node-analysis method, (2) the state-variable approach, (3) the (k-tree) topological technique, and (4) the flow graph technique. In the following paragraphs, we shall describe four analysis programs, each of which typifies a particular analysis technique. Each of these programs is included here for one or more of the following reasons: (a) it is very well known, (b) it is widely used, (c) it is well tested and appears to be free of errors, (d) it contains good error-checking routines, (e) it has extensive diagnostic checks for inputs, (f) it is well documented, (g) it is known for its speed, (h) it has special features.

In each of the next four sections, we shall describe briefly a program in the form of a short user's guide and illustrate the application of the program with examples including the necessary inputs and the corresponding computer outputs. However, any detailed discussion of the program will not be attempted. The interested reader will be referred to literature elsewhere.

10.2 THE ELECTRONIC CIRCUIT ANALYSIS PROGRAM (ECAP)

In this section we shall present a program called "*Electronic Circuit Analysis Program*" (ECAP) which is developed by International Business Machines Corporation (IBM) originally for the IBM 1620 system and is now available for use with other systems including IBM 7094, IBM 1130, and IBM 360.*

ECAP, based on nodal analysis, is composed of four related programs, namely the input language, the dc-analysis, the ac-analysis, and the transient analysis programs. The input language program may be considered by the user as the main program which interprets the input data, calls one of the three analysis programs, and specifies the desired outputs. The dc-analysis program determines the dc-solutions of a linear network with options for providing various types of outputs such as node voltages, branch currents, worst-case solutions, and automatic variation of parameter values by a "modify" routine. The ac-analysis program provides the sinusoidal steady-state solution of a network at a given frequency, which may be varied automatically using a "modify" routine. The transient analysis program determines the transient response of a network to a given excitation, which may be a dc-source or an arbitrary time-varying source.

To use ECAP or any other general-purpose circuit analysis program, the user must first learn the "input language" of the program. In other words, he must know how to enter information on network topology, element values, calculations required, and output desired. Table 10.2.1 shows the types of circuit elements acceptable to each of the three analysis programs of ECAP. MKS units are assumed as indicated in Table 10.2.2. ECAP also requires that every branch in the circuit to be analyzed takes the form of the "standard circuit branch" as shown in Fig. 10.2.1. Each branch in a circuit must contain one non-zero passive (R, L, or C) element. This rules out the use of ideal sources.† A value of "0" for any other quantity in a branch will be assumed by ECAP unless otherwise specified.

The following general procedure may be used as a guide for the application of ECAP for circuit analysis:

1. Decide if a dc, ac, or transient solution is required.
2. Draw an ECAP equivalent circuit using the ECAP standard circuit branch representation (Fig. 10.2.1).

---

* For information on the availability of various versions of ECAP, the reader is referred to IBM Corporation, Program Information Department, Hawthorn, New York 10532.

† To simulate an ideal voltage source, one can add a very small resistance in series with the source.

Table 10.2.1 Circuit Elements Allowed in ECAP.

Circuit elements	Analysis programs		
	dc	ac	Transient
Resistor or conductor	×	×	×
Capacitor		×	×
Inductor		×	×
Mutual inductance		×	
Independent voltage sources:			
(a) Constant	×		×
(b) Sinusoidal		×	
(c) Time dependent			×
Independent current sources:			
(a) Constant	×	×	×
(b) Sinusoidal		×	
(c) Time dependent			×
Dependent current source	×	×	×
Switch			×

Table 10.2.2 Units Assumed in ECAP.

Variables	Units	Variables	Units
Resistance	ohms	Voltage	volts
Conductance	mhos	Current	amperes
Capacitance	farads	Time	seconds
Inductance	henries	Frequency	hertz
Mutual inductance	henries	Phase angle	degrees

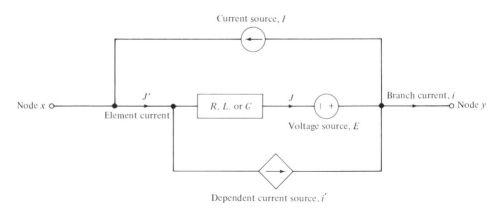

Fig. 10.2.1 Standard circuit branch used in ECAP.

3. Number the nodes arbitrarily in ascending order: 0, 1, 2, ..., with 0 assigned to the reference node.

4. Number the branches arbitrarily in ascending order: 1, 2, ...

5. Assign a positive current direction to each branch.

6. Assuming that punched cards are used, prepare the data cards in the manner as described in the following paragraphs.

All cards must be coded within columns 1 through 72 inclusively. Columns 73–80 are ignored by ECAP as they are normally used only for sequencing purposes. The first 72 columns of a card are divided into two "data fields." The first data field includes columns 1 through 5 and it is used for identifying the type and number of a circuit data card as will be explained later in this section. The second data field includes columns 7 through 72 and is used for circuit data description such as node numbers of a branch, and type and value of an element. Column 6 must be left blank unless the card is a "continuation card" in the case that one card is insufficient for entering all the circuit data associated with a branch or describing a time-dependent nonperiodic source, etc.; in such a case, an asterisk "*" is coded in column 6 on each continuation card. Within each of the two data fields, blanks are ignored by ECAP so that they may be used to increase readability.

A deck of cards necessary to execute ECAP, referred to as an "ECAP input deck," consists of the following types of cards: system control, comment, circuit data, solution control, output specification, and command. These types of cards are each explained as follows.

System Control Cards

Each computer installation requires a certain set of system control cards depending upon the type of data processing equipment, accounting procedure of the installation, etc.

Comment Cards

ECAP will print out any statement specified in a card with a C in column 1. The user may use these comment cards to document any information in connection with an ECAP problem, such as problem title, names of the users, contract numbers, etc.

Circuit Data Cards

Circuit data cards are used to describe the network topology and element values. They are classified into four types: the B-cards, the T-cards, the M-cards, and the S-cards. In each card, columns 1 through 5 are used to specify the type and number of the card, whereas columns 7 through 72 are for describing the topology and values of the branch associated with the cards. Each type of these cards will be described separately.

The B-Cards. A Branch card describes the topology and values of the elements of a standard branch with the format as shown in Fig. 10.2.2. In this figure, *Bbb* may be coded anywhere within the first data field (columns 1 through 5). The order of

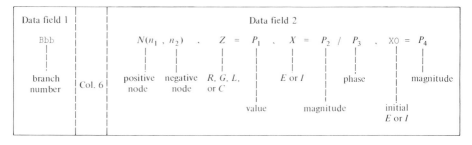

Fig. 10.2.2 *B*-card format.

Table 10.2.3 Availability of BETA and GM.

From-branch element	dc		ac		Transient	
	BETA	GM	BETA	GM	BETA	GM
R	Yes	No	Yes	Yes	Yes	Yes
L	No	No	No	Yes	Yes	No
C	No	No	No	Yes	Yes	No

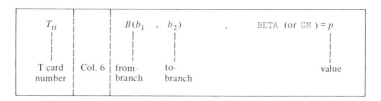

Fig. 10.2.3 *T*-card format.

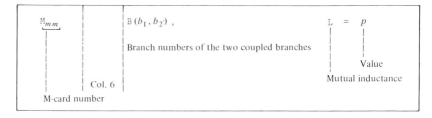

Fig. 10.2.4 *M*-card format.

the nodes indicates that an assumed positive current flows from n_1 to n_2. Z indicates the type of elements with a value P_1. X may be either E or I, representing a voltage- or current-source, respectively, which is specified by magnitude P_2 alone (if dc-analysis is used), or by both magnitude P_2 and phase P_3 separated by a slash "/". However, if both voltage and current sources are present in a branch, the data $X = P_2/P_3$ must be entered twice, one representing the voltage source and the other, the current source. In the case that Z is C (or L), the corresponding initial voltage across C (or the initial current through L) is specified by E0 (or I0) in the place of $X0$ with the magnitude given by P_4.

The order of the entries in the second data field of the B-card is arbitrary, but the entries must be separated by commas.

The T-Cards. A T-card describes the relationship between the dependent current source in a branch, called the "*to-branch*," and the current through (or the voltage across) any other branch, called the "*from-branch*," by means of specifying the current gain BETA in the relation $i_{\text{to-branch}} = \text{BETA} \cdot i_{\text{from-branch}}$ (or the transconductance GM in $i_{\text{to-branch}} = \text{GM} \cdot V_{\text{from-branch}}$). The format for a T-card is shown in Fig. 10.2.3. The availability of BETA or GM for use in the three analysis programs of ECAP is indicated in Table 10.2.3.

Remarks. In entering input data for both the B- and the T-cards, one should exercise caution about whether to use negative signs for the sources as well as BETA and GM. In the standard circuit branch shown in Fig. 10.2.1, the direction of the current through the passive element J is always taken to be the reference direction. If the polarities of a source do not agree with those schematically specified in the standard branch, a negative value should be entered in the input data. Thus, for example, if the dependent current source and the branch current J do not have the same direction, then a negative value must be used for either the BETA or the GM describing this dependent source.

The sign convention for initial conditions can be treated as follows: If the initial current in a coil has the same direction as that of the current J, then a positive value is used for I0; in like manner, a positive value is used for the initial voltage across a capacitor E0 if the current J traverses from the negative to the positive polarity of E0.

A T-card must not precede the B-card which defines the from-branch. Even though T-cards and B-cards may be mixed with each other, they must be entered *sequentially* beginning with 1.

The M-cards. An M-card describes the relationship between two mutually coupled inductive branches. The format for the M-cards is given in Fig. 10.2.4, in which the letter L must be used to indicate the mutual inductance with its value given by p while the order of the branch numbers is arbitrary. A negative value must be used for p if one branch has the dot at the positive node and the other at the negative node. Again, M-card numbers must be assigned sequentially beginning with 1.

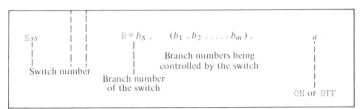

```
B 5          N(3 , 4) , R = (1, 100)

B 6          N(7 , 8) , R = 50
  .
  .
  .
S 1          B = 6  , (5) , ON
```

Fig. 10.2.5 Illustration for switch coding.

```
Sss     │ │        B = b_S .    (b₁ . b₂ . . . . . b_m) .               a
        │ │                        Branch numbers being
        │ │ │                      controlled by the switch
        Switch number    │
                Branch number
                 of the switch                                    ON or OFF
```

Fig. 10.2.6
S-card format.

The S-cards. An *S*-card is used to describe a "switch." Any branch may be designated as a switch if the branch current is continuously monitored for current
direction, a change in which will cause some parameter in a certain branch (called
the controlled branch) to "switch" to another value. A parameter in a controlled
branch will be affected by a switch controlling that branch only if it has two specified
values.

A simple example of the use of *S*-card is shown in Fig. 10.2.5, in which two
values, namely 1 and 100 ohms, have been assigned to the resistance R_5 in branch 5.
The *S*-card designated as S1 indicates that (a) the branch B6 is assigned as the
switch; (b) the branch B5 is to be controlled, i.e., it is the controlled branch; and
(c) the term ON requires R in B5 to assume the first value (1 ohm) when the current in
B6 is positive. If, on the other hand, OFF were used instead of ON, then R in B6 will
assume the first value (1 ohm) when the current in B6 is negative.

Thus, in general, if a switch is originally designated as ON in the *S*-card, then the
first of the two values of every parameter controlled by this switch is taken when
the current through the switch is positive and the second value is assumed when the
current through the switch is negative. The opposite effects will be true if the switch
is originally designated as OFF in the *S*-card. Thus, the ON or OFF designation, along
with the actual current direction in the switch, determines the value(s) assumed for
the parameter(s) in the controlled branch.

The format for the *S*-card is given in Fig. 10.2.6, in which b_S is the branch number of the branch designated as a switch and the letter *a* is either ON or OFF as
specified by the user. It should be noted that *S*-cards must also be numbered
sequentially beginning with 1.

Solution Control Cards

Solution control statements, coded in the second data field, may be divided into
two types as follows.

Type I. A type-I solution control card contains three items: a control word, an
equality sign, and a value. This type of card is used to specify values such as the

"frequency" to be used in the AC program and the "time step" for the transient program. As an illustration, a statement

```
FREQUENCY = 1000
```

will specify the frequency to be 1000 hertz.

Type II. A type-II solution control card contains only one solution word (such as "sensitivity analysis") which is used to indicate a specific calculation not included in a so-called "nominal solution."*

It should be noted that ECAP can recognize only the first two letters of any word (alphabetic mnemonics) except the word BETA, which must be spelled out completely. Thus, FR is sufficient for *FR*equency, TR for *TR*ansient analysis, and so on.

Output Specification Cards

An output specification card, coded in the second data field, is used to indicate a specific desired output. Such a card will take a format similar to the following statement:

```
PRINT, NV, BA
```

In this case, all the node voltages (NV) and branch currents (BA) will then be printed out.

Command Cards

At least two command cards, coded in the second data field, are necessary in each ECAP input deck. The first command card indicates which of the three analysis programs is to be used. For example, a card with the statement DC at the beginning of the deck indicates that the dc program will be used to analyze the circuit described in the deck. The second command card with the statement EX (for *EX*ecute) is placed at the end of the deck to start the calculation. The use of other command cards will be discussed later in the section.

The order of various types of cards in an ECAP input deck should, in general, be arranged as follows.

1. System control cards
2. Command card (DC, AC, or TR analysis)
3. Data cards (The cards within each type must be sequential.)
    a) *B*-cards
    b) *T*-cards
    c) *M*-cards
    d) *S*-cards

4. Solution control cards
5. Output specification cards
6. Command card (EX)

---

* A nominal solution in ECAP includes routine calculations such as node voltages, branch currents, element powers, etc.

Fig. 10.2.7 Numerical formats used in ECAP.

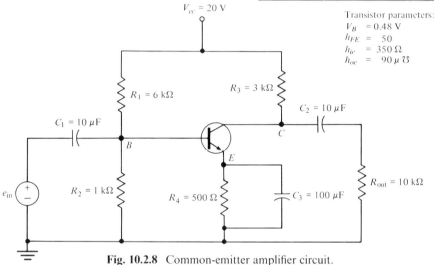

Fig. 10.2.8 Common-emitter amplifier circuit.

7. Another command card with END punched in the second data field indicating the termination of the ECAP computation

In addition, comment cards may be used as discussed previously and may appear anywhere after the system control cards in the input deck.

The numerical formats acceptable to ECAP are shown in Fig. 10.2.7 in which an X represents any digit and YY signifies a power of ten (with a value between -99 and $+99$) in the standard E format in FORTRAN.

The following examples will illustrate the coding methods for ECAP we have discussed thus far.

Example 10.2.1 We shall determine the dc-solution for the common-emitter amplifier given in Fig. 10.2.8. The ECAP equivalent circuit for the dc-analysis is shown in Fig. 10.2.9 with the nodes and branches numbered and current directions assigned as indicated. Note that branches 5 and 6 in Fig. 10.2.9 represent the dc-equivalent circuit of the transistor in the amplifier circuit so that nodes 1, 2, and 3 correspond to, respectively, the base, the collector, and the emitter of the transistor. Following the coding method discussed earlier in the section, the coding for the input deck is given in Fig. 10.2.10.

Through the use of appropriately prepared system control cards for the particular computer installation, the input deck as depicted in Fig. 10.2.10 will produce the ECAP output as shown in Fig. 10.2.11, which represents the nominal solution of the given common-emitter amplifier circuit determined by the dc-analysis program of ECAP.

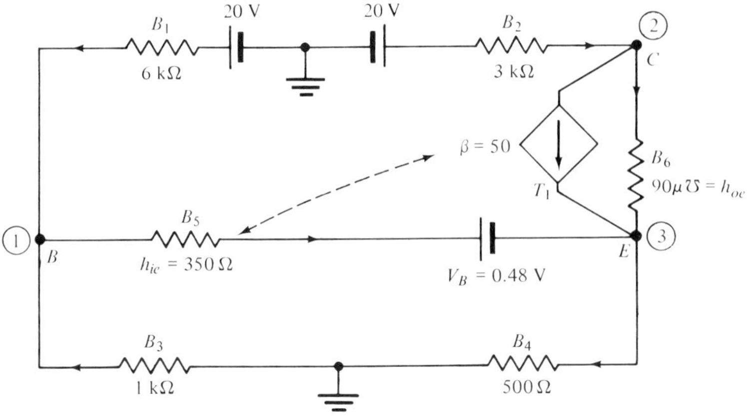

Fig. 10.2.9 DC-equivalent circuit of common-emitter amplifier.

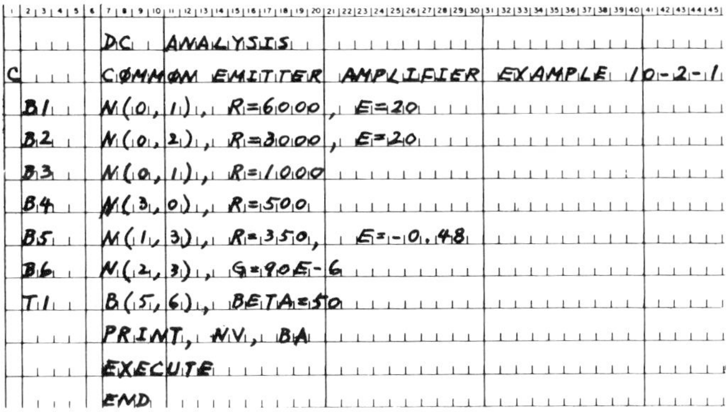

```
      DC ANALYSIS
C     CØMMØN EMITTER AMPLIFIER EXAMPLE 10-2-1
 B1   N(0,1), R=6000, E=20
 B2   N(0,2), R=3000, E=20
 B3   N(0,1), R=1000
 B4   N(3,0), R=500
 B5   N(1,3), R=350, E=-0.48
 B6   N(2,3), G=90E-6
 T1   B(5,6), BETA=50
      PRINT, NV, BA
      EXECUTE
      END
```

Fig. 10.2.10 Input data for Example 10.2.1.

```
      DC  ANALYSIS
C     COMMON EMITTER AMPLIFIER  EXAMPLE 10-2-1
 B1   N(0,1), R=6000, E=20
 B2   N(0,2), R=3000, E=20
 B3   N(0,1), R=1000
 B4   N(3,0), R=500
 B5   N(1,3), R=350,   E=-0.48
 B6   N(2,3), G=90E-6
 T1   B(5,6), BETA=50
      PRINT, NV, BA
      EXECUTE

NODE VOLTAGES

NODES                 VOLTAGES

 1-  3       0.27870557D 01     0.65746692D 01     0.22784369D 01

BRANCH CURRENTS

BRANCHES              CURRENTS

 1-  4       0.28688240D-02     0.44751079D-02    -0.27870555D-02      0.45568734D-02
 5-  6       0.81768289D-04     0.44750751D-02
```

Fig. 10.2.11 Computer output for Example 10.2.1.

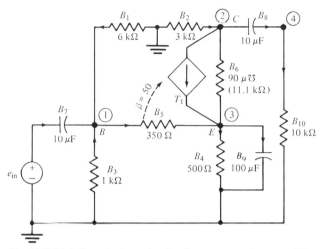

Fig. 10.2.12 AC-equivalent circuit of common-emitter amplifier.

```
           AC   ANALYSIS
C   COMMON  EMITTER  AMPLIFIER  EXAMPLE  10-2-2
B1      N(0, 1),  R=6000
B2      N(0, 2),  R=3000
B3      N(0, 1),  R=1000
B4      N(3, 0),  R=500
B5      N(1, 3),  R=350
B6      N(2, 3),  R=11.1E3
B7      N(0, 1),  C=10E-6,  E=25E-3
B8      N(2, 4),  C=10E-6
B9      N(3, 0),  C=100E-6
B10     N(4, 0),  R=10E3
T1      B(5, 6),  BETA=50
        FREQUENCY=2000
        PRINT,  NV,  BA
        EXECUTE
        END
```

Fig. 10.2.13 Input data for Example 10.2.2.

Example 10.2.2 As an illustration for coding the input data for the ac-analysis of ECAP, we shall obtain the ac-solution of the common-emitter amplifier circuit shown in Fig. 10.2.8.

Similar to the dc-case in the last example, we first draw the ac-equivalent circuit as shown in Fig. 10.2.12. The coding for the input deck is given in Fig. 10.2.13, where we are assuming a frequency of 2000 hertz for the sinusoidal voltage source. Note that a solution control card is used to specify the frequency for the ac-analysis. Figure 10.2.14 shows the nominal solution for this example.

```
          AC   ANALYSIS
   C      COMMON EMITTER AMPLIFIER  EXAMPLE 10-2-2
   B1     N(0,1),  R=6000
   B2     N(0,2),  R=3000
   B3     N(0,1),  R=1000
   B4     N(3,0),  R=500
   B5     N(1,3),  R=350
   B6     N(2,3),  R=11.1E3
   B7     N(0,1),  C=10E-6,  E=25E-3
   B8     N(2,4),  C=10E-6
   B9     N(3,0),  C=100E-6
   B10    N(4,0),  R=10E3
   T1     B(5,6),  BETA=50
          FREQUENCY=2000
          PRINT, NV, BA
          EXECUTE


   FREQ =  0.19999998E 04


          NODES                NODE VOLTAGES


   MAG    1-  4   0.24933290E-01  0.67725258E 01  0.23918259E-02  0.67725239E 01
   PHA            0.18178825E 01 -0.17268253E 03 -0.82580917E 02 -0.17263696E 03




          BRANCHES             BRANCH CURRENTS


   MAG    1-  4   0.41555486E-05  0.22575073E-02  0.24933281E-04  0.47836511E-05
   PHA           -0.17818208E 03  0.73174372E 01 -0.17818208E 03 -0.82580917E 02

   MAG    5-  8   0.70897950E-04  0.29347595E-02  0.99891622E-04  0.67725219E-03
   PHA            0.73226662E 01  0.73279591E 01  0.57219868E 01 -0.17263696E 03

   MAG    9- 10   0.30056536E-02  0.67725219E-03
   PHA            0.74190226E 01 -0.17263696E 03
```

Fig. 10.2.14 Computer output for Example 10.2.2.

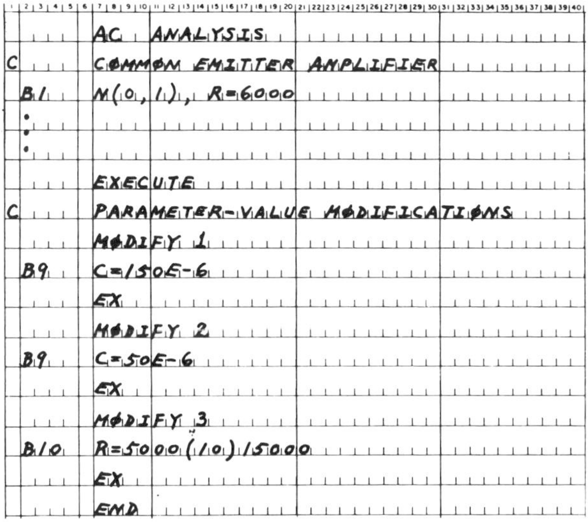

Fig. 10.2.15 Parameter-value modifications for Example 10.2.3.

In the ac- or dc-analysis we often desire to vary one or more parameters of the network under consideration. This may be accomplished by using the parameter modification capability of ECAP as the next example will illustrate.

```
      MODIFY 1
 B9   C=150E-6
      EX
```

FREQ = 0.19999998E 04

        NODES   NODE VOLTAGES

   MAG   1- 4   0.24951003E-01   0.67953329E 01   0.15999207E-02   0.67953310E 01
   PHA   0.18259115E 01 -0.17450395E 03 -0.84432770E 02 -0.17445837E 03

        BRANCHES   BRANCH CURRENTS

   MAG   1- 4   0.41584999E-05   0.22651097E-02   0.24950990E-04   0.31998406E-05
   PHA   -0.17817406E 03   0.54959898E 01 -0.17817406E 03 -0.84432770E 02

   MAG   5- 8   0.71136688E-04   0.29446427E-02   0.10020369E-03   0.67953300E-03
   PHA   0.55023794E 01   0.55065117E 01   0.44351006E 01 -0.17445837E 03

   MAG   9- 10   0.30157776E-02   0.67953300E-03
   PHA   0.55672045E 01 -0.17445837E 03

      MODIFY 2
 B9   C=50E-6
      EX

FREQ = 0.19999998E 04

        NODES   NODE VOLTAGES

   MAG   1- 4   0.24882920E-01   0.66643496E 01   0.47072209E-02   0.66643476E 01
   PHA   0.17791281E 01 -0.16731799E 03 -0.77125336E 02 -0.16727242E 03

        BRANCHES   BRANCH CURRENTS

   MAG   1- 4   0.41471530E-05   0.22214488E-02   0.24882916E-04   0.94144452E-05
   PHA   -0.17822083E 03   0.12681953E 02 -0.17822083E 03 -0.77125336E 02

   MAG   5- 8   0.69765534E-04   0.28878832E-02   0.98424774E-04   0.66643464E-03
   PHA   0.12683699E 02   0.12692475E 02   0.94852304E 01 -0.16727242E 03

   MAG   9- 10   0.29576339E-02   0.66643464E-03
   PHA   0.12874646E 02 -0.16727242E 03
```

**Fig. 10.2.16**  Computer output for Example 10.2.3.


**Example 10.2.3**  Suppose that we wish to determine (a) the ac-solution for the network shown in Fig. 10.2.8 for several values of $C_3$, say, 50, 100, and 150 $\mu$F; and (b) the ac-solution for the same circuit with $C_3 = 50$ $\mu$F and the value of $R_{out}$ varied from 5 to 15 k$\Omega$ in 10 equally spaced steps.

The input deck for the desired computations is coded as illustrated in Fig. 10.2.15. In this figure all the statements up to the first EXECUTE are identical to those for Example 10.2.2 as shown in Fig. 10.2.13 for the ac-solution of the same circuit with a particular set of parameter values. The command card MODIFY 1 used in the ac- (or dc-) analysis dictates that the ac- (or dc-) analysis is to be repeated with modified parameter values. In our case, all the parameter values remain the same
```

```
        MODIFY 3
   B10  R=5000(10)15000
        EX

R    =  0.50000000E 04

FREQ =  0.19999998E 04


        NODES              NODE VOLTAGES

MAG    1-  4   0.24879821E-01   0.55878754E 01   0.48539937E-02   0.55878687E 01
PHA            0.17759161E 01  -0.16699577E 03  -0.76779465E 02  -0.16690457E 03



        BRANCHES           BRANCH CURRENTS

MAG    1-  4   0.41466365E-05   0.18626242E-02   0.24879817E-04   0.97079910E-05
PHA           -0.17822406E 03   0.13004219E 02  -0.17822406E 03  -0.76779465E 02

MAG    5-  8   0.69672213E-04   0.29801971E-02   0.98304168E-04   0.11175736E-02
PHA            0.13026280E 02   0.13038414E 02   0.97239180E 01  -0.16690457E 03

MAG    9- 10   0.30498537E-02   0.11175736E-02
PHA            0.13220513E 02  -0.16690457E 03

R    =  0.60000000E 04

FREQ =  0.19999998E 04


        NODES              NODE VOLTAGES

MAG    1-  4   0.24880733E-01   0.59058857E 01   0.48106574E-02   0.59058809E 01
PHA            0.17768745E 01  -0.16708920E 03  -0.76881744E 02  -0.16701320E 03



        BRANCHES           BRANCH CURRENTS

MAG    1-  4   0.41467883E-05   0.19686276E-02   0.24880719E-04   0.96213198E-05
PHA           -0.17822307E 03   0.12910751E 02  -0.17822307E 03  -0.76881744E 02

       5-  8   0.697000⁰⁰      ?9529405E-02   0.983401⁷ᵉ⁻      ³1343E-03
               0.129⁰         ³4084E 02   0.965³'          ³?0E 03
```

Fig. 10.2.16 Computer output for Example 10.2.3 (*concluded*).

except that in branch 9 the value of C is changed to 150 μF. Similarly, the command card MODIFY 2 causes another nominal ac-solution with another change in parameter values as indicated. Finally, the command card MODIFY 3 causes still another parameter-value variation, in which the resistance in branch 10 is "iterated" in ten equal steps from 5kΩ to 15kΩ, and for each iteration the nominal ac-solution will appear in the computer output. It is important to note that in the last modification C = 50E − 6 is assumed for branch 9 since ECAP keeps and assumes all parameter values used in the most recent execution unless a value is specifically modified. Figure 10.2.16 shows the output for Example 10.2.3.

Parameter modification is available only in the dc- or ac-analysis program; in the latter, the modify routine is especially useful for frequency iteration. For example, a frequency card

$$\text{FR} = p_1(p_2)p_3$$

will cause a solution at each of the following frequencies:

$$p_1, p_1p_2, p_1p_2^2, p_1p_2^3, \ldots, p_1p_2^k,$$

where p_1, p_2, p_3 are positive numbers and $p_1p_2^k \geq p_3$. Thus, the frequency iteration is logarithmic (or multiplicative). For an additive frequency iteration, the frequency card is in the form

$$\text{FR} = p_1(+p_2)p_3$$

which requests analysis to be performed for the frequency range from p_1 to p_3 in p_2 frequency steps; that is, analyses at frequencies

$$p_1, p_1 + \left[\frac{p_3 - p_1}{p_2}\right], p_1 + 2\left[\frac{p_3 - p_1}{p_2}\right], p_1 + 3\left[\frac{p_3 - p_1}{p_2}\right], \ldots, p_3.$$

The transient analysis program provides the user with time response of a network (both linear and nonlinear) subject to arbitrary driving inputs. If the network under consideration is nonlinear, a piecewise linear equivalent circuit through the use of switches must first be drawn before the transient analysis program can be used for obtaining the solution. Before introducing the coding method for the transient analysis program, we shall discuss some of the techniques that must be used in the transient analysis of ECAP such as piecewise linear modeling of nonlinear elements, as well as the format for describing a time-dependent function, and solution control statements.

The use of a switch in the piecewise linear modeling of a nonlinear element can perhaps be best described by considering a simple example.

Example 10.2.4   Figure 10.2.17 shows a simple nonlinear network in which the characteristic of the diode D is depicted in Fig. 10.2.18. To draw a piecewise linear equivalent circuit, the given diode characteristic is approximated by means of two straight-line segments which intersect at the voltage designated as V_D ($=0.5$ volts). The branch representing the piecewise linear model of the diode is shown in Fig. 10.2.19 in which the branch is assigned as a switch that controls the value of its own resistance. If the branch voltage $e < 0.5$, then the branch current i is negative and the switch will be OFF so that $R = 10^7\Omega$—a value equal to the inverse of the slope for the "nonconducting mode" of the diode characteristic. Similarly, if $e > 0.5$, then i becomes positive and the switch will be ON so that $R = 10\Omega$, the inverse of the slope for the "conducting mode" of the diode characteristic. Thus, except in the neighborhood of $e = 0.5$, this proposed model of the diode represents the device fairly accurately. With this diode model the ECAP equivalent circuit for the given network can be drawn as shown in Fig. 10.2.20. In this figure, a resistance of 0.1Ω is inserted in branch 1 due to the fact that ECAP does not accept any branch with zero impedance.

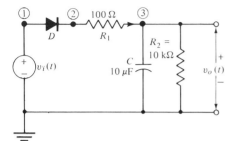

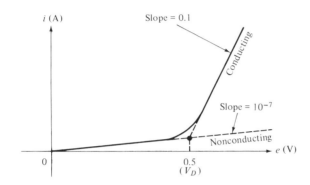

Fig. 10.2.17   Network for Example 10.2.4.

Fig. 10.2.18   Diode characteristic for Example 10.2.4.

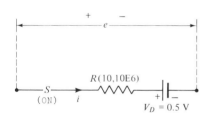

Fig. 10.2.19   Piecewise linear diode model.

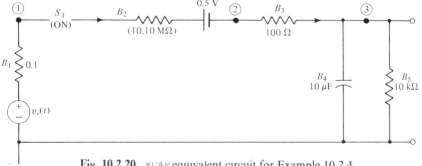

Fig. 10.2.20   ECAP equivalent circuit for Example 10.2.4.

Next, we shall discuss several Type-I solution control cards which are needed in the input deck for the transient analysis program. They are described as follows:

a) *Time step* (in seconds) indicates the time interval the program will use for each step of calculation in the transient solution. Thus, a solution control card with the statement

$$\text{TIME STEP} = 0.1\text{E} - 3$$

punched in the second data field will cause a solution at every 10^{-4} sec.

b) *Output interval* indicates the frequency with which the solution is to be printed in the output. For instance, a card with the statement

$$\text{OUTPUT INTERVAL} = 10$$

punched in the second data field will cause the output to print a set of solutions after every 10 steps of calculations have been performed.

c) *Final time* indicates the time of termination of calculation in the problem. Thus a card with the statement

$$\text{FINAL TIME} = .01$$

punched in the second data field will cause the computation of the solution to terminate after $t = 0.01$ sec.

In the transient analysis program, the user can specify a time-dependent voltage—or current—source in any branch. This is done by including a card placed immediately after the B-card of the branch, to which this source belongs. The format used for this purpose depends on the type of the source described as follows.

1. *Nonperiodic source.* The input format for coding the data of a nonperiodic voltage—or current—source is

$$X\text{bb} \qquad (m), s_0, s_1, s_2, \ldots, s_n$$

where $X = \text{E}$ (or I) designates a voltage—or current—source; bb denotes the number identifying the branch containing the source; m is the number of time steps between any pair of source values specified (linear interpolation between two specified points is assumed by ECAP); $s_0, s_1, s_2, \ldots$ are values of the source specified at time intervals of "$m \times$ (time step)." $X\text{bb}$ is to be punched in the first data field and the remaining statement is to be punched in the second data field.

As an example, the data card for a certain voltage source may take the following form.

$$\text{E1} \qquad (10), 0, 10, 20, 30, 20, 10, 0, -10, 0, 10, 20, 30$$

This statement indicates the following facts: The voltage source is in branch 1;

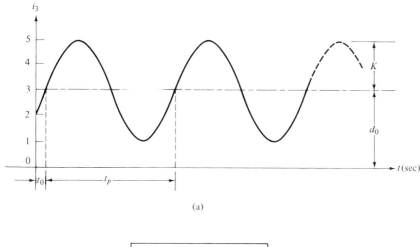

(a)

```
B 3        N ( , ) , . . .

I 3        SIN (.08) , 2 , 3 . .01
```

(b)

Fig. 10.2.21 Illustration for sinusoidal source data card: (a) current source in branch 3, (b) appropriate data cards.

the values of this source are given for the time intervals equal to ten times the TIME STEP (not specified); the initial value of the source is zero, and it changes its value at every 10 time steps with the final value of 30 at the 110th time step. This final value will be assumed constant by ECAP for t larger than 110 time steps.

2. *Periodic source.* The format for coding the data of a periodic source is

$$X\text{bb} \qquad P(m), s_0, s_1, s_2, \ldots, s_n$$

where P indicates that the source is periodic and Xbb, m, $s_0, s_1, \ldots, s_n$ are as defined for a nonperiodic source. However, since $s_0, s_1, \ldots, s_n$ describe one full period of the source, s_n must be equal in value to s_0.

3. *Sinusoidal source.* A special class of periodic sources is the sinusoidal source. The format is given as follows:

$$X\text{bb} \quad \text{SIN} \ (t_p), K, d_0, t_0$$

where SIN indicates that the source is sinusoidal; t_p is the period of the source (in seconds); K is the amplitude of the sinusoid; d_0 is the dc-offset of the source; t_0 is the time offset of the source; and Xbb has the same meaning as in the previous cases. Figure 10.2.21 illustrates how various terms can be used in the sinusoidal source card.

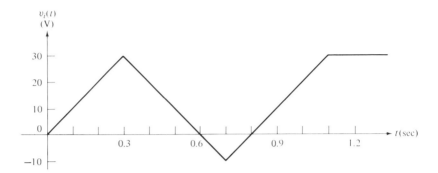

Fig. 10.2.22 Waveform of input voltage for network of Fig. 10.2.20.

```
       TRANSIENT ANALYSIS
C      DIODE WITH RC FILTER EXAMPLE 10-2-5
B1     N(0,1), R=0.1
E1     (1,0),0,10,20,30,20,10,0,-10,0,10,20,30
B2     N(1,2), R=(10,10E6), E=-.5
B3     N(2,3), R=100
B4     N(3,0), C=10E-6
B5     N(3,0), R=10E3
S1     B=2,(2),ON
       TIME STEP =.01
       OUTPUT INTERVAL =10
       FINAL TIME =3
       PRINT,VOLTAGES,CURRENTS
       EXECUTE
       END
```

Fig. 10.2.23 Input data for Example 10.2.5.

Example 10.2.5 As an example for introducing the coding techniques necessary to perform a transient analysis, consider again the network depicted in Fig. 10.2.20. The waveform for the input voltage $v_i(t)$ is assumed to be that shown in Fig. 10.2.22.

For a time step of 0.1 sec, an output interval of 10 time steps, and a final time of 3 sec, the ECAP statements for the transient analysis of the given network are shown in Fig. 10.2.23. It should be noted that in the E1 card, a value was used to describe $v_i(t)$ for every 10 time steps with the last one given at the 110th time step (i.e., $t = 1.1$ sec) having a value of 30. The last value of 30 is taken by ECAP to

```
       TRANSIENT ANALYSIS
C      DIODE WITH RC FILTER  EXAMPLE 10-2-5
 B1    N(0,1),  R=0.1
 E1    (10),0,10,20,30,20,10,0,-10,0,10,20,30
 B2    N(1,2),  R=(10,10E6),  E=-.5
 B3    N(2,3),  R=100
 B4    N(3,0),  C=10F-6
 B5    N(3,0),  R=10E3
 S1    B=2,(2),ON
       TIME STEP =.01
       OUTPUT INTERVAL = 10
       FINISH TIME = 3
       PRINT, VOLTAGES, CURRENTS
       EXECUTE


   T =   0.0

NODES                   NODE VOLTAGES

  1-  3      0.45409100E-03    -0.45413667E 00    -0.45409062E-04

BRANCHES              ELEMENT CURRENTS

  1-  4     -0.454091210-02   -0.454091210-02   -0.454091210-02   -0.454090750-02
  5-  5     -0.454090640-08


   T =   0.9999990E-01

NODES                   NODE VOLTAGES

  1-  3      0.99998074E 01     0.94806280E 01     0.92888288E 01

BRANCHES              ELEMENT CURRENTS

  1-  4      0.191799290-02    0.191799290-02    0.191799290-02    0.989110060-03
  5-  5      0.928882810-03


   T =   0.1999998E 00

NODES                   NODE VOLTAGES

  1-  3      0.19999695E 02     0.19470627E 02     0.19179916E 02

BRANCHES              ELEMENT CURRENTS

  1-  4      0.290710270-02    0.290710270-02    0.290710270-02    0.989110060-03
  5-  5      0.191799260-02


   T =   0.2999997E 00

NODES                   NODE VOLTAGES

  1-  3      0.29999603E 02     0.29460648E 02     0.29071014E 02

BRANCHES              ELEMENT   CENTS

           0.38°                        ° 02
```

Fig. 10.2.24 Output for Example 10.2.5

be the value of $v_i(t)$ for $t > 1.1$ sec. Figure 10.2.24 contains the output for this example.

The basic applications of the three analysis programs in ECAP have been briefly discussed by means of the preceding examples. The formats for input, solution control, output specifications, and commands discussed here are standard, not only for the various versions of ECAP provided by the IBM Corporation, but also for many modified versions of ECAP available in commercial time-share services.

Table 10.2.4 Output Available in ECAP (× indicates availability).

Ouput indicator	Description of output	Analysis		
		DC	AC	TR
NV or VOLTAGES	Node voltages	×	×	×
CA or CURRENTS	Element currents	×	×	×
CV	Element voltages	×	×	×
BA	Branch currents	×	×	×
BV	Branch voltages	×	×	×
BP	Element power loss	×	×	×
SENSITIVITY	Sensitivity and partial derivatives	×		
WORST CASE	Worst-case calculations	×		
STANDARD	Standard deviation calculations	×		
MISCELLANEOUS	Nodal admittance matrix	×	×	×
	Equivalent current vector	×	×	×
	Nodal impedance matrix	×		

Limitations and capabilities of ECAP depend on the particular version of the program used and thus the coverage of such topics is beyond the scope of this text. The interested reader is referred to literature elsewhere.* For the purpose of reference, outputs obtainable from each of the three analysis programs in ECAP are listed in Table 10.2.4. Some of the terms listed in this table will be discussed briefly as follows.

Sensitivity. When the sensitivity analysis is requested in the dc-analysis program, ECAP will calculate the partial derivative of every node voltage with respect to every circuit parameter and also calculate the change in every node voltage for a one-percent change in every circuit parameter in the form

$$\frac{\partial e_i}{\partial P_j} \frac{P_j}{100},$$

where the subscript i refers to the ith node, the subscript j refers to the jth branch, and P is a circuit parameter such as R, β, E, or I.

Standard deviation. The standard deviation analysis provides as output the printout of an approximation to the standard deviation of every node voltage using a three-percent parameter tolerance. A card with STANDARD DEVIATION (or simply ST) entered in the field 2 in the input deck will request that the standard deviation be calculated and printed for every node voltage.

Miscellaneous output. If the word MISCELLANEOUS (or simply MI) is entered in the PRINT statement card of the input deck such as PR, NV, CA, MI, then the printout will

---

* For example, see IBM 1620 ECAP user's manual or Jensen and Lieberman [JE 1].

include, in addition to all the node voltages and element currents, the nodal conductance matrix, the equivalent current vector, as well as the nodal impedance matrix (available only in dc-analysis). These are the quantities used by ECAP to solve the node equations of a given network.

The miscellaneous outputs available in the analysis programs are useful as an intermediate check for some problems. The various terms listed in Table 10.2.4, including the miscellaneous outputs, are discussed in detail in the appendix of the IBM 1620 user's manual.

Worst case. The worst-case analysis provides as output both the minimum and maximum values for every node voltage specified due to variation of one or more circuit parameters. The variation of a circuit parameter such as R, E, I, or BETA can be specified in one of the two ways:

a) The statement R = 80(0.1) indicates that the resistance in the branch has a nominal value of 80Ω with a 10% tolerance. That is, it has a minimum value of 72Ω and a maximum value of 88Ω.

b) A statement such as E = 80 (72, 88) specifies a voltage source of a nominal voltage of 80 V and the minimum and maximum values of 72 and 88 volts, respectively.

To request for the worst-case analysis on the node voltages at nodes n_1, n_2, and n_3, for example, the input deck must contain a solution control card in the form

                 WORST CASE, N1, N2, N3 (or WO, N1, N2, N3)

With this card, ECAP will provide the worst-case analysis for the node voltages identified as N1, N2, and N3. However, if node numbers are not written following the word WORST CASE (or simply WO), then the worst-case node voltages will be calculated and printed for all nodes.

10.3 THE CORNELL NETWORK ANALYSIS PROGRAM (CORNAP)

There exists today a number of well-known circuit analysis programs based on the state variable method. In this section we shall present a program, written by C. Pottle of Cornell University, which is originally referred to as the "Pottle program" and later called "the *COR*nell *N*etwork *A*nalysis *P*rogram" (CORNAP).* Given the topology and the element values of a linear time-invariant network, CORNAP will generate a set of state equations and determine the specified transfer functions as well as frequency and time responses.

In Chapter 3, Section 3.7, we discussed a method for the formulation of state equations. Basically, this method consists of the following steps.

---

* For information on the availability of CORNAP, the reader is advised to contact Professor C. Pottle, Department of Electrical Engineering, Cornell University, Ithaca, New York.

a) select a tree from the network, including as tree branches

  (i) all the known voltage sources,

  (ii) as many capacitors as possible,

  (iii) as few inductors as possible, and

  (iv) no current sources;

b) obtain the fundamental cutset (KCL) equations and the fundamental circuit (KVL) equations with respect to the chosen tree to form the desired state equations; and

c) express the state equations in "normal form."

This general procedure is used in CORNAP for the formulation of state equations.*

The types of circuit elements acceptable to CORNAP are: resistance, inductance, mutual inductance, capacitance, mutual capacitance, and the four types of dependent sources, namely, voltage-/current-controlled voltage/current sources. Assuming that the data for executing CORNAP are to be entered in card form, the input deck is divided into the following groups of cards: title, circuit data, solution control, and command cards, in addition to the necessary system control cards, which are different for different computer facilities. These groups of cards will next be described.

System Control Cards
System control cards appropriate for the particular computer facility must be included.

Title Card
The first card after the necessary system control cards will be read by CORNAP and the contents of this card will be printed on every section of the output. Thus the user may use this card for recording or accounting purposes.

Circuit Data Cards
Network topology and element values are described by the circuit data cards, using one card for each branch in any arbitrary order. The items and their corresponding column locations for each card in this group are presented in the order of their appearance in the card as follows.

Column 1. This column contains a letter, describing the type of element in a branch, as defined by

> V — voltage source
> I — current source
> R — resistor
> L — inductor
> C — capacitor
> M — mutual element (inductance or capacitance)
> K — mutual element (coefficient of coupling)

---

* See F. F. Kuo and J. F. Kaiser [KU 5] Chapter 3 (by C. Pottle).

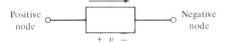

Positive node ⎯ Negative node

+ v −

Fig. 10.3.1 Node-designation convention of a branch for CORNAP.

Columns 3–6. These columns contain one to four alphameric characters indicating the "branch name"—any name assigned to the branch by the user. The characters in this data field must be left-justified. For example, if the branch name is R2 then R2 should be punched in columns 3 and 4.

Columns 8–9 *and* 11–12. All the nodes of the network must be numbered consecutively in ascending order with 00 assigned to the ground node. Columns 8–9 and 11–12 contain a pair of two-digit integers designating, respectively, the node number of the positive node and that of the negative node of the branch as illustrated in Fig. 10.3.1. If this card is for a mutual element, these columns are left blank.

Columns 14–31. These columns contain either the value of the element or the "strength" (the gain) of the dependent source. If the element is an independent source, these columns are left blank. The value may be entered in the form of a string of digits with a decimal point or in exponential form; in the case of the latter, the exponent part must be right-justified in the field. For example, if the value is 0.0052, it may be punched anywhere within the field as it is given or 5.2E-03 may be used with E-03 punched in Columns 28–31.

Columns 33–36. (a) If the branch is a controlled source, these columns contain the name of the controlling branch; (b) if the "branch" is a mutual element, these columns contain the name of one of the two elements involved (for example, either L_1 or L_2 of a transformer), while the name of the other element is previously entered in Columns 3–6. These names must be left-justified.

Column 38. A letter V (or I) in this column will indicate that the branch is a voltage- (or current-) controlled dependent source. (The name of the controlling branch is entered in Columns 33–36.)

Columns 40–41. In these columns one of the symbols +V, − V, +I, and − I will be used to designate that the branch is to be taken as an output port so that the voltage across the element or the current through the element is considered to be the output quantity. The + or − sign in front of the letter, V or I, indicates, respectively, whether the polarity of the output agrees with, or is opposite to, the convention given in Fig. 10.2.1.

Solution Control Cards

A solution control card is used to indicate the type of solution desired, the step size to be used, the time or frequency interval in which the solution is requested, etc. These cards are described in detail as follows.

The "Scale" Card. The first solution control card immediately following the circuit data cards indicates in Columns 14–31 and 44–61, respectively, the frequency (radians/second) and the impedance level (ohms) about which the circuit is designed to operate. The program will use these numbers as scale factors to scale the given network to operate about the "normalized" frequency of 1 rad/sec and the impedance level of 1Ω.

The "Response Option" Card(s). One or more of this type of solution control card may be included in the input deck for controlling the calculations and the print-outs of the frequency and/or time responses. The items and their corresponding column locations in each of these cards are given next.

Column 1. In this column, a letter F (or T) indicates that the frequency (or time) response is desired.

Columns 3–6. If these columns contain the branch name of an independent source, the response to this source alone as the input will be determined. If these columns are left blank, the response to each and every independent source as input will be calculated. (The output port at which the responses are to be determined will be specified later in Columns 33–36.)

Columns 8–9. These columns specify the frequency scale to be used for frequency-response calculations. "Blanks" in these columns indicate a linear scale, while a positive integer indicates the number of decades of the frequency variable to be included in the logarithmic scale.

Columns 10–12. These columns provide the space for entering a three-digit integer which indicates the number of solution points to be printed out in the following manner:

1. For the time response, this integer specifies the total number of solution points required.
2. For the frequency response, this integer specifies the number of solution points desired if a linear scale is used, or the number of points per decade if a logarithmic scale is used.

Columns 14–31. This field contains a number indicating the increment in frequency (linear scale) or in time between adjacent solution points to be printed out. This number may be entered either in the form of a string of digits with a decimal point or in exponential form in which the exponent part must be right-justified in the field.

Columns 33–36. If this field contains the branch name of a passive element, that branch is taken as an output port so that the response(s) at the port will be calculated. If this field is left blank, responses at each and every output port defined in the circuit data cards will be determined.

Column 38. A blank in this column indicates that the solution values which appear in the printout correspond to the unscaled network. Otherwise, any nonblank entry in this column indicates that the values are to remain scaled.

Column 41. This column indicates either the frequency scale to be used in the frequency response or the type of time response to be calculated as explained in the following:

1. For the frequency response, a blank in this column signifies that the frequency scale is to be in hertz; otherwise, a nonblank character indicates that the scale is in radians per second.
2. For the time response, a blank in this column will make the request for both the impulse and step responses; otherwise, a nonblank character indicates that the input(s) will be specified in the next card(s). The format used for entering the value of the source at regular intervals is the same as described for the time increment in Columns 14–31 previously.

Columns 44–61. This field contains a number indicating one of the following:

1. the frequency at which the desired frequency response is to begin, or
2. the integration step size to be used in the solution for the time response, or
3. the sampling interval of the sampled source.*

Remarks. (a) The print interval specified in Columns 14–31 must be a multiple (≥ 2) of the sampling interval. (b) If this field is left blank, the program will automatically choose an appropriate step size for the integration. (c) The format used in this field is identical to that used in Columns 14–31 of this card.

As indicated earlier, if a nonblank character appears in Column 41 of a time-response solution control card, the input(s) will be specified in the subsequent card(s), which will be described as follows. The samples of the input signal are entered 6 to a card, each occupying 12 columns (real notation or right-justified exponential notation).

Command Card

Following the group of solution control cards, a command card containing one word beginning in Column 1 indicates the following two options:

1. MORE indicates that the input deck for another problem begins in the next card so that the program will repeat the same process of computation to solve the next problem.
2. CANCEL indicates that no more problems are to be solved, and hence the computation terminates.

---

* That is, the time interval at which the values of the given (continuous) source are picked to obtain the so-called "sampled" source, the values of which are called *samples*.

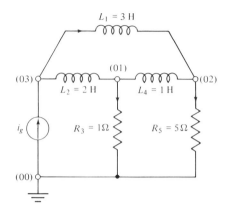

Fig. 10.3.2 Network for Example 10.3.1.

EXAMPLE 10-3-1 NETWORK WITH DEGENERATE CUTSET				
R	R3	01	00 1.0	
R	R5	02	00 5.0	+V
L	L1	03	02 3.0	
L	L2	03	01 2.0	
L	L4	01	02 1.0	
I	IG	00	03	
			1.0	1.0

Fig. 10.3.3 Input coding for Example 10.3.1.

It should be noted that all the cards for the input deck must be arranged in the same group order in which they are presented above.

We shall next illustrate the input formats and some of the capabilities of CORNAP in the following examples.

Example 10.3.1 We shall determine the state equations of the network shown in Fig. 10.3.2 which is identical to the network used in Example 3.7.3, Chapter 3.

With the nodes numbered and the current directions assigned, we obtain the input coding as shown in Fig. 10.3.3. The corresponding computer output is given in Fig. 10.3.4, from which we write the state equations as

$$\begin{bmatrix} \dot{i}_2 \\ \dot{i}_4 \end{bmatrix} = \begin{bmatrix} -\frac{6}{5} & \frac{6}{5} \\ 6 & -6 \end{bmatrix} \begin{bmatrix} i_2 \\ i_4 \end{bmatrix} + \begin{bmatrix} 1 \\ -5 \end{bmatrix} [i_g] + \begin{bmatrix} \frac{3}{5} \\ 0 \end{bmatrix} [\dot{i}_g]. \tag{10.3.1}$$

Comparing (10.3.1) with (3.7.28) in Chapter 3, we can see that if the element values given in Fig. 10.3.2 are substituted in (3.7.28), it will reduce to (10.3.1).

EXAMPLE 10-3-1 NETWORK WITH DEGENERATE CUTSET

THE GRAPH OF THIS NETWORK IS DESCRIBED BY THE FOLLOWING BRANCHES

BRANCH TYPE	NAME	NODE NOS.	ELEMENT VALUE	BRANCH NAME	CON- TROL	OUT PUT
R	R3	1 0	1.000000E 00			
R	R5	2 0	5.000000E 00			+V
L	L1	3 2	3.000000E 00			
L	L2	3 1	2.000000E 00			
L	L4	1 2	1.000000E 00			
I	IG	0 3	0.0			

THIS NETWORK HAS BEEN SCALED FOR COMPUTATION BY THE FOLLOWING FACTORS

FREQUENCY 1.0000000D 00 RADIANS/SEC.   IMPEDANCE 1.0000000D 00 OHMS

EXAMPLE 10-3-1 NETWORK WITH DEGENERATE CUTSET

TRANSFER FUNCTION CRITICAL FREQUENCIES (SCALED)

OUTPUT VARIABLE - V R5
SOURCE VARIABLE -   IG

GAIN CONSTANT IS   2.0000000D 00

POLE POSITIONS			ZERO POSITIONS		
REAL PART	IMAGINARY PART	ORDER	REAL PART	IMAGINARY PART	ORDER
-7.2000000D 00	0.0	1	-3.0000000D 00	0.0	1
0.0	0.0	1	0.0	0.0	1

EXAMPLE 10-3-1 NETWORK WITH DEGENERATE CUTSET

THE (SCALED) ENTRIES OF THE A MATRIX ARE

, STATE
VARIABLES   STATE
                                VARIABLES

	L2	L4
L2	-1.2000000D 00	1.2000000D 00
L4	6.0000000D 00	-6.0000000D 00

THE (SCALED) ENTRIES OF THE B MATRIX ARE

, STATE
VARIABLES   SOURCE
                                VARIABLES

	IG
L2	1.0000000D 00
L4	-5.0000000D 00

THE (SCALED) ENTRIES OF THE B1 MATRIX ARE

, STATE
VARIABLES   SOURCE
                                VARIABLES

	IG
L2	6.0000000D-01
L4	0.0

THE (SCALED) ENTRIES OF THE C MATRIX ARE

,OUTPUT
VARIABLES   STATE
                                VARIABLES

	L2	L4
V R5	-5.0000000D 00	5.0000000D 00

THE (SCALED) ENTRIES OF THE D MATRIX ARE

,OUTPUT
VARIABLES   SOURCE
                                VARIABLES

	IG
V R5	5.0000000D 00

THE (SCALED) ENTRIES OF THE D1 MATRIX ARE

,OUTPUT
VARIABLES   SOURCE
                                VARIABLES

	IG
V R5	0.0

Fig. 10.3.4 Output for Example 10.3.1.

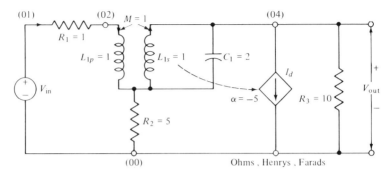

Fig. 10.3.5 Network for Example 10.3.2.

EXAMPLE	1.0	-3.-2				
R	R1	0.1	0.2	1..0		
R	R2	0.3	0.0	5.0		
R	R3	0.4	0.0	1.0..0		+V
L	L1P	0.2	0.3	1..0		
L	L1S	0.3	0.4	1..0		
M	L1P			1..0	L1S	
C	C1	0.3	0.4	2..0		
V	VIN	0.1	0.0			
I	ID	0.4	0.0	-5..0	L1S V	
			1..			1..
F			5.0	..0.2.5		
F			2	2.5		

Fig. 10.3.6 Input coding for Example 10.3.2.

Example 10.3.2   We shall determine the frequency response of the network shown in Fig. 10.3.5.

To cover the significant frequency range adequately, we shall request for the frequency response in both the linear and logarithmic scales as follows: (a) from 0 to 1.25 Hz in a linear scale, and (b) from 0 to 100 Hz in a logarithmic scale. The input coding and the computer output are given in Fig. 10.3.6 and Fig. 10.3.7, respectively.

Example 10.3.3   Consider again the network in Fig. 10.3.5. We shall obtain the step and impulse responses of the network. The input coding is given in Fig. 10.3.8 in which we request for the step and impulse responses in the time interval between 0 and 2.5 seconds. In addition, as a check for the step response, we enter a step function input for the time interval between 0 and 2.0 seconds. The computer output is shown in Fig. 10.3.9. A comparison between the two step responses shows that the results from the "internal" and the "external" step inputs are the same.

EXAMPLE   10-3-2   449

THE GRAPH OF THIS NETWORK IS DESCRIBED BY THE FOLLOWING BRANCHES

BRANCH TYPE	NAME	NODE NOS.	ELEMENT VALUE	BRANCH NAME	CON-TROL	OUT PUT
R	R1	1 2	1.000000E 00			
R	R2	3 0	5.000000E 00			
R	R3	4 0	1.000000E 01			+V
L	LIP	2 3	1.000000E 00			
L	LIS	3 4	1.000000E 00			
M	LIP	0 0	1.000000E 00	LIS		
C	C1	3 4	2.000000E 00			
I	ID	4 0	-5.000000E 00	LIS	V	
V	VIN	1 0	0.0			

THIS NETWORK HAS BEEN SCALED FOR COMPUTATION BY THE FOLLOWING FACTORS
FREQUENCY 1.0000000D 00 RADIANS/SEC.   IMPEDANCE 1.0000000D 00 OHMS

EXAMPLE   10-3-2

TRANSFER FUNCTION CRITICAL FREQUENCIES (SCALED)

OUTPUT VARIABLE - V R3
SOURCE VARIABLE -   VIN

GAIN CONSTANT IS   7.6923077D-01

POLE POSITIONS			ZERO POSITIONS		
REAL PART	IMAGINARY PART	ORDER	REAL PART	IMAGINARY PART	ORDER
-4.3544045D 00	0.0	1	-4.7957408D 00	0.0	1
-1.1482626D-01	0.0	1	-1.0425918D-01	0.0	1

THIS NETWORK HAS BEEN SCALED FOR COMPUTATION BY THE FOLLOWING FACTORS
FREQUENCY 1.0000000D 00 RADIANS/SEC.   IMPEDANCE 1.0000000D 00 OHMS

UNSCALED FREQ RESPONSE

FREQUENCY HZ	MAGNITUDE	MAGNITUDE DB	REAL PART	IMAGINARY PART	PHASE FRAC OF PI	DELAY SEC
0.0	7.6923E-01	-2.2789E 00	7.6923E-01	0.0	0.0	-5.4132E 00
2.5000E-02	8.2079E-01	-1.7154E 00	8.2007E-01	3.4421E-02	0.01335	7.5905E-01
5.0000E-02	8.3800E-01	-1.5351E 00	8.3777E-01	1.9591E-02	0.00744	6.0064E-01
7.5000E-02	8.4216E-01	-1.4921E 00	8.4210E-01	9.6170E-03	0.00363	3.8317E-01
1.0000E-01	8.4327E-01	-1.4806E 00	8.4327E-01	2.7741E-03	0.00105	2.7905E-01
1.2500E-01	8.4329E-01	-1.4804E 00	8.4329E-01	-2.4654E-03	-0.00093	2.2310E-01
1.5000E-01	8.4278E-01	-1.4858E 00	8.4275E-01	-6.7767E-03	-0.00256	1.8872E-01
1.7500E-01	8.4193E-01	-1.4945E 00	8.4186E-01	-1.0479E-02	-0.00396	1.6504E-01
2.0000E-01	8.4085E-01	-1.5057E 00	8.4073E-01	-1.3739E-02	-0.00520	1.4713E-01
2.2500E-01	8.3959E-01	-1.5187E 00	8.3942E-01	-1.6649E-02	-0.00631	1.3257E-01
2.5000E-01	8.3820E-01	-1.5331E 00	8.3797E-01	-1.9265E-02	-0.00732	1.2010E-01
2.7500E-01	8.3669E-01	-1.5487E 00	8.3641E-01	-2.1624E-02	-0.00823	1.0900E-01
3.0000E-01	8.3510E-01	-1.5652E 00	8.3477E-01	-2.3751E-02	-0.00905	9.8885E-02
3.2500E-01	8.3345E-01	-1.5824E 00	8.3305E-01	-2.5664E-02	-0.00980	8.9530E-02
3.5000E-01	8.3174E-01	-1.6002E 00	8.3129E-01	-2.7381E-02	-0.01048	8.0803E-02
3.7500E-01	8.3001E-01	-1.6184E 00	8.2950E-01	-2.8914E-02	-0.01109	7.2627E-02
4.0000E-01	8.2825E-01	-1.6368E 00	8.2770E-01	-3.0275E-02	-0.01164	6.4959E-02
4.2500E-01	8.2648E-01	-1.6553E 00	8.2588E-01	-3.1477E-02	-0.01213	5.7772E-02
4.5000E-01	8.2472E-01	-1.6738E 00	8.2408E-01	-3.2530E-02	-0.01256	5.1047E-02
4.7500E-01	8.2297E-01	-1.6923E 00	8.2229E-01	-3.3445E-02	-0.01294	4.4770E-02
5.0000E-01	8.2124E-01	-1.7106E 00	8.2053E-01	-3.4233E-02	-0.01327	3.8930E-02
5.2500E-01	8.1954E-01	-1.7286E 00	8.1880E-01	-3.4902E-02	-0.01356	3.3512E-02
5.5000E-01	8.1787E-01	-1.7463E 00	8.1710E-01	-3.5464E-02	-0.01381	2.8503E-02
5.7500E-01	8.1623E-01	-1.7637E 00	8.1544E-01	-3.5927E-02	-0.01401	2.3886E-02
6.0000E-01	8.1464E-01	-1.7807E 00	8.1383E-01	-3.6299E-02	-0.01419	1.9644E-02
6.2500E-01	8.1309E-01	-1.7972E 00	8.1227E-01	-3.6588E-02	-0.01433	1.5760E-02
6.5000E-01	8.1158E-01	-1.8133E 00	8.1075E-01	-3.6804E-02	-0.01444	1.2213E-02
6.7500E-01	8.1013E-01	-1.8290E 00	8.0928E-01	-3.6951E-02	-0.01452	8.9853E-03
7.0000E-01	8.0871E-01	-1.8441E 00	8.0787E-01	-3.7038E-02	-0.01458	6.0556E-03
7.2500E-01	8.0735E-01	-1.8588E 00	8.0650E-01	-3.7071E-02	-0.01462	3.4044E-03
7.5000E-01	8.0603E-01	-1.8729E 00	8.0518E-01	-3.7054E-02	-0.01464	1.0122E-03
7.7500E-01	8.0477E-01	-1.8866E 00	8.0392E-01	-3.6994E-02	-0.01464	-1.1400E-03
8.0000E-01	8.0355E-01	-1.8998E 00	8.0270E-01	-3.6896E-02	-0.01462	-3.0704E-03
8.2500E-01	8.0237E-01	-1.9125E 00	8.0153E-01	-3.6763E-02	-0.01459	-4.7967E-03
8.5000E-01	8.0124E-01	-1.9247E 00	8.0040E-01	-3.6599E-02	-0.01454	-6.3355E-03
8.7500E-01	8.0015E-01	-1.9365E 00	7.9933E-01	-3.6409E-02	-0.01449	-7.7024E-03
9.0000E-01	7.9911E-01	-1.9479E 00	7.9829E-01	-3.6196E-02	-0.01442	-8.9123E-03
9.2500E-01	7.9811E-01	-1.9587E 00	7.9730E-01	-3.5962E-02	-0.01435	-9.9790E-03
9.5000E-01	7.9715E-01	-1.9692E 00	7.9635E-01	-3.5710E-02	-0.01426	-1.0915E-02
9.7500E-01	7.9623E-01	-1.9793E 00	7.9544E-01	-3.5443E-02	-0.01417	-1.1733E-02
1.0000E 00	7.9534E-01	-1.9889E 00	7.9456E-01	-3.5164E-02	-0.01408	-1.2443E-02
1.0250E 00	7.9449E-01	-1.9982E 00	7.9373E-01	-3.4873E-02	-0.01398	-1.3056E-02
1.0500E 00	7.9368E-01	-2.0071E 00	7.9292E-01	-3.4573E-02	-0.01387	-1.3580E-02
1.0750E 00	7.9290E-01	-2.0157E 00	7.9215E-01	-3.4266E-02	-0.01376	-1.4026E-02
1.1000E 00	7.9215E-01	-2.0239E 00	7.9142E-01	-3.3952E-02	-0.01365	-1.4399E-02
1.1250E 00	7.9143E-01	-2.0318E 00	7.9071E-01	-3.3633E-02	-0.01353	-1.4708E-02
1.1500E 00	7.9074E-01	-2.0394E 00	7.9003E-01	-3.3311E-02	-0.01341	-1.4959E-02
1.1750E 00	7.9007E-01	-2.0467E 00	7.8938E-01	-3.2986E-02	-0.01329	-1.5157E-02
1.2000E 00	7.8944E-01	-2.0537E 00	7.8876E-01	-3.2659E-02	-0.01317	-1.5309E-02
1.2250E 00	7.8883E-01	-2.0604E 00	7.8816E-01	-3.2331E-02	-0.01305	-1.5419E-02
1.2500E 00	7.8824E-01	-2.0668E 00	7.8759E-01	-3.2002E-02	-0.01293	-1.5492E-02

Fig. 10.3.7 Output for Example 10.3.2.

```
EXAMPLE    10-3-2
                TRANSFER FUNCTION CRITICAL FREQUENCIES (SCALED)
OUTPUT VARIABLE - V R3
SOURCE VARIABLE -    VIN

GAIN CONSTANT IS      7.6923077D-01

           POLE POSITIONS                              ZERO POSITIONS
   REAL PART       IMAGINARY PART   ORDER       REAL PART       IMAGINARY PART   ORDER
-4.3544045D 00    0.0                 1      -4.7957408D 00   0.0                 1
-1.1482626D-01    0.0                 1      -1.0425918D-01   0.0                 1

THIS NETWORK HAS BEEN SCALED FOR COMPUTATION BY THE FOLLOWING FACTORS
FREQUENCY 1.0000000D 00 RADIANS/SEC.     IMPEDANCE 1.0000000D 00 OHMS

              UNSCALED FREQ RESPONSE

FREQUENCY    MAGNITUDE    MAGNITUDE      REAL          IMAGINARY    PHASE        DELAY
  HZ                        DB           PART           PART       FRAC OF PI     SEC
1.0000E 00   7.9534E-01  -1.9889E 00   7.9456E-01   -3.5164E-02   -0.01408   -1.2443E-02
1.0965E 00   7.9225E-01  -2.0228E 00   7.9152E-01   -3.3996E-02   -0.01366   -1.4350E-02
1.2023E 00   7.8938E-01  -2.0543E 00   7.8871E-01   -3.2629E-02   -0.01316   -1.5321E-02
1.3183E 00   7.8675E-01  -2.0832E 00   7.8614E-01   -3.1109E-02   -0.01259   -1.5530E-02
1.4454E 00   7.8438E-01  -2.1095E 00   7.8383E-01   -2.9481E-02   -0.01197   -1.5159E-02
1.5849E 00   7.8226E-01  -2.1330E 00   7.8177E-01   -2.7790E-02   -0.01131   -1.4373E-02
1.7378E 00   7.8038E-01  -2.1539E 00   7.7995E-01   -2.6071E-02   -0.01064   -1.3319E-02
1.9055E 00   7.7874E-01  -2.1722E 00   7.7835E-01   -2.4356E-02   -0.00996   -1.2115E-02
2.0893E 00   7.7730E-01  -2.1882E 00   7.7697E-01    2.2672E-02   -0.00929   -1.0854E-02
2.2909E 00   7.7606E-01  -2.2021E 00   7.7578E-01   -2.1038E-02   -0.00863   -9.6039E-03
2.5119E 00   7.7500E-01  -2.2140E 00   7.7475E-01   -1.9470E-02   -0.00800   -8.4100E-03
2.7542E 00   7.7409E-01  -2.2242E 00   7.7388E-01   -1.7977E-02   -0.00739   -7.3015E-03
3.0200E 00   7.7332E-01  -2.2329E 00   7.7314E-01   -1.6566E-02   -0.00682   -6.2939E-03
3.3113E 00   7.7266E-01  -2.2403E 00   7.7251E-01   -1.5240E-02   -0.00628   -5.3931E-03
3.6308E 00   7.7210E-01  -2.2465E 00   7.7198E-01   -1.4001E-02   -0.00577   -4.5983E-03
3.9811E 00   7.7163E-01  -2.2518E 00   7.7153E-01   -1.2847E-02   -0.00530   -3.9044E-03
4.3652E 00   7.7124E-01  -2.2562E 00   7.7115E-01   -1.1776E-02   -0.00486   -3.3039E-03
4.7863E 00   7.7091E-01  -2.2599E 00   7.7083E-01   -1.0786E-02   -0.00445   -2.7876E-03
5.2481E 00   7.7063E-01  -2.2631E 00   7.7057E-01   -9.8720E-03   -0.00408   -2.3464E-03
5.7544E 00   7.7040E-01  -2.2657E 00   7.7035E-01   -9.0301E-03   -0.00373   -1.9711E-03
6.3096E 00   7.7021E-01  -2.2679E 00   7.7016E-01   -8.2560E-03   -0.00341   -1.6531E-03
6.9183E 00   7.7001E-01  -2.2697E 00   7.7001E-01   -7.5451E-03   -0.00312   -1.3845E-03
7.5858E 00   7.6991E-01  -2.2712E 00   7.6988E-01   -6.8930E-03   -0.00285   -1.1582E-03
8.3177E 00   7.6979E-01  -2.2725E 00   7.6977E-01   -6.2955E-03   -0.00260   -9.6791E-04
9.1202E 00   7.6970E-01  -2.2736E 00   7.6968E-01   -5.7484E-03   -0.00238   -8.0826E-04
1.0000E 01   7.6962E-01  -2.2745E 00   7.6960E-01   -5.2478E-03   -0.00217   -6.7450E-04
1.0965E 01   7.6956E-01  -2.2752E 00   7.6954E-01   -4.7900E-03   -0.00198   -5.6256E-04
1.2023E 01   7.6950E-01  -2.2758E 00   7.6949E-01   -4.3716E-03   -0.00181   -4.6899E-04
1.3183E 01   7.6946E-01  -2.2763E 00   7.6945E-01   -3.9892E-03   -0.00165   -3.9083E-04
1.4454E 01   7.6942E-01  -2.2767E 00   7.6941E-01   -3.6399E-03   -0.00151   -3.2559E-04
1.5849E 01   7.6939E-01  -2.2771E 00   7.6938E-01   -3.3210E-03   -0.00137   -2.7117E-04
1.7378E 01   7.6936E-01  -2.2774E 00   7.6935E-01   -3.0298E-03   -0.00125   -2.2579E-04
1.9055E 01   7.6934E-01  -2.2777E 00   7.6933E-01   -2.7639E-03   -0.00114   -1.8798E-04
2.0893E 01   7.6932E-01  -2.2779E 00   7.6932E-01   -2.5213E-03   -0.00104   -1.5647E-04
2.2909E 01   7.6930E-01  -2.2780E 00   7.6930E-01   -2.2999E-03   -0.00095   -1.3023E-04
2.5119E 01   7.6929E-01  -2.2782E 00   7.6929E-01   -2.0979E-03   -0.00087   -1.0838E-04
2.7543E 01   7.6928E-01  -2.2783E 00   7.6928E-01   -1.9135E-03   -0.00079   -9.0183E-05
3.0200E 01   7.6927E-01  -2.2784E 00   7.6927E-01   -1.7453E-03   -0.00072   -7.5038E-05
3.3113E 01   7.6927E-01  -2.2785E 00   7.6926E-01   -1.5919E-03   -0.00066   -6.2433E-05
3.6308E 01   7.6926E-01  -2.2786E 00   7.6926E-01   -1.4520E-03   -0.00060   -5.1942E-05
3.9811E 01   7.6926E-01  -2.2786E 00   7.6925E-01   -1.3243E-03   -0.00055   -4.3213E-05
4.3652E 01   7.6925E-01  -2.2786E 00   7.6925E-01   -1.2078E-03   -0.00050   -3.5949E-05
4.7863E 01   7.6925E-01  -2.2787E 00   7.6925E-01   -1.1016E-03   -0.00046   -2.9905E-05
5.2481E 01   7.6924E-01  -2.2787E 00   7.6924E-01   -1.0047E-03   -0.00042   -2.4877E-05
5.7545E 01   7.6924E-01  -2.2787E 00   7.6924E-01   -9.1633E-04   -0.00038   -2.0694E-05
6.3096E 01   7.6924E-01  -2.2788E 00   7.6924E-01   -8.3572E-04   -0.00035   -1.7214E-05
6.9184E 01   7.6924E-01  -2.2788E 00   7.6924E-01   -7.6220E-04   -0.00032   -1.4319E-05
7.5859E 01   7.6924E-01  -2.2788E 00   7.6924E-01   -6.9515E-04   -0.00029   -1.1911E-05
8.3177E 01   7.6923E-01  -2.2788E 00   7.6923E-01   -6.3399E-04   -0.00026   -9.9072E-06
9.1202E 01   7.6923E-01  -2.2788E 00   7.6923E-01   -5.7822E-04   -0.00024   -8.2408E-06
1.0000E 02   7.6923E-01  -2.2788E 00   7.6923E-01   -5.2734E-04   -0.00022   -6.8546E-06
```

Fig. 10.3.7   Output for Example 10.3.2 (*concluded*).

10.4 THE LINEAR NETWORK ANALYSIS COMPUTER PROGRAM (CALAHAN)

The two programs, ECAP and CORNAP, presented in Sections 10.2 and 10.3 are respectively based on the node-analysis and the state-variable methods which are the two approaches most commonly used in computer-aided network analysis and design. In this section, we shall discuss a third program called the "Linear Network Analysis Computer Program," the original version* of which was written by D. A. Calahan and hence is often referred to as the CALAHAN analysis program.

---

* See D. A. Calahan [CA 1] for the original version of this program.

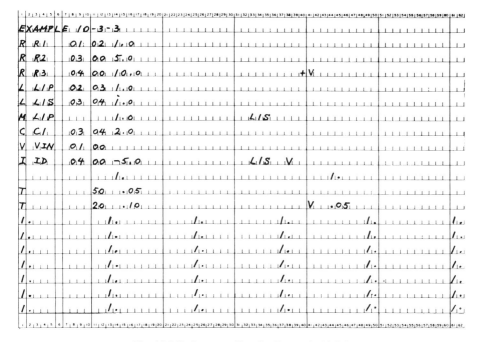

Fig. 10.3.8 Input coding for Example 10.3.3.

```
EXAMPLE    10-3-3

THE GRAPH OF THIS NETWORK IS DESCRIBED BY THE FOLLOWING BRANCHES

BRANCH      NODE    ELEMENT      BRANCH CON-  OUT
TYPE NAME   NOS.    VALUE        NAME  TROL   PUT

 R    R1     1  2   1.000000E 00
 R    R2     3  0   5.000000E 00
 R    R3     4  0   1.000000E 01              +V
 L    LIP    2  3   1.000000E 00
 L    LIS    3  4   1.000000E 00
 M    LIP    0  0   1.000000E 00  LIS
 C    C1     3  4   2.000000E 00
 I    ID     4  0  -5.000000E 00  LIS   V
 V    VIN    1  0   0.0

THIS NETWORK HAS BEEN SCALED FOR COMPUTATION BY THE FOLLOWING FACTORS
 FREQUENCY 1.0000000D 00 RADIANS/SEC.     IMPEDANCE 1.0000000D 00 OHMS


EXAMPLE    10-3-3

         TRANSFER FUNCTION CRITICAL FREQUENCIES (SCALED)

 OUTPUT VARIABLE - V R3
 SOURCE VARIABLE -   VIN

 GAIN CONSTANT IS    7.6923077D-01

          POLE POSITIONS                      ZERO POSITIONS
   REAL PART    IMAGINARY PART   ORDER    REAL PART    IMAGINARY PART   ORDER

-4.3544045D 00  0.0               1    -4.7957408D 00  0.0               1
-1.1482626D-01  0.0               1    -1.0425918D-01  0.0               1
THIS NETWORK HAS BEEN SCALED FOR COMPUTATION BY THE FOLLOWING FACTORS
FREQUENCY 1.0000000D 00 RADIANS/SEC.     IMPEDANCE 1.0000000D 00 OHMS
```

Fig. 10.3.9 Output for Example 10.3.3.

UNSCALED TIME RESPONSE

INTEGRATION STEP SIZE 1.6667E-02 SEC

TIME	STEP	IMPULSE
0.0	7.6923D-01	3.3136D-01
5.0000D-02	7.8408D-01	2.6483D-01
1.0000D-01	7.9593D-01	2.1132D-01
1.5000D-01	8.0538D-01	1.6829D-01
2.0000D-01	8.1290D-01	1.3369D-01
2.5000D-01	8.1886D-01	1.0587D-01
3.0000D-01	8.2358D-01	8.3497D-02
3.5000D-01	8.2729D-01	6.5514D-02
4.0000D-01	8.3019D-01	5.1059D-02
4.5000D-01	8.3244D-01	3.9441D-02
5.0000D-01	8.3417D-01	3.0106D-02
5.5000D-01	8.3548D-01	2.2606D-02
6.0000D-01	8.3645D-01	1.6583D-02
6.5000D-01	8.3716D-01	1.1748D-02
7.0000D-01	8.3764D-01	7.8673D-03
7.5000D-01	8.3796D-01	4.7552D-03
8.0000D-01	8.3813D-01	2.2610D-03
8.5000D-01	8.3819D-01	2.6371D-04
9.0000D-01	8.3816D-01	-1.3339D-03
9.5000D-01	8.3806D-01	-2.6101D-03
1.0000D 00	8.3791D-01	-3.6278D-03
1.0500D 00	8.3770D-01	-4.4376D-03
1.1000D 00	8.3747D-01	-5.0803D-03
1.1500D 00	8.3720D-01	-5.5886D-03
1.2000D 00	8.3691D-01	-5.9889D-03
1.2500D 00	8.3660D-01	-6.3023D-03
1.3000D 00	8.3628D-01	-6.5458D-03
1.3500D 00	8.3595D-01	-6.7333D-03
1.4000D 00	8.3561D-01	-6.8757D-03
1.4500D 00	8.3526D-01	-6.9818D-03
1.5000D 00	8.3491D-01	-7.0589D-03
1.5500D 00	8.3455D-01	-7.1127D-03
1.6000D 00	8.3420D-01	-7.1477D-03
1.6500D 00	8.3384D-01	-7.1677D-03
1.7000D 00	8.3348D-01	-7.1757D-03
1.7500D 00	8.3312D-01	-7.1740D-03
1.8000D 00	8.3276D-01	-7.1646D-03
1.8500D 00	8.3241D-01	-7.1491D-03
1.9000D 00	8.3205D-01	-7.1287D-03
1.9500D 00	8.3169D-01	-7.1044D-03
2.0000D 00	8.3134D-01	-7.0770D-03
2.0500D 00	8.3099D-01	-7.0471D-03
2.1000D 00	8.3063D-01	-7.0154D-03
2.1500D 00	8.3028D-01	-6.9821D-03
2.2000D 00	8.2994D-01	-6.9477D-03
2.2500D 00	8.2959D-01	-6.9124D-03
2.3000D 00	8.2925D-01	-6.8764D-03
2.3500D 00	8.2890D-01	-6.8399D-03
2.4000D 00	8.2856D-01	-6.8031D-03
2.4500D 00	8.2822D-01	-6.7660D-03
2.5000D 00	8.2788D-01	-6.7288D-03

UNSCALED TIME RESPONSE

INTEGRATION STEP SIZE 1.0000E-01 SEC
EXTERNAL INPUT SIGNAL SAMPLING INTERVAL 5.0000E-02 SEC

TIME	INPUT	OUTPUT
0.0	1.0000D 00	7.6923D-01
1.0000D-01	1.0000D 00	7.9594D-01
2.0000D-01	1.0000D 00	8.1291D-01
3.0000D-01	1.0000D 00	8.2359D-01
4.00C0D-01	1.0000D 00	8.3020D-01
5.0000D-01	1.0000D 00	8.3419D-01
6.0000D-01	1.0000D 00	8.3647D-01
7.0000D-01	1.0000D 00	8.3767D-01
8.0000D-01	1.0000D 00	8.3816D-01
9.0000D-01	1.0000D 00	8.3819D-01
1.0000D 00	1.0000D 00	8.3794D-01
1.1000D 00	1.0000D 00	8.3750D-01
1.2000D 00	1.0000D 00	8.3694D-01
1.3000D 00	1.0000D 00	8.3631D-01
1.4000D 00	1.0000D 00	8.3564D-01
1.5000D 00	1.0000D 00	8.3495D-01
1.6000D 00	1.0000D 00	8.3424D-01
1.7000D 00	1.0000D 00	8.3352D-01
1.8000D 00	1.0000D 00	8.3281D-01
1.9000D 00	1.0000D 00	8.3209D-01
2.0000D 00	1.0000D 00	8.3138D-01

Fig. 10.3.9 Output for Example 10.3.3 (*concluded*).

The CALAHAN program employs *topological formulas* (namely, formulas developed using topological properties) for determining network functions such as driving-point and transfer immittances and voltage (current) ratio transfer functions. As in almost every method, the topological approach in network analysis has advantages and disadvantages, which will be discussed in a later section. However, we shall present briefly, in the next few paragraphs, the basic theory behind the topological approach used in the CALAHAN program.

In Chapter 2, we defined a tree of a connected graph G of v vertices as a connected subgraph T of G, which contains all of the v vertices but none of the circuits of G (refer to Section 2.2). We also showed that any tree of G must contain the same number $(v - 1)$ of tree-branches (or simply branches). If, for a given tree T, any one of the $(v - 1)$ branches is removed from T but all of the v vertices are retained, a two-piece circuitless (proper) subgraph T_2 results. This subgraph T_2 consists of two separately connected subgraphs, which together contain all of the v vertices of G with the possibility that either (or both, in a trivial case) subgraph may be an isolated vertex. Subgraph T_2 is known as a 2-*tree* of G. Obviously, all of the 2-trees of G must contain exactly $(v - 2)$ branches, since each of them may be obtained from a certain tree of G less one (appropriate) branch. Similarly, a k-*tree* of G (k being a positive integer $\leq v$) may be defined as a circuitless subgraph of G consisting of k separately connected subgraphs, all of which together contain all the v vertices and $(v - k)$ branches. The k-*tree admittance product** of a k-tree is the product of the admittances of all of the $(v - k)$ branches of the k-tree. Evidently, the definitions of a k-tree and the corresponding k-tree admittance product include the definition of a tree and its *tree admittance product* as a special case where $k = 1$. The following example will illustrate the above newly defined topological terms.

Example 10.4.1 Consider the passive RLC network N consisting of five elements as shown in Fig. 10.4.1(a). The corresponding graph G is depicted in Fig. 10.4.1(b). A tree T containing edges e_1, e_2, and e_4 is shown in Fig. 10.4.1(c). Removing edge e_2 from T, we obtain the 2-tree $T_2^{(1)}$ consisting of edges e_1 and e_4. If, however, edge e_1 is removed from T instead of e_2, we obtain another 2-tree $T_2^{(2)}$ which contains edges e_2 and e_4 along with an isolated vertex, v_2. Both $T_2^{(1)}$ and $T_2^{(2)}$ are illustrated in (d) and (e), respectively, of Fig. 10.4.1. A 3-tree T_3 may be obtained from $T_2^{(1)}$ by removing an edge, say, e_4. In this case, T_3 consists of edge e_1 and two isolated vertices; namely, v_3 and v_4. When edge e_1 is removed from T_3, a 4-tree T_4 results which consists of four isolated vertices and no edges as shown in Fig. 10.4.1(g). Since G contains 4 vertices (with $v = 4$), T_4 is the only 4-tree possible and there exist no k-trees with $k \geq 5$ because of the condition that $k \leq v = 4$, as stated in the definition.

---

* In the case that the k-tree consists of isolated vertices only, the corresponding k-tree admittance product is, by definition, equal to unity.

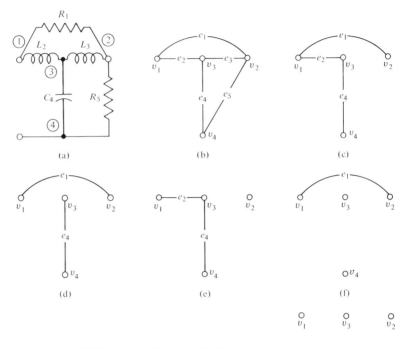

Fig. 10.4.1 (a) Network N, (b) graph G, (c) tree T, (d) 2-tree $T_2^{(1)}$, (e) 2-tree $T_2^{(2)}$, (f) 3-tree T_3, and (g) 4-tree T_4.

From (a) and (b) of Fig. 10.4.1, the admittances $y_i(i = 1, 2, \ldots, 5)$ of the elements of N are given by

$$y_1 = \frac{1}{R_1}, \quad y_2 = \frac{1}{sL_2}, \quad y_3 = \frac{1}{sL_3}, \quad y_4 = sC_4, \quad y_5 = \frac{1}{R_5}.$$

$$(10.4.1)$$

Hence, the tree admittance product $T(y)$ of T is

$$T(y) = y_1 y_2 y_4 = \left(\frac{1}{R_1}\right)\left(\frac{1}{sL_2}\right)(sC_4) = \frac{C_4}{R_1 L_2}.$$

$$(10.4.2)$$

Similarly, the 2-tree admittance products $T_2^{(1)}(y)$ for $T_2^{(1)}$ and $T_2^{(2)}(y)$ for $T_2^{(2)}$ are respectively

$$T_2^{(1)}(y) = y_1 y_4 = s\frac{C_4}{R_1}$$

$$(10.4.3)$$

and

$$T_2^{(2)}(y) = y_2 y_4 = \frac{C_4}{L_2}.$$

$$(10.4.4)$$

In like manner, the 3-tree admittance product $T_3(y)$ for T_3 is

$$T_3(y) = y_1 = \frac{1}{R_1}. \tag{10.4.5}$$

Finally, since 4-tree T_4 contains only four isolated vertices and no edges, the corresponding 4-tree admittance product is, by definition

$$T_4(y) \triangleq 1.$$

This completes the example. In the subsequent paragraphs, we shall give a brief discussion on some of the topological formulas that are based on the so-called k-tree approach.*

In Chapter 5, we defined the driving-point and transfer functions of a network (Table 5.3.1), and showed that these network functions may be expressed as rational functions of the complex frequency variables. We shall now show by means of an example that these functions may also be expressed in terms of the determinant $\Delta(\equiv \det Y_n)$ of the node-admittance matrix Y_n (or simply the node-determinant) and/or its cofactors.

Example 10.4.2   Consider the same network N as used in Example 10.4.1. If a current driver $I_1(s)$ is connected between nodes 1 and 4, and the node voltages $V_1(s)$, $V_2(s)$, $V_3(s)$ are defined as shown in Fig. 10.4.2 with node 4 as reference, we obtain the following matrix equation for the node analysis:

$$\begin{bmatrix} \left(\dfrac{1}{R_1} + \dfrac{1}{sL_2} \right) & \left(-\dfrac{1}{R_1} \right) & \left(-\dfrac{1}{sL_2} \right) \\[2ex] \left(-\dfrac{1}{R_1} \right) & \left(\dfrac{1}{R_1} + \dfrac{1}{sL_3} + \dfrac{1}{R_5} \right) & \left(-\dfrac{1}{sL_3} \right) \\[2ex] \left(-\dfrac{1}{sL_2} \right) & \left(-\dfrac{1}{sL_3} \right) & \left(\dfrac{1}{sL_2} + \dfrac{1}{sL_3} + sC_4 \right) \end{bmatrix} \begin{bmatrix} V_1(s) \\[2ex] V_2(s) \\[2ex] V_3(s) \end{bmatrix}$$

$$= \begin{bmatrix} I_1(s) \\[2ex] 0 \\[2ex] 0 \end{bmatrix} \tag{10.4.6}$$

or

$$Y_n V_n = I_n, \tag{10.4.7}$$

where the correspondences between the terms in (10.4.6) and (10.4.7) are evident upon direct comparison.

---

* The interested reader is referred to S. P. Chan [CH 1] for all the proofs and further discussion of this topic.

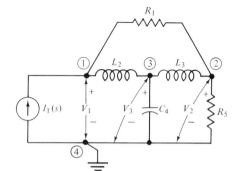

Fig. 10.4.2 Network for Example 10.4.2.

Solving for the ratio $V_1(s)/I_1(s)$ from (10.4.6), we obtain the expression for the driving-point impedance $Z_d(s)$ of N:

$$Z_d(s) = \frac{V_1(s)}{I_1(s)} = \frac{\Delta_{11}}{\Delta}, \tag{10.4.8}$$

where

$$\Delta = \det Y_n = \begin{vmatrix} \left(\dfrac{1}{R_1} + \dfrac{1}{sL_2}\right) & \left(-\dfrac{1}{R_1}\right) & \left(-\dfrac{1}{sL_2}\right) \\[2ex] \left(-\dfrac{1}{R_1}\right) & \left(\dfrac{1}{R_1} + \dfrac{1}{sL_3} + \dfrac{1}{R_5}\right) & \left(-\dfrac{1}{sL_3}\right) \\[2ex] \left(-\dfrac{1}{sL_2}\right) & \left(-\dfrac{1}{sL_3}\right) & \left(\dfrac{1}{sL_2} + \dfrac{1}{sL_3} + sC_4\right) \end{vmatrix} \tag{10.4.9}$$

is the node-determinant, and

$$\Delta_{11} = \begin{vmatrix} \left(\dfrac{1}{R_1} + \dfrac{1}{sL_3} + \dfrac{1}{R_5}\right) & \left(-\dfrac{1}{sL_3}\right) \\[2ex] \left(-\dfrac{1}{sL_3}\right) & \left(\dfrac{1}{sL_2} + \dfrac{1}{sL_3} + sC_4\right) \end{vmatrix} \tag{10.4.10}$$

is the 1, 1-cofactor of Δ. In a similar manner, it can be readily shown that the voltage-ratio transfer function $V_2(s)/V_1(s)$ is given by

$$\frac{V_2(s)}{V_1(s)} = \frac{\dfrac{V_2(s)}{I_1(s)}}{\dfrac{V_1(s)}{I_1(s)}} = \frac{\dfrac{\Delta_{12}}{\Delta}}{\dfrac{\Delta_{11}}{\Delta}} = \frac{\Delta_{12}}{\Delta_{11}}, \tag{10.4.11}$$

where

$$\Delta_{12} = (-1)^{1+2} \begin{vmatrix} \left(-\dfrac{1}{R_1}\right) & \left(-\dfrac{1}{sL_3}\right) \\[2ex] \left(-\dfrac{1}{sL_2}\right) & \left(\dfrac{1}{sL_2} + \dfrac{1}{sL_3} + sC_4\right) \end{vmatrix} \qquad (10.4.12)$$

is the 1, 2-cofactor of Δ.

Next, assuming the network under consideration is connected and relaxed (zero initial condition), we shall state three topological theorems which relate Δ and its cofactors to their corresponding topological quantities.

Theorem 10.4.1 Let N be a passive network without mutual inductances. The determinant Δ of the node admittance matrix Y_n is equal to the sum of all the tree-admittance products of N. That is,

$$\Delta = \det Y_n = \sum_i T^{(i)}(y), \qquad (10.4.13)$$

where $T^{(i)}(y)$ is the tree-admittance product of the ith tree $T^{(i)}$ in the summation.

Theorem 10.4.2 Let Δ be the node determinant of a passive network N with $n + 1$ nodes and without mutual inductances. Also let the reference node be denoted by $1'$. Then the j, j-cofactor Δ_{jj} of Δ is equal to the sum of all the 2-tree-admittance products $T_{2_{j,1'}}(y)$ of N, each of which contains node j in one connected part (connected subgraph) and node $1'$ in the other. That is,

$$\Delta_{jj} = \sum_k T^{(k)}_{2_{j,1'}}(y), \qquad (10.4.14)$$

where the summation is taken over all the 2-tree-admittance products of the form $T_{2_{j,1'}}(y)$.

Theorem 10.4.3 The i, j-cofactor, Δ_{ij}, of Δ of a passive network N with $n + 1$ nodes (with node $1'$ as the reference node) and without mutual inductances is given by

$$\Delta_{ij} = \sum_k T^{(k)}_{2_{ij,1'}}(y), \qquad (10.4.15)$$

where the summation is taken over all the 2-tree-admittance products of the form $T_{2_{ij,1'}}(y)$ with each of the products containing nodes i and j in one connected part and the reference node $1'$ in the other.

For convenience, we define the following shorthand notation:

(a) $\qquad V(Y) \triangleq \sum_i T^{(i)}(y) =$ sum of all tree-admittance products $\qquad (10.4.16)$

and

(b) $\qquad W_{j,r}(Y) \triangleq \sum_k T^{(k)}_{2_{j,r}}(y) =$ sum of all 2-tree-admittance products
$\qquad\qquad\qquad\qquad\qquad$ with node j and the reference node r
$\qquad\qquad\qquad\qquad\qquad$ contained in different parts. $\qquad (10.4.17)$

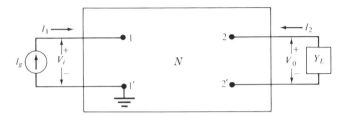

Fig. 10.4.3 A loaded passive two-port N.

Thus, for instance, we can express the driving-point impedance $Z_d(s)$ of a passive network without mutual inductances in topological terms in light of (10.4.8), (10.4.13), (10.4.14), (10.4.16), and (10.4.17):

$$Z_d(s) = \frac{W_{1,1'}(Y)}{V(Y)}.$$

(10.4.18)

In a two-port network N, there are four nodes to be specified, namely nodes 1 and 1' at the input port $(1, 1')$, and nodes 2 and 2' at the output port $(2, 2')$, as illustrated in Fig. 10.4.3.

With very little effort, it can be shown that, in general, the following relationship holds:

$$W_{ij,1'}(Y) = W_{ijk,1'}(Y) + W_{ij,k1'}(Y)$$

(10.4.19)

or simply,

$$W_{ij,1'} = W_{ijk,1'} + W_{ij,k1'},$$

(10.4.20)

where $i, j, k,$ and 1' are the four terminals of N with 1' denoting the datum node. The symbol $W_{ijk,1'}$ denotes the sum of all the 2-tree-admittance products, each containing nodes $i, j,$ and k in one connected part and node 1' (reference) in the other.

Before stating the topological formulas for a two-port network, we need the following theorem.

Theorem 10.4.4 With the same hypothesis and notation as stated earlier,

$$\Delta_{12} - \Delta_{12'} = W_{12,1'2'}(Y) - W_{12',1'2}(Y)$$

(10.4.21)

It is interesting to note that (10.4.21) is stated by Percival* in the following descriptive fashion:

$$\Delta_{12} - \Delta_{12'} = W_{12,1'2'} - W_{12',1'2} = \begin{pmatrix} 1 \circ\!\!-\!\!-\!\!-\!\!\circ 2 \\ 1' \circ\!\!-\!\!-\!\!-\!\!\circ 2' \end{pmatrix} - \begin{pmatrix} 1 \circ\!\!\!\!\!\times\!\!\!\!\!\circ 2 \\ 1' \circ\!\!\!\!\!\times\!\!\!\!\!\circ 2' \end{pmatrix}$$

(10.4.22)

which illustrates the two types of 2-trees involved in the formula.

In the next theorem, we shall state the four transfer functions listed in Table 5.3.1 in terms of topological quantities.

---

* For his original work, see R. H. Percival [PE 1].

Theorem 10.4.5 With the same notation as before,

(a)
$$Z_{21}(s) = \frac{W_{12,1'2'} - W_{12',1'2}}{V}$$

(b)
$$Y_{21}(s) = Y_L\left(\frac{W_{12,1'2'} - W_{12',1'2}}{W_{1,1'}}\right)$$

(10.4.23)

(c)
$$G_{21}(s) = \frac{W_{12,1'2'} - W_{12',1'2}}{W_{1,1'}}$$

(d)
$$\alpha_{21}(s) = Y_L\left(\frac{W_{12,1'2'} - W_{12',1'2}}{V}\right)$$

Other topological formulas may also be obtained following a similar approach. However, they will not be discussed here due to limited space; the interested reader is referred to literature elsewhere.* The following example will serve to illustrate the ideas involved in the theorems stated above.

Example 10.4.3 The network N used in Example 10.4.1 may be considered as a bridged-T network with a resistive load R_5. To compute the voltage-ratio $G_{21}(s)$ topologically, according to (10.4.23(c)), we need to evaluate the terms $W_{12,1'2'}$, $W_{12',1'2}$, and $W_{1,1'}$ from N. Thus, in light of (10.4.22), we obtain by inspecting the network N in Fig. 10.4.1 and noting that node 4 is now identified as both the reference node $1'$ and node $2'$ for a two-port:

$$W_{12,1'2'} = y_1 y_2 + y_1 y_3 + y_1 y_4 + y_2 y_3$$
$$= \frac{1}{sR_1 L_2} + \frac{1}{sR_1 L_3} + s\frac{C_4}{R_1} + \frac{1}{s^2 L_2 L_3},$$
$$W_{12',1'2} = 0,$$

and

$$W_{1,1'} = y_1 y_2 + y_1 y_3 + y_1 y_4 + y_2 y_3 + y_2 y_5 + y_3 y_4 + y_3 y_5 + y_4 y_5$$
$$= \frac{1}{sR_1 L_2} + \frac{1}{sR_1 L_3} + s\frac{C_4}{R_1} + \frac{1}{s^2 L_2 L_3} + \frac{1}{sL_2 R_5} + \frac{C_4}{L_3} \mid \frac{1}{sL_3 R_5} + s\frac{C_4}{R_5}$$

so that

$$G_{21} = \frac{W_{12,1'2'} - W_{12',1'2}}{W_{1,1'}}$$

$$= \frac{\dfrac{1}{sR_1 L_2} + \dfrac{1}{sR_1 L_3} + s\dfrac{C_4}{R_1} + \dfrac{1}{s^2 L_2 L_3}}{\dfrac{1}{sR_1 L_2} + \dfrac{1}{sR_1 L_3} + s\dfrac{C_4}{R_1} + \dfrac{1}{s^2 L_2 L_3} + \dfrac{C_4}{L_3} + \dfrac{1}{sL_3 R_5} + s\dfrac{C_4}{R_5} + \dfrac{1}{sL_2 R_5}}$$

---

* See, for example, S. P. Chan [CH 1], or Seshu and Reed [SE 1].

which, upon simplification, reduces to

$$G_{21} = \frac{(L_2 L_3 C_4 R_5)s^2 + (L_2 + L_3)R_5 s + R_1 R_5}{(R_1 L_2 L_3 C_4)s^3 + (L_2 L_3 C_4 R_5 + R_1 L_2 C_4 R_5)s^2 + (L_3 R_5 + L_2 R_5 + R_1 L_3 + R_1 L_2)s + R_1 R_5}.$$

(10.4.24)

Equation (10.4.24) is seen to be correct by comparison with the ratio of (10.4.12) to (10.4.10).

Other transfer functions in Table 5.3.1 may be similarly obtained using (10.4.23) with $Y_L = 1/R_5$. The remaining task is left as an exercise.

It should be noted that (10.4.24) is referred to as the expression of G_{21} in *symbolic form* for the obvious reason. Symbolic expressions are useful in sensitivity analysis for computing various partial derivatives. They also provide physical insight to the circuit designer since the elements are symbolically related to the network functions so as to reveal the effects of certain elements (which are of particular interest to the designer) on the behavior of the network under study in terms of a specified (transfer) function. It is for these reasons that a program such as the CALAHAN program, developed by using the (k-tree) topological approach (which is especially suited for generating symbolic expressions), should be studied in connection with the computer-aided analysis and design.

The CALAHAN program accepts two different forms of input for describing the problem. The *circuit input*, similar to the input of CORNAP or ECAP, describes the topology and element values of the network to be analyzed. The *network function input* supplies the coefficients of the numerator and denominator polynomials of a network function. The outputs available from the program are the following:

1. coefficients of the network function (for circuit input);

2. poles and zeros of the network function;

3. frequency response giving both the table of the magnitude, phase, and delay versus frequency and the corresponding printer plots;

4. time response to an arbitrary input including both the table of computed values of the output at a specified time increment and the corresponding printer plot;

5. repeated outputs of items (1) through (4) with the value of a chosen element as a variable parameter;

6. a network function in symbolic form (for circuit input).

In the case of the circuit input, the user may request for one of the six types of network functions listed in Table 10.4.1. In this table, an integer value is assigned to each network function under the solution indicator KEY 1 indicating the particular network function desired. The symbols used in this table are illustrated in Fig. 10.4.4. Another solution indicator is the KEY 2, whose value indicates a particular set of outputs desired as given in Table 10.4.2.

Although the input format for the CALAHAN program is somewhat similar to that for CORNAP, the solution controls in CALAHAN are not completely separated from the rest of the input information. The coding methods for the input deck are

Table 10.4.1 Definitions of `KEY 1`.

Value of KEY 1	Network function	Defining ratio
1	Voltage-ratio transfer function	$\left.\dfrac{V_2}{V_1}\right|_{I_2=0}$
2	Open-circuit driving-point impedance	$\left.\dfrac{V_1}{I_1}\right|_{I_2=0}$
3	Open-circuit transfer impedance	$\left.\dfrac{V_2}{I_1}\right|_{I_2=0}$
4	Short-circuit driving-point admittance	$\left.\dfrac{I_1}{V_1}\right|_{V_2=0}$
5	Short-circuit transfer admittance	$\left.\dfrac{I_2}{V_1}\right|_{V_2=0}$
6	Current-ratio transfer function	$\left.\dfrac{I_2}{I_1}\right|_{V_2=0}$

Table 10.4.2 Definitions of `KEY 2`.

Value of KEY 2	Desired outputs
1	(Network function) coefficients, poles, and zeros
2	Coefficients, poles, zeros, and frequency response
3	Coefficients, poles, zeros, and time response
4	Coefficients, poles, zeros, time and frequency responses
5	Symbolic form of network functions

Fig. 10.4.4 A 2-port pertaining to the definitions of `KEY 1`.

described for two separate cases depending on whether the input is in the form of a network function or in the form of network topology and element values (called the circuit form).

Case A: Network Function Input

Card I. This is a title card for problem identification and accounting purposes. The contents in this card will be printed on the output.

Card II. A negative integer of `KEY 2` is entered in Columns 1–2 in this card. The negative sign shows that the input is a network function and the absolute value of this integer is used as the value of Key 2 as defined in Table 10.4.2.

Card III. Two two-digit integers, $n + 1$ and $m + 1$, are entered in Columns 1–2 and 4–5, respectively, where n is the degree of the numerator and m is the degree of the denominator of the network function.

Cards IV. Two cards in this group are used to enter first the numerator coefficients and then the denominator coefficients of the network function, beginning with the constant term of each polynomial in the ascending order. The format used in these cards is 8F10.0. The program assumes that there are no missing terms in the numerator and denominator polynomials. Thus, a zero must be entered to account for each missing term in the polynomials.

Cards V. If the frequency response is requested in KEY 2, two cards in this group are needed for solution control. (If time response is requested instead, these cards can be omitted.)

    Card V-1. In Column 1 an integer will indicate the type of frequency scale to be used as follows: 1 means linear scale, and 2 means logarithmic scale. In Columns 3–5, a three-digit integer is entered requesting either the total number of solution points if the linear scale is used or the number of solution points per decade if the logarithmic scale is chosen.

    Card V-2. The second card in this group contains two numbers indicating the lower and upper limits of the frequencies (in Hz) to be used for the frequency response. These numbers are entered beginning in Column 1 using the format of 2F10.0.

Cards VI. If time response is requested in KEY 2, two or more cards in this group are included in the input deck. (If frequency response is requested instead, cards VI can be omitted.)

    Card VI-1. A positive integer is entered in Columns 1–2 indicating the number of cards that follow to describe the input pulses.

    Cards VI-2, 3, ... Each card in this subgroup contains five numbers to describe one input pulse using the format of 5F10.0. These numbers describe the initial value, initial time, final value, final time of the pulse, and the time increment (to be used for each step of calculation within the interval from the initial time to the final time of the pulse), in that order.

    This completes the input deck for the program. However, of course, appropriate system control cards must be included since they are always required in any computer installation.

    Next, if the input is in circuit form we have the following case.

Case B: Circuit Input

For circuit input the nodes of the network must first be numbered sequentially beginning with 1, and a current direction is assigned to each branch in the same way as in CORNAP.

Card I. It is identical to that in Case A.

Card II. The integer 1 entered in Column 2 indicates that the input is in the circuit form.

Card III. This card contains eleven two-digit integers representing the following data: the number of passive elements (entered in Columns 1–2); the number of controlled sources (Columns 4–5); the number of nodes (Columns 7–8); the positive input node number (Columns 10–11); the negative input node number (Columns 13–14); the positive output node number (Columns 16–17); the negative output node number (Columns 19–20); KEY 1 (Columns 22–23); KEY 2 (Columns 25–26); the "element number"* of the element to be varied (Columns 28–29); and the number of values of the varied element (Columns 31–32).

Cards IV. This group of cards describes the topology and the values of the passive elements in the network using one card per element with data entered in the following manner: the positive node number of the element (Columns 1–2); the negative node number of the element (Columns 4–5); the letter R, L, or C indicating the element kind (Column 7); the value of the element or the initial value of the element if it is to be varied [using the F10.0 format (Columns 9–18)].

Cards V. This group of cards describes the topology and values of the active elements (controlled sources) and the mutually-coupled element contained in the circuit using one card for each element. Only voltage-controlled current sources are allowed in the program. The formats used are as follows: Columns 1–2 and 4–5 contain the node numbers of the positive and negative nodes, respectively, of the controlled source, using the I2 format. Columns 7–8 and 10–11 contain the positive and negative node numbers of the controlling branch using the I2 format.

Column 13 contains a letter, G or M, indicating that the element is represented by the transconductance G or mutual inductance M.

Columns 15–24 contain the value of the transconductance or mutual inductance in the F10.0 format.

Cards VI. This group of cards is necessary only if there is an element to be varied as specified in Card III. The values of the variable element are described in one or more cards, in eight fields each, using the format of 8F10.0.

Cards VII. This group of cards is identical to group V in Case A discussed previously.

Cards VIII. This group of cards is identical to group VI in Case A discussed previously.

This completes the description of the input deck for Case B. More than one problem may be solved in each computer run by including two or more sets of problem description data cards in an input deck.

Next, we shall illustrate the use of the program for network analysis by means of the following examples.

---

* The "element number" *k* used in connection with the branch cards in card group IV (to be described next) denotes that the element to be varied appears in the *k*th card in that group.

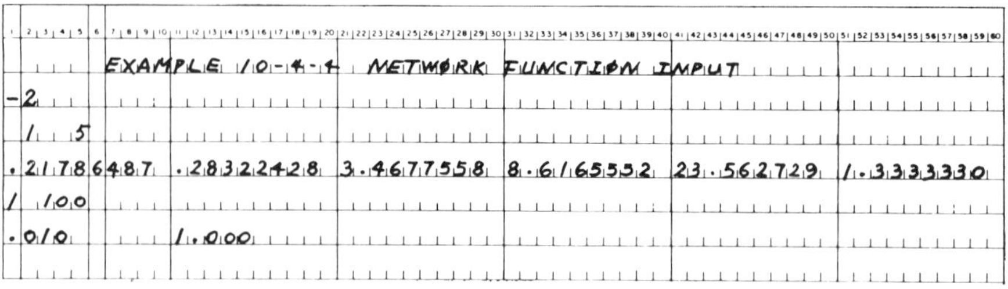

EXAMPLE 10-4-4   NETWORK FUNCTION INPUT
-2
1.5
.21786487 .28322428 3.4677558 8.6165552 23.562729 1.3333330
1.0,0
.0/0   1.000

Fig. 10.4.5 Input coding for Example 10.4.4.

EXAMPLE 10-4-4   NETWORK FUNCTION INPUT

NUMERATOR COEFFICIENTS, BEGINNING WITH THE CONSTANT TERM

2.17864811E-01

ZEROS

    NONE

DENOMINATOR COEFFICIENTS, BEGINNING WITH THE CONSTANT TERM

2.83224225E-01   3.46775436E 00   8.61655426E 00   2.35627289E 01   1.33333206E 00

POLES

REAL PART	IMAG PART	REAL PART	IMAG PART
-1.3254201E-01	3.2492495E-01	-1.3254201E-01	-3.2492495E-01
-9.9666834E-02	-3.1974423E-14	-1.7307312E 01	-7.7399709E-10

FREQ (HZ)	GAIN (DB)	PHASE(DEG)	DELAY (SEC)	FREQ (HZ)	GAIN (DB)	PHASE(DEG)	DELAY (SEC)
0.0100000	-3.5331717	-40.391	9.5635500	0.0199000	-5.5999670	-68.989	6.8625183
0.0298000	-7.1270447	-91.994	6.3753767	0.0397000	-8.1593189	-116.390	7.5175543
0.0496000	-9.5488405	-145.814	8.7720232	0.0595000	-12.2958136	-175.830	7.6256247
0.0694000	-16.0306396	-198.666	5.2084074	0.0792999	-19.8207550	-213.786	3.4182606
0.0892000	-23.2849731	-223.890	2.3452129	0.0990999	-26.3785460	-231.007	1.7009878
0.1090000	-29.1452179	-236.293	1.2944326	0.1188999	-31.6388092	-240.393	1.0237427
0.1288000	-33.9058380	-243.686	0.8348297	0.1386999	-35.9834747	-246.405	0.6977016
0.1486000	-37.9010925	-248.700	0.5948756	0.1585000	-39.6818848	-250.673	0.5156624
0.1684000	-41.3446672	-252.396	0.4532447	0.1782998	-42.9039154	-253.919	0.4031135
0.1881999	-44.3725586	-255.280	0.3621884	0.1980999	-45.7606506	-256.509	0.3283058
0.2079998	-47.0767822	-257.626	0.2999088	0.2178999	-48.3282013	-258.651	0.2758524
0.2278000	-49.5211792	-259.597	0.2552778	0.2376999	-50.6609802	-260.474	0.2375329
0.2475999	-51.7523041	-261.293	0.2221109	0.2574999	-52.7992096	-262.060	0.2086158
0.2673998	-53.8052063	-262.782	0.1967343	0.2772999	-54.7794985	-263.464	0.1862111
0.2871998	-55.7068176	-264.110	0.1768492	0.2970999	-56.6077271	-264.726	0.1684750
0.3069999	-57.4783783	-265.312	0.1600532	0.3168998	-58.3208313	-265.874	0.1541694
0.3267999	-59.1368713	-266.412	0.1480283	0.3367000	-59.9281769	-266.929	0.1424492
0.3465998	-60.6961670	-267.428	0.1373641	0.3564999	-61.4422760	-267.909	0.1327152
0.3663999	-62.1677399	-268.374	0.1286519	0.3762999	-62.8736572	-268.825	0.1245325
0.3861999	-63.5611115	-269.262	0.1209191	0.3960998	-64.2310638	-269.687	0.1175802
0.4059999	-64.8843994	89.899	0.1144875	0.4158999	-65.5219574	89.496	0.1116167
0.4257998	-66.1444550	89.103	0.1089463	0.4356999	-66.7526703	88.719	0.1064570
0.4455999	-67.3472290	88.344	0.1041324	0.4554998	-67.9287415	87.977	0.1019574
0.4653999	-68.4978180	87.617	0.0999189	0.4753000	-69.0549622	87.265	0.0980052
0.4851999	-69.6006927	86.918	0.0962055	0.4950999	-70.1354828	86.579	0.0945103
0.5049999	-70.6597290	86.245	0.0929114	0.5148998	-71.1739655	85.916	0.0914009
0.5247999	-71.6784668	85.593	0.0899721	0.5346998	-72.1736908	85.275	0.0886185
0.5445999	-72.6598969	84.961	0.0873346	0.5544999	-73.1374817	84.652	0.0861152
0.5643998	-73.6067963	84.347	0.0849558	0.5742999	-74.0680237	84.046	0.0838516
0.5842000	-74.5215607	83.750	0.0827993	0.5940998	-74.9676056	83.456	0.0817948
0.6039999	-75.4064331	83.167	0.0808353	0.6138999	-75.8382416	82.880	0.0799174
0.6237999	-76.2632904	82.597	0.0790387	0.6336999	-76.6818390	82.317	0.0781965
0.6436000	-77.0940247	82.039	0.0773883	0.6534998	-77.5001068	81.765	0.0766122
0.6633999	-77.9002380	81.493	0.0758664	0.6732998	-78.2945404	81.224	0.0751486
0.6831999	-78.6833344	80.958	0.0744573	0.6930999	-79.0666962	80.693	0.0737911
0.7029998	-79.4447479	80.432	0.0731480	0.7128999	-79.8176422	80.172	0.0725272
0.7228000	-80.1856232	79.915	0.0719271	0.7326999	-80.5487823	79.659	0.0713466
0.7425999	-80.9072266	79.406	0.0707848	0.7524999	-81.2610321	79.155	0.0702404
0.7623999	-81.6104584	78.905	0.0697128	0.7722999	-81.9552206	78.658	0.0692007
0.7821998	-82.2963257	78.412	0.0687034	0.7920999	-82.6330414	78.168	0.0682203
0.8019999	-82.9657745	77.926	0.0677506	0.8118999	-83.2945862	77.685	0.0672932
0.8217999	-83.6195679	77.446	0.0668479	0.8317000	-83.9408417	77.209	0.0664140
0.8415998	-84.2584686	76.973	0.0659911	0.8514999	-84.5725708	76.738	0.0655783
0.8613999	-84.8832397	76.505	0.0651754	0.8712999	-85.1904602	76.274	0.0647818
0.8811999	-85.4944305	76.043	0.0643971	0.8911000	-85.7951508	75.815	0.0640208
0.9009999	-86.0927582	75.587	0.0636524	0.9108999	-86.3872528	75.361	0.0632920
0.9207999	-86.6787720	75.136	0.0629387	0.9306999	-86.9673004	74.912	0.0625926
0.9405999	-87.2529907	74.690	0.0622529	0.9504999	-87.5358124	74.469	0.0619198
0.9603999	-87.8158722	74.248	0.0615928	0.9703000	-88.0932617	74.030	0.0612715
0.9801999	-88.3680115	73.812	0.0609559	0.9900999	-88.6401672	73.595	0.0606457

Fig. 10.4.6 Output for Example 10.4.4.

Example 10.4.4 We shall use CALAHAN to plot the magnitude, phase, and delay as functions of frequency for the function

$$H(s) = \frac{0.21786487}{0.28322428 + 3.4677553s + 8.6165552s^2 + 23.562729s^3 + 1.3333330s^4}.$$

The input coding is illustrated in Fig. 10.4.5 in which we should note that the coefficients of the numerator and denominator polynomials are entered in the ascending order beginning with the numerator. The outputs are shown in Fig. 10.4.6 in which the plots are in a linear scale as requested.

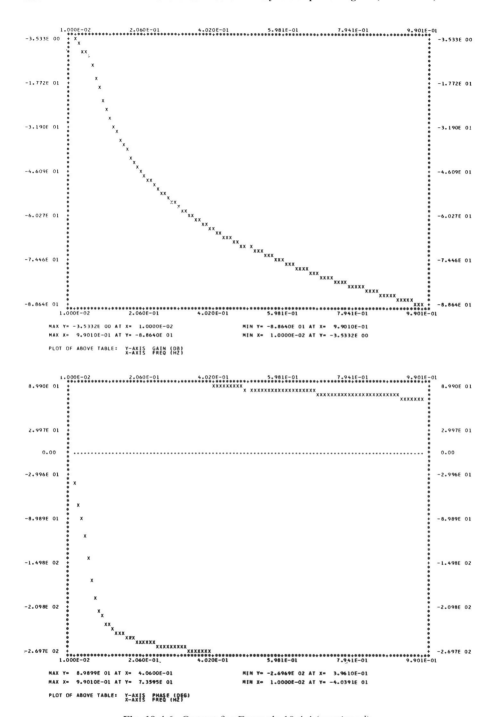

Fig. 10.4.6 Output for Example 10.4.4 (*continued*).

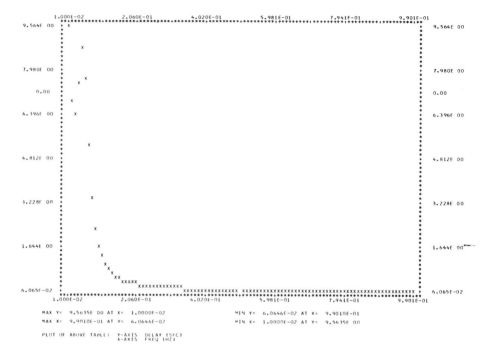

Fig. 10.4.6 Output for Example 10.4.4 (*concluded*).

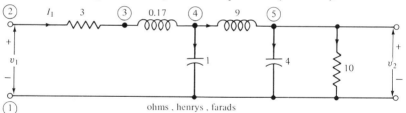

ohms , henrys , farads

Fig. 10.4.7 Network for Example 10.4.5.

Fig. 10.4.8 Input coding for Example 10.4.5.

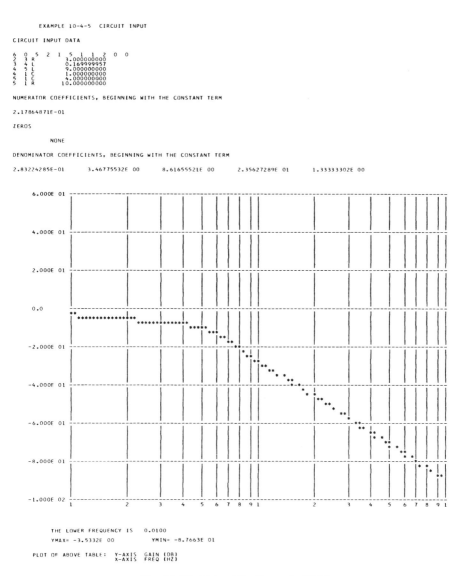

```
                    EXAMPLE 10-4-5   CIRCUIT INPUT

   CIRCUIT INPUT DATA

   6   0   5   2   1   5   1   1   2   0   0
   2   3   R              3.000000000
   3   4   L              0.169999957
   4   5   L              9.000000000
   4   1   C              1.000000000
   5   1   C              4.000000000
   5   1   R             10.000000000

   NUMERATOR COEFFICIENTS, BEGINNING WITH THE CONSTANT TERM

   2.17864871E-01

   ZEROS

            NONE

   DENOMINATOR COEFFICIENTS, BEGINNING WITH THE CONSTANT TERM

   2.83224285E-01      3.46775532E 00      8.61655521E 00      2.35627289E 01      1.33333302E 00
```

```
   THE LOWER FREQUENCY IS    0.0100
   YMAX= -3.5332E 00          YMIN= -8.7663E 01

   PLOT OF ABOVE TABLE:   Y-AXIS   GAIN (DB)
                          X-AXIS   FREQ (HZ)
```

Fig. 10.4.9 Output for Example 10.4.5.

Example 10.4.5 To illustrate the logarithmic plot, we shall obtain the frequency response of the network given in Fig. 10.4.7. The input coding and the corresponding computer output are shown in Fig. 10.4.8 and Fig. 10.4.9, respectively. A comparison between the printed output on the "numerator coefficients" and the "denominator coefficients" in Examples 10.4.4 and 10.4.5 reveals the fact that the transfer function $H(s)$ used in Example 10.4.4 and the function obtained in Example 10.4.5 are essentially the same. The output plots are not identical due to different

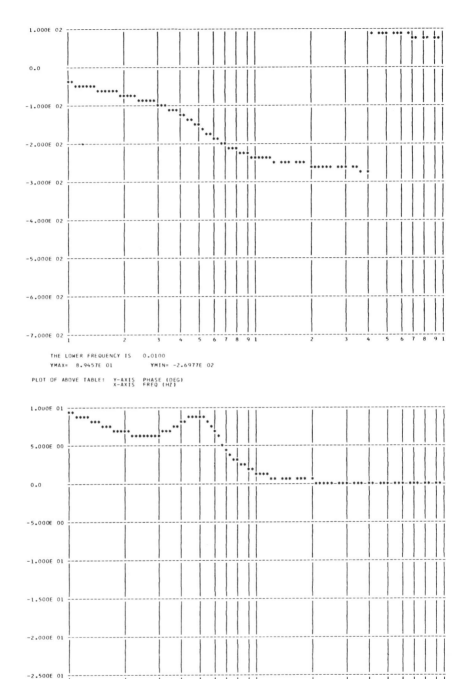

Fig. 10.4.9 Output for Example 10.4.5 (*concluded*).

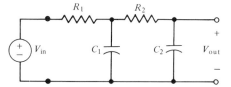

Fig. 10.5.1 An *RC* filter network.

frequency ranges used in the two examples. By comparing the "delay" plots in the two outputs, we can realize the importance of an appropriate choice of the frequency range for obtaining the complete information on a particular characteristic.

From the results of the last two examples, we can appreciate the convenience of having the plotting capability in the computed data.

10.5 THE NETWORK ANALYSIS FOR SYSTEM APPLICATION PROGRAM (NASAP)

The idea of using flow graphs to represent linear systems and networks was introduced in Chapter 7, Section 7.7. We illustrated that, given a set of independent linear algebraic equations (obtained from a network or system), we can construct a flow graph. By identifying the input and output nodes in the flow graph, we can obtain a transfer function by the application of certain "gain" formulas. A program based on the concept of flow graphs will be discussed next.

The program called "Network *A*nalysis for *S*ystems *A*pplications *P*rogram" (NASAP)* is originally developed by the Electronics Research Center of the National Aeronautics and Space Administration. This program employs the flow-graph technique to determine transfer functions, sensitivity functions, and worst-case squared functions. In the following paragraphs, we shall briefly describe the procedure used in NASAP for the construction of the flow graph from a given network and the method for generating a desired transfer function.

Since the configuration of a flow graph representing a network depends on the set of equations used, an algorithm for generating a set of independent network equations must be included in a program based on the flow graph approach. In NASAP, this algorithm begins by choosing a tree including all the voltage sources. Next, the cutset equations are obtained by expressing the currents in the tree-branches in terms of the currents in the chords or links. With these cutset equations, we construct a flow graph by following a set of rules (to be discussed later). Finally, the so-called "Shannon–Happ formula" is applied to determine the transfer function. This procedure is best explained with the aid of an example.

Consider the *RC* filter network shown in Fig. 10.5.1 and recognize that the problem is to determine the voltage transfer ratio, i.e.,

$$T(s) = \frac{V_{out}(s)}{V_{in}(s)}.$$

---

* There exist several versions of NASAP. The discussion presented in this section is based mainly on the version developed by L. P. McNamee *et al*, of the University of California at Los Angeles (see [MC 3]).

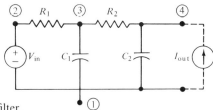

Fig. 10.5.2 NASAP equivalent circuit of the RC filter.

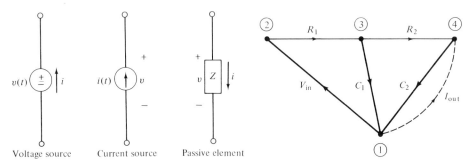

Fig. 10.5.3 Voltage polarities and current direction conventions.

Fig. 10.5.4 Directed graph of the network in Fig. 10.5.2.

For the purpose of identifying the output port, the program assumes a "current generation,"* I_{out}, with zero magnitude across the output terminals as depicted in Fig. 10.5.2, which is called the NASAP *equivalent circuit* of the given RC filter. Next, a directed graph for the equivalent circuit is drawn as shown in Fig. 10.5.4, in which the assigned orientations follow the conventions given in Fig. 10.5.3. From the directed graph, select a tree such that it includes all the voltage sources. Let this tree be the one represented by the heavy strokes in Fig. 10.5.4. Then the KCL-equations expressing the branch currents in terms of link currents are given by

$$I_{v_{\text{in}}} = I_{R_1},$$
$$I_{C_1} = I_{R_1} - I_{R_2}, \qquad (10.5.1)$$
$$I_{C_2} = I_{R_2} + I_{\text{out}}.$$

With the cutset equations, a flow graph can be constructed in the following steps:

1. Draw a partial flow graph to represent the KCL-(cutset) equations with both the branch-current and link-current variables taken as nodes. Arrange these nodes so that they lie on a horizontal line resulting in a *current flow graph*—a part of the

---

*A current generator with zero magnitude connected across the output terminals of any two-port does not inject any current into (or draw any current from) the network and thus the electrical properties of the network remain unaltered.

Fig. 10.5.5 Current flow graph representing (10.5.2).

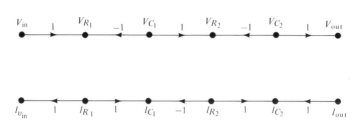

Fig. 10.5.6 Voltage flow graph generated from a current flow graph.

final flow graph. Thus applying the current relations in (10.5.1), we obtain the current flow graph for the simple network of Fig. 10.5.2, as depicted in Fig. 10.5.5.

2. Taking the mirror image of the current flow graph yields the corresponding *voltage flow graph* which, for our simple example, is drawn as shown in Fig. 10.5.6. Comparing the two graphs in Fig. 10.5.6, it is evident that the orientation of every transmittance in the voltage flow graph is opposite to that of its counterpart in the current flow graph. However, the signs of the voltage transmittances are determined according to the following rules: If one of the two elements that the transmittance is associated with is an active element while the other is passive, the sign for that transmittance is taken to be the same as that of the corresponding transmittance in the current flow graph; otherwise the opposite sign is used.

3. Next, draw a vertical line connecting each pair of the voltage and the corresponding current nodes for each passive element. If the element is a tree-branch, the orientation from the current node to the voltage node is assigned to the vertical line with an impedance as its transmittance. (In NASAP the tree-branches are referred to as the "voltage-type elements" with their voltage-current relations represented by the symbol $V \leftarrow i * z$ meaning that in the flow graph the orientation is directed from the current node to the voltage node with impedance as its transmittance.) On the other hand, if the element is a link (called the "current-type element"), the direction for the vertical line is taken from the voltage node to the current node with an admittance assigned as its transmittance (i.e., in NASAP notation, $i \leftarrow v * 1/z$).

4. The relations for dependent sources are drawn in next. Thus, for example, a relation such as $I_a = \beta I_b$ can be implemented by drawing a line directed from node I_b to node I_a with a transmittance β.

5. Finally, a line with a transmittance P is drawn from the output node to the input node (in NASAP notation, Input $\leftarrow P * $ Output). Thus, in our example, a line with a transmittance P is drawn originating from the node V_{out} to the node V_{in}.

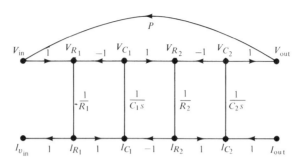

Fig. 10.5.7 NASAP flow graph for the network in Fig. 10.5.2.

Following the above steps, the final flow graph for the network of Fig. 10.5.1 is obtained as shown in Fig. 10.5.7.

It should be noted that the flow graph obtained using this procedure is unique once a tree is specified or chosen.

Before stating the formula used in NASAP for the determination of the transfer function from a flow graph, we need the following definitions:

A *first-order loop* is defined as a directed loop, and the *value* of the loop (or simply the *loop value*) is defined as the product of all the branch transmittances of the loop.

An *Nth-order loop* is defined as a set of N disjoint first-order loops with no common nodes (that is, N nontouching directed loops). Its value is the product of the values of all the included first-order loops.

A *zeroth-order loop* is defined to have a value of 1 and has no physical significance in a flow graph.

The Shannon–Happ formula used in NASAP is given as

$$H = 1 - \sum_{j=1}^{Q_1} L_j(1) + \sum_{j=1}^{Q_2} L_j(2) - \sum_{j=1}^{Q_3} L_j(3) + \cdots + (-1)^R \sum_{j=1}^{Q_R} L_j(R) = 0$$

or,

$$H = \sum_{r=0}^{R} \left\{ \sum_{j=1}^{Q_r} (-1)^r L_j(r) \right\} = 0, \tag{10.5.2\dagger}$$

where $L_j(r)$ is the value of the jth loop of order r, R is the order of the highest-order loop(s), Q_r is the total number of rth-order loops in the flow graph with $Q_0 \triangleq 1$. The transfer function $T(s)$, the ratio of the output transform to the input transform, is defined as

$$T(s) = \frac{1}{P}, \tag{10.5.3}$$

where P, the transmittance assigned to the added path in the flow graph from the

† Refer to [MC 3].

output node to the input node, can be determined by applying the Shannon–Happ formula in (10.5.2).

In our example, we obtain from Fig. 10.5.7 the following loops and their corresponding values:

(a) First-order loops   Value

$$L_1: \left(\frac{1}{R_1}\right)(1)\left(\frac{1}{C_1 s}\right)(-1) \qquad = -\frac{1}{R_1 C_1 s}$$

$$L_2: \left(\frac{1}{C_1 s}\right)(1)\left(\frac{1}{R_2}\right)(-1) \qquad = -\frac{1}{R_2 C_1 s}$$

$$L_3: \left(\frac{1}{R_2}\right)(1)\left(\frac{1}{C_2 s}\right)(-1) \qquad = -\frac{1}{R_2 C_2 s}$$

$$L_4: (P)(1)\left(\frac{1}{R_1}\right)(1)\left(\frac{1}{C_1 s}\right)(1)\left(\frac{1}{R_2}\right)(1)\left(\frac{1}{C_2 s}\right)(1) = \frac{P}{R_1 R_2 C_1 C_2 s^2}$$

(b) Second-order loops   Value

$$L_1 L_3: \left(\frac{-1}{R_1 C_1 s}\right)\left(\frac{-1}{R_2 C_2 s}\right) = \frac{1}{R_1 R_2 C_1 C_2 s^2}$$

With these loop values, (10.5.2) can be expressed as

$$H = 1 - \left[\frac{-1}{R_1 C_1 s} + \frac{-1}{R_2 C_1 s} + \frac{-1}{R_2 C_2 s} + \frac{P}{R_1 R_2 C_1 C_2 s^2}\right] + \frac{1}{R_1 R_2 C_1 C_2 s^2} = 0$$

which reduces to

$$H = \frac{R_1 R_2 C_1 C_2 s^2 + (R_1 C_1 + R_2 C_2 + R_1 C_2)s - P + 1}{R_1 R_2 C_1 C_2 s^2} = 0. \quad (10.5.4)$$

Solving for P, we obtain

$$P = R_1 R_2 C_1 C_2 s^2 + (R_1 C_1 + R_2 C_2 + R_1 C_2)s + 1.$$

Finally, with (10.5.3) we get the transfer function

$$\frac{V_{out}}{V_{in}} = T = \frac{1}{P} = \frac{1}{R_1 R_2 C_1 C_2 s^2 + (R_1 C_1 + R_2 C_2 + R_1 C_2)s + 1}. \quad (10.5.5)$$

It is interesting to note that by applying Mason's gain formula to the flow graph of Fig. 10.5.7 with the "added branch, P," removed, we will obtain the same result as in (10.5.5). The reader is invited to verify this statement.

The program NASAP(UCLA) contains three subprograms. These are: (a) NASAP, which selects a tree for minimum computation time and proceeds to solve for the transfer function as discussed previously; (b) TREE, which allows the user to specify a tree and then continues with the solution; (c) CIRCUIT, which selects a tree so that optimum accuracy is obtained for the solution at a specified frequency.

The format for the input deck of NASAP(UCLA) depends on which of the subprograms the user intends to use for his problem. In general, an input deck consists of four groups of cards. *Group I* is optional and is composed of up to ten "title cards," the contents of which (if included) will be printed on the output. *Group II* consists of one command card, indicating which of the three subprograms is to be executed. This card contains one of the following three words beginning in Column 1: NASAP, TREE, or CIRCUIT. *Group III* consists of the circuit description cards, specifying the network topology and element values. The formats for this group of cards differ in the three subprograms and will be discussed in three parts as follows.

Subprogram NASAP

One card is used to describe each element, in the format

NAME	POSITIVE	NEGATIVE	VALUE	UNIT	DEPENDENCY
	NODE	NODE		(optional)	(if appropriate)

where the terms are described in the following paragraphs:

a) NAME of an element must begin with a letter, V, I, R, L, or C, indicating the type of element as specified in Table 10.5.1, and followed by up to eleven alphanumeric characters. That is, an element name may not occupy more than a total of twelve columns.

b) POSITIVE NODE and NEGATIVE NODE are the numbers assigned to the corresponding nodes of the element in integer form.

c) VALUE is the element value in any format commonly used in FORTRAN.

d) UNIT is the unit of the element the user may wish to specify. The units acceptable to NASAP and their corresponding codes are given in Table 10.5.2. If no unit is specified, NASAP assumes MKS units.

e) DEPENDENCY contains the element name of the controlling element (previously described) of the dependent source (described by this card), preceded by a letter V or I indicating whether the source is voltage-controlled or current controlled.

In a circuit description card, in addition to the requirement that the element name may not occupy more than twelve columns, two consecutive items must be separated by at least one blank column. Except for these two restrictions, the data in the card are "in free field formats." That is, each field is not restricted to a particular column group.

The last of the circuit description cards must contain the word, OUTPUT or END.

Table 10.5.1 Element Types and Assigned Names for NASAP.

Element type	Element name
Voltage source	V — — — — — — — — — —
Current source	I — — — — — — — — — —
Resistor	R — — — — — — — — — —
Inductor	L — — — — — — — — — —
Capacitor	C — — — — — — — — — —

Table 10.5.2 Units and Codes Used in NASAP.

Element	Scale-factor code	Units multiplier
Resistor	K	10^3
	M	10^6
Inductor	MH	10^{-3}
	UH	10^{-6}
Capacitor	UF	10^{-6}
	PF	10^{-12}

Table 10.5.3 Element Types and Assigned Names for TREE.

Element type		Element name
Independent voltage source		E — — — — — — — — — —
Independent current source		J — — — — — — — — — —
Resistor	as a branch	RE — — — — — — — — — —
	as a link	RJ — — — — — — — — — —
Inductor	as a branch	LE — — — — — — — — — —
	as a link	LJ — — — — — — — — — —
Capacitor	as a branch	CE — — — — — — — — — —
	as a link	CJ — — — — — — — — — —
Dependent voltage source		DE — — — — — — — — — —
Dependent current source		— — — — — — — — — —

Subprogram TREE

If the user intends to specify the tree to be used, the subprogram TREE in NASAP/70 can be used. The command card that follows the title card(s) contains the word TREE. The element types and their corresponding element names acceptable to TREE are listed in Table 10.5.3. Again, an element name may occupy up to a total of twelve columns in a card. One card is used to describe each element in the format.

$$\text{NAME} \quad \begin{pmatrix} \text{POSITIVE} & - & \text{NEGATIVE} \\ \text{NODE} & & \text{NODE} \end{pmatrix} \quad = \quad \text{VALUE} \quad \begin{array}{c} \text{UNIT} \\ \text{(optional)} \end{array}$$

Table 10.5.4 Units and Codes Used in Subprograms TREE and CIRCUIT.

Element	Scale-factor code	Units multiplier
Resistor	K	10^3
	M	10^6
Inductor	MH	10^{-3}
	MMH	10^{-6}
Capacitor	MF	10^{-6}
	MMF or PF	10^{-12}

where each term is the same as in the format used in subprogram NASAP discussed previously, except for the case of dependent sources. For a dependent source, the element name must consist of two parts separated by a slash. The first part contains the name of the *controlled* element, and the second part consists of the name of the *controlling* element preceded by a letter V or I indicating whether the source is voltage-controlled or current-controlled. The following example will illustrate the format for a dependent-source data card:

$$DE2/IRE5 \ (3-6) = 1.$$

which indicates: (a) the element with the name DE2 is a dependent voltage source; (b) it is controlled by the current in element RE5; (c) the positive and negative nodes of the source are nodes 3 and 6, respectively; and (d) the value of the dependency (i.e. the ratio of the value of the dependent source to the value of the controlling variable) is 1.

Table 10.5.4 shows the units and their corresponding codes used in TREE. Again, if no unit is given, TREE will assume **MKS** units. With the exception for the required format as indicated, all the data are again free field formats, and blanks and commas may be used to improve readability since they are ignored by subprogram TREE.

While TREE allows the user to specify a particular tree to be used for the determination of a flow graph, it usually results in a longer computation time as compared to that required by subprogram NASAP for solving the same network problem, since the latter subprogram is designed to select automatically a tree which will generate a flow graph with a minimum number of loops. Thus, unless the user has a special reason for choosing a particular tree, it would be much simpler to use subprogram NASAP to perform the entire computation including the choice of a tree.

Subprogram CIRCUIT

The subprogram CIRCUIT is designed to select a tree so that optimum accuracy will be obtained in the solution for a specified frequency. The format for the command card following the optional title card(s) is

$$CIRCUIT \ (FREQUENCY)$$

which specifies the subprogram CIRCUIT to be used for the calculations and the value of FREQUENCY in Hz at which the solution accuracy is to be optimized. If the value of frequency is not specified here (i.e., only the word CIRCUIT is included), 1 Hz is assumed.

The rest of the input data is identical to that for TREE except that CIRCUIT does not distinguish between branch and link elements. Thus, the only passive element names are R, L, and C as in subprogram NASAP.

The fourth group or *Group IV* of data cards used in each subprogram of NASAP (UCLA) is a number of "solution control" and "output request" cards.

Output requests must appear in the following order, using one card for each item:

1. Sensitivity request
2. Worst-case analysis
3. Transfer function
4. Roots
5. Plots

We shall now discuss the meanings of these items and their required formats.

Sensitivity request. The first card after the OUTPUT (or END) card must be the sensitivity card (if requested) using the format

$$\text{SENSITIVITY} = \text{ELEMENT NAME}$$

where ELEMENT NAME is the name of the element with respect to which the sensitivity of the transfer function is to be calculated. The output will include the printout of the change in the transfer function (to be specified later) due to variation of the value of the element.

Worst-case analysis. A worst-case analysis can be requested for a network with up to twenty elements, excluding independent sources. To do so, a card must first be included after the END (or OUTPUT) card in the format

$$\text{WORST}$$

Immediately after the WORST card, "tolerance" cards should be included to describe the tolerance of each element using the format

$$\text{TOL} = \text{ELEMENT NAME} = \text{VALUE}$$

where the word VALUE contains a number indicating the *relative tolerance*, $\Delta x/x$, of the element with a nominal value of x and a change of value Δx. If no "tolerance" card is provided for an element, the value of $\Delta x/x = 0.1$ will be assumed by the program. The worst-case analysis will be performed using the tolerance value of *each* element to determine the maximum change of the transfer function, which is specified in the next card.

Table 10.5.5 Available `OPTION 1`.

OPTION 1	Resulting output plot
TYPE = IMPULSE	Impulse response
TYPE = STEP	Step response
TYPE = SINE	Response to sine wave input
TYPE = EXPONENTIAL	Response to exponential input
TYPE = PULSE	Response to pulse (train) input
TYPE = FREQUENCY	Frequency response
TYPE = SENSITIVITY	Sensitivity plot
TYPE = WORST	Worst-case magnitude

The output for a worst-case analysis gives the square of the worst-case tolerance in the same form of the output(s) (to be requested) for the transfer function. When a worst-case analysis is requested, no "sensitivity" card can be included in the input deck.

Transfer function. The desired transfer function is requested in a card following either the "sensitivity" card or the "tolerance" card if one of these is present. The format used is

TYPE NAME 1/TYPE NAME 2

`TYPE NAME 1` is an element name which has already appeared in the input deck, preceded by a letter `V` or `I` to indicate whether the desired dependent variable is a voltage or a current. `TYPE NAME 2` is an element name of an independent source. For example, the card

VR2/VEO

indicates that the desired transfer function is the ratio of the voltage across the element R_2 to the input voltage source of e_o. Thus, the output will be a rational function in s.

Roots. A card with the word

ROOTS

following the "transfer function" card will request for the poles and zeros of the transfer function.

Plots. There are three types of plots available: time response, frequency response, and sensitivity plots. Each "plot request" is indicated in a single card in the format

PLOT (OPTION 1/OPTION 2/...)

where `OPTION 1` defines the type of plot and may contain one of the items shown in Table 10.5.5, in which the available options of plot types are listed. The "fre-

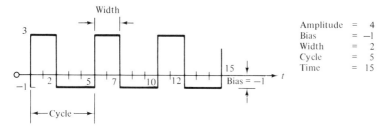

Fig. 10.5.8 A pulse train illustrating associated terms.

quency," "sensitivity," and "worst" plots include both magnitude and phase plots versus frequency.

Each of the remaining options (namely OPTION 2, OPTION 3, · · ·) are separated by a slash and contain various information depending on the type of plot requested in OPTION 1. These options for entering related information for the plot are briefly described as follows.

AMPLITUDE = a number; indicates the amplitude of an input wave

DENSITY   = an integer, k; specifies that the plotted output will contain one plotted point after every k solution points calculated

STEP   = a number (in sec); defines the time interval between calculated points

TIME   = a number (in sec); specifies the time interval of the input to be considered

FREQUENCY = a number (in Hz); indicates the frequency of a sinusoidal input and this option is used only when SINE is specified for OPTION 1.

CONSTANT = a number; specifies the constant b in an exponential input, e^{bt}.

BIAS   = a number; is used only with PULSE to indicate the amount of bias assigned to a pulse train. This is used together with the AMPLITUDE as illustrated in Fig. 10.5.8.

WIDTH   = a number (in sec); indicates the duration of a pulse in a pulse train as defined in Fig. 10.5.8

CYCLE   = a number (in sec); defines the period of a pulse train (Fig. 10.5.8)

ELEMENT   = element name; indicates the name of an element, corresponding to which a sensitivity plot is requested.

FROM = x/TO = y indicates that the range of a frequency plot is from x to y.

Since all information related to a plot must be included in a single card, the program allows abbreviations and omission of some of the options as follows.

a) Only the first two letters of a word in an option are necessary.

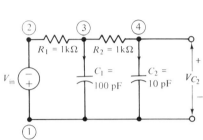

```
 | 2 | 3 | 4 | 5 | 6 | 7 | 8 | 9 |10|11|12|13|14|15|16|17|18|19|20|21|22|23|24|25|26|27|28|29|30|
EXAMPLE 1/10-5-1/                              NASAP
NASAP
VIM      1   2    1.
C1       3   1    1E+2    PF
C2       4   1    10.     PF
R1       2   3    1.      K
R2       3   4    1000.
OUTPUT
VC2/VVIM
EXECUTE
```

Fig. 10.5.9 Network for Example 10.5.1.

Fig. 10.5.10 Input coding for Example 10.5.1.

b) The program assumes the *default value** if a required option is not specified in a PLOT card. The available default values in the program are

$$TYPE = IMPULSE$$
$$TIME = 1$$
$$STEP = .01$$
$$AMPLITUDE = 1$$
$$DENSITY = 1$$
$$CONSTANT = 1$$
$$FREQUENCY = 1$$

The options may appear in any order on a PLOT card, and several plots (one for each plot) may be requested in a problem.

The last card of the input deck for a problem contains a word

$$EXECUTE$$

Several different problems may be included in the same run. A card with the word

$$STOP$$

included at the end of the input deck will terminate the execution of the program.

In the following examples, we shall illustrate the use of NASAP (UCLA).

Example 10.5.1   We shall use NASAP to determine the transfer function V_{c_2}/V_{in} of the network shown in Fig. 10.5.9 (the same network in Fig. 10.5.1).

---

* The default value of a variable is the value assumed by the program when it is not specified by the user.

```
****NASAP****

    NETWORK ANALYSIS AND SYSTEMS APPLICATION PROGRAM
THIS VERSION WAS DEVELOPED AT UCLA ENGR. DEPT.


EXAMPLE 10-5-1        NASAP
NASAP
VIN   1  2   1.
C1    3  1   1E+2   PF
C2    4  1   10.    PF
R1    2  3   1.     K
R2    3  4   1000.
OUTPUT


FLOWGRAPH
FROM          TO          S          VALUE
  2            7         -1        1.00000031E 10
  3            8         -1        1.00000006E 11
  9            4          0        9.99999931E-04
 10            5          0        9.99999931E-04
  4            1          0        1.00000000E 00
  6            9          0        1.00000000E 00
  4            2          0        1.00000000E 00
  7            9          0       -1.00000000E 00
  5            2          0       -1.00000000E 00
  7           10          0        1.00000000E 00
  5            3          0        1.00000000E 00
  8           10          0       -1.00000000E 00




TRANSFER   FUNCTION


 - 1.000E 00    + 0.0         + 0.0


 ------------------------------------------------

                                                    2
   1.000E 15   + 1.200E 08S  + 1.000E 00S·

THE FUNCTION FACTOR =    -1.000E 15
```

Fig. 10.5.11 Output for Example 10.5.1.

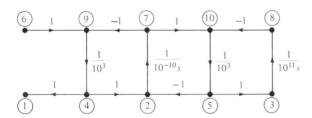

Fig. 10.5.12 Flow graph for Example 10.5.1.

After numbering the nodes and assigning the current directions in the network, the NASAP input deck can be coded as shown in Fig. 10.5.10. The computer output is given in Fig. 10.5.11 from which we obtain the transfer function

$$\frac{V_{c_2}}{V_{in}} = \frac{10^{15}}{1 \times 10^{15} + 1.2 \times 10^8 s + s^2}.$$

The flow graph of the network obtained from the output is shown in Fig. 10.5.12, which is identical to the flow graph in Fig. 10.5.7 when the zero-magnitude current generator I_{out} and the feedback path P are added to Fig. 10.5.12.

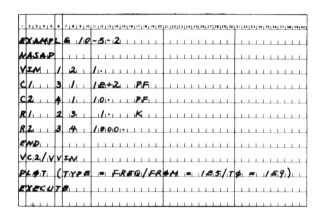

Fig. 10.5.13 Input coding for Example 10.5.2.

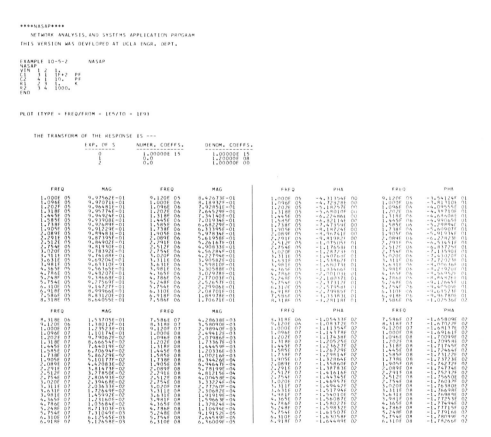

Fig. 10.5.14 Output for Example 10.5.2.

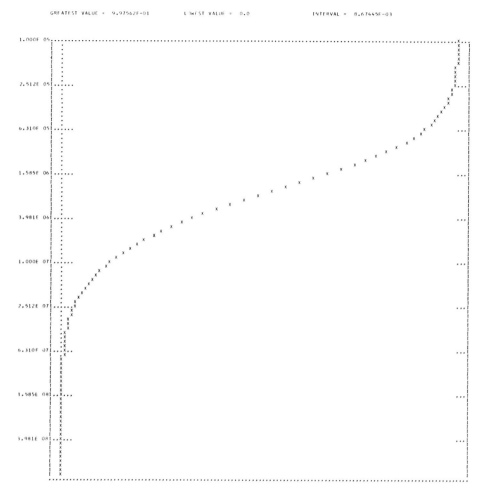

GREATEST VALUE = 9.97562E-01   LOWEST VALUE = 0.0   INTERVAL = 8.67445E-03

Fig. 10.5.14 Output for Example 10.5.2 (*continued*).

Example 10.5.2 To illustrate the plotting capability of NASAP, we shall obtain the frequency response of the network shown in Fig. 10.5.9. The input coding and the corresponding output are shown in Fig. 10.5.13 and Fig. 10.5.14, respectively.

Example 10.5.3 Suppose that we wish to determine the step response of the same network in Fig. 10.5.9. In addition to the "circuit description cards" and the "solution control cards" as given in Fig. 10.5.10, an additional output request card in the form of

$$PLOT \ (TYPE = STEP/TIME = 1 \ E - 6)$$

must be included just before the EXECUTE card in the input deck. The computer output is shown in Fig. 10.5.15.

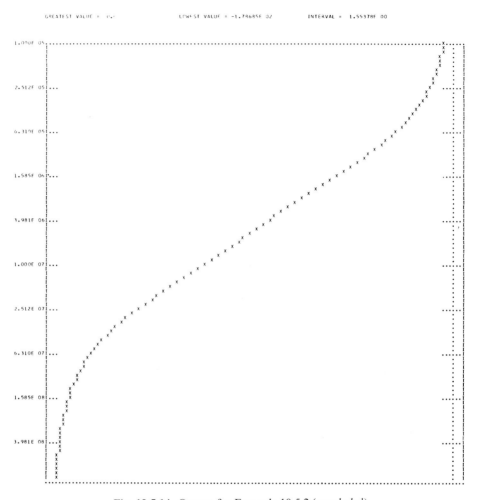

Fig. 10.5.14 Output for Example 10.5.2 (*concluded*).

As mentioned earlier in this section, the subprogram TREE allows the user to specify the tree for the solution. However, in general, subprogram NASAP should be used since it is designed to select a tree which results in optimum solution time. This is primarily due to the fact that the tree selected by the subprogram NASAP will result in a flow graph in which the number of loops is minimum. The reader is invited to verify this fact.

10.6 SUMMARY AND REMARKS

In the previous sections in this chapter we introduced four computer programs for network analysis. We used each of these programs as an example to illustrate the computer implementation of a particular analysis method.

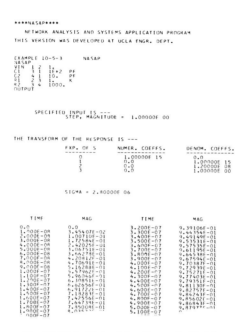

Fig. 10.5.15 Output for Example 10.5.3.

In Section 10.2 we presented ECAP as an example for the computer application of the nodal analysis, which is one of the two most well-known analysis techniques to electrical and electronics engineers.* ECAP is very well documented and tested. Although the solution speed of ECAP compares unfavorably to some other existing programs using nodal analysis, it is a suitable program to be presented here at an introductory level for computer-aided network analysis because of the following reasons. First, ECAP has a very extensive diagnostics routine which checks virtually every statement in the input deck for errors and then prints out error messages; this is a valuable help to a first-time user. Another reason is that ECAP is well known and available in major commercial time-share systems with some modifications. The program also contains many special features such as "nodal current checks" for all the solutions at the user's specified accuracy and various other user's options as discussed in the User's Manual. One obvious disadvantage is the lack of plotting routine. However, a printer plotting routine can be added easily to the program.

Although the state-variable method has become available as a tool for circuit analysis only recently, many programs based on this concept are already in existence today. CORNAP, presented in Section 10.3, is an example. For the types of solutions available from CORNAP, its solution time is quite short compared to most other programs that are based on the state-variable approach.

---

* The other one is, of course, the loop analysis.

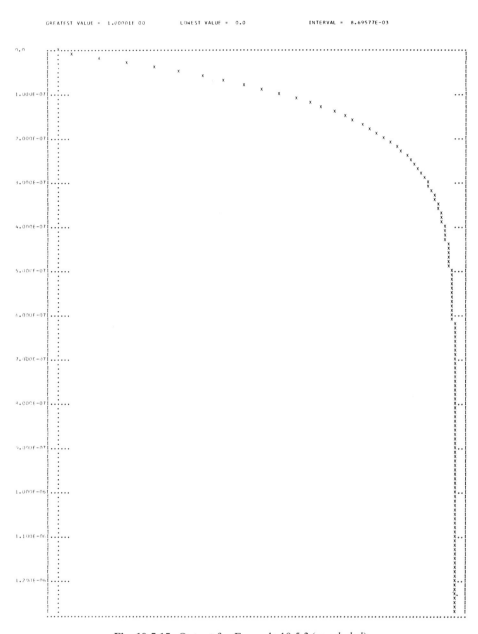

Fig. 10.5.15 Output for Example 10.5.3 (*concluded*).

The CALAHAN program, discussed in Section 10.4, determines network functions using topological formulas. Because of the necessity for determining all the trees and some of the 2-trees and 3-trees, the size of the network that can be solved by using CALAHAN is limited to about 15 nodes and 30 elements. This limitation is due primarily to the large amount of time required to generate the k-trees. A replacement of the tree-finding routine by a faster algorithm can improve this situation somewhat. In spite of the disadvantage of slow solution speed for larger networks, this program offers a unique output, namely the symbolic form* of network functions, which is especially useful in performing sensitivity analysis.

The NASAP program, presented in Section 10.5, employs the flow graph technique for the calculation of the transfer function. The program offers a wide variety of output options to the user. The input format is relatively simple. Certain versions of NASAP (similar to the subprogram NASAP discussed in this chapter) are available in some commercial time-share systems.

Computer implementations of the k-tree topological method and the flow graph technique are still not very popular. Most of the analysis programs that exist today are based on either the nodal analysis or the state-variable method. Other programs recently developed, using the nodal analysis method, include TRAC and BIAS -3. Programs CIRCUS and SCEPTRE are two well-known programs based on the state-variable technique. BIAS -3 is a program for the nonlinear dc-analysis of bipolar transistor circuits. TRAC, CIRCUS, and SCEPTRE are especially designed for determining the transient radiation effects on electronic circuits. Although these may be called "special-purpose programs," their applications are quite general. In fact, these programs are known for their extensive modeling capabilities which enable the user to analyze electronic circuits composed of various types of nonlinear elements.

REFERENCES

1. *1620 Electronic Circuit Analysis Program (ECAP) (1620-EE-02X) User's Manual*, IBM Corporation, 1965.

2. R. W. Jensen and M. D. Lieberman, *IBM Electronic Circuit Analysis Program*, Prentice-Hall, 1968.

3. D. A. Calahan, "Linear Network Analysis and Realization Digital Computer Programs: An Instruction Manual," *University of Illinois Bulletin* 472, Vol. 62, No. 58, Feb., 1965.

4. C. Pottle, "State-Space Techniques for General Active Network Analysis," in *System Analysis by Digital Computer*, F. F. Kuo and J. F. Kaiser, Eds., John Wiley & Sons, 1966, Chapter 3.

5. C. Pottle, "A 'Textbook' Computerized State-Space Network Analysis Algorithm," *Trans. IEEE*, Vol. CT-16, Nov. 1969, pp. 566–568.

6. L. P. McNamee and H. Potash, *A User's Guide and Programmer's Manual for NASAP*, Report No. 68–38, University of California, Los Angeles, 1968.

---

* The symbolic form of output from CALAHAN is given in the form of tree-admittance products (e.g., Y1 Y2 Y3, Y1 Y3 Y4 etc.).

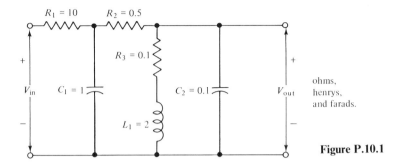

ohms,
henrys,
and farads.

Figure P.10.1

PROBLEMS

10.1 Using ECAP, determine
a) the frequency response, and
b) the step response of the network shown in Fig. P.10.1.

10.2 Repeat Problem 10.1, using CORNAP.

10.3 Repeat Problem 10.1, using CALAHAN.

10.4 Repeat Problem 10.1, using NASAP (UCLA).

10.5 Repeat Problem 10.1 for the network shown in Fig. P.10.5.

10.6 Repeat Problem 10.1 for the network of Fig. P.10.6.

10.7 A 4-MHz inverter network is shown in Fig. P.10.7.
a) Draw an ac-equivalent circuit of this network using the hybrid π model for the 2N708 transistor.
b) With proper scaling, determine the bandwidth of the network (by obtaining the frequency response), using ECAP.

10.8 Using ECAP, determine the time response of the network shown in Fig. P.10.7 to the input $v_{in}(t)$ given in Fig. P.10.8.

10.9 Repeat Problem 10.5, using CORNAP.

10.10 Repeat Problem 10.6, using CORNAP.

10.11 Repeat Problem 10.7, using CORNAP.

10.12 Repeat Problem 10.8, using CORNAP, with only one input pulse.

10.13 Repeat Problem 10.5, using CALAHAN.

10.14 Repeat Problem 10.6, using CALAHAN.

10.15 Repeat Problem 10.7, using CALAHAN.

10.16 Repeat Problem 10.8, using CALAHAN.

10.17 Repeat Problem 10.5, using NASAP (UCLA).

10.18 Repeat Problem 10.6, using NASAP (UCLA).

10.19 Repeat Problem 10.7, using NASAP (UCLA).

10.20 Repeat Problem 10.8, using NASAP (UCLA).

10.21 Obtain the response $v_2(t)$ of the network shown in Fig. P.10.5 to the input $v_1(t)$ depicted in Fig. P.10.21. Use any suitable program.

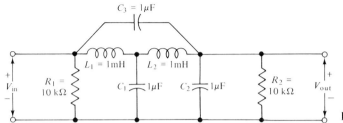

Figure P.10.5

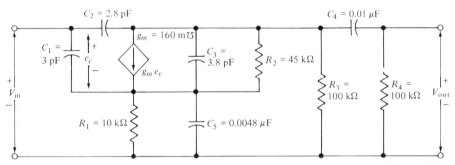

Figure P.10.6

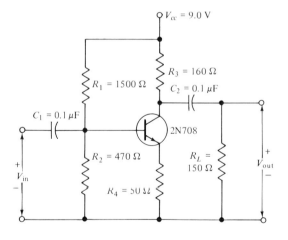

Figure P.10.7

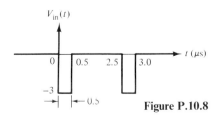

Figure P.10.8

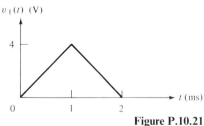

Figure P.10.21

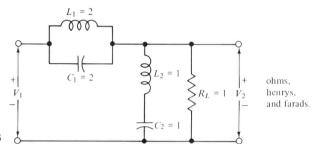

Figure P.10.23

10.22 Using an appropriate program to determine the effect of varying the values of (a) C_1, (b) C_2 on the frequency response, $V_{out}(s)/V_{in}(s)$, of the network shown in Fig. P.10.1.

10.23 a) Use an appropriate program to plot the frequency characteristics of the voltage-ratio transfer function, V_2/V_1, of the network shown in Fig. P.10.23.

b) Check the solution in part (a) analytically (hand calculation).

c) For the same network, investigate the sensitivity of the transfer function V_2/V_1 to the variation of values of C_1 and C_2.

10.24 a) Using CORNAP, determine the impulse response of the network of Fig. P.10.23.

b) Verify the computed result in part (a) by hand calculation.

10.25 Repeat Problem 10.24 part (a), using NASAP.

CHAPTER 11

SELECTED TOPICS IN SYSTEM ANALYSIS

11.1 INTRODUCTION

In the discussion of network responses (Chapter 5), we found that, depending upon the values of the natural frequencies of a given network, the magnitude of the network response may increase without bound as $t \rightarrow \infty$ even though the source input is bounded in magnitude. This unboundedness of the network response simply means that the network under consideration is unstable. In addition to various definitions of stability to be discussed from the system point of view, it is intended in this chapter to present a brief introduction to some of the relatively new concepts in system theory, controllability, and observability. Using the state-space approach (Chapters 7 and 8) these concepts not only provide more insight into the physical structure of a system, which is unobtainable otherwise, but also serve as a useful tool in the study of optimal control systems.

11.2 CONTROLLABILITY OF LINEAR SYSTEMS

Consider the two-port RLC network depicted in Fig. 11.2.1 in which the switches S_1 and S_2 are suddenly closed at $t = 0$. Suppose that our problem is to determine, for a known set of parameter values, a voltage waveform $v_s(t)$ so that the output variable, say, $v_2(t) = i_2 R_2$, be driven to zero at $t = T$ seconds and be kept constant at zero thereafter in the absence of subsequent disturbances.

The problem just posed is a practical control problem of regulating the output. To determine whether this problem has a solution, we need to perform an analysis of the system, which necessitates the establishment of a mathematical model. Of course, we must assume that the mathematical model used in the analysis represents accurately the behavior of the physical system. Using the state-space approach with $i_1(t)$ and $v_c(t)$ as state variables, that is, $x_1 \triangleq i_1(t)$ and $x_2 \triangleq v_c(t)$, the mathematical model of the system is given by

$$\begin{bmatrix} \dot{x}_1 \\ \dot{x}_2 \end{bmatrix} = \begin{bmatrix} -\dfrac{R_1}{L} & 0 \\ 0 & \dfrac{1}{CR_2} \end{bmatrix} \begin{bmatrix} x_1 \\ x_2 \end{bmatrix} + \begin{bmatrix} \dfrac{1}{L} \\ 0 \end{bmatrix} [v_s] \qquad (11.2.1)$$

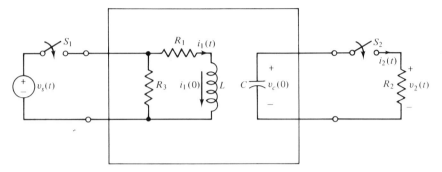

Fig. 11.2.1 A two-port RLC network.

and

$$[v_2] = \begin{bmatrix} 0 & 1 \end{bmatrix} \begin{bmatrix} x_1 \\ x_2 \end{bmatrix} = [x_2]. \tag{11.2.2}$$

Now we have the mathematical model. The solution of (11.2.1) enables us to determine exactly how the input $v_s(t)$ influences the state of the system. However, the initial state, $x_1(0)$ and $x_2(0)$, must be known before the solution can be obtained. In general, these initial conditions may not be measured directly as is the case for $x_1(t) = i_1(t)$ in this network, so that one must determine if the initial state can be calculated from the knowledge of the measured quantities, $v_s(t)$ and $v_2(t)$, in this system. Equation (11.2.2) provides the relation between the output and the state variables. If $x_1(0)$ and $x_2(0)$ can be determined by solving (11.2.2)—a condition which we refer to as the observability condition—then one can proceed to determine exactly how the input $v_s(t)$ influences the state and hence the output of the system. In this simple problem, we can easily conclude that the system is not observable since $x_1(t)$ is missing in (11.2.2) and hence $x_1(0)$ cannot be determined by solving (11.2.2).

Having pointed out the important physical implication of the concept of observability, we turn to the other aspect of the problem—the concept of controllability. Suppose that the system under study was observable. Then the initial state, $x_1(0)$ and $x_2(0)$, could have been determined and hence one could proceed to find out, by investigating the solution of (11.2.1), if a control $v_s(t)$ for $0 < t < T$ exists so that the state vector $\mathbf{x}(T) = 0$ and hence that $v_2(t) = 0$ for $t \geq T$ if $v_s(t) = 0$ for $t > T$. If it does, the system is said to be a controllable system. Of course, in this simple problem, we can easily conclude that the system is not controllable since, by inspection of (11.2.1), $v_s(t)$ does not influence the state variable x_2 in any manner.

In the above discussion it is evident that the output of a system can be regulated only if the system is both controllable and observable. Furthermore, this simple problem illustrates that a combination of experience and intuition sometimes reveals the feasibility of a workable system. However, as the complexity of the system increases, it is unlikely that the solution of a problem can be determined on the basis of intuition alone. One must, instead, also make use of mathematical

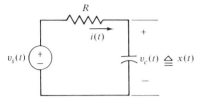

Fig. 11.2.2 The RC network N.

tools to assist him in analyzing the mathematical model which represents the behavior of a given physical system. The concept of controllability is taken up in this section and that of observability will be discussed in Section 11.3.

We shall limit our discussion on controllability to linear time-invariant systems only. Roughly speaking, the term controllability means that it is possible to drive the system from any given initial state to any other finite state in some finite time. A system in which the state vector is independent of the input is said to be uncontrollable. In the following discussion we shall separate systems into two groups: (1) linear time-invariant continuous systems, and (2) linear discrete-time systems.

Linear time-invariant continuous systems

Let a system be described by the matrix equation

$$\dot{\mathbf{x}} = \mathbf{A}\mathbf{x} + \mathbf{B}\mathbf{u} \tag{11.2.3}$$

where $\mathbf{x}$ is the state vector, $\mathbf{u}$ is the input or control vector, and $\mathbf{A}$ and $\mathbf{B}$ are constant matrices of orders $n \times n$ and $n \times r$, respectively.

Definition 11.2.1   The system (11.2.3) is said to be *controllable* if, for any time t_0, it is possible by means of a suitably selected control vector $\mathbf{u}$ to transfer the system from any initial state $\mathbf{x}(t_0)$ to any other state $\mathbf{x}(t_1)$ in a finite time interval $T = t_1 - t_0$.*

Let us illustrate this concept by considering the following simple examples.

Example 11.2.1   The state equation for the RC network N of Fig. 11.2.2 was found to be

$$\dot{x}(t) = -\frac{1}{RC} x(t) + \frac{1}{RC} v_s(t), \tag{11.2.4}$$

where $x(t) \triangleq v_c(t)$, the voltage across the capacitor as shown [see (7.2.3)]. The solution of (11.2.4) is given by (7.2.6) which is repeated here for convenience:

$$x(t) = e^{-(1/RC)(t-t_0)}x(t_0) + \int_{t_0}^{t} e^{-(1/RC)(t-\tau)}\left(\frac{1}{RC}\right) v_s(\tau)\,d\tau. \tag{11.2.5}$$

---

* Strictly speaking, the system satisfying the property state above is said to be *completely state controllable*. Thus, for a system described by $\dot{\mathbf{x}} = \mathbf{A}\mathbf{x} + \mathbf{B}\mathbf{u}$ and $\mathbf{y} = \mathbf{C}\mathbf{x}$, if in the above definition $\mathbf{x}(t_0)$ and $\mathbf{x}(t_1)$ are replaced by the output $\mathbf{y}(t_0)$ and any final output $\mathbf{y}(t_1)$, respectively, then the system is said to be *completely output controllable*.

This system is controllable if the capacitor voltage can be changed to any specified value $x(t_1)$ from an arbitrarily chosen initial value $x(t_0)$ in a finite time interval $T = t_1 - t_0$. This is equivalent to requiring the voltage source $v_s(t)$ to satisfy the equality

$$x(t_1) - e^{-(1/RC)(t_1 - t_0)}x(t_0) = \frac{1}{RC}e^{-(1/RC)t_1}\int_{t_0}^{t_1} e^{(1/RC)\tau}v_s(\tau)\,d\tau. \qquad (11.2.6)$$

There is no unique solution to (11.2.4); that is, a finite number of functions $v_s(t)$ can be found to satisfy (11.2.6). For convenience, let $v_s(t) = A$ for $t_0 \le t \le t_1$. Then the unknown constant A can be determined by solving (11.2.6), or

$$x(t_1) - e^{-(1/RC)(t_1 - t_0)}x(t_0) = e^{-(1/RC)t_1}\left[e^{(1/RC)\tau}A\right]_{t_0}^{t_1}$$

which gives

$$A = \frac{[x(t_1) - e^{-(1/RC)(t_1 - t_0)}x(t_0)]}{1 - e^{-(1/RC)(t_1 - t_0)}}. \qquad (11.2.7)$$

Therefore, the RC network is a controllable system.

Example 11.2.2 Consider the second-order system described by the equation $\dot{\mathbf{x}} = \mathbf{A}\mathbf{x} + \mathbf{B}u$ or

$$\begin{bmatrix} \dot{x}_1 \\ \dot{x}_2 \end{bmatrix} = \begin{bmatrix} a_{11} & a_{12} \\ a_{21} & a_{22} \end{bmatrix}\begin{bmatrix} x_1 \\ x_2 \end{bmatrix} + \begin{bmatrix} b_1 \\ b_2 \end{bmatrix}u. \qquad (11.2.8)$$

Let λ_1 and λ_2 be the distinct eigenvalues of $\mathbf{A}$. Then, as discussed in Section 7.4, by means of the linear nonsingular transformation

$$\mathbf{x} = \mathbf{S}\mathbf{z} \qquad (11.2.9)$$

the system (11.2.8) can be reduced to the following one such that the new state variables z_1 and z_2 are completely decoupled from each other. That is,

$$\dot{\mathbf{z}} = \mathbf{S}^{-1}\mathbf{A}\mathbf{S}\mathbf{z} + \mathbf{S}^{-1}\mathbf{B}u$$
$$= \mathbf{\Lambda}\mathbf{z} + \mathbf{E}u \qquad (11.2.10)$$

where $\mathbf{\Lambda} = \mathbf{S}^{-1}\mathbf{A}\mathbf{S}$ and $\mathbf{E} = \mathbf{S}^{-1}\mathbf{B}$. In terms of components, (11.2.10) becomes

$$\dot{z}_1 = \lambda_1 z_1 + e_1 u$$
$$\dot{z}_2 = \lambda_2 z_2 + e_2 u. \qquad (11.2.11)$$

Integrating these equations between t_0 and t_1 yields

$$z_1(t_1) = e^{\lambda_1(t_1 - t_0)}z_1(t_0) + e^{\lambda_1 t_1}\int_{t_0}^{t_1} e^{\lambda_1 \tau}e_1 u(\tau)\,d\tau$$

$$\qquad (11.2.12)$$

$$z_2(t_1) = e^{\lambda_2(t_1 - t_0)}z_2(t_0) + e^{\lambda_2 t_1}\int_{t_0}^{t_1} e^{\lambda_2 \tau}e_2 u(\tau)\,d\tau.$$

The system (11.2.8) will be controllable if the system (11.2.11) is. This fact can be appreciated by considering the zero-input response of the nth-order system described by

$$\dot{\mathbf{x}} = \mathbf{A}\mathbf{x} + \mathbf{B}u, \qquad \mathbf{x}(0) = \mathbf{x}_0,$$

which as discussed in Appendix A.1, Section A.1.5 can always be expressed in the form

$$\mathbf{x} = \alpha_1 \mathbf{u}_1 e^{\lambda_1 t} + \alpha_2 \mathbf{u}_2 e^{\lambda_2 t} + \cdots + \alpha_n \mathbf{u}_n e^{\lambda_n t}$$

with $\mathbf{u}_1, \mathbf{u}_2, \ldots, \mathbf{u}_n$ being the normalized eigenvectors associated with the *distinct* eigenvalues $\lambda_1, \lambda_2, \ldots, \lambda_n$, respectively. The scalar constants $\alpha_1, \alpha_2, \ldots, \alpha_n$ can be determined once the initial state $\mathbf{x}_0$ is specified. The vector term $\mathbf{u}_i e^{\lambda_i t}$ $(i = 1, \ldots, n)$ is usually referred to as the ith *natural mode* associated with the eigenvalue λ_i. In terms of the so-called *canonical coordinates* $z_1, \ldots, z_n$, the state equation becomes

$$\mathbf{z} = \Lambda \mathbf{z} + \mathbf{E}u$$

or

$$\dot{z}_i = \lambda_i z_i + e_i u, \qquad i = 1, \ldots, n$$

which yields the zero-input response

$$z_i = z_i(0) e^{\lambda_i t}, \qquad i = 1, \ldots, n.$$

Thus, each z_i represents a natural mode of the system. Therefore, if every one of the z_i's is controllable, then all the natural modes are controllable and hence the original system itself is also controllable, since $\mathbf{x}$ and $\mathbf{z}$ are related by a linear non-singular transformation.

For the system (11.2.11) to be controllable, we should be able to find a control function $u(t)$ such that the two equations in (11.2.12) are satisfied for any arbitrary values assigned to $z_1(t_0), z_1(t_1), z_2(t_0), z_2(t_1), t_0$, and t_1. Equivalently, we have to find an input $u(t)$ satisfying the following equations:

$$\beta_1 = \int_{t_0}^{t_1} e^{\lambda_1 \tau} e_1 u(\tau) \, d\tau$$

$$(11.2.13)$$

$$\beta_2 = \int_{t_0}^{t_1} e^{\lambda_2 \tau} e_2 u(\tau) \, d\tau$$

with β_1 and β_2 being arbitrary constants. Let $u(t)$ be a piecewise constant function with unknown constants A and B as shown in Fig. 11.2.3. Then, (11.2.13) can be integrated to yield

$$\frac{\lambda_1 \beta_1}{e_1} = e^{\lambda_1 (t_m - t_0)} A + e^{\lambda_1 (t_1 - t_m)} B$$

$$(11.2.14)$$

$$\frac{\lambda_2 \beta_2}{e_2} = e^{\lambda_2 (t_m - t_0)} A + e^{\lambda_2 (t_1 - t_m)} B$$

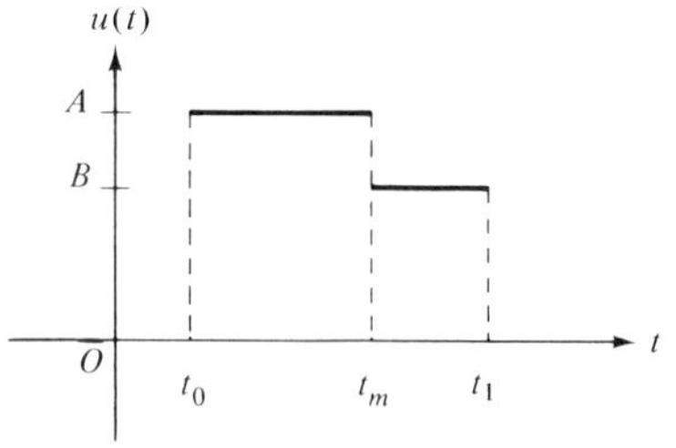

Fig. 11.2.3 A possible choice of $u(t)$ for controlling the system (11.2.11).

from which the constants A and B can be uniquely determined since, for distinct values λ_1 and λ_2, the determinant

$$\begin{vmatrix} e^{\lambda_1(t_m - t_0)} & e^{\lambda_1(t_1 - t_m)} \\ e^{\lambda_2(t_m - t_0)} & e^{\lambda_2(t_1 - t_m)} \end{vmatrix}$$

is not equal to zero. Thus, we have shown that the system (11.2.11) [and hence (11.2.8)] is controllable if both constants e_1 and e_2 are nonzero.

In light of the discussion in the preceding examples, the condition of controllability can be generalized to include an nth-order system regardless of whether or not the eigenvalues are distinct, as stated in the following theorem.

Theorem 11.2.1*   The necessary and sufficient condition for the controllability of a system described by the equation

$$\dot{\mathbf{x}} = \mathbf{A}\mathbf{x} + \mathbf{B}\mathbf{u},$$

with $\mathbf{x}$, $\mathbf{u}$, $\mathbf{A}$, and $\mathbf{B}$ being, respectively, the n-dimensional state vector, the r-dimensional control vector, and the $n \times n$, and $n \times r$ constant matrices, is that the composite $n \times (nr)$ matrix $\mathbf{P}$ defined by

$$\mathbf{P} \triangleq [\mathbf{B} \quad \mathbf{A}\mathbf{B} \quad \cdots \quad \mathbf{A}^{n-1}\mathbf{B}] \tag{11.2.15}$$

be of rank n.

Remark. If the system in Theorem 11.2.1 consists of distinct eigenvalues and if $\mathbf{S}$ is a nonsingular matrix which diagonalizes $\mathbf{A}$, then the linear transformation $\mathbf{x} = \mathbf{S}\mathbf{z}$ will reduce the system to

$$\dot{\mathbf{z}} = \mathbf{S}^{-1}\mathbf{A}\mathbf{S}\mathbf{z} + \mathbf{S}^{-1}\mathbf{B}\mathbf{u}$$

$$= \mathbf{\Lambda}\mathbf{z} + \mathbf{E}\mathbf{u},$$

where $\mathbf{\Lambda} \triangleq \mathbf{S}^{-1}\mathbf{A}\mathbf{S}$, a diagonal matrix with the eigenvalues of $\mathbf{A}$ as its diagonal elements, and $\mathbf{E} \triangleq \mathbf{S}^{-1}\mathbf{B}$. It can be readily shown† that the necessary and sufficient condition for controllability as stated in Theorem 11.2.1 is equivalent to the

---

* For a proof, see, for example, L. A. Zadeh and C. A. Desoer [ZA 1], p. 499.

† For proof, refer to E. G. Gilbert [GI 1], pp. 128–151.

condition that no rows of the matrix $\mathbf{E}$ be zero. We shall now illustrate the application of this theorem by means of a simple example.

Example 11.2.3 Let us investigate the controllability of the system described by

$$
\begin{bmatrix} \dot{x}_1 \\ \dot{x}_2 \end{bmatrix} = \begin{bmatrix} 3 & -\frac{2}{3} \\ 4 & -\frac{1}{3} \end{bmatrix} \begin{bmatrix} x_1 \\ x_2 \end{bmatrix} + \begin{bmatrix} 1 \\ 3 \end{bmatrix} [u]. \tag{11.2.16}
$$

In this system

$$
\mathbf{A} = \begin{bmatrix} 3 & -\frac{2}{3} \\ 4 & -\frac{1}{3} \end{bmatrix}, \qquad \mathbf{B} = \begin{bmatrix} 1 \\ 3 \end{bmatrix},
$$

and the matrix $\mathbf{P}$ is given by

$$
\mathbf{P} = [\mathbf{B} \quad \mathbf{AB}] = \begin{bmatrix} 1 & 1 \\ 3 & 3 \end{bmatrix}
$$

which is of rank 1. Hence the system is not a controllable system.

The fact that the system is noncontrollable can be readily confirmed by subjecting it to a canonical transformation. It can be shown that the eigenvalues of the system are $\lambda_1 = 1$ and $\lambda_2 = \frac{5}{3}$. Thus, using the linear transformation $\mathbf{x} = \mathbf{Sz}$ with

$$
\mathbf{S} = \begin{bmatrix} 1 & 1 \\ 3 & 2 \end{bmatrix},
$$

the system (11.2.16) reduces to $\dot{\mathbf{z}} = \mathbf{S}^{-1}\mathbf{A}\mathbf{S}\mathbf{z} + \mathbf{S}^{-1}\mathbf{B}u = \Lambda\mathbf{z} + \mathbf{S}^{-1}\mathbf{B}u$ or

$$
\begin{bmatrix} \dot{z}_1 \\ \dot{z}_2 \end{bmatrix} = \begin{bmatrix} -2 & 1 \\ 3 & -1 \end{bmatrix} \begin{bmatrix} 3 & -\frac{2}{3} \\ 4 & -\frac{1}{3} \end{bmatrix} \begin{bmatrix} 1 & 1 \\ 3 & 2 \end{bmatrix} \begin{bmatrix} z_1 \\ z_2 \end{bmatrix} + \begin{bmatrix} -2 & 1 \\ 3 & -1 \end{bmatrix} \begin{bmatrix} 1 \\ 3 \end{bmatrix} [u]
$$

$$
= \begin{bmatrix} 1 & 0 \\ 0 & \frac{5}{3} \end{bmatrix} \begin{bmatrix} z_1 \\ z_2 \end{bmatrix} + \begin{bmatrix} 1 \\ 0 \end{bmatrix} [u]
$$

which, in terms of components, yields

$$
\dot{z}_1 = z_1 + u \qquad \text{and} \qquad \dot{z}_2 = \tfrac{5}{3}z_2. \tag{11.2.17}
$$

It is clear from (11.2.17) that the state variable z_2, associated with the eigenvalues $\lambda_2 = \frac{5}{3}$, cannot be controlled by the function u. This fact can be appreciated further by calculating the transfer functions $[x_1(s)]/u(s)$ and $[x_2(s)]/u(s)$. Without loss of generality, let $\mathbf{x}(t_0) = \mathbf{x}(0) = 0$ since the controllability of a system will not be affected by its initial conditions. Then the Laplace transform of the state vector $\mathbf{x}$ is

$$
\mathbf{X}(s) = [s\mathbf{I} - \mathbf{A}]^{-1}\mathbf{B}U(s) \tag{11.2.18}
$$

which, with the numerical values properly substituted, can be written as

$$\mathbf{X}(s) = \frac{\text{Adjoint}\,[s\mathbf{I} - \mathbf{A}]}{\det\,[s\mathbf{I} - \mathbf{A}]}\,\mathbf{B}U(s)$$

$$= \frac{\begin{bmatrix} s - \frac{5}{3} \\ 3(s - \frac{5}{3}) \end{bmatrix} U(s)}{(s - 1)(s - \frac{5}{3})} \tag{11.2.19}$$

and, in terms of state variables, we obtain

$$X_1(s) = \frac{(s - \frac{5}{3})U(s)}{(s - 1)(s - \frac{5}{3})} = \frac{U(s)}{s - 1}$$

$$X_2(s) = \frac{3(s - \frac{5}{3})U(s)}{(s - 1)(s - \frac{5}{3})} = \frac{3U(s)}{s - 1}. \tag{11.2.20}$$

An inspection of (11.2.20) indicates that if there were no cancellation of the factor $(s - \frac{5}{3})$, each of the state variables would, in general, consist of three terms in the t-domain: one associated with the control function u, and the other two involving e^t and $e^{5t/3}$, the natural modes of the system associated with the two eigenvalues $\lambda_1 = 1$ and $\lambda_2 = \frac{5}{3}$, respectively. Furthermore, the complete suppression of the natural mode associated with λ_2 in this example suggests a correspondence between the condition "the rank of P being less than n" and the condition "a cancellation of a common factor in the matrix product $[s\mathbf{I} - \mathbf{A}]^{-1}\mathbf{B}$." (Indeed, it can be readily established that these two conditions are equivalent.) This correspondence can be stated in the form of the following theorem.

Theorem 11.2.2* A necessary and sufficient condition that the system

$$\dot{\mathbf{x}} = \mathbf{A}\mathbf{x} + \mathbf{B}u, \qquad u = \text{scalar} \tag{11.2.21}$$

be controllable is that there be no cancellation in $[s\mathbf{I} - \mathbf{A}]^{-1}\mathbf{B}$.

Linear Discrete-Time Systems

Using the same notations established in Chapter 8, a discrete-time, linear time-invariant system can be represented by the matrix equation of the form

$$\mathbf{x}(k + 1) = \mathbf{A}\mathbf{x}(k) + \mathbf{B}\mathbf{u}(k), \tag{11.2.22}$$

where $\mathbf{x}(k)$, and $\mathbf{u}(k)$ are, respectively, the n-dimensional state and the r-dimensional input (control) vectors at the time instant kT, and $\mathbf{A}$ and $\mathbf{B}$ are constant matrices of proper orders. In addition, let the matrix $\mathbf{A}$ be a nonsingular matrix.

For convenience, the initial time t_0 can be chosen zero. The system (11.2.20) is said to be (completely) *controllable* if a control vector $\mathbf{u}$ can be found such that the state $\mathbf{x}(t)$ be transferred from any given initial state $\mathbf{x}(0)$ to zero for $t \geq kT$, i.e., $\mathbf{x}(t) = 0$ for $t \geq kT$.

---

* A proof of this theorem can be found, for example, in Ogata [OG 1], pp. 389–398.

The discussion on the controllability for continuous systems carries over directly to discrete-time systems as evidenced by referring to the following theorem.

Theorem 11.2.3*   The system (11.2.20) is controllable if and only if the matrix $\mathbf{Q}$ defined by

$$\mathbf{Q} \triangleq [\mathbf{B} \quad \mathbf{A}^{-1}\mathbf{B} \quad \cdots \quad \mathbf{A}^{-n+1}\mathbf{B}] \tag{11.2.23}$$

is of rank n. (Note that since $\mathbf{A}$ is nonsingular the condition just stated is equivalent to requiring that the rank of the matrix $\mathbf{P} \triangleq [\mathbf{B} \quad \mathbf{AB} \quad \cdots \quad \mathbf{A}^{n-1}\mathbf{B}]$ be n.)

11.3 OBSERVABILITY OF LINEAR SYSTEMS

The concepts of controllability and observability are closely related to each other. Controllability means that for any given initial state a suitably chosen input can be found so that the system is brought to any final state within a finite interval of time. An observable system is one in which, based on the measurements of the available output vector over a finite time interval, it is sufficient to determine the initial state $\mathbf{x}(t_0)$. More specifically, we can state the definition of observability as follows.

Definition   A system is said to *observable* at time t_0 if, with the system initially at the state $\mathbf{x}(t_0)$, it is possible to identify this state $\mathbf{x}(t_0)$ by observing the output of a system over a finite time interval $t_0 \leq t \leq t_0 + T$.

We begin our discussion on observability by considering the following second-order system described by

$$\begin{bmatrix} \dot{x}_1 \\ \dot{x}_2 \end{bmatrix} = \begin{bmatrix} 3 & -\frac{2}{3} \\ 4 & -\frac{1}{3} \end{bmatrix} \begin{bmatrix} x_1 \\ x_2 \end{bmatrix} + \begin{bmatrix} 1 \\ 3 \end{bmatrix} [u] \tag{11.3.1}$$

with the output vector $\mathbf{y}$ expressed in terms of $\mathbf{x}$ as

$$\begin{bmatrix} y_1 \\ y_2 \end{bmatrix} = \begin{bmatrix} 1 & -\frac{1}{3} \\ 3 & -1 \end{bmatrix} \begin{bmatrix} x_1 \\ x_2 \end{bmatrix}, \tag{11.3.2}$$

where the system (11.3.1) is the same as that used in Example 11.2.3. Using the linear transformation $\mathbf{x} = \mathbf{Sz}$ with

$$\mathbf{S} = \begin{bmatrix} 1 & 1 \\ 3 & 2 \end{bmatrix},$$

we found that the state equation in terms of the variables z_1 and z_2 was given by

$$\begin{bmatrix} \dot{z}_1 \\ \dot{z}_2 \end{bmatrix} = \begin{bmatrix} 1 & 0 \\ 0 & \frac{5}{3} \end{bmatrix} \begin{bmatrix} z_1 \\ z_2 \end{bmatrix} + \begin{bmatrix} 1 \\ 0 \end{bmatrix} [u].$$

---

* A proof of this theorem can be found, for example, in Ogata [OG 1], pp. 383–384.

Substituting $\mathbf{x} = \mathbf{Sz}$ into (11.3.2), we find, accordingly, that

$$\begin{bmatrix} y_1 \\ y_2 \end{bmatrix} = \begin{bmatrix} 1 & -\frac{1}{3} \\ 3 & -1 \end{bmatrix} \begin{bmatrix} 1 & 1 \\ 3 & 2 \end{bmatrix} \begin{bmatrix} z_1 \\ z_2 \end{bmatrix}$$

$$= \begin{bmatrix} 0 & \frac{1}{3} \\ 0 & 1 \end{bmatrix} \begin{bmatrix} z_1 \\ z_2 \end{bmatrix}$$

or

$$y_1 = \tfrac{1}{3} z_2$$
$$y_2 = z_2. \tag{11.3.3}$$

Obviously the system under consideration is not observable since the variable z_1 is missing in the above system (11.3.3) and hence it cannot be ascertained from the measurements of the available output.

The result of the above simple problem suggests that if no component of $\mathbf{z}$ is missing in the expression for the output vector, then the system is an observable system. Indeed, this is the necessary and sufficient condition for observability. Thus, let us examine the general system described by

$$\dot{\mathbf{x}} = \mathbf{Ax} + \mathbf{Bu},$$
$$\mathbf{y} = \mathbf{Cx}. \tag{11.3.4}$$

If the eigenvalues of this system are distinct, then a linear transformation $\mathbf{x} = \mathbf{Sz}$ (with $\mathbf{S}$ being a nonsingular matrix which diagonalizes $\mathbf{A}$) will lead to the (matrix) differential equation for the transformed vector $\mathbf{z}$

$$\dot{\mathbf{z}} = \mathbf{S}^{-1}\mathbf{ASz} + \mathbf{S}^{-1}\mathbf{Bu}$$
$$= \mathbf{\Lambda z} + \mathbf{Eu} \tag{11.3.5}$$

and

$$\mathbf{y} = \mathbf{CSz} = \mathbf{Fz},$$

where $\mathbf{\Lambda} \triangleq \mathbf{S}^{-1}\mathbf{AS}$, $\mathbf{E} \triangleq \mathbf{S}^{-1}\mathbf{B}$, and $\mathbf{F} \triangleq \mathbf{CS}$. It can be shown* that *if no column of the matrix $\mathbf{F}$ is zero, then the system* (11.3.4) *(with distinct eigenvalues) is observable.*

The necessary and sufficient condition for observability of a system such that its eigenvalues are not necessarily distinct can be stated in the following theorems.

Theorem 11.3.1 The linear time-invariant continuous system

$$\dot{\mathbf{x}} = \mathbf{Ax} + \mathbf{Bu}$$
$$\mathbf{y} = \mathbf{Cx},$$

where $\mathbf{x}$, $\mathbf{u}$, and $\mathbf{y}$ are the state, the control, and the output vectors of dimensions,

---

* For proof, see E. G. Gilbert [GI 1].

n, r, and m, respectively, is observable if and only if the $n \times (nm)$-matrix $\mathbf{P}$,*

$$\mathbf{P} \triangleq [\mathbf{C}^T \quad \mathbf{A}^T\mathbf{C}^T \quad \cdots \quad (\mathbf{A}^T)^{n-1}\mathbf{C}^T] \tag{11.3.6}$$

has the rank n.

Theorem 11.3.2 The linear discrete-time system described by

$$\mathbf{x}(k+1) = \mathbf{A}\mathbf{x}(k) + \mathbf{B}\mathbf{u}(k)$$

and

$$\mathbf{y}(k) = \mathbf{C}\mathbf{x}(k)$$

where $\mathbf{x}(k)$ is the state vector, $\mathbf{u}(k)$ the control vector, and $\mathbf{y}(k)$ the output vector at the kth time instant (i.e., $t = kT$) with dimensions n, r, and m, respectively, and where $\mathbf{A}, \mathbf{B}$, and $\mathbf{C}$ are constant matrices with $\mathbf{A}$ being a nonsingular matrix, is observable if and only if the $n \times (nm)$-matrix $\mathbf{P}$,†

$$\mathbf{P} = [\mathbf{C}^T \quad \mathbf{A}^T\mathbf{C}^T \quad \cdots \quad (\mathbf{A}^T)^{n-1}\mathbf{C}^T] \tag{11.3.7}$$

has rank n.

Example 11.3.1 Consider the system

$$\begin{bmatrix} \dot{x}_1 \\ \dot{x}_2 \end{bmatrix} = \begin{bmatrix} 0 & 2 \\ 1 & 1 \end{bmatrix} \begin{bmatrix} x_1 \\ x_2 \end{bmatrix} + \begin{bmatrix} 0 \\ 1 \end{bmatrix} u \tag{11.3.8}$$

$$y = x_2.$$

To determine whether the system is observable, let us rewrite the second equation of (11.3.8) as

$$y = [0 \quad 1] \begin{bmatrix} x_1 \\ x_2 \end{bmatrix}.$$

Thus, $\mathbf{C} = [0 \quad 1]$, and

$$\mathbf{A}^T\mathbf{C}^T = \begin{bmatrix} 0 & 1 \\ 2 & 1 \end{bmatrix} \begin{bmatrix} 0 \\ 1 \end{bmatrix} = \begin{bmatrix} 1 \\ 1 \end{bmatrix}.$$

Hence the matrix

$$\mathbf{P} = [\mathbf{C}^T \quad \mathbf{A}^T\mathbf{C}^T]$$

$$= \begin{bmatrix} 0 & 1 \\ 1 & 1 \end{bmatrix}$$

is of rank 2, so that the system is observable.

---

* The condition (11.3.6) is based on the assumption that the elements of $\mathbf{A}$ and $\mathbf{C}$ are real; otherwise (11.3.6) should be replaced by $\mathbf{P} = [\mathbf{C}^* \quad \mathbf{A}^*\mathbf{C}^* \quad \cdots \quad (\mathbf{A}^*)^{n-1}\mathbf{C}^*]$ where $\mathbf{C}^*$ is the conjugate transpose of $\mathbf{C}$, etc. For proof of theorem see K. Ogata [OG 1], Chapter 7.

† See footnote for Theorem 11.3.1 for assumptions on the elements of $\mathbf{A}$ and $\mathbf{C}$.

Example 11.3.2   As another example, consider again the system (11.3.8) of which the eigenvalues are $\lambda_1 = -1$ and $\lambda_2 = 2$. Thus using the linear transformation

$$\begin{bmatrix} x_1 \\ x_2 \end{bmatrix} = \begin{bmatrix} 1 & 1 \\ -\frac{1}{2} & 1 \end{bmatrix} \begin{bmatrix} z_1 \\ z_2 \end{bmatrix},$$

we find that the system (11.3.8) reduces to

$$\begin{bmatrix} \dot{z}_1 \\ \dot{z}_2 \end{bmatrix} = \begin{bmatrix} \frac{2}{3} & -\frac{2}{3} \\ \frac{1}{3} & \frac{2}{3} \end{bmatrix} \begin{bmatrix} 0 & 2 \\ 1 & 1 \end{bmatrix} \begin{bmatrix} 1 & 1 \\ -\frac{1}{2} & 1 \end{bmatrix} \begin{bmatrix} z_1 \\ z_2 \end{bmatrix} + \begin{bmatrix} \frac{2}{3} & -\frac{2}{3} \\ \frac{1}{3} & \frac{2}{3} \end{bmatrix} \begin{bmatrix} 0 \\ 1 \end{bmatrix} u$$

$$= \begin{bmatrix} -1 & 0 \\ 0 & 2 \end{bmatrix} \begin{bmatrix} z_1 \\ z_2 \end{bmatrix} + \begin{bmatrix} -\frac{2}{3} \\ \frac{2}{3} \end{bmatrix} u$$

and

$$y = \begin{bmatrix} 0 & 1 \end{bmatrix} \begin{bmatrix} 1 & 1 \\ -\frac{1}{2} & 1 \end{bmatrix} \begin{bmatrix} z_1 \\ z_2 \end{bmatrix}.$$

$$= \begin{bmatrix} -\frac{1}{2} & 1 \end{bmatrix} \begin{bmatrix} z_1 \\ z_2 \end{bmatrix}$$

from which we obtain $\mathbf{F} = \mathbf{CS} = \begin{bmatrix} -\frac{1}{2} & 1 \end{bmatrix}$.   Since $\mathbf{F}$ has no zero column, the system under consideration is observable.

11.4 BASIC CONCEPTS OF STABILITY

As discussed in the preceding sections, the analysis or design of a dynamic system necessitates the establishment of a mathematical model which, in the case of a continuous system, is usually in the form of differential equations.   Once the differential equations for the motion of a physical system have been set up, it is necessary to obtain the solution by integration for a given set of initial conditions. And then, since all physical systems are constantly exposed to various kinds of disturbances, the problem of stability of motion is of great importance in the design of a system. As a matter of fact, control systems must be designed so that they are always stable.

  In the present section, we shall introduce some of the well-known concepts of stability and their implications. The general stability requirements, as well as the effects of inaccuracies of components on the behavior of a system, will be discussed in the later sections.

  The stability of a linear system can be defined in terms of its input and output quantities.   In words, a linear system is said to be *stable* if for any bounded input the output is bounded. This stability concept, based on the input–output relationship, is usually referred to as *stability in the sense of b. i. b. o.* (bounded input implies bounded output).

Fig. **11.4.1** A single-input–single-output system.

Let us observe the physical implication of the above definition by considering a single-input–single-output system as depicted in Fig. 11.4.1, where $e_i(t)$ and $r_o(t)$ are, respectively, the input and the output of the system.

Without loss of generality, let the system be initially in zero state. Then, the response $r_o(t)$ is given by (see Chapter 5, Section 5.2):*

$$r_o(t) = \int_0^\infty h(\tau)e_i(t - \tau)\,d\tau, \qquad (11.4.1)$$

where $h(t)$ is the impulse response of the system. If the system is stable b. i. b. o., then r_o is bounded for every bounded e_i; that is, if $|e_i(t)| \le M_1 < \infty$, then $|r_o(t)| \le M_2 < \infty$. Hence, (11.4.1) can be written as

$$|r_o(t)| = \left| \int_0^\infty h(\tau)e_i(t - \tau)\,d\tau \right|$$

$$\le \int_0^\infty |h(\tau)e_i(t - \tau)|\,d\tau$$

$$\le M_1 \int_0^\infty |h(\tau)|\,d\tau$$

$$\le M_1 M_3 = M_2, \qquad (11.4.2)$$

where $\int_0^\infty |h(\tau)|\,d\tau \triangleq M_3$ and $M_1 M_3 \triangleq M_2$. Thus, it is evident that the system is stable b. i. b. o. if the impulse response $h(t)$ is absolutely integrable, i.e.,

$$\int_0^\infty |h(\tau)|\,d\tau = M_3 < \infty. \qquad (11.4.3)$$

Next, we shall investigate how the requirement on the impulse response for stability can be related to the restrictions on the poles of the transfer function $H(s) \triangleq \mathscr{L}\{h(t)\}$ as discussed in Chapter 5 (Section 5.4).

In terms of the impulse response, we write

$$H(s) = \int_0^\infty h(t)e^{-st}\,dt. \qquad (11.4.4)$$

---

* In (11.4.1), the upper limit ∞ instead of t is used. This change of limit is correct since $e_i(t) = 0$ for $t < 0$ is assumed.

Taking the absolute value on both sides of (11.4.4) yields

$$|H(s)| = \left| \int_0^\infty h(t)e^{-st}\, dt \right| \leq \int_0^\infty |h(t)||e^{-\sigma t}||e^{-j\omega t}|\, dt \qquad (11.4.5)$$

or

$$|H(s)| \leq \int_0^\infty |h(t)|e^{-\sigma t}\, dt,$$

where σ is the real part of s; i.e., $s \triangleq \sigma + j\omega$.

Let $s_i = \sigma_i + j\omega_i$ be a pole of $H(s)$ which does not lie on the left-hand side of the s-plane. Then, with $s = s_i$, (11.4.5) reduces to

$$\infty \leq \int_0^\infty |h(t)|e^{-\sigma_i t}\, dt$$

$$\leq \int_0^\infty |h(t)|\, dt, \qquad (11.4.6)$$

since $\sigma_i \geq 0$ and $e^{-\sigma_i t} \leq 1$. Comparing the two expressions (11.4.3) and (11.4.6), we arrive at the very interesting conclusion; that is, if one or more poles of $H(s)$ are in the right half or on the $j\omega$-axis of the s-plane, the requirement on the impulse response is violated and hence the system is unstable. Therefore, *the poles of the transfer function must have negative real parts if the system is to be stable b. i. b. o.*

The bounded-input–bounded-output stability concept can be applied equally well to a general multiple-input–multiple-output linear system described by the vector differential equation

$$\dot{\mathbf{x}} = \mathbf{A}\mathbf{x} + \mathbf{B}\mathbf{u}, \qquad (11.4.7)$$

where, as before, $\mathbf{x}$ and $\mathbf{u}$ are, respectively, the n-dimensional ouput and the r-dimensional input vectors, and where $\mathbf{A}$ and $\mathbf{B}$ are matrices of appropriate orders. In this case, the system is said to be stable b.i.b.o. if, for every bounded input $\mathbf{u}$ i.e., the norm* of $\mathbf{u}$ is bounded ($\|\mathbf{u}\| < M_1 < \infty$), the output $\mathbf{x}$ is also bounded, i.e., $\|\mathbf{x}\| < M_2 < \infty$. In other words, the boundedness of a vector is expressed in terms of its norm (instead of the absolute value for a scalar quantity). It can be shown by using the same procedure as in the scalar case that the stability of the system is again independent of the input.†

In the absence of an input, a linear system can be represented by (11.4.7) with $\mathbf{u}$ set to zero, or

$$\dot{\mathbf{x}} = \mathbf{A}\mathbf{x}, \qquad (11.4.8)$$

where the state vector $\mathbf{x}$ is again considered as the output of the force-free system.

---

* Refer to Appendix A.1 for definition of norm.
† See Problem 11.4.

The system (11.4.8) is said to be stable if for any well-defined initial state $x(0) = x_0$, the solution $x(t)$ is bounded (in norm) for all $t > 0$. The stability just defined is usually referred to as *bounded stability*. The system (11.4.8) is said to be *asymptotically stable* if it is (bounded) stable and if every solution $x(t)$ tends toward 0 as $t \to \infty$. The last definition of stability is referred to as *asymptotic stability*.

The conditions for stability and asymptotic stability can be stated in the form of a theorem.

Theorem 11.4.1   Let $\Phi(t)$ be the state-transition matrix of the system (11.4.8). Then the system (11.4.8) is stable if and only if the state-transition matrix $\Phi(t)$ is bounded for all $t > 0$; that is $\|\Phi(t)\| \leq M < \infty$. The system is asymptotically stable if and only if

$$\lim_{t \to \infty} \|\Phi(t)\| = 0.$$

The proof of the theorem is very straightforward and is left as an exercise.*

Remark. It is evident from the above discussion that asymptotic stability always implies bounded stability and, of course, the converse is not true. Furthermore, it can be shown that asymptotic stability also implies stability b. i. b. o.†

11.5 STABILITY ANALYSIS

In this section we shall briefly discuss several well-known stability criteria for linear time-invariant systems. Stability analysis of linear time-invariant systems is important enough to merit its own discussion because the general concept of stability may be understood more easily as an extension of the linear case. Also the local stability of a certain class of nonlinear systems depends on whether their linearized counterparts are stable. In addition, many physical systems can be accurately approximated by linear time-invariant mathematical models.

As stated in Problem 11.6, if all the eigenvalues of matrix A have negative real parts, then the system (11.4.8) is asymptotically stable at the origin $x = 0$. The eigenvalues of A are the roots of the characteristic equation

$$|A - \lambda I| = 0. \tag{11.5.1}$$

The determination of the roots of an algebraic equation can become very cumbersome when the degree of the equation is large. Fortunately, the asymptotic stability of the system does not depend upon the knowledge of the exact locations of the roots of the characteristic equation. Instead, we only need to test if all of the eigenvalues of A have negative real parts. The *Routh-Hurwitz criterion* that we shall discuss next will enable us to perform such a test.

---

* See Problem 11.5.
† See Problems 11.6 and 11.7.

Routh-Hurwitz Criterion*

Let the characteristic equation of the system (11.4.8) be given by

$$a_n s^n + a_{n-1} s^{n-1} + \cdots + a_1 s + a_0 = 0, \qquad (11.5.2)$$

where the coefficients a_i, $i = 0, 1, \ldots, n$, are real numbers. The Routh-Hurwitz criterion determines the number of roots of (11.5.2) that have either positive real parts or zero real parts.

We shall assume that all the coefficients a_i are positive since this assumption is a necessary (but not sufficient) condition for the negativeness of the real parts of the roots of an algebraic equation.†

Once the above assumption is satisfied, the coefficients of the characteristic equation are arranged in the pattern shown in the following array

s^n	a_n	a_{n-2}	$a_{n-4} \cdots$
s^{n-1}	a_{n-1}	a_{n-3}	$a_{n-5} \cdots$
s^{n-2}	b_{n-1}	b_{n-3}	$b_{n-5} \cdots$
s^{n-3}	c_{n-1}	c_{n-3}	$c_{n-5} \cdots$
$\vdots$	$\vdots$	$\vdots$	$\vdots$

where the terms $s^n, s^{n-1}, \ldots, s^1, s^0$, inserted to the left of the vertical line, will be necessary only when the given characteristic equation has at least one pair of roots that are negative to each other. The constants b_i are evaluated by means of the formulas

$$b_{n-1} = -\frac{1}{a_{n-1}} \begin{vmatrix} a_n & a_{n-2} \\ a_{n-1} & a_{n-3} \end{vmatrix} \qquad (11.5.3)$$

$$b_{n-3} = -\frac{1}{a_{n-1}} \begin{vmatrix} a_n & a_{n-4} \\ a_{n-1} & a_{n-5} \end{vmatrix}, \qquad (11.5.4)$$

$$b_{n-5} = -\frac{1}{a_{n-1}} \begin{vmatrix} a_n & a_{n-6} \\ a_{n-1} & a_{n-7} \end{vmatrix}, \qquad (11.5.5)$$

etc., and, in like manner, the c_i constants are determined by using the elements of

---

* In 1877, Routh developed the method for determining the number of roots not having negative real parts. In 1895, Hurwitz independently obtained essentially the same method in a different form.

† For example, see E. A. Guillemin [GU 2], pp. 395–396.

the two previous rows; that is,

$$c_{n-1} = -\frac{1}{b_{n-1}} \begin{vmatrix} a_{n-1} & a_{n-3} \\ b_{n-1} & b_{n-3} \end{vmatrix}, \tag{11.5.6}$$

$$c_{n-3} = -\frac{1}{b_{n-1}} \begin{vmatrix} a_{n-1} & a_{n-5} \\ b_{n-1} & b_{n-5} \end{vmatrix}, \tag{11.5.7}$$

$$c_{n-5} = -\frac{1}{b_{n-1}} \begin{vmatrix} a_{n-1} & a_{n-7} \\ b_{n-1} & b_{n-7} \end{vmatrix}, \tag{11.5.8}$$

and so forth. This pattern of determining the elements of a row by using the elements of the two previous rows as indicated in Eqs. (11.5.3) through (11.5.8) is continued until a row of zeros is obtained. Normally, for a characteristic equation of degree n, a total of $n + 1$ rows results. Then the process is terminated. *The Routh-Hurwitz criterion states that the system* (11.4.8) *is asymptotically stable if and only if all the elements in the first column of the array* (i.e., a_n, a_{n-1}, b_{n-1}, ...) *are positive. Otherwise, the number of roots of the characteristic equation with positive real parts is equal to the number of changes of sign of the elements in the same column.*

In the above procedure, we encounter difficulty if one of the elements in the first column of the array is zero (since a division by zero will result). This difficulty is usually avoided by multiplying the characteristic equation by a factor $(s + a)$ with $a > 0$, except when the given equation has (at least) one pair of roots negative to each other. The presence of such a pair of roots, $\pm (\sigma_a + j\omega_a)$, of course, amounts to the fact that the system cannot be asymptotically stable. The following examples will illustrate the ideas involved.

Example 11.5.1 Let us apply the Routh-Hurwitz criterion to determine whether the third-order system having the characteristic equation

$$s^4 + 6s^3 + 12s^2 + 10s + 3 = 0 \tag{11.5.9}$$

is asymptotically stable. In light of (11.5.3) through (11.5.8)* we can establish the following array

s^4	1	12	3
s^3	6	10	
s^2	10.33	3	
s^1	8.26		
s^0	3		

Since all the coefficients in the first column are positive, the system under consideration is asymptotically stable.

---

* In calculating the constant b's and c's, if an element is not available in a row, a value of zero is used as in the case for determining the constant 3 in the third row.

Example 11.5.2 Consider next the system with the characteristic equation given by

$$s^4 + 5s^3 + 3s^2 + 15s + 1 = 0. \tag{11.5.10}$$

Again we arrange the array as follows

s^4	1	3	1
s^3	$\cancel{5}^1$	$\cancel{15}^3$	
s^2	0	1	
s^1			
s^0			

(The result remains unchanged if all the elements of one row are multiplied by the same positive number.)

The process cannot be continued since a zero appears in the first column of the third row. To avoid this difficulty, let us multiply our original equation (11.5.10) by a factor, say, $(s + 1)$, since this multiplication will not modify the locations of the roots of the characteristic equation. Thus, we obtain the following equation

$$s^5 + 6s^4 + 8s^3 + 18s^2 + 16s + 1 = 0, \tag{11.5.11}$$

and repeat the process of establishing the array. As a result, we find

s^5	1	8	16
s^4	6	18	1
s^3	5	15.83	
s^2	-1	1	
s^1	20.83		
s^0	1		

There are two changes of sign in this case. Therefore, the system under consideration is unstable.

Example 11.5.3 Again, let us apply the Routh-Hurwitz criterion to check the asymptotic stability of a system with the characteristic equation given by

$$s^5 + 4s^4 + 6s^3 + 6s^2 + 5s + 2 = 0. \tag{11.5.12}$$

Arranging the coefficients, we obtain

s^5	1	6	5	
s^4	$\cancel{4}^2$	$\cancel{6}^3$	$\cancel{2}^1$	(After division by 2)
s^3	$\cancel{4.5}^1$	$\cancel{4.5}^1$		(After division by 4.5)
s^2	1	1		
s^1	0			
s^0				

With a zero in the first column in the array, we modify, as in Example 11.5.4, the given equation by multiplying it by $(s + 1)$, yielding

$$s^6 + 5s^5 + 10s^4 + 12s^3 + 11s^2 + 7s + 2 = 0, \qquad (11.5.13)$$

and test if the zero in the first column will disappear.

Repeating the same process, we obtain

s^6	1	10	11	2
s^5	5	12	7	
s^4	7.6	9.6	2	
s^3	5.69^1	5.69^1	(After division by 5.69)	
s^2	2	2	($F(s) = 2s^2 + 2$)	
s^1	0			
s^0				

Again, we face the same problem as before; that is, a zero appears in the first column. In general, if a row of zeros results in the array development, it indicates that the original characteristic equation contains at least a pair of roots which are negative to each other. Such a situation arises when the corresponding elements of two preceding rows differ by a common multiple.

To remedy this situation, we first form the *auxiliary polynomial*

$$F(s) \triangleq 2s^2 + 2,$$

where the coefficients and the degree of $F(s)$ are obtained from the elements of the row indicated by s^2 on the left-hand side of the vertical line as shown. Next, differentiate $F(s)$, yielding

$$\frac{dF(s)}{ds} = 4s, \qquad (11.5.14)$$

and replace the zero by the coefficient 4 of the polynomial $dF(s)/ds$. The process is then continued until the last element associated with the term s^0 is obtained. Thus, the complete array is given by

s^6	1	10	11	2
s^5	5	12	7	
s^4	7.6	9.6	2	
s^3	5.69^1	5.69^1		
s^2	2	2		
s^1	4	(After the zero is replaced by		
s^0	2	the coefficient of dF/ds)		

It is seen that there is no change in sign in the first column of the new array indicating that the original equation (11.5.12) does not have any root with positive real part. A by-product of the above procedure is that the pair of roots that are negative to each other can be easily obtained by factoring the auxiliary polynomial. In our example, we find that

$$F(s) = 2(s + j)(s - j),$$

yielding a pair of roots of the original equation; that is,

$$s = \pm j,$$

which are purely imaginary numbers. In fact, (11.5.12) can be expressed in factored form as

$$(s + 1)^2(s + 2)(s + j)(s - j) = 0. \qquad (11.5.15)$$

With a pair of characteristic roots lying on the $j\omega$-axis, the system is not asymptotically stable.

The procedure presented in Example 11.5.3 can now be summarized for the general case as follows. If a row of zeros occurs in the row identified by, say, s^{k-1}, the elements of the preceding row will be examined. Let these elements be, from left to right, $d_{n-1}, d_{n-3}, \ldots$. Form the auxiliary polynomial $F(s)$:

$$F(s) = d_{n-1}s^k + d_{n-3}s^{k-2} + \cdots + d_1 s \qquad \text{(for k odd)}$$

or

$$F(s) = d_{n-1}s^k + d_{n-3}s^{k-2} + \cdots + d_0 \qquad \text{(for k even)}.$$

Next, differentiate $F(s)$, yielding

$$\frac{dF(s)}{ds} = kd_{n-1}s^{k-1} + (k - 2)d_{n-3}s^{k-3} + \cdots,$$

and then replace the row of zeros by the coefficients of $dF(s)/ds$ with, from left to right, the first element being kd_{n-1}, the second element being $(k - 2)d_{n-3}$, etc. This process of array development is continued until the last element is computed.

The roots that are negative to each other can then be obtained by factoring the auxiliary polynomial $F(s)$.

In the next few paragraphs we shall present another important stability criterion which is based on a graphical approach.

Nyquist Criterion

Consider a typical single-input–single-output feedback system as shown in Fig. 11.5.1, where $G(s) \triangleq R_o(s)/E(s)$ is the transfer function of the dynamic unit (the system to be controlled) and $H(s) \triangleq Y(s)/R_o(s)$ is the transfer function of the

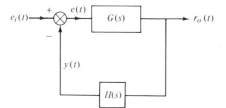

Fig. 11.5.1 A typical feedback system.

feedback element. Referring to Fig. 11.5.1, the following relations can be readily established:

$$E(s) = E_i(s) - Y(s), \qquad (11.5.16)$$

$$Y(s) = H(s)R_o(s), \qquad (11.5.17)$$

$$R_o(s) = G(s)E(s). \qquad (11.5.18)$$

Upon elimination of both $E(s)$ and $Y(s)$ in (11.5.16) through (11.5.18), the so-called *closed-loop transfer function* $R_o(s)/E_i(s)$ can be readily determined as

$$\frac{R_o(s)}{E_i(s)} = \frac{G(s)}{1 + G(s)H(s)}. \qquad (11.5.19)$$

The asymptotic stability of the system is assured if none of the roots of the characteristic equation

$$1 + G(s)H(s) = 0 \qquad (11.5.20)$$

lies in the right-half or on the $j\omega$-axis of the s-plane.

The Nyquist criterion is a criterion for determining the stability of the closed-loop system depicted in Fig. 11.5.1 by examining the frequency response of the *open-loop transfer function*

$$G(j\omega)H(j\omega) = G(s)H(s)\big|_{s=j\omega}.$$

The product function $G(s)H(s)$ represents the transfer function around the loop from $E(s)$ to $Y(s)$. In any physical system, the transmission from $E(s)$ to $Y(s)$ at high frequencies is limited by the presence of losses, series inductive effects, as well as shunt capacitive effects. Therefore physical limitations demand that *the degree of the numerator polynomial is not greater than that of the denominator polynomial of the rational function $G(s)H(s)$*, which is also the condition necessary for the validity of the Nyquist criterion.

The procedure for applying the Nyquist stability criterion can be performed in the following steps:

1. Make a polar plot of the open-loop function $G(j\omega)H(j\omega)$ for all the values of ω from $-\infty$ to $+\infty$. Note that the plot for the negative frequencies is the conjugate of the plot drawn for positive frequencies. An example of a simple Nyquist diagram is depicted in Fig. 11.5.2.

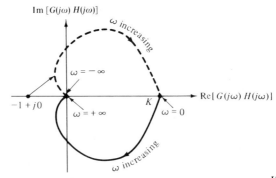

Fig. 11.5.2 A simple Nyquist diagram with $G(s)H(s) = \dfrac{K}{(1 + sT_1)(1 + sT_2)}$.

2. Draw a vector from the point $-1 + j0$ to the plot and observe how this vector rotates as ω increases from $-\infty$ to $+\infty$. Let N_r be the number of times this vector rotates about the point $-1 + j0$ in *counterclockwise* direction.

3. Determine the number of poles of $G(j\omega)H(j\omega)$ whose real parts are positive. Let this number be N_p.

The Nyquist stability criterion can be stated as follows: *The feedback system of Fig. 11.5.1 is asymptotically stable if (a) the locus of $G(j\omega)H(j\omega)$ does not pass through the point of $-1 + j0$ and (b) if the number of counterclockwise encirclements about $-1 + j0$ is equal to the number of poles of $G(s)H(s)$ with positive real parts, that is, $N_r = N_p$.*

The requirements (a) and (b) ensure that the polynomial $1 + G(s)H(s)$ in (11.5.20) will not have, respectively, purely imaginary roots and roots with positive real parts. The following examples illustrate how the stability criterion can be applied.

Example 11.5.4 Consider the open-loop transfer function

$$G(s)H(s) = \frac{K}{(1 + sT_1)(1 + sT_2)} \tag{11.5.21}$$

from which it can be shown that when $\omega = 0$,

(a) $$G(j\omega)H(j\omega) = K, \tag{11.5.22}$$

and when $\omega \to \infty$,

(b) $$G(j\omega)H(j\omega) \to 0 \,\underline{/-180°}. \tag{11.5.22}$$

The polar plot of $G(j\omega)H(j\omega)$ is shown in Fig. 11.5.2 from which we find

$$N_r = 0.$$

_____

* For a detailed discussion, refer to, for example, Chestnut and Mayer [CH 3], Chapter 6.

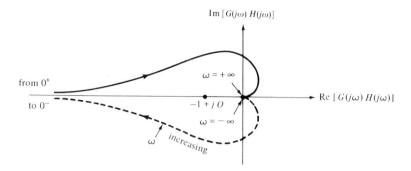

Fig. 11.5.3   Polar plot for (11.5.23).

Since $G(s)H(s)$ has no poles with positive real parts (that is; $N_p = 0$), both conditions (a) and (b) of the Nyquist criterion are satisfied and hence the system is asymptotically stable.

Example 11.5.5   The construction of the polar plot for the transfer function

$$G(s)H(s) = \frac{K}{s^2(1 + sT_1)(1 + sT_2)} \qquad (11.5.23)$$

can be facilitated by first noting that as $\omega \to 0^+$,

$$G(j\omega)H(j\omega) \to \infty \, \underline{/-180°},$$

and as $\omega \to \infty$,

$$G(j\omega)H(j\omega) \to 0 \, \underline{/-360°}.$$

The polar plot is sketched in Fig. 11.5.3, from which we obtain

$$N_r = -1,$$

one revolution in the clockwise direction as ω varies from $-\infty$ to $+\infty$. Therefore, the system is unstable.

    This section constitutes a brief account of some of the well-known stability criteria. The first, the Routh-Hurwitz criterion, is an analytical method, while the second, the Nyquist criterion, is based on a graphical approach. Another commonly used graphical technique is the *root-locus method* which investigates the variation of locations of the poles of a closed-loop transfer function in the s-plane under the influence of changing parameter values. It is used mainly for synthesis purposes. Due to space limitation, this method as well as other interesting ones will not be treated here. The interested reader is referred to literature on automatic control systems.

11.6 SENSITIVITY ANALYSIS

In addition to the problem of stability, another problem of practical importance in the design of a dynamic system is the knowledge of the influence on the system's performance due to variations in system parameters. This problem is known as the sensitivity problem. The extreme importance of this sensitivity problem results from the fact that the actual physical system and the mathematical model which the designer analyzes or specifies are not identical. Furthermore, the components of a system may change in value as a result of temperature variation, aging, manufacturing tolerances, as well as many other factors. These changes in parameter values will cause the actual behavior of a given physical system to be different from that of an idealized one.

The definition of sensitivity function was first introduced in feedback systems by Bode.* His definition of *sensitivity function* S_p^T, denoting the sensitivity of the system transfer function T to the variation of a parameter P, is defined as

$$S_p^T = \frac{dT/T}{dP/P}.$$

For linear time-varying and nonlinear systems, the definition given above cannot be applied to the evaluation of system deterioration.

Due to space limitation, we shall present, in this section, a brief discussion on two important types of sensitivity functions—sensitivity of the response of a system in the time-domain and that of a transfer function—as a result of variation of system parameters.

Sensitivity Function of the System's Response in *t*-domain

Consider the linear time-invariant system of Fig. 11.6.1. Let α be a parameter of the system with a nominal value α_0; that is, $\alpha = \alpha_0$. Then $r_o(t, \alpha_0)$ represents the nominal response of the system. If the actual value of α in the physical system differs from its nominal value α_0, the actual response $r_o(t,\alpha)$ can be expanded into a Taylor series:

$$r_o(t, \alpha) = r_o(t, \alpha_0) + \left[\frac{\partial r_o(t, \alpha)}{\partial \alpha}\right]_{\alpha = \alpha_0} (\alpha - \alpha_0) + \frac{1}{2}\left[\frac{\partial^2 r_o(t, \alpha)}{\partial \alpha^2}\right]_{\alpha = \alpha_0} (\alpha - \alpha_0)^2 + \cdots.$$

$$(11.6.1)$$

For small variation of α about its nominal value α_0, the change in system performance, designated as $\delta r_o(t, \alpha)$, is predominated by the linear term of the Taylor series; that is,

$$\delta r_o(t, \alpha) \triangleq r_o(t, \alpha) - r_o(t, \alpha_0) \cong \left[\frac{\partial r_o(t, \alpha)}{\partial \alpha}\right]_{\alpha = \alpha_0} \delta\alpha, \qquad (11.6.2)$$

where $\delta\alpha \triangleq \alpha - \alpha_0$. The coefficient $\partial r_o(t, \alpha)/\partial \alpha$ is called the *sensitivity function*.

---

* H. Bode [BO 1].

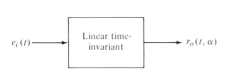

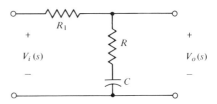

Fig. 11.6.1 A linear time-invariant system
with parameter α.

Fig. 11.6.2 A compensating network.

If the transfer function of the system is denoted by $H(s, \alpha)$, then we have

$$R_o(s, \alpha) = H(s, \alpha)E_i(s), \qquad (11.6.3)$$

where $R_o(s, \alpha)$ and $E_i(s)$ are, respectively, the Laplace transforms of $r_o(t, \alpha)$ and $e_i(t)$, and the response $r_o(t, \alpha)$ is given by

$$r_o(t, \alpha) = \mathscr{L}^{-1}\{H(s, \alpha)E_i(s)\}. \qquad (11.6.4)$$

Differentiating both sides of (11.6.4) with respect to α yields

$$\left[\frac{\partial r_o(t, \alpha)}{\partial \alpha}\right]_{\alpha = \alpha_0} = \left[\frac{\partial}{\partial \alpha}\mathscr{L}^{-1}\{H(s, \alpha)E_i(s)\}\right]_{\alpha = \alpha_0}$$

$$= \mathscr{L}^{-1}\left\{\left[\frac{\partial H(s, \alpha)}{\partial \alpha}\right]_{\alpha = \alpha_0} E_i(s)\right\}. \qquad (11.6.5)$$

In the last expression of (11.6.5), the order of differentiation and integration has been interchanged since the inverse Laplace transform is governed by the inversion integral (see Appendix A.2). This interchange of differentiation and integration is always justified provided that the parameter variation does not alter the degree of the denominator polynomial of the system transfer function $H(s, \alpha)$.* In like manner, higher order partial derivatives of $r_o(t, \alpha)$ can be obtained by differentiating both sides of (11.6.5) repeatedly with respect to α, or

$$\left[\frac{\partial^n r_o(t, \alpha)}{\partial \alpha^n}\right]_{\alpha = \alpha_0} - \mathscr{L}^{-1}\left\{\left[\frac{\partial^n H(s, \alpha)}{\partial \alpha^n}\right]_{\alpha = \alpha_0} E_i(s)\right\}, \qquad n - 1, 2, \dots . \quad (11.6.6)$$

Example 11.6.1 Consider the RC network shown in Fig. 11.6.2 which can be used as a compensating network to improve the stability of a system, say, a high-gain dc-amplifier.

Let the input be a unit-step function and we wish to determine the sensitivity function to a change in value of R_1. For the sake of demonstrating how the sensitivity function can be calculated, let $R = 1$ ohm, $C = 1$ farad, and R_1 be the

---

* For a detailed discussion on the convergence of inversion integrals, refer to S. S. L. Chang [CH 2], pp. 195–198.

parameter subject to variation, which has a nominal value of 1 ohm. The transfer function

$$H(s, R_1) \triangleq \frac{V_o(s, R_1)}{V_i(s)}$$

is given by

$$H(s, R_1) = \frac{V_o(s, R_1)}{V_i(s)} = \frac{1 + sCR}{1 + sC(R + R_1)} \tag{11.6.7}$$

which, after substituting the numerical values of R and C, reduces to

$$H(s, R_1) = \frac{1 + s}{1 + s(1 + R_1)}. \tag{11.6.8}$$

Differentiating both sides of (11.6.8) and then setting R_1 to its nominal value, we find that

$$\left[\frac{\partial H(s, R_1)}{\partial R_1}\right]_{R_1 = 1} = -\frac{(1 + s)s}{[1 + s(1 + R_1)]^2}\bigg|_{R_1 = 1}$$

$$= -\frac{1}{4}\frac{s^2 + s}{(s + \frac{1}{2})^2} \tag{11.6.9}$$

From (11.6.5) and (11.6.9), we have

$$\left[\frac{\partial v_o(t, R_1)}{\partial R_1}\right]_{R_1 = 1} = \mathscr{L}^{-1}\left\{-\frac{1}{4}\frac{s^2 + s}{s(s + \frac{1}{2})^2}\right\}$$

$$= -\frac{1}{4}(1 + \frac{1}{2}t)e^{-(1/2)t}, \quad t > 0. \tag{11.6.10}$$

The result in (11.6.10) can be readily verified by just finding the unit-step response which is computed to be

$$v_o(t, R_1) = 1 - \frac{R_1}{1 + R_1}e^{-(1/1 + R_1)t}. \tag{11.6.11}$$

Differentiating $v_o(t, R_1)$ with respect to R_1 and then setting R_1 to unity yields the sensitivity function (11.6.10).

In the next few paragraphs, we shall discuss briefly the effects of parameter variations upon the locations of poles and zeros of a closed-loop system.

Effects of Parameter Variations on the Zeros and Poles of a Transfer Function

Consider again the closed-loop system of Fig. 11.5.1. In terms of its zeros z_1 and poles p_k, the closed-loop function is represented by

$$\frac{R_o(s)}{E_i(s)} = \frac{K\prod_i(s - z_i)}{\prod_k(s - p_k)} \tag{11.6.12}$$

which can also be expressed in terms of $G(s)$ and $H(s)$ as

$$\frac{R_o(s)}{E_i(s)} = \frac{G(s)}{1 + G(s)H(s)}. \tag{11.6.13}$$

Combining (11.6.12) and (11.6.13), we obtain

$$\frac{\prod_k (s - p_k)}{K \prod_i (s - z_i)} = \frac{1 + G(s)H(s)}{G(s)}$$

or

$$\frac{\prod_k (s - p_k)}{K \prod_i (s - z_i)} = G^{-1}(s) + H(s). \tag{11.6.14}$$

Assuming that both the zeros and poles of the closed-loop transfer function are distinct—a usual situation for practical systems, then for small variations in parameter values α_j, the change in a pole location, say, p_k is given by

$$\delta p_k = \sum_j \frac{\partial p_k}{\partial \alpha_j} \delta \alpha_j. \tag{11.6.15}$$

To evaluate $\partial p_k / \partial \alpha_j$, $k = 1, 2, \ldots$, we take the natural logarithm of both sides of (11.6.14) yielding

$$\sum_k \log(s - p_k) - \sum_i \log(s - z_i) - \log K = \log[G^{-1}(s) + H(s)]. \tag{11.6.16}$$

Next, differentiating (11.6.16) with respect to α_j, multiplying the result by $-(s - p_i)$, and then setting $s = p_i$, we get

$$\left[\sum_k \frac{\frac{\partial p_k}{\partial \alpha_j}}{(s - p_k)} (s - p_i) - \sum_i \frac{\frac{\partial z_i}{\partial \alpha_j}}{(s - z_i)} (s - p_i) \right]_{s = p_i} = - \left[\frac{\frac{\partial G^{-1}(s)}{\partial \alpha_j} + \frac{\partial H(s)}{\partial \alpha_j}}{G^{-1}(s) + H(s)} (s - p_i) \right]_{s = p_i}$$

or

$$\frac{\partial p_i}{\partial \alpha_j} = - \left[(s - p_i) \frac{R_o(s)}{E_i(s)} \right]_{s = p_i} \left[\frac{\partial G^{-1}(p_i)}{\partial \alpha_j} + \frac{\partial H(p_i)}{\partial \alpha_j} \right]. \tag{11.6.17}$$

We note that the term

$$\left[(s - p_i) \frac{R_o(s)}{E_i(s)} \right]_{s = p_i}$$

in (11.6.17) is simply the expression for the residue of $R_o(s)/E_i(s)$ associated with the pole p_i, which we denote as K_i. Thus, (11.6.17) reduces to

$$\frac{\partial p_i}{\partial \alpha_j} = K_i \left[\frac{\partial G^{-1}(p_i)}{\partial \alpha_j} + \frac{\partial H(p_i)}{\partial \alpha_j} \right], \qquad i, j = 1, 2, \ldots. \tag{11.6.18}$$

Once the transfer functions $G(s)$ and $H(s)$ are given, $\partial p_i/\partial \alpha_j$ can be evaluated and the change in the pole location about its nominal value can be determined with the aid of (11.6.15). The *zero sensitivity*, i.e., the variation of the location of a zero of the closed-loop transfer function $R_o(s)/E_i(s)$ with respect to a change in parameter value, can be developed in exactly the same manner. However, the detailed derivation is left as an exercise.*

The sensitivity study presented above is based on the assumption that the change in parameter is very small so that the effect on the system performance can be accurately determined by calculating the first variation in the Taylor-series expansion. The approach used for determining the sensitivity function in the transient response can be readily extended to both linear time-varying, as well as nonlinear, systems with multiple inputs and outputs. Since the purpose of the section is intended to give the reader a brief introduction to this important and interesting topic, other concepts such as gain and phase sensitivities, peak response sensitivity, sensitivity function for large variation of a system, just to name a few, are beyond our scope. The reader interested in pursuing this topic further is referred to literature elsewhere.†

11.7 INTRODUCTION TO OPTIMAL CONTROL SYSTEMS

Little attention was paid to the subject of optimal control systems prior to 1950. During the latter part of the decade of the 'fifties, the analysis and synthesis of optimal control systems has been extensively studied both in the United States and in Russia. Since 1950 a considerable body of theory has been developed. Pontryagin's maximum principle and Bellman's dynamic programming are well known in the field of control systems. Other methods based on the calculus of variations were also modified so that they could be applied to optimal control systems. Direct effective computational techniques such as the method of steepest descent for solving optimal control problems were not developed until recently.

In the present section, we intend to provide a brief introduction to the interesting topic of optimal control systems. The mathematical formulation of an optimal control system can be described in the following paragraphs.

The behavior of an optimal control system is in general governed by a set of differential equations of the form

$$\dot{x}_i = f_i(x_1, \ldots, x_n, u_1, \ldots, u_r, t) \qquad i = 1, 2, \ldots, n \qquad (11.7.1)$$

with initial conditions $x_i(t_0) = x_{i0}$. In vector form, (11.7.1) can be expressed as

$$\dot{\mathbf{x}} = \mathbf{f}(\mathbf{x}, \mathbf{u}, t), \qquad \mathbf{x}(t_0) = \mathbf{x}_0, \qquad (11.7.2)$$

where $\mathbf{x}$ is the n-dimensional state vector describing the instantaneous behavior of the system, and $\mathbf{u}$ is the r-dimensional control vector.

---

* See Problem 11.10.

† For example, S. S. L. Chang [CH 2], Chapter 1, and S. K. Mitra [MI 1], Chapter 5.

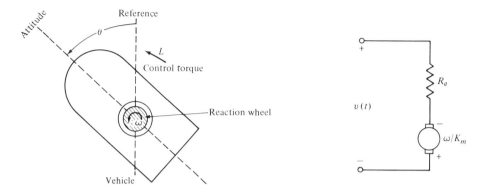

Fig. 11.7.1 Vehicle and reaction wheel with a dc-servomotor.

The functions f_i, $i = 1, 2, \ldots, n$, are, in general, nonlinear functions. It is assumed that these functions, as well as their first partial derivatives with respect to x_j, $j = 1, 2, \ldots, n$, are defined and continuous in some domain Γ of the space $\mathbf{R}$ of variables $x_1, \ldots, x_n, u_1, \ldots, u_r, t$. In other words, the space $\mathbf{R}$ can be taken to be the Cartesian product* $\mathbf{R} = \mathbf{X} \times \mathbf{U} \times \mathbf{T}$. If, in addition, the control vector $\mathbf{u}$ and its partial derivatives $\partial u_k/\partial x_j$, $k = 1, \ldots, r$, $j = 1, \ldots, n$, are continuous, then for each $\mathbf{u}$ there exists a unique solution of the system such that $\mathbf{x}(t)$ is defined and continuous on some interval containing the initial time t_0 and satisfies the initial condition $\mathbf{x}(t_0) = \mathbf{x}_0$.

The performance of the system is measured by the *performance index* which can be represented by an integral of the form

$$J(\mathbf{u}) = \int_{t_0}^{t_1} f_0(\mathbf{x}, \mathbf{u})\, dt, \tag{11.7.3}$$

where t_1 is the terminal time.

For the system described by (11.7.2) and (11.7.3), we can state an optimal control problem as follows. Find a control $\mathbf{u}$ such that the performance index (11.7.3) is optimized. Depending upon exactly what the performance index is, (11.7.3) may be either minimized or maximized. If, for example, (11.7.3) represents the cost of operation of a factory, then, of course, its minimization is desired. On the other hand, (11.7.3) must be maximized if it represents the profit margin for manufacturing a product. Let us illustrate how a practical optimal control system can be formulated by means of the following example.

Example 11.7.1 We shall investigate an optimal control system which can be considered as a single-axis model of a vehicle with a reaction wheel as depicted in Fig. 11.7.1. It is assumed that the control torque L is provided by a dc, armature-

---

* Given two sets $\mathbf{A}$ and $\mathbf{B}$, the set of all ordered pairs (a, b) such that a and b are elements of $\mathbf{A}$ and $\mathbf{B}$, respectively, is called the *Cartesian product* of $\mathbf{A}$ and $\mathbf{B}$, and is denoted by $\mathbf{A} \times \mathbf{B}$. In like manner, the Cartesian product of sets $\mathbf{A}$, $\mathbf{B}$, $\mathbf{C}$, and $\mathbf{D}$ is denoted by $\mathbf{A} \times \mathbf{B} \times \mathbf{C} \times \mathbf{D}$.

controlled, permanent-magnet servomotor and that the motor is driven by an ideal voltage source $v(t)$. The problem is to determine the control $v(t)$ such that for a given time interval T the vehicle be transferred from some fixed initial attitude to some fixed final attitude and, at the same time, the energy delivered to the servomotor be kept to a minimum.

Let the vehicle attitude and wheel velocity with respect to the vehicle be designated as θ and ω, respectively. Then the equations of motion can be written as

$$J\ddot{\theta} = L$$
$$I_w(\dot{\omega} + \ddot{\theta}) = -L, \tag{11.7.4}$$

where J and I_w are moments of inertia of the vehicle and the wheel, respectively. Assuming that the vehicle and the wheel are initially nonrotating, (11.7.4) can be reduced to

$$I_w(\dot{\omega} + \ddot{\theta}) = -J\ddot{\theta}$$
$$\dot{\omega} = -\frac{I_w + J}{I_w}\ddot{\theta}$$

or

$$\omega = -\frac{J}{I}\dot{\theta}, \tag{11.7.5}$$

where

$$I = \frac{I_w}{(I_w/J) + 1}.$$

The torque L is given by

$$L = K_L i = K_L\left(\frac{v + K_m\omega}{R_a}\right), \tag{11.7.6}$$

where K_L and K_m are the torque and speed constants of the motor. Substituting (11.7.4) and (11.7.5) into (11.7.6), we find that

$$J\ddot{\theta} = \frac{K_L v}{R_a} - \frac{K_L K_m}{R_a I}\dot{\theta}$$

or

$$\tau_m^2\ddot{\theta} + \tau_m\dot{\theta} = \frac{\tau_m}{K_m}\left(\frac{I}{J}\right)v, \tag{11.7.7}$$

where $\tau_m = IR_a/(K_L K_m)$ is the motor time constant. For convenience, let us define

$$u(t) = v(t), \tag{11.7.8}$$

$$\tau = \frac{t}{\tau_m}, \qquad \text{the normalized time,} \tag{11.7.9}$$

and
$$x(t) = \frac{K_m\theta}{\tau_m(I/J)}. \tag{11.7.10}$$

By means of these definitions, (11.7.7) reduces to

$$\frac{d^2x}{d\tau^2} + \frac{dx}{d\tau} = u(\tau). \tag{11.7.11}$$

Let $x_1 = x$ and $x_2 = dx_1/d\tau$. Then (11.7.11) can be written in vector form as

$$\begin{bmatrix} \dot{x}_1 \\ \dot{x}_2 \end{bmatrix} = \begin{bmatrix} 0 & 1 \\ 0 & -1 \end{bmatrix} \begin{bmatrix} x_1 \\ x_2 \end{bmatrix} + \begin{bmatrix} 0 \\ 1 \end{bmatrix} [u(\tau)] \tag{11.7.12}$$

with initial conditions $x_1(0) = x_{10}$ and $x_2(0) = 0$.

The power delivered to the servomotor is given by

$$P = vi = v\frac{(v + K_m\omega)}{R_a}$$
$$= \frac{u(\tau)}{R_a}[u(\tau) - x_2(\tau)]. \tag{11.7.13}$$

Then the control problem is to find $u(\tau)$ such that it transfers the system (11.7.12) from the initial state $(x_{10}, 0)$ to the final state $(0, 0)$ and at the same time, minimizes the energy delivered to the servomotor, for a given time T, which is given by

$$J(u) = \frac{1}{R_a}\int_0^T u(u - x_2)\, d\tau. \tag{11.7.14}$$

It can be shown that for the given time T the energy represented by (11.7.14) is minimized if the control function $u(\tau)$ assumes the form

$$u(x_1, x_2) = x_2 + \frac{36}{T^3}x_{10}\frac{2x_1 - x_{10}}{2x_2 - (6/T)x_{10}}. \tag{11.7.15}$$

The derivation of (11.7.15) depends on the knowledge of the calculus of variations which is unsuited to our purpose, since the theory on this subject is beyond the scope of this text. However, a number of books have been written on the subject of optimal control systems in recent years and the interested reader may refer to literature elsewhere.*

11.8 SUMMARY

The topics covered in this chapter represent some of the recent developments in the field of modern system theory. In Sections 11.2 and 11.3, the concepts of controllability and observability were introduced. These two concepts can be con-

---

* See L. S. Pontryagin, et al. [PO 1], and M. Athens and P. Falb [AT 2], for example.

sidered as duals to each other. Roughly speaking, controllability investigates the possibility of steering the state by means of adjusting the input while observability studies the possibility of determining the state by measuring the output. These important concepts of Kalman are very important in the study of both the theoretical and practical aspects of control systems. In fact, controllability and observability are often interpreted as both necessary and sufficient conditions for the existence of a solution to an optimal control problem.

Some of the important concepts of stability were discussed in Section 11.4, and two stability criteria—Routh-Hurwitz and Nyquist—were given in Section 11.5. The Routh-Hurwitz criterion is an analytical approach, while the Nyquist criterion is a graphical one. Needless to say, the topic of stability is extremely important in a wide spectrum of different fields. For example, in the design of a feedback control system, among many usual performance specifications that the system must satisfy, the most important requirement is that the system must be stable at all times.

The sensitivity analysis discussed in Section 11.6 deals mainly with the deterioration of a system's performance due to variations in system parameters. The stability and sensitivity of dynamic systems are closely related. For example, with the definition of sensitivity function given in Section 11.6, it is clear that for an nth-order linear differential equation with constant coefficients the differential equation of the sensitivity function has the same characteristic equation as that of the original differential equation.

Finally, in Section 11.7, the topic of optimal control was briefly touched.

REFERENCES

1. K. Ogata, *State Space Analysis of Control Systems*, Prentice-Hall, 1967, Chapters 7, 8, and 9.

2. L. A. Zadeh and C. A. Desoer, *Linear System Theory*, McGraw-Hill, 1963, Chapters 7 and 11.

3. S. S. L. Chang, *Synthesis of Optimum Control Systems*, McGraw-Hill, 1961, Chapter 8.

4. S. K. Mitra, *Analysis and Synthesis of Linear Active Networks*, John Wiley & Sons, 1969, Chapter 5.

5. L. S. Pontryagin, V. G. Boltyanskii, R. V. Gramkrelidze, and E. F. Mishchenko, *The Mathematical Theory of Optimal Processes*, Interscience, 1962.

6. M. Athens and P. Falb, *Optimal Control: An Introduction to the Theory and Its Applications*, McGraw-Hill, 1966.

PROBLEMS

11.1 A second-order system is characterized by the equation

$$\begin{bmatrix} \dot{x}_1 \\ \dot{x}_2 \end{bmatrix} = \begin{bmatrix} 2 & 1 \\ -1 & 3 \end{bmatrix} \begin{bmatrix} x_1 \\ x_2 \end{bmatrix} + \begin{bmatrix} 0 \\ 1 \end{bmatrix} [u].$$

Is the system controllable?

11.2 Consider the second-order system of Problem 11.1. If the output $y(t)$ is related to the state vector by the relation

$$y = \begin{bmatrix} 1 & 3 \end{bmatrix} \begin{bmatrix} x_1 \\ x_2 \end{bmatrix}$$

a) is the system output controllable?
b) is it an observable system?

11.3 The motion of a second-order system is governed by the equation

$$\begin{bmatrix} \dot{x} \\ \dot{x} \end{bmatrix} = \begin{bmatrix} 3 & 1 \\ \alpha & 2 \end{bmatrix} \begin{bmatrix} x_1 \\ x_2 \end{bmatrix} + \begin{bmatrix} 1 \\ 3 \end{bmatrix} [u],$$

where α is a parameter of the system. Is the system controllable for all finite values of α?

11.4 Show that the condition for stability b.i.b.o. for the system $\dot{\mathbf{x}} = \mathbf{A}\mathbf{x} + \mathbf{B}u$ is that

$$\int_0^\infty \| \Phi(t - \tau) \| \, d\tau \le M < \infty.$$

11.5 Verify Theorem 11.4.1.

11.6 Show that if the linear time-invariant system

$$\dot{\mathbf{x}} = \mathbf{A}\mathbf{x} + \mathbf{B}u$$

is asymptotically stable, it is also stable b.i.b.o.

11.7 By means of an example, show that the converse of the statement in Problem 11.6 is not true.

11.8 Show that if all the eigenvalues of $\mathbf{A}$ have negative real parts, the system $\dot{\mathbf{x}} = \mathbf{A}\mathbf{x}$ is asymptotically stable.

11.9 Use the Routh-Hurwitz criterion to determine whether the systems having the characteristic equations in (a) and (b) are asymptotically stable. If any roots lie on the $j\omega$-axis, determine the locations of these roots.

a) $s^4 + 2s^3 + 2s^2 + 2s + 1 = 0$;   b) $s^4 + 7s^3 + 17s^2 + 17s + 6 = 0$

11.10 Derive an expression for the change in location of a zero of a closed-loop transfer function

$$\frac{R_o(s)}{E_i(s)} = \frac{G(s)}{1 + G(s)H(s)}$$

in terms of the system components $G(s)$ and $H(s)$.

MATRIX ANALYSIS

A.1.1 INTRODUCTION

The principal objective of this appendix is twofold: first, to review some of the basic properties of matrix algebra and, second, to introduce a few topics in matrix analysis which find extensive applications in engineering and other applied sciences. To limit the size of this appendix, rigorous mathematical proofs are omitted since they can be found in most standard texts.

A.1.2 DEFINITIONS

A *matrix* **A** is defined as a rectangular array of numbers as shown in (A.1.2.1),

$$\mathbf{A} = \begin{bmatrix} a_{11} & a_{12} & \cdots & a_{1n} \\ a_{21} & a_{22} & \cdots & a_{2n} \\ \vdots & & & \\ a_{m1} & a_{m2} & \cdots & a_{mn} \end{bmatrix}, \tag{A.1.2.1}$$

where the *elements* a_{ij} can be (real or complex) numbers or functions. Since **A** consists of m rows and n columns, **A** is called an $m \times n$ (m by n) matrix or a rectangular matrix of *order* $m \times n$. If, however, m is equal to n, **A** becomes a *square matrix* of *order* n because **A** has exactly n rows and n columns;

$$\mathbf{A} = \begin{bmatrix} a_{11} & a_{12} & \cdots & a_{1n} \\ a_{21} & a_{22} & \cdots & a_{2n} \\ \vdots & & & \\ a_{n1} & a_{n2} & \cdots & a_{nn} \end{bmatrix}. \tag{A.1.2.2}$$

If, in (A.1.2.2), all the elements $a_{ij}(i \neq j)$ are equal to zero, then the matrix **A** reduces to

$$\mathbf{A} = \begin{bmatrix} a_{11} & 0 & 0 & \cdots & 0 \\ 0 & a_{22} & 0 & \cdots & 0 \\ & & \ddots & & \\ 0 & 0 & 0 & \cdots & a_{nn} \end{bmatrix} \tag{A.1.2.3}$$

and is called a *diagonal matrix*. If, in addition, all the elements of **A** in (A.1.2.3) are equal to unity, the corresponding matrix is called the *unit matrix* of order n, denoted by I_n.* The *trace* of a square matrix of order n is defined to be the algebraic sum of all the elements on the main diagonal of **A**, viz.,

$$\text{Tr } \mathbf{A} \triangleq \sum_{i=1}^{n} a_{ii}, \qquad (\text{A.1.2.4})$$

where Tr **A** denotes the trace of **A**. If the elements of a square matrix **A** satisfy the relation $a_{ij} = a_{ji}$ for every pair of i and j, **A** is referred to as a *symmetric matrix*. The matrices

$$\mathbf{A} = \begin{bmatrix} 1 & -1 & 2 \\ -1 & 2 & 5 \\ 2 & 5 & 3 \end{bmatrix},$$

$$\mathbf{B} = \begin{bmatrix} 2 & 0 & 0 \\ 0 & 1 & 0 \\ 0 & 0 & 3 \end{bmatrix}$$

are examples of symmetric and diagonal matrices, respectively, and Tr **A** = Tr **B** = 6.

The matrix **B** is said to be the *transpose* of a matrix **A**, **B** = **A**^{T}, if the elements of **B** and **A** satisfy the conditions

$$b_{ij} = a_{ji} \qquad \text{for every pair of } i \text{ and } j. \qquad (\text{A.1.2.5})$$

For example, the two matrices

$$\mathbf{A} = \begin{bmatrix} 2 & 4 & -2 \end{bmatrix}, \qquad \mathbf{B} = \begin{bmatrix} 2 \\ 4 \\ -2 \end{bmatrix}$$

are transposes of each other. A matrix having a single row is commonly known as a *row vector*, and one having a single column is called a *column vector*. For example, the matrices **A** and **B** given above are, respectively, a row vector and a column vector. An inspection of (A.1.2.5) and the above example indicate the following properties:

1. The transpose of a symmetric matrix is the symmetric matrix itself.

2. The transpose of a column matrix is a row matrix, and vice versa.

3. The transpose of the transpose of a matrix is the matrix itself.

---

* Or simply **I** if no ambiguity arises, or when the order of the unit matrix need not be emphasized.

A *triangular matrix* is a square matrix whose elements below (or above) the main diagonal are all zeros. Thus the matrices

$$C = \begin{bmatrix} 1 & 1 & 2 \\ 0 & 2 & 1 \\ 0 & 0 & 3 \end{bmatrix},$$

$$D = \begin{bmatrix} 1 & 0 & 0 \\ 1 & 2 & 0 \\ 3 & 1 & 3 \end{bmatrix}$$

are examples of triangular matrices, with C referred to as an *upper-triangular* and D a *lower-triangular* matrix.

A.1.3 MATRIX OPERATIONS

Equality, Addition, and Subtraction

Two matrices A and B are said to be *equal* if they are of the same order and the relation $a_{ij} = b_{ij}$ is satisfied for every pair of i and j. Two matrices can be added together if they are of the same order. Explicitly, if

$$A = \begin{bmatrix} a_{11} & a_{12} & \cdots & a_{1n} \\ a_{21} & a_{22} & \cdots & a_{2n} \\ \vdots & & & \\ a_{m1} & a_{m2} & \cdots & a_{mn} \end{bmatrix}, \quad B = \begin{bmatrix} b_{11} & b_{12} & \cdots & b_{1n} \\ b_{21} & b_{22} & \cdots & b_{2n} \\ \vdots & & & \\ b_{m1} & b_{m2} & \cdots & b_{mn} \end{bmatrix}, \quad \text{(A.1.3.1)}$$

then the *sum* C is given by

$$C = A + B = \begin{bmatrix} c_{11} & c_{12} & \cdots & c_{1n} \\ c_{21} & c_{22} & \cdots & c_{2n} \\ \vdots & & & \\ c_{m1} & c_{m2} & \cdots & c_{mn} \end{bmatrix}$$

$$= \begin{bmatrix} (a_{11} + b_{11}) & (a_{12} + b_{12}) & \cdots & (a_{1n} + b_{1n}) \\ (a_{21} + b_{21}) & (a_{22} + b_{22}) & \cdots & (a_{2n} + b_{2n}) \\ \vdots & & & \\ (a_{m1} + b_{m1}) & (a_{m2} + b_{m2}) & \cdots & (a_{mn} + b_{mn}) \end{bmatrix}. \quad \text{(A.1.3.2)}$$

In exactly the same manner, the *difference* $A - B$ is defined as the matrix whose elements are the differences of corresponding elements of A and B.

Scalar Multiple of a Matrix

The *product* of a matrix *multiplied by a scalar* is a matrix whose elements are the corresponding elements of the given matrix multiplied by the scalar; thus

$$C = \alpha A = \alpha \begin{bmatrix} a_{11} & a_{12} & \cdots & a_{1n} \\ a_{21} & a_{22} & \cdots & a_{2n} \\ \vdots & & & \\ a_{m1} & a_{m2} & \cdots & a_{mn} \end{bmatrix} = \begin{bmatrix} \alpha a_{11} & \alpha a_{12} & \cdots & \alpha a_{1n} \\ \alpha a_{21} & \alpha a_{22} & \cdots & \alpha a_{2n} \\ \vdots & & & \\ \alpha a_{m1} & \alpha a_{m2} & \cdots & \alpha a_{mn} \end{bmatrix}. \quad (A.1.3.3)$$

Product of Two Matrices

The *product of* two matrices A and B (say, AB) is defined to be a matrix $C = AB = [c_{ij}]$ whose elements are determined by the relation

$$c_{ij} = \sum_{k=1}^{n} a_{ik} b_{kj}, \quad (A.1.3.4)$$

where n denotes the number of columns of A as well as the number of rows of B. The relation (A.1.3.4) indicates that the product AB exists (i.e., A and B are said to be *conformable* for multiplication in the order AB) only if the number of columns of A is the same as the number of rows of B, since the running index k in (A.1.3.4) refers to the columns of A as well as the rows of B. As an example, the product AB of the matrices

$$A = \begin{bmatrix} a_{11} & a_{12} & a_{13} \\ a_{21} & a_{22} & a_{23} \end{bmatrix}, \quad B = \begin{bmatrix} b_{11} & b_{12} \\ b_{21} & b_{22} \\ b_{31} & b_{32} \end{bmatrix}$$

is given by

$$AB = \begin{bmatrix} (a_{11}b_{11} + a_{12}b_{21} + a_{13}b_{31}) & (a_{11}b_{12} + a_{12}b_{22} + a_{13}b_{32}) \\ (a_{21}b_{11} + a_{22}b_{21} + a_{23}b_{31}) & (a_{21}b_{12} + a_{22}b_{22} + a_{23}b_{32}) \end{bmatrix}.$$

Similarly, the product BA is

$$BA = \begin{bmatrix} (b_{11}a_{11} + b_{12}a_{21}) & (b_{11}a_{12} + b_{12}a_{22}) & (b_{11}a_{13} + b_{12}a_{23}) \\ (b_{21}a_{11} + b_{22}a_{21}) & (b_{21}a_{12} + b_{22}a_{22}) & (b_{21}a_{13} + b_{22}a_{23}) \\ (b_{31}a_{11} + b_{32}a_{21}) & (b_{31}a_{12} + b_{32}a_{22}) & (b_{31}a_{13} + b_{32}a_{23}) \end{bmatrix}.$$

An inspection of the above example reveals a very important fact: *the commutative law, in general, does not hold for multiplication of matrices*, i.e.,

$$AB \neq BA. \quad (A.1.3.5)$$

To avoid any possible confusion involving products of matrices, we shall use the terms *premultiplication* and *postmultiplication*. Thus, in the product AB, B is said to be *premultiplied* by A; or, equivalently, A is *postmultiplied* by B.

The rules of operations of matrices with respect to addition, multiplication, and scalar multiple of a matrix can be summarized as follows:

1. Commutative laws:

$$\mathbf{A} + \mathbf{B} = \mathbf{B} + \mathbf{A},$$

$$\mathbf{AB} \neq \mathbf{BA}.$$

(A.1.3.6)

2. Distributive laws:

$$\mathbf{A}(\mathbf{B} + \mathbf{C}) = \mathbf{AB} + \mathbf{AC},$$

$$a(\mathbf{B} + \mathbf{C}) = a\mathbf{B} + a\mathbf{C}, \qquad \text{for scalar } a.$$

(A.1.3.7)

3. Associative laws:

$$(\mathbf{AB})\mathbf{C} = \mathbf{A}(\mathbf{BC}),$$

$$(a\mathbf{B})\mathbf{C} = a(\mathbf{BC}), \qquad \text{for scalar } a$$

(A.1.3.8)

$$(\mathbf{A} + \mathbf{B}) + \mathbf{C} = \mathbf{A} + (\mathbf{B} + \mathbf{C}).$$

Matrix Inversion

Let $\mathbf{A}$ be a square matrix of order n. Then the *determinant of* $\mathbf{A}$, denoted by det $\mathbf{A}$, is the determinant of n^2 elements arranged in n rows and n columns in exactly the same way as the elements appeared in $\mathbf{A}$; that is, if

$$\mathbf{A} = \begin{bmatrix} a_{11} \cdots a_{1n} \\ \vdots \\ a_{n1} \cdots a_{nn} \end{bmatrix},$$

then

$$\det \mathbf{A} = \begin{vmatrix} a_{11} \cdots a_{1n} \\ \vdots \\ a_{n1} \cdots a_{nn} \end{vmatrix}.$$

(A.1.3.9)*

A square matrix $\mathbf{A}$ is said to be *nonsingular* if the determinant of $\mathbf{A}$ is nonzero; otherwise it is called a *singular matrix*. If a given matrix $\mathbf{A}$ is nonsingular, then it is said to have an *inverse* $\mathbf{A}^{-1}$ such that the following relation holds:

$$\mathbf{AA}^{-1} = \mathbf{A}^{-1}\mathbf{A} = \mathbf{I},$$

(A.1.3.10)

where $\mathbf{I}$ is the unit (*identity*) matrix. It can be readily shown that the inverse matrix $\mathbf{A}^{-1}$ is given by

$$\mathbf{A}^{-1} = \left(\frac{1}{\det \mathbf{A}}\right)[A_{ij}]^T = \left(\frac{1}{\det \mathbf{A}}\right) \begin{bmatrix} A_{11} & A_{21} & \cdots & A_{n1} \\ A_{12} & A_{22} & \cdots & A_{n2} \\ \vdots \\ A_{1n} & A_{2n} & \cdots & A_{nn} \end{bmatrix},$$

(A.1.3.11)

---

* Methods for evaluating determinants and cofactors are usually discussed in a college algebra text or any introductory text on matrices and determinants. For example, refer to F. E. Hohn [HO 1].

where A_{ij} is the *cofactor** of a_{ij}, defined by

$$A_{ij} = (-1)^{i+j} \times \left\{ \begin{array}{c} \text{determinant of the matrix obtained by deleting} \\ \text{the } i\text{th row and } j\text{th column of } \mathbf{A}. \end{array} \right\}$$

and

$$[A_{ij}]^T = \begin{bmatrix} A_{11} & A_{21} & \cdots & A_{n1} \\ A_{12} & A_{22} & \cdots & A_{n2} \\ \vdots & & & \\ A_{1n} & A_{2n} & \cdots & A_{nn} \end{bmatrix} \triangleq \mathscr{A}. \qquad \text{(A.1.3.12)}$$

The matrix $\mathscr{A} \triangleq [A_{ij}]^T$ defined above is called the *adjoint* of the given matrix $\mathbf{A}$. Thus, combining (A.1.3.11) and (A.1.3.12), we obtain

$$\mathbf{A}^{-1} = \frac{\mathscr{A}}{\det \mathbf{A}}. \qquad \text{(A.1.3.13)}$$

As an example, consider the following matrix:

$$\mathbf{A} = \begin{bmatrix} 1 & 0 & -1 \\ -1 & 1 & 0 \\ 0 & 1 & 2 \end{bmatrix}.$$

The inverse of $\mathbf{A}$ can be found by means of (A.1.3.12) and (A.1.3.13):

$$\mathbf{A}^{-1} = \frac{\mathscr{A}}{\det \mathbf{A}} = \frac{1}{3} \times \begin{bmatrix} 2 & -1 & 1 \\ 2 & 2 & 1 \\ -1 & -1 & 1 \end{bmatrix} = \begin{bmatrix} \frac{2}{3} & -\frac{1}{3} & \frac{1}{3} \\ \frac{2}{3} & \frac{2}{3} & \frac{1}{3} \\ -\frac{1}{3} & -\frac{1}{3} & \frac{1}{3} \end{bmatrix}$$

Differentiation and Integration

Consider a matrix $\mathbf{A}(t) = [a_{ij}(t)]$ whose elements $a_{ij}(t)$ are functions of t. Then the *derivative* of $\mathbf{A}(t)$ with respect to t is given by

$$\frac{d}{dt}\mathbf{A}(t) = \frac{d}{dt}[a_{ij}(t)] = \left[\frac{d}{dt}a_{ij}(t)\right]. \qquad \text{(A.1.3.14)}$$

Of course, in (A.1.3.14), the existence of the derivatives of $a_{ij}(t)$ is assumed. Similarly, the *integral* of $\mathbf{A}(t)$ is defined by

$$\int \mathbf{A}(t)\,dt = \left[\int a_{ij}(t)\,dt\right]. \qquad \text{(A.1.3.15)}$$

An inspection of (A.1.3.14) and (A.1.3.15) reveals that the derivative or the integral of $\mathbf{A}(t)$ is obtained by differentiating or integrating each element of $\mathbf{A}(t)$,

---

* Methods for evaluating determinants and cofactors are usually discussed in a college algebra text or any introductory text on matrices and determinants. For example, refer to F. E. Hohn [HO 1].

respectively. As an example, the derivative of

$$\mathbf{A}(t) = \begin{bmatrix} 2t & 3t^2 \\ 4 & 6t \end{bmatrix}$$

is simply the matrix

$$\frac{d}{dt}\mathbf{A}(t) = \begin{bmatrix} 2 & 6t \\ 0 & 6 \end{bmatrix}.$$

It can be readily shown that the following relations always hold:

$$\frac{d}{dt}(\mathbf{AB}) = \frac{d\mathbf{A}}{dt}\mathbf{B} + \mathbf{A}\frac{d\mathbf{B}}{dt},$$

$$\frac{d}{dt}(\alpha\mathbf{A}) = \frac{d\alpha}{dt}\mathbf{A} + \alpha\frac{d\mathbf{A}}{dt}, \qquad (A.1.3.16)$$

$$\frac{d}{dt}(\mathbf{A} + \mathbf{B}) = \frac{d\mathbf{A}}{dt} + \frac{d\mathbf{B}}{dt}.$$

Matrix Partitioning

In working out a matrix product, it is sometimes convenient to subdivide or *partition* matrices into *submatrices*. Matrix partitioning is especially useful when matrices have a large number of rows and columns. Such a partitioned matrix may be written

$$\mathbf{A} = \begin{bmatrix} \mathbf{A}_{11} & \mathbf{A}_{12} \\ \mathbf{A}_{21} & \mathbf{A}_{22} \end{bmatrix}, \qquad (A.1.3.17)$$

where $\mathbf{A}_{11}$, $\mathbf{A}_{12}$, $\mathbf{A}_{21}$, and $\mathbf{A}_{22}$ are submatrices.

Thus a matrix $\mathbf{A}$ of order $m \times n$ can be partitioned into groups of p and q rows and r and s columns as long as the following relations are satisfied:

$$p + q = m, \qquad r + s = n. \qquad (A.1.3.18)$$

If a second matrix $\mathbf{B}$ of order $n \times 1$ is also partitioned such that $\mathbf{B}$ may be written as

$$\mathbf{B} = \begin{bmatrix} \mathbf{B}_{11} & \mathbf{B}_{12} \\ \mathbf{B}_{21} & \mathbf{B}_{22} \end{bmatrix}, \qquad (A.1.3.19)$$

then the product $\mathbf{AB}$ can also be expressed in partitioned form with submatrices as its elements, viz.,

$$\begin{aligned} \mathbf{AB} &= \begin{bmatrix} \mathbf{A}_{11} & \mathbf{A}_{12} \\ \mathbf{A}_{21} & \mathbf{A}_{22} \end{bmatrix}\begin{bmatrix} \mathbf{B}_{11} & \mathbf{B}_{12} \\ \mathbf{B}_{21} & \mathbf{B}_{22} \end{bmatrix} \\ &= \begin{bmatrix} \mathbf{A}_{11}\mathbf{B}_{11} + \mathbf{A}_{12}\mathbf{B}_{21} & \mathbf{A}_{11}\mathbf{B}_{12} + \mathbf{A}_{12}\mathbf{B}_{22} \\ \mathbf{A}_{21}\mathbf{B}_{11} + \mathbf{A}_{22}\mathbf{B}_{21} & \mathbf{A}_{21}\mathbf{B}_{12} + \mathbf{A}_{22}\mathbf{B}_{22} \end{bmatrix}. \end{aligned} \qquad (A.1.3.20)$$

It should be noted that the products of submatrices in (A.1.3.20) must be conformable for multiplication. The following simple example will demonstrate the ideas involved.

Example A.1.3.1 Let us evaluate the product **AB** of the following matrices

$$A = \begin{bmatrix} 2 & 1 & 0 & 0 \\ 3 & 2 & 0 & 0 \end{bmatrix}, \qquad B = \begin{bmatrix} 2 \\ 1 \\ 6 \\ 5 \end{bmatrix}.$$

The two matrices can be partitioned as follows:

$$A = \begin{bmatrix} 2 & 1 & 0 & 0 \\ 3 & 2 & 0 & 0 \end{bmatrix} = [A_{11} \quad A_{12}], \qquad B = \begin{bmatrix} 2 \\ 1 \\ 6 \\ 5 \end{bmatrix} = \begin{bmatrix} B_{11} \\ B_{21} \end{bmatrix},$$

where

$$A_{11} = \begin{bmatrix} 2 & 1 \\ 3 & 2 \end{bmatrix}, \qquad A_{12} = \begin{bmatrix} 0 & 0 \\ 0 & 0 \end{bmatrix};$$

$$B_{11} = \begin{bmatrix} 2 \\ 1 \end{bmatrix}, \qquad B_{21} = \begin{bmatrix} 6 \\ 5 \end{bmatrix}.$$

Now the product **AB** can be expressed in partitioned form as

$$AB = [A_{11} \quad A_{12}]\begin{bmatrix} B_{11} \\ B_{21} \end{bmatrix} = [A_{11}B_{11} + A_{12}B_{21}].$$

But,

$$A_{11}B_{11} = \begin{bmatrix} 2 & 1 \\ 3 & 2 \end{bmatrix}\begin{bmatrix} 2 \\ 1 \end{bmatrix} = \begin{bmatrix} 5 \\ 8 \end{bmatrix}$$

$$A_{12}B_{21} = \begin{bmatrix} 0 & 0 \\ 0 & 0 \end{bmatrix}\begin{bmatrix} 6 \\ 5 \end{bmatrix} = \begin{bmatrix} 0 \\ 0 \end{bmatrix}.$$

Finally,

$$AB = \begin{bmatrix} 5 \\ 8 \end{bmatrix}.$$

In the above example, the elements of A_{12} are all equal to zero. *A matrix whose elements are all zero is called a zero or null matrix.* Thus the product $A_{12}B_{21}$ in the above example is also a null matrix.

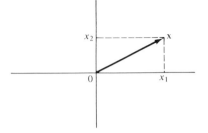

Fig. A.1.4.1 Illustration of a position vector.

A.1.4 VECTORS AND VECTOR SPACES

In this section we present a brief discussion of vectors and vector spaces. Because it is intended as a review, we shall not attempt to cover any topic in great detail. In fact, only those items which are necessary for the understanding of the text material are introduced here. The interested reader should consult standard texts on this subject for proofs or discussions of any topics in depth.

Vector

In general, a position vector in a plane can be described by means of its components x_1 and x_2 in the form

$$\mathbf{x} = \begin{bmatrix} x_1 \\ x_2 \end{bmatrix}. \tag{A.1.4.1}$$

These two components define uniquely the exact position of the vector $\mathbf{x}$ as depicted in Fig. A.1.4.1. Since it takes a set of two numbers to define the vector $\mathbf{x}$ in (A.1.4.1), it is customary to call $\mathbf{x}$ a 2-dimensional vector, or simply a 2-vector. In exactly the same manner an *n-dimensional vector* (*n-vector*) can be represented by an ordered set of *n* numbers, i.e.,

$$\mathbf{x} = \begin{bmatrix} x_1 \\ x_2 \\ \vdots \\ x_n \end{bmatrix}. \tag{A.1.4.2}$$

If a vector has zeros as its elements, then it is called a *null vector*.

Vector Space

A set **R** of elements called vectors is said to be a *linear vector space* if the following algebraic laws are satisfied:

A1. If $\mathbf{x}$ and $\mathbf{y}$ are any two vectors in **R**, then their sum $\mathbf{x} + \mathbf{y}$ is also in **R**.

A2. $\mathbf{x} + \mathbf{y} = \mathbf{y} + \mathbf{x}$ (Commutative law).

A3. For vectors $\mathbf{x}, \mathbf{y}, \mathbf{z}$ in **R**:

$$\mathbf{x} + (\mathbf{y} + \mathbf{z}) = (\mathbf{x} + \mathbf{y}) + \mathbf{z}, \quad \text{(Associative law)}.$$

A4. There exists an element **0** in **R** such that

$$\mathbf{x} + \mathbf{0} = \mathbf{x}, \qquad \text{for all } \mathbf{x} \text{ in } \mathbf{R}.$$

A5. For every **x** in **R**, there is a vector $-\mathbf{x}$ in **R** such that

$$\mathbf{x} + (-\mathbf{x}) = \mathbf{0}.$$

B1. For an arbitrary number α and an element **x** in **R**, there is defined an element $\alpha\mathbf{x}$ in **R**.

B2. For arbitrary numbers α, β, and any vector **x** in **R**:

$$\alpha(\beta\mathbf{x}) = (\alpha\beta)\mathbf{x}.$$

B3. For vectors **x**, **y** in **R** and an arbitrary number α:

$$\alpha(\mathbf{x} + \mathbf{y}) = \alpha\mathbf{x} + \alpha\mathbf{y}.$$

B4. For arbitrary numbers α, β and any vector **x** in **R**:

$$(\alpha + \beta)\mathbf{x} = \alpha\mathbf{x} + \beta\mathbf{x}.$$

B5. There exists an element **I** in **R** such that

$$\mathbf{I} \cdot \mathbf{x} = \mathbf{x}, \qquad \text{for all } \mathbf{x} \text{ in } \mathbf{R}.$$

Depending on whether all complex numbers or only real numbers are admitted, we distinguish between complex and real linear vector spaces. Since, in most applications, the real vector spaces are sufficient, we shall consider only real linear spaces unless stated otherwise. If a vector space **R** is a set of all finite n-dimensional vectors, the space **R** is referred to as an *n-dimensional vector space* V_n. In addition to operations of addition and multiplication in a linear space, there is usually an operation of passage to a limit, which can be easily established by introducing a *norm* into the space.

A linear space **R** is said to be *normed* if to each **x** in **R** a nonnegative number $\|\mathbf{x}\|$, called the *norm* of **x**, exists with the following properties:

1. $\|\mathbf{x}\| = 0$ if and only if $\mathbf{x} = \mathbf{0}$.

2. $\|\alpha\mathbf{x}\| = |\alpha| \cdot \|\mathbf{x}\|$, where $|\alpha|$ is the absolute value of the arbitrary number α.

$$(\text{A.1.4.3})$$

3. $\|\mathbf{x} + \mathbf{y}\| \leq \|\mathbf{x}\| + \|\mathbf{y}\|$ (triangular inequality).

Actually the concept of a normed linear space is not new at all. The simplest example of a normed space is the real line. In this case, the norm is simply the absolute value of the real number. Geometrically, the norm of a vector can be regarded as the *length* of that vector. Perhaps it should be pointed out that there are different definitions of norms commonly used in the literature. However, in this book we shall use the following definition: In a finite n-dimensional space, *the norm of a vector* $\mathbf{x}$ is defined as

$$\|\mathbf{x}\| \triangleq \left\{ \sum_{i=1}^{n} x_i^2 \right\}^{1/2}. \qquad (\text{A.1.4.4})$$

A linear vector space with the norm defined in (A.1.4.4) is called *Euclidean n-space.*

Linear Manifold and Subspace

In a normed linear space $\mathbf{R}$, a linear manifold $\mathbf{L}$ is any set of elements in $\mathbf{R}$ satisfying the following condition: For any two vectors $\mathbf{x}, \mathbf{y}$ in $\mathbf{L}$ and arbitrary numbers α and β, the vector $\alpha\mathbf{x} + \beta\mathbf{y}$ is also an element of $\mathbf{L}$. Since we consider only finite dimensional vector spaces, a subspace* can be regarded as the same as a linear manifold.

Linear Independence and Basis

A set of vectors $\mathbf{x}_1, \mathbf{x}_2, \ldots, \mathbf{x}_n$ are said to be *linearly independent* if the linear equation

$$c_1\mathbf{x}_1 + c_2\mathbf{x}_2 + \cdots + c_n\mathbf{x}_n = \mathbf{0}$$

or

$$\sum_{i=1}^{n} c_i\mathbf{x}_i = \mathbf{0} \qquad (\text{A.1.4.5})$$

is satisfied only when all the constants c_i are equal to zero; otherwise the set of vectors is said to be *linearly dependent.* An inspection of (A.1.4.5) reveals that if a set of vectors is linearly dependent, then any one of the vectors in the set can be expressed as a linear combination of the others. As an example, let us investigate whether the set of vectors

$$\mathbf{x}_1 = \begin{bmatrix} 1 \\ 1 \\ 0 \end{bmatrix}, \qquad \mathbf{x}_2 = \begin{bmatrix} 1 \\ 2 \\ 1 \end{bmatrix}$$

forms an independent set. Applying (A.1.4.5), we obtain

$$c_1\begin{bmatrix} 1 \\ 1 \\ 0 \end{bmatrix} + c_2\begin{bmatrix} 1 \\ 2 \\ 1 \end{bmatrix} = \begin{bmatrix} 0 \\ 0 \\ 0 \end{bmatrix}$$

which is equivalent to the following set of equations:

$$c_1 + c_2 = 0,$$

$$c_1 + 2c_2 = 0,$$

$$c_2 = 0.$$

---

* Rigorously speaking, a subspace of the space $\mathbf{R}$ is a *closed* linear manifold. Here, the term "subspace" is used in the same sense as a "subset" as defined in set theory.

These three equations are satisfied only if $c_1 = c_2 = 0$; therefore the vectors $\mathbf{x}_1$ and $\mathbf{x}_2$ given above are linearly independent.

A set of vectors is said to *span* the space $\mathbf{R}$ if every vector in $\mathbf{R}$ can be expressed as a linear combination of the vectors of that set. As an example, the vectors $\mathbf{x}_1$ and $\mathbf{x}_2$ given above do not span the 3-dimensional space $\mathbf{V}_3$ because the vector, say,

$$\mathbf{x}_3 = \begin{bmatrix} 0 \\ 0 \\ 1 \end{bmatrix},$$

cannot be expressed as a linear combination of $\mathbf{x}_1$ and $\mathbf{x}_2$.

A set $\mathbf{L}$ of vectors of a linear vector space $\mathbf{R}$ is said to be a *basis* of $\mathbf{R}$ if the set $\mathbf{L}$ (a) spans the space $\mathbf{R}$, and (b) is also a linearly independent set.

Example A.1.4.1   The vectors

$$\mathbf{x}_1 = \begin{bmatrix} 1 \\ 0 \\ 0 \end{bmatrix}, \qquad \mathbf{x}_2 = \begin{bmatrix} 0 \\ 1 \\ 0 \end{bmatrix}, \qquad \mathbf{x}_3 = \begin{bmatrix} 0 \\ 0 \\ 1 \end{bmatrix}$$

form a basis of the 3-dimensional space $\mathbf{V}_3$. It is obvious that any arbitrary vector

$$\mathbf{x} = \begin{bmatrix} a_1 \\ a_2 \\ a_3 \end{bmatrix}$$

can be expressed as a linear combination of $\mathbf{x}_1$, $\mathbf{x}_2$, and $\mathbf{x}_3$:

$$\mathbf{x} = a_1\mathbf{x}_1 + a_2\mathbf{x}_2 + a_3\mathbf{x}_3$$

and also that the relation

$$c_1\mathbf{x}_1 + c_2\mathbf{x}_2 + c_3\mathbf{x}_3 = 0$$

is satisfied only when $c_1 = c_2 = c_3 = 0$.

The number of vectors in a basis of a space $\mathbf{R}$ is called the *dimension* of $\mathbf{R}$. Thus, if the number of elements in a basis of $\mathbf{R}$ is n, then $\mathbf{R}$ is said to be an n-dimensional vector space.

Inner Product

The *inner product* or *scalar product* of two vectors $\mathbf{x}$ and $\mathbf{y}$ denoted by $(\mathbf{x}, \mathbf{y})$ is a real number such that the following axioms are satisfied:

a) $(\mathbf{x}, \mathbf{y}) = (\mathbf{y}, \mathbf{x})$,
b) $(c\mathbf{x}, \mathbf{y}) = (\mathbf{x}, c\mathbf{y}) = c(\mathbf{x}, \mathbf{y})$,
c) $(\mathbf{x} + \mathbf{y}, \mathbf{z} + \mathbf{w}) = (\mathbf{x}, \mathbf{z}) + (\mathbf{x}, \mathbf{w}) + (\mathbf{y}, \mathbf{z}) + (\mathbf{y}, \mathbf{w})$,
d) $(\mathbf{x}, \mathbf{x}) > 0$ for $\mathbf{x} \neq 0$.

(A.1.4.6)

It should be pointed out that there are many different definitions of inner product. However, for n-dimensional vector space $\mathbf{V}_n$, we shall use the following definition: the inner product of two vectors $\mathbf{x}$ and $\mathbf{y}$ is defined by the relation

$$(\mathbf{x}, \mathbf{y}) = \mathbf{x}^T \mathbf{y} = \sum_{i=1}^{n} x_i y_i. \tag{A.1.4.7}$$

Of course, it can be easily verified that all the axioms in (A.1.4.6) are satisfied by the above definition. It is interesting to note that if, in (A.1.4.7), $\mathbf{y}$ is replaced by $\mathbf{x}$, we obtain

$$(\mathbf{x}, \mathbf{x}) = \sum_{i=1}^{n} x_i^2 \tag{A.1.4.8}$$

which, upon comparison with (A.1.4.4), can be written as

$$(\mathbf{x}, \mathbf{x}) = \|\mathbf{x}\|^2. \tag{A.1.4.9}$$

Expression (A.1.4.9) reveals that the inner product $(\mathbf{x}, \mathbf{x})$ can be considered as representing the square of the length of the vector $\mathbf{x}$.

Another interesting point we should point out is that for any two vectors $\mathbf{x}, \mathbf{y}$ the Schwarz inequality is always valid; that is,

$$|(\mathbf{x}, \mathbf{y})| \leq \|\mathbf{x}\|\,\|\mathbf{y}\| \tag{A.1.4.10}$$

which can be established as follows. For any arbitrary number α, we form

$$(\alpha \mathbf{x} + \mathbf{y}, \alpha \mathbf{x} + \mathbf{y}) = \|\alpha \mathbf{x} + \mathbf{y}\|^2$$
$$= \alpha^2 \|\mathbf{x}\|^2 + 2\alpha(\mathbf{x}, \mathbf{y}) + \|\mathbf{y}\|^2 \geq 0. \tag{A.1.4.11}$$

The last expression of (A.1.4.11) is a quadratic in α which will be nonnegative only if its discriminant is nonpositive, or

$$4(\mathbf{x}, \mathbf{y})^2 - 4\|\mathbf{x}\|^2\|\mathbf{y}\|^2 \leq 0. \tag{A.1.4.12}$$

Upon taking the positive square root, we finally arrive at the Schwarz inequality.

Orthogonality of Vectors

Two vectors $\mathbf{x}$ and $\mathbf{y}$ are said to be *orthogonal* to each other if their inner product $(\mathbf{x}, \mathbf{y})$ is zero. For example, it is easy to verify that the vectors

$$\mathbf{x}_1 = \begin{bmatrix} -1 \\ 1 \\ 0 \end{bmatrix}, \qquad \mathbf{x}_2 = \begin{bmatrix} 0 \\ 0 \\ 1 \end{bmatrix}, \qquad \mathbf{x}_3 = \begin{bmatrix} 1 \\ 1 \\ 0 \end{bmatrix} \tag{A.1.4.13}$$

form an orthogonal set for $(\mathbf{x}_1, \mathbf{x}_2) = (\mathbf{x}_2, \mathbf{x}_3) = (\mathbf{x}_3, \mathbf{x}_1) = 0$.

In general, a set of vectors $\mathbf{x}_1, \mathbf{x}_2, \ldots, \mathbf{x}_r$ is said to be an *orthonormal set* if the vectors satisfy the following conditions: For all $i, j = 1, \ldots, r$

a) $(\mathbf{x}_i, \mathbf{x}_j) = 0 \qquad (i \neq j)$

b) $(\mathbf{x}_i, \mathbf{x}_i) = 1.$
$$\tag{A.1.4.14}$$

It can be shown readily that, in an n-dimensional space, the set of vectors $\mathbf{x}_1, \mathbf{x}_2, \ldots, \mathbf{x}_n$ defined by

$$\mathbf{x}_1 = \begin{bmatrix} 1 \\ 0 \\ 0 \\ \vdots \\ 0 \end{bmatrix}, \mathbf{x}_2 = \begin{bmatrix} 0 \\ 1 \\ 0 \\ \vdots \\ 0 \end{bmatrix}, \ldots, \mathbf{x}_n = \begin{bmatrix} 0 \\ 0 \\ \vdots \\ 0 \\ 1 \end{bmatrix} \qquad (A.1.4.15)$$

satisfies the conditions in (A.1.4.14) and is therefore an orthonormal set.

A set of orthogonal vectors $\mathbf{x}_1, \mathbf{x}_2, \ldots, \mathbf{x}_r$ can always be normalized into a set or orthonormal vectors $\mathbf{y}_1, \mathbf{y}_2, \ldots, \mathbf{y}_r$ by means of the relation

$$\mathbf{y}_i = \frac{\mathbf{x}_i}{\|\mathbf{x}_i\|}. \qquad (A.1.4.16)$$

Example A.1.4.2 Let us normalize the set of orthogonal vectors in (A.1.4.13) by first determining their norms. Using (A.1.4.4), we obtain

$$\|\mathbf{x}_1\| = \sqrt{(-1)^2 + 1^2} = \sqrt{2},$$
$$\|\mathbf{x}_2\| = \sqrt{1^2} = 1,$$
$$\|\mathbf{x}_3\| = \sqrt{1^2 + 1^2} = \sqrt{2}.$$

Next, we form the vectors $\mathbf{y}_1, \mathbf{y}_2$, and $\mathbf{y}_3$ by means of (A.1.4.16):

$$\mathbf{y}_1 = \begin{bmatrix} \dfrac{-1}{\sqrt{2}} \\ \dfrac{1}{\sqrt{2}} \\ 0 \end{bmatrix}, \qquad \mathbf{y}_2 = \begin{bmatrix} 0 \\ 0 \\ 1 \end{bmatrix}, \qquad \mathbf{y}_3 = \begin{bmatrix} \dfrac{1}{\sqrt{2}} \\ \dfrac{1}{\sqrt{2}} \\ 0 \end{bmatrix}$$

which, of course, form an orthonormal set.

A.1.5 ELEMENTARY MATRIX ANALYSIS

In this section we introduce some of the basic tools in matrix analysis. Specifically, we discuss the following: linear transformations, elementary transformation of a matrix, eigenvalues and eigenvectors, similarity transformations, quadratic forms, functions of a matrix, and vector differential equations.

Linear Transformation

Consider the matrix equation

$$\mathbf{y} = \mathbf{Ax}, \qquad (A.1.5.1)$$

where $\mathbf{x}$ is an n-vector, $\mathbf{y}$ is an m-vector, and $\mathbf{A}$ is a matrix of order $m \times n$.

It is evident that in (A.1.5.1) matrix $\mathbf{A}$ transforms every vector $\mathbf{x}$ in an n-space $\mathbf{V}_n$ into a vector $\mathbf{y}$ in an m-space $\mathbf{V}_m$. Equation (A.1.5.1) represents a *linear transformation* T which can be described as a process that maps a vector $\mathbf{x}$ into a vector $\mathbf{y} = T(\mathbf{x})$ of $\mathbf{V}_m$ such that the following relationship holds:

$$T(\alpha_1 \mathbf{x}_1 + \alpha_2 \mathbf{x}_2) = T(\alpha_1 \mathbf{x}_1) + T(\alpha_2 \mathbf{x}_2)$$
$$= \alpha_1 T(\mathbf{x}_1) + \alpha_2 T(\mathbf{x}_2), \qquad (A.1.5.2)$$

where α_1 and α_2 are arbitrary constants. The first equality in (A.1.5.2) expresses the *property of additivity* which indicates that the transformation of the sum is equal to the sum of transformations. The second equality describes the *property of homogeneity* which states that if the vector *before* the transformation is multiplied by an arbitrary constant, then the vector *after* the transformation is also multiplied by the same constant. In other words, linearity implies both additivity and homogeneity.

Elementary Transformations

Suppose that a matrix $\mathbf{A}$ of order $m \times n$ is a linear transformation from a vector space $\mathbf{V}_n$ to a vector space $\mathbf{V}_m$. Frequently, in engineering applications, it is required to find two nonsingular square matrices $\mathbf{P}$ and $\mathbf{Q}$ of orders m and n, respectively, so that the given matrix $\mathbf{A}$ can be reduced to the following form

$$\mathbf{PAQ} = \begin{bmatrix} \mathbf{I}_r & \mathbf{0} \\ \mathbf{0} & \mathbf{0} \end{bmatrix}, \qquad (A.1.5.3)$$

where $\mathbf{I}_r$ is an identity matrix of order r with $r \leq m$ and $r \leq n$. The matrices $\mathbf{P}$ and $\mathbf{Q}$ have the effects of changing the bases of the vector spaces $\mathbf{V}_m$ and $\mathbf{V}_n$, respectively, and can be obtained by a sequence of *elementary transformations* of the following types:

1. Interchange of any two rows (or columns),

2. Multiplication of all the elements of a row (or column) by a nonzero constant,

3. Addition of all the elements of a row (or column) multiplied by a nonzero constant to the corresponding elements of another row (or column).

A square matrix of order n obtained from the identity matrix of the same order through (repeated) applications of elementary transformations is called an *elementary matrix*. Since there are three types of elementary transformations, there are three fundamental types of elementary matrices.

If the desired transformation is intended to affect the *rows* of a matrix $\mathbf{A}$, then $\mathbf{A}$ is *premultiplied* by an elementary matrix. If, on the other hand, $\mathbf{A}$ is *postmultiplied* by an elementary matrix, then the columns of $\mathbf{A}$ will be modified.

As a rule, each type of elementary matrix is obtained from the identity matrix, by performing upon it the same elementary transformation that we intend to perform upon the matrix $\mathbf{A}$. The following example will demonstrate the ideas involved.

Example A.1.5.1 Consider the matrices

$$\mathbf{S}_1 = \begin{bmatrix} 0 & 0 & 1 \\ 0 & 1 & 0 \\ 1 & 0 & 0 \end{bmatrix}, \qquad \mathbf{S}_2 = \begin{bmatrix} 1 & k & 0 \\ 0 & 1 & 0 \\ 0 & 0 & 1 \end{bmatrix}, \qquad \mathbf{S}_3 = \begin{bmatrix} 1 & 0 & 0 \\ 0 & k & 0 \\ 0 & 0 & 1 \end{bmatrix}.$$

$\mathbf{S}_1$ is obtained from $\mathbf{I}_3$ by interchanging the first and third rows (columns) of $\mathbf{I}_3$. If a matrix $\mathbf{A}$ is premultiplied (postmultiplied) by $\mathbf{S}_1$, the result is the matrix obtained from $\mathbf{A}$ by interchanging its first and third rows (columns). If $\mathbf{A}$ is pre-multiplied by $\mathbf{S}_2$, the resultant matrix will have the same elements as $\mathbf{A}$ with the exception of the first row, the elements of which are equal to the sum of the elements of the second row multiplied by k and the corresponding elements of the first row of $\mathbf{A}$. Likewise, the columns are affected similarly when $\mathbf{A}$ is postmultiplied by $\mathbf{S}_2$. Finally, if $\mathbf{A}$ is premultiplied (postmultiplied) by $\mathbf{S}_3$, the product matrix is the same as $\mathbf{A}$ except that the elements of the second row (column) are equal to the elements of the same row (column) of $\mathbf{A}$ multiplied by the constant k.

Note that elementary matrices are always nonsingular and that the inverse of an elementary matrix is again an elementary matrix. An immediate application of elementary transformations is the formulation of a systematic procedure for determining the rank of a matrix.

The *rank* of a matrix $\mathbf{A}$ of order $m \times n$ is defined as the maximum number of linearly independent rows (or columns) of $\mathbf{A}$. An equivalent definition can be stated as follows. The rank of a matrix $\mathbf{A}$ is equal to the order of the highest non-singular submatrix of $\mathbf{A}$.

Since the rank of a matrix $\mathbf{A}$ will not be altered if it is multiplied by a non-singular matrix, this implies that once $\mathbf{A}$ is reduced to the form (A.1.5.3), the rank of $\mathbf{A}$ is immediately known—the order r of the identity matrix.

Eigenvalues and Eigenvectors
In many engineering applications, it is quite common to arrive at the problem of determining the values of λ such that the system of homogeneous equations

$$(a_{11} - \lambda)u_1 + a_{12}u_2 + \cdots + a_{1n}u_n = 0$$
$$a_{21}u_1 + (a_{22} - \lambda)u_2 + \cdots + a_{2n}u_n = 0$$
$$\vdots \qquad\qquad\qquad\qquad\qquad \text{(A.1.5.4)}$$
$$a_{n1}u_1 + a_{n2}u_2 + \cdots + (a_{nn} - \lambda)u_n = 0$$

have nontrivial solutions. Writing (A.1.5.4) in vector form, we obtain

$$[\mathbf{A} - \lambda\mathbf{I}]\mathbf{u} = \mathbf{0}, \qquad\qquad\qquad \text{(A.1.5.5)}$$

where $\mathbf{u}$ is the n-vector having $u_1, u_2, \ldots, u_n$ as its components.

In order that $\mathbf{u}$ has a nontrivial solution it is necessary and sufficient that the system determinant of (A.1.5.4) or (A.1.5.5) vanishes, i.e.,

$$\det[\mathbf{A} - \lambda\mathbf{I}] = 0 \qquad\qquad\qquad \text{(A.1.5.6)}$$

which is called the *characteristic equation of* **A**. An expansion of the left-hand member of (A.1.5.6) yields a polynomial of degree n in λ,

$$g(\lambda) = \det[\mathbf{A} - \lambda\mathbf{I}] \qquad (A.1.5.7)$$

which is known as the *characteristic polynomial of* **A**. The zeros of the characteristic equation (A.1.5.6) are referred to as the *eigenvalues* (*characteristic values*) of **A**. Let $\lambda_1, \lambda_2, \ldots, \lambda_n$ be the n eigenvalues of **A** which may not be distinct. Then, for any $\lambda_i (i = 1, \ldots, n)$, (A.1.5.6) is satisfied and hence the equation

$$[\mathbf{A} - \lambda_i\mathbf{I}]\mathbf{u} = \mathbf{0} \qquad (i = 1, \ldots, n) \qquad (A.1.5.8)$$

has nontrivial solutions for **u**. Any vector **u** satisfying (A.1.5.8) is called an *eigenvector* (*characteristic vector*) of **A** associated with the eigenvalue λ_i.

Example A.1.5.2 Determine the eigenvalues and eigenvectors of **A**, where

$$\mathbf{A} = \begin{bmatrix} 0 & -1 \\ 2 & -3 \end{bmatrix}.$$

The characteristic polynomial of **A** is

$$g(\lambda) = \det[\mathbf{A} - \lambda\mathbf{I}] = \begin{vmatrix} -\lambda & -1 \\ 2 & -3 - \lambda \end{vmatrix}$$
$$= \lambda^2 + 3\lambda + 2$$

and the corresponding characteristic equation $g(\lambda) = 0$ yields $\lambda_1 = -1$ and $\lambda_2 = -2$. To determine the eigenvectors, let

$$\mathbf{u}_1 = \begin{bmatrix} u_{11} \\ u_{12} \end{bmatrix} \qquad \text{and} \qquad \mathbf{u}_2 = \begin{bmatrix} u_{21} \\ u_{22} \end{bmatrix}$$

be the eigenvectors of **A** associated with λ_1 and λ_2, respectively.
  Then, for $\lambda_1 = -1$, we have

$$[\mathbf{A} - \lambda_1\mathbf{I}]\mathbf{u}_1 = \mathbf{0}$$

or,

$$\begin{bmatrix} 1 & -1 \\ 2 & -2 \end{bmatrix}\begin{bmatrix} u_{11} \\ u_{12} \end{bmatrix} = \mathbf{0}$$

which yields the single independent equation

$$u_{11} - u_{12} = 0$$

or,

$$\mathbf{u}_1 = u_{11}\begin{bmatrix} 1 \\ 1 \end{bmatrix}$$

with u_{11} being an arbitrary number. For $\lambda_2 = -2$, we have

$$[\mathbf{A} - \lambda_2\mathbf{I}]\mathbf{u}_2 = \mathbf{0}$$

or,

$$\begin{bmatrix} 2 & -1 \\ 2 & -1 \end{bmatrix}\begin{bmatrix} u_{21} \\ u_{22} \end{bmatrix} = \mathbf{0}$$

which gives the equation

$$2u_{21} - u_{22} = 0$$

or,

$$\mathbf{u}_2 = u_{21}\begin{bmatrix} 1 \\ 2 \end{bmatrix}.$$

Again u_{21} is an arbitrary number. It should be pointed out that, for each eigenvalue, there exists an infinite number of eigenvectors. For instance, in the above example, the eigenvector $\mathbf{u}_1$ is proportional to

$$\begin{bmatrix} 1 \\ 1 \end{bmatrix},$$

and hence by assigning different values to the arbitrary number u_{11}, we find that different eigenvectors will result. However, we should emphasize the fact that there will be only one linearly independent eigenvector corresponding to each eigenvalue. This one-to-one correspondence between eigenvalues and eigenvectors is, of course, valid only when the eigenvalues of $\mathbf{A}$ are distinct. Thus, for convenience, the two linearly independent eigenvectors in the above example can be chosen to be

$$\mathbf{u}_1 = \begin{bmatrix} 1 \\ 1 \end{bmatrix} \quad \text{and} \quad \mathbf{u}_2 = \begin{bmatrix} 1 \\ 2 \end{bmatrix}$$

by setting both arbitrary constants u_{11} and u_{21} to unity. The results of the above example can be generalized into the following theorem.

Theorem A.1.5.1 Let the eigenvalues $\lambda_1, \lambda_2, \ldots, \lambda_n$ of a matrix, $\mathbf{A}$, of order n be all *distinct*. Then the corresponding nonzero eigenvectors $\mathbf{u}_1, \mathbf{u}_2, \ldots, \mathbf{u}_n$ are linearly independent.*

If the eigenvalues of a matrix are not all distinct, the eigenvectors can still be obtained in exactly the same manner as described above. However, in such a case, the number of eigenvectors, in general, will not be the same as the order of the matrix $\mathbf{A}$; instead it will be equal to the number of *distinct* eigenvalues of $\mathbf{A}$.

---

* For a complete proof, refer to F. E. Hohn [HO 1], Chapter 8.

Similarity Transformation—Diagonal Form

Suppose that the eigenvalues of a given matrix **A** are all distinct. Then a non-singular matrix **C** can be formed by taking the eigenvectors of **A** as its columns, since these eigenvectors are linearly independent by Theorem A.1.5.1. This non-singular matrix **C** can be used to transform the matrix **A** into a *diagonal matrix* Λ with the eigenvalues of **A** as its elements; i.e.,

$$\mathbf{S}^{-1}\mathbf{A}\mathbf{S} = \Lambda = \begin{bmatrix} \lambda_1 & 0 & \cdots & 0 \\ 0 & \lambda_2 & \cdots & 0 \\ & & \ddots & \\ 0 & 0 & \cdots & \lambda_n \end{bmatrix}. \tag{A.1.5.9}$$

A transformation defined by

$$\mathbf{B} = \mathbf{T}^{-1}\mathbf{A}\mathbf{T} \tag{A.1.5.10}$$

is called a *similarity transformation* and matrix **A** is said to be *similar* to matrix **B**. A very important property of similar matrices is that they have the same set of eigenvalues.

The above ideas can be demonstrated by the following examples.

Example A.1.5.3 Transform the matrix **A** into diagonal form, where

$$\mathbf{A} = \begin{bmatrix} 0 & -1 \\ 2 & -3 \end{bmatrix}.$$

The eigenvalues and eigenvectors of **A** are $\lambda_1 = -1$, $\lambda_2 = -2$, and

$$\mathbf{u}_1 = \begin{bmatrix} 1 \\ 1 \end{bmatrix}, \qquad \mathbf{u}_2 = \begin{bmatrix} 1 \\ 2 \end{bmatrix}$$

respectively. Next, the matrix can be formed by taking $\mathbf{u}_1$ and $\mathbf{u}_2$ as its columns; thus

$$\mathbf{S} = \begin{bmatrix} 1 & 1 \\ 1 & 2 \end{bmatrix}.$$

The inverse $\mathbf{S}^{-1}$ of **S** is given by

$$\mathbf{S}^{-1} = \begin{bmatrix} 2 & -1 \\ -1 & 1 \end{bmatrix}.$$

Hence, by means of (A.1.5.9), we obtain

$$\Lambda = \begin{bmatrix} 2 & -1 \\ -1 & 1 \end{bmatrix} \begin{bmatrix} 0 & -1 \\ 2 & -3 \end{bmatrix} \begin{bmatrix} 1 & 1 \\ 1 & 2 \end{bmatrix} = \begin{bmatrix} -1 & 0 \\ 0 & -2 \end{bmatrix}$$

which, of course, is a diagonal matrix with the eigenvalues of **A** as its diagonal elements.

Example A.1.5.4 Repeat the above example for the matrix $\mathbf{A}$ given by

$$\mathbf{A} = \begin{bmatrix} 0 & 1 \\ -2 & -2 \end{bmatrix}.$$

Here the eigenvalues and eigenvectors can be readily verified to be

$$\lambda_1 = -1 + j, \quad \lambda_2 = -1 - j, \quad \text{and} \quad \mathbf{u}_1 = \begin{bmatrix} 1 \\ -1 + j \end{bmatrix}, \quad \mathbf{u}_2 = \begin{bmatrix} 1 \\ -1 - j \end{bmatrix}.$$

The matrix $\mathbf{S}$ and its inverse $\mathbf{S}^{-1}$ are given by

$$\mathbf{S} = \begin{bmatrix} 1 & 1 \\ -1 + j & -1 - j \end{bmatrix}, \quad \mathbf{S}^{-1} = \begin{bmatrix} \dfrac{1-j}{2} & \dfrac{-j}{2} \\ \dfrac{1+j}{2} & \dfrac{j}{2} \end{bmatrix}.$$

Consequently,

$$\Lambda = \mathbf{S}^{-1}\mathbf{A}\mathbf{S} = \begin{bmatrix} -1 + j & 0 \\ 0 & -1 - j \end{bmatrix}.$$

The preceding two examples reveal the important fact that this procedure always yields a diagonal form regardless of whether the eigenvalues of $\mathbf{A}$ are real or complex as long as they are distinct. The above discussion can be stated formally as a theorem.

Theorem A.1.5.2 If the eigenvalues $\lambda_1, \lambda_2, \ldots, \lambda_n$ of a matrix $\mathbf{A}$ are distinct, then $\mathbf{A}$ can be transformed into a diagonal matrix composed of the eigenvalues $\lambda_1, \lambda_2, \ldots, \lambda_n$ as its diagonal elements by means of a similarity transformation

$$\Lambda = \mathbf{S}^{-1}\mathbf{A}\mathbf{S} = \text{diag}(\lambda_1, \lambda_2, \ldots, \lambda_n), \tag{A.1.5.11}$$

where $\mathbf{S}$ is a matrix whose columns are made up of the eigenvectors of $\mathbf{A}$, $\mathbf{S} = [\mathbf{u}_1 \ \mathbf{u}_2 \ \cdots \ \mathbf{u}_n]$, and $\text{diag}(\lambda_1, \lambda_2, \ldots, \lambda_n)$ is a diagonal matrix with the eigenvalues of $\mathbf{A}$ as its diagonal elements.*

It is interesting to note that, for any *real symmetric* matrix (i.e., a matrix of real elements with the property $a_{ij} = a_{ji}$ for all i and j), the eigenvalues are always real. Another important property of real symmetric matrices is that the eigenvectors associated with distinct eigenvalues are always orthogonal. Let us demonstrate these interesting points by considering the following example.

Example A.1.5.5 Given that

$$\mathbf{A} = \begin{bmatrix} 1 & \sqrt{3} & 0 \\ \sqrt{3} & -1 & 0 \\ 0 & 0 & 1 \end{bmatrix},$$

---

* For the complete proof, refer to E. A. Guillemin [GU 2], Chapter 3.

determine the eigenvalues and eigenvectors of $\mathbf{A}$ and then test the orthogonality property of the eigenvectors.

It can be readily shown that the eigenvalues of $\mathbf{A}$ are given by $\lambda_1 = 1, \lambda_2 = 2$, and $\lambda_3 = -2$. A set of eigenvectors corresponding to these eigenvalues is found to be

$$\mathbf{u}_1 = \begin{bmatrix} 0 \\ 0 \\ 1 \end{bmatrix}, \qquad \mathbf{u}_2 = \begin{bmatrix} \sqrt{3} \\ 1 \\ 0 \end{bmatrix}, \qquad \mathbf{u}_3 = \begin{bmatrix} 1 \\ -\sqrt{3} \\ 0 \end{bmatrix}.$$

Next, let us apply (A.1.4.7) to determine if $\mathbf{u}_1$, $\mathbf{u}_2$, and $\mathbf{u}_3$ are mutually orthogonal; we have

$$(\mathbf{u}_1, \mathbf{u}_2) = 0(\sqrt{3}) + 0(1) + 1(0) = 0,$$

$$(\mathbf{u}_1, \mathbf{u}_3) = 0(1) + 0(-\sqrt{3}) + 1(0) = 0,$$

$$(\mathbf{u}_2, \mathbf{u}_3) = \sqrt{3}(1) + 1(-\sqrt{3}) + 0(0) = 0.$$

As expected, they are indeed mutually orthogonal. We now state the afore-mentioned fundamental properties of real symmetric matrices as a theorem.

Theorem A.1.5.3 The eigenvalues of a real symmetric matrix are real and the eigenvectors associated with distinct eigenvalues are orthogonal to each other.*

Similarity Transformation—Jordan Canonical Form

We have just illustrated that a square matrix $\mathbf{A}$ with distinct eigenvalues can always be reduced to a diagonal form by means of a similarity transformation. However, for systems with multiple eigenvalues (i.e., at least two of the eigenvalues of $\mathbf{A}$ are identical), $\mathbf{A}$ (of order n) can be reduced to a diagonal form only if there exist n linearly independent eigenvectors. It can be shown that any square matrix can be reduced by means of a similarity transformation to a *Jordan canonical form*† which can be considered as a generalization of the diagonal form. Before we define the Jordan canonical form and underline the procedure for the reduction of a matrix to a Jordan canonical form, we shall first demonstrate the ideas involved by means of the following example.

Example A.1.5.6 The characteristic equation of the matrix

$$\mathbf{A} = \begin{bmatrix} 0 & 0 & 1 \\ 1 & 0 & -3 \\ 0 & 1 & 3 \end{bmatrix}$$

is found to be $(\lambda - 1)^3 = 0$ which yields the three identical eigenvalues $\lambda_1 = \lambda_2 = \lambda_3 = 1$.

* For the proof, refer to R. Bellman [BE 1], Chapter 3.
† For example, refer to B. Friedman [FR 2], Chapter 2.

To determine the eigenvectors, we write for $\lambda_1 = 1$

$$[\mathbf{A} - \lambda_1\mathbf{I}]\mathbf{u}_1 = \mathbf{0},$$

or

$$\begin{bmatrix} -1 & 0 & 1 \\ 1 & -1 & -3 \\ 0 & 1 & 2 \end{bmatrix} \begin{bmatrix} u_{11} \\ u_{12} \\ u_{13} \end{bmatrix} = \mathbf{0}$$

which yields the following linearly independent equations

$$-u_{11} + u_{13} = 0$$

$$u_{12} + 2u_{13} = 0.$$

Hence,

$$\mathbf{u}_1 = u_{13} \begin{bmatrix} 1 \\ -2 \\ 1 \end{bmatrix}$$

which indicates that there is only one linearly independent eigenvector associated with the eigenvalue of unity. In other words, the given matrix **A** cannot be reduced to a diagonal form since we need three linearly independent eigenvectors to form the matrix **S**. The next logical step would be to reduce **A** to a Jordan canonical form. For convenience, let $u_{13} = 1$ so that

$$\mathbf{u}_1 = \begin{bmatrix} 1 \\ -2 \\ 1 \end{bmatrix}.$$

Then, we proceed to determine the second linearly independent vector $\mathbf{u}_2$ by means of the following relationship:

$$\mathbf{A}\mathbf{u}_2 = \lambda_1\mathbf{u}_2 + \mathbf{u}_1$$

or

$$[\mathbf{A} - \lambda_1\mathbf{I}]\mathbf{u}_2 = \mathbf{u}_1.$$

In terms of components, we have

$$\begin{bmatrix} -1 & 0 & 1 \\ 1 & -1 & -3 \\ 0 & 1 & 2 \end{bmatrix} \begin{bmatrix} u_{21} \\ u_{22} \\ u_{23} \end{bmatrix} = \begin{bmatrix} 1 \\ -2 \\ 1 \end{bmatrix}$$

which yields the linearly *independent* equations

$$-u_{21} + u_{23} = 1$$

$$u_{22} + 2u_{23} = 1.$$

Thus, by letting $u_{23} = 1$, we see that $\mathbf{u}_2$ becomes

$$\mathbf{u}_2 = \begin{bmatrix} 0 \\ -1 \\ 1 \end{bmatrix}.$$

Repeating the same process, we can find the third linearly independent vector $\mathbf{u}_3$ by solving

$$[\mathbf{A} - \lambda_1 \mathbf{I}]\mathbf{u}_3 = \mathbf{u}_2$$

which yields

$$\mathbf{u}_3 = u_{33}\begin{bmatrix} 1 \\ -1 \\ 1 \end{bmatrix} = \begin{bmatrix} 1 \\ -1 \\ 1 \end{bmatrix},$$

where u_{33} is taken to be unity for convenience. As before, let the matrix $\mathbf{S}$ be formed in such a way that the vectors $\mathbf{u}_1$, $\mathbf{u}_2$, and $\mathbf{u}_3$ appear as its columns

$$\mathbf{S} = [\mathbf{u}_1 \quad \mathbf{u}_2 \quad \mathbf{u}_3] = \begin{bmatrix} 1 & 0 & 1 \\ -2 & -1 & -1 \\ 1 & 1 & 1 \end{bmatrix};$$

and its inverse $\mathbf{S}^{-1}$ is

$$\mathbf{S}^{-1} = \frac{1}{-1}\begin{bmatrix} 0 & 1 & 1 \\ 1 & 0 & -1 \\ -1 & -1 & -1 \end{bmatrix} = \begin{bmatrix} 0 & -1 & -1 \\ -1 & 0 & 1 \\ 1 & 1 & 1 \end{bmatrix}.$$

Next, we form the following matrix

$$\mathbf{J} \triangleq \mathbf{S}^{-1}\mathbf{A}\mathbf{S} = \begin{bmatrix} 0 & -1 & -1 \\ -1 & 0 & 1 \\ 1 & 1 & 1 \end{bmatrix}\begin{bmatrix} 0 & 0 & 1 \\ 1 & 0 & -3 \\ 0 & 1 & 3 \end{bmatrix}\begin{bmatrix} 1 & 0 & 1 \\ -2 & -1 & -1 \\ 1 & 1 & 1 \end{bmatrix} = \begin{bmatrix} 1 & 1 & 0 \\ 0 & 1 & 1 \\ 0 & 0 & 1 \end{bmatrix}.$$

It should be pointed out that the diagonal elements of the matrix $\mathbf{J}$ are the eigenvalues of $\mathbf{A}$, which in this case are equal to unity, and that the elements immediately to the right of the principal diagonal (*superdiagonal* elements) are always equal to one. In fact, the matrix $\mathbf{J}$ is called the Jordan canonical form of $\mathbf{A}$.

Remark. The vectors $\mathbf{u}_1$, $\mathbf{u}_2$, and $\mathbf{u}_3$ so obtained in the above example are always linearly independent* and $\mathbf{u}_2$ and $\mathbf{u}_3$ are sometimes called *generalized eigenvectors*

---

* For proof, see B. Friedman [FR 2], Chapter 2.

of $\mathbf{A}$ associated with the eigenvalue λ_1. Note that the Jordan canonical form representation may not be unique for a given matrix depending upon the number of linearly independent eigenvectors the matrix possesses. For example, if a third-order matrix has λ_1 as the eigenvalue of *multiplicity* three (i.e., three identical eigenvalues of the same value), it may be reduced, by means of a similarity transformation, to one of the three forms

$$\mathbf{J}_1 = \begin{bmatrix} \lambda_1 & 0 & 0 \\ 0 & \lambda_1 & 0 \\ 0 & 0 & \lambda_1 \end{bmatrix}, \quad \mathbf{J}_2 = \begin{bmatrix} \lambda_1 & 0 & 0 \\ 0 & \lambda_1 & 1 \\ 0 & 0 & \lambda_1 \end{bmatrix}, \quad \mathbf{J}_3 = \begin{bmatrix} \lambda_1 & 1 & 0 \\ 0 & \lambda_1 & 1 \\ 0 & 0 & \lambda_1 \end{bmatrix}$$

depending upon whether $\mathbf{A}$ has, respectively, three, two, or one linearly independent eigenvectors.

We are now in a position to generalize the above procedure for transforming an arbitrary square matrix $\mathbf{A}$ to a Jordan canonical form through a similarity transformation. We shall first state the following theorem without proof.*

Theorem A.1.5.4 Every matrix $\mathbf{A}$ of order n can be transformed to a *Jordan canonical form* $\mathbf{J}$ by means of a similarity transformation:

$$\mathbf{J} = \mathbf{S}^{-1}\mathbf{A}\mathbf{S}, \tag{A.1.5.12}$$

where

$$\mathbf{J} = \begin{bmatrix} \mathbf{J}_{k_1}(\lambda_1) & 0 & \cdots & 0 \\ 0 & \mathbf{J}_{k_2}(\lambda_2) & \cdots & 0 \\ \vdots & & & \\ 0 & 0 & \cdots & \mathbf{J}_{k_r}(\lambda_r) \end{bmatrix} \tag{A.1.5.13}$$

with $k_1 + k_2 + \cdots + k_r = n$ and λ_i being the eigenvalues of $\mathbf{A}$ not necessarily distinct. The *Jordan block* $\mathbf{J}_{k_i}(\lambda_i)$ $(i = 1, \ldots, r)$ denotes a $k_i \times k_i$-matrix of the form

$$\mathbf{J}_{k_i}(\lambda_i) = \begin{bmatrix} \lambda_i & 1 & 0 & \cdots & 0 & 0 \\ 0 & \lambda_i & 1 & \cdots & 0 & 0 \\ 0 & 0 & \lambda_i & \cdots & 0 & 0 \\ & & & \ddots & & \\ 0 & 0 & 0 & \cdots & \lambda_i & 1 \\ 0 & 0 & 0 & \cdots & 0 & \lambda_i \end{bmatrix}. \tag{A.1.5.14}$$

where $\mathbf{J}_1(\lambda) = \lambda$.

A close examination of the above theorem reveals the following facts:

1. The number of the Jordan blocks is equal to the number of *eigenvectors* of $\mathbf{A}$.

---

* For proof, see B. Friedman, *op. cit.*

2. The elements along the principal diagonal of $\mathbf{J}$, $\lambda_1, \lambda_2, \ldots, \lambda_r$, are the eigenvalues of $\mathbf{A}$ and they need not be distinct.

3. In each of the Jordan blocks $\mathbf{J}_{k_i}(\lambda_i)$, the elements immediately to the right of the principal diagonal (the superdiagonal elements) are always equal to unity.

In view of (A.1.5.12), it is obvious that, for a given matrix $\mathbf{A}$, once matrix $\mathbf{S}$ is found, a Jordan canonical form can be determined immediately by simple matrix multiplications. Consequently, our next step is to determine matrix $\mathbf{S}$. To be specific, let $\mathbf{A}$ be an $n \times n$-matrix which consists of p equal eigenvalues λ_1 as well as other distinct eigenvalues $\lambda_{p+1}, \lambda_{p+2}, \ldots, \lambda_n$. We shall consider the following two cases: (1) there are p linearly independent eigenvectors of $\mathbf{A}$ associated with the eigenvalue λ_1, and (2) there is only one linearly independent eigenvector for λ_1. Of course, there are other cases where the number of linearly independent eigenvectors is less than p but greater than one. However, no particular difficulty will be involved in such cases since techniques similar to the one in case 2 can be employed to determine the matrix $\mathbf{S}$.

Case 1. There are p linearly independent eigenvectors for λ_1. Together with the $(n - p)$ linearly independent eigenvectors for the distinct eigenvalues λ_{p+1}, $\lambda_{p+2}, \ldots, \lambda_n$, we have a total of n eigenvectors in this case. Hence, if we denote these n eigenvectors by $\mathbf{u}_1, \mathbf{u}_2, \ldots, \mathbf{u}_n$ with the first p eigenvectors associated with λ_1 and the remaining $(n - p)$ eigenvectors $\mathbf{u}_{p+1}, \ldots, \mathbf{u}_n$ associated with $\lambda_{p+1}, \ldots, \lambda_n$, respectively, the Jordan canonical form will become a diagonal matrix, viz.,

$$\mathbf{J} = \mathbf{S}^{-1}\mathbf{A}\mathbf{S} = \text{diag}(\lambda_1, \ldots, \lambda_1, \lambda_{p+1}, \ldots, \lambda_n) \qquad (A.1.5.15)$$

with the matrix $\mathbf{S}$ consisting of the eigenvectors as its columns:

$$\mathbf{S} = [\mathbf{u}_1 \quad \mathbf{u}_2 \quad \cdots \quad \mathbf{u}_n]. \qquad (A.1.5.16)$$

Example A.1.5.7   Consider the matrix

$$\mathbf{A} = \begin{bmatrix} 7 & 4 & -4 \\ 4 & 7 & -4 \\ -1 & -1 & 4 \end{bmatrix}.$$

The characteristic equation of $\mathbf{A}$ is

$$g(\lambda) = \det[\mathbf{A} - \lambda\mathbf{I}]$$
$$= -\lambda^3 + 18\lambda^2 - 81\lambda + 108 = 0$$

or

$$(\lambda - 3)^2(\lambda - 12) = 0$$

which yields two distinct eigenvalues: $\lambda_1 = 3$ of multiplicity two, and $\lambda_3 = 12$ of multiplicity one. To determine the eigenvectors, we write, for $\lambda_1 = 3$,

$$[\mathbf{A} - \lambda_1\mathbf{I}]\mathbf{u}_1 = \mathbf{0}$$

or

$$\begin{bmatrix} 4 & 4 & -4 \\ 4 & 4 & -4 \\ -1 & -1 & 1 \end{bmatrix} \begin{bmatrix} u_{11} \\ u_{12} \\ u_{13} \end{bmatrix} = \mathbf{0}$$

which yields only one linearly independent equation

$$u_{11} + u_{12} - u_{13} = 0$$

or

$$u_{13} = u_{11} + u_{12}.$$

There are two linearly independent eigenvectors associated with $\lambda_1 = 3$. Since we have one equation with three unknowns, two of them can be assigned arbitrarily. Thus $\mathbf{u}_1$ can be obtained, for instance, by setting $u_{11} = 1$ and $u_{12} = 0$; that is,

$$\mathbf{u}_1 = \begin{bmatrix} 1 \\ 0 \\ 1 \end{bmatrix}.$$

Likewise, if we set $u_{11} = 0$ and $u_{12} = 1$, we obtain

$$\mathbf{u}_2 = \begin{bmatrix} 0 \\ 1 \\ 1 \end{bmatrix}.$$

For the third eigenvector, we write

$$[\mathbf{A} - \lambda_3 \mathbf{I}]\mathbf{u}_3 = \mathbf{0}$$

or

$$\begin{bmatrix} -5 & 4 & -4 \\ 4 & -5 & -4 \\ -1 & -1 & -8 \end{bmatrix} \begin{bmatrix} u_{31} \\ u_{32} \\ u_{33} \end{bmatrix} = \mathbf{0}$$

which gives the following linearly independent equations

$$-5u_{31} + 4u_{32} - 4u_{33} = 0$$
$$-u_{31} - u_{32} - 8u_{33} = 0.$$

Hence, by setting $u_{32} = 4$, $\mathbf{u}_3$ becomes

$$\mathbf{u}_3 = \begin{bmatrix} 4 \\ 4 \\ -1 \end{bmatrix}.$$

Our next step is to form the matrix $\mathbf{S}$ and to calculate $\mathbf{S}^{-1}$

$$\mathbf{S} = [\mathbf{u}_1 \quad \mathbf{u}_2 \quad \mathbf{u}_3] = \begin{bmatrix} 1 & 0 & 4 \\ 0 & 1 & 4 \\ 1 & 1 & -1 \end{bmatrix}, \quad \mathbf{S}^{-1} = \begin{bmatrix} \frac{5}{9} & -\frac{4}{9} & \frac{4}{9} \\ -\frac{4}{9} & \frac{5}{9} & \frac{4}{9} \\ \frac{1}{9} & \frac{1}{9} & -\frac{1}{9} \end{bmatrix}.$$

Therefore, applying (A.1.5.15), we finally obtain

$$\mathbf{J} = \mathbf{S}^{-1}\mathbf{AS}$$

$$= \begin{bmatrix} \frac{5}{9} & -\frac{4}{9} & \frac{4}{9} \\ -\frac{4}{9} & \frac{5}{9} & \frac{4}{9} \\ \frac{1}{9} & \frac{1}{9} & -\frac{1}{9} \end{bmatrix} \begin{bmatrix} 7 & 4 & -4 \\ 4 & 7 & -4 \\ -1 & -1 & 4 \end{bmatrix} \begin{bmatrix} 1 & 0 & 4 \\ 0 & 1 & 4 \\ 1 & 1 & -4 \end{bmatrix}$$

$$= \begin{bmatrix} \frac{5}{9} & -\frac{4}{9} & \frac{4}{9} \\ -\frac{4}{9} & \frac{5}{9} & \frac{4}{9} \\ \frac{1}{9} & \frac{1}{9} & -\frac{1}{9} \end{bmatrix} \begin{bmatrix} 3 & 0 & 48 \\ 0 & 3 & 48 \\ 3 & 3 & -12 \end{bmatrix} = \begin{bmatrix} 3 & 0 & 0 \\ 0 & 3 & 0 \\ 0 & 0 & 12 \end{bmatrix}.$$

Case 2. There is only one linearly independent eigenvector associated with λ_1. Together with the $(n - p)$ linearly independent eigenvectors for the distinct eigenvalues $\lambda_{p+1}, \lambda_{p+2}, \ldots, \lambda_n$, we have a total of $(n - p + 1)$ linearly independent eigenvectors. To form the matrix $\mathbf{S}$, we need $(p - 1)$ more linearly independent vectors $\mathbf{u}_2, \mathbf{u}_3, \ldots, \mathbf{u}_p$ which can be easily computed according to the following relationship:

$$(\mathbf{A} - \lambda_1\mathbf{I})\mathbf{u}_{k+1} = \mathbf{u}_k \qquad (k = 1, 2, \ldots, p - 1). \tag{A.1.5.17}$$

Of course, the eigenvectors $\mathbf{u}_1, \mathbf{u}_{p+1}, \mathbf{u}_{p+2}, \ldots, \mathbf{u}_n$ associated with $\lambda_1, \lambda_{p+1}, \lambda_{p+2}, \ldots, \lambda_n$, respectively, can be found in the usual manner, i.e.,

$$(\mathbf{A} - \lambda_i\mathbf{I})\mathbf{u}_i = \mathbf{0}, \qquad i = 1, p + 1, p + 2, \ldots, n. \tag{A.1.5.18}$$

From (A.1.5.17) and (A.1.5.18), we obtain the following set of equations

$$\mathbf{Au}_1 = \lambda_1\mathbf{u}_1$$

$$\mathbf{Au}_2 = \lambda_1\mathbf{u}_2 + \mathbf{u}_1$$

$$\mathbf{Au}_3 = \lambda_1\mathbf{u}_3 + \mathbf{u}_2$$

$$\vdots$$

$$\mathbf{Au}_p = \lambda_1\mathbf{u}_p + \mathbf{u}_{p-1} \tag{A.1.5.19}$$

$$\mathbf{Au}_{p+1} = \lambda_{p+1}\mathbf{u}_{p+1}$$

$$\vdots$$

$$\mathbf{Au}_n = \lambda_n\mathbf{u}_n.$$

Let us define, again, the matrix $\mathbf{S}$ as the matrix consisting of the vectors $\mathbf{u}_1, \mathbf{u}_2, \ldots, \mathbf{u}_n$ as its columns; i.e.,

$$\mathbf{S} = [\mathbf{u}_1 \quad \mathbf{u}_2 \quad \cdots \quad \mathbf{u}_n]. \tag{A.1.5.20}$$

Then the set of equations in (A.1.5.19) reduces to

$$\mathbf{AS} = \mathbf{SJ} \quad \text{or} \quad \mathbf{J} = \mathbf{S}^{-1}\mathbf{AS},* \tag{A.1.5.21}$$

where $\mathbf{J}$ is the Jordan canonical form and is given by

$$\mathbf{J} = \begin{bmatrix} \lambda_1 & 1 & 0 & \cdots & 0 & 0 & 0 & 0 & \cdots & 0 \\ 0 & \lambda_1 & 1 & \cdots & 0 & 0 & 0 & 0 & \cdots & 0 \\ 0 & 0 & \lambda_1 & \cdots & 0 & 0 & 0 & 0 & \cdots & 0 \\ \vdots & & & & & & & & & \\ 0 & 0 & 0 & \cdots & \lambda_1 & 1 & 0 & 0 & \cdots & 0 \\ 0 & 0 & 0 & \cdots & 0 & \lambda_1 & 0 & 0 & \cdots & 0 \\ 0 & 0 & 0 & \cdots & 0 & 0 & \lambda_{p+1} & 0 & \cdots & 0 \\ 0 & 0 & 0 & \cdots & 0 & 0 & 0 & \lambda_{p+2} & \cdots & 0 \\ \vdots & & & & & & & & & \\ 0 & 0 & 0 & \cdots & 0 & 0 & 0 & 0 & \cdots & \lambda_n \end{bmatrix} \tag{A.1.5.22}$$

Example A.1.5.8   Consider the matrix

$$\mathbf{A} = \begin{bmatrix} -2 & 1 \\ -1 & 0 \end{bmatrix}.$$

The characteristic equation is

$$\lambda^2 + 2\lambda + 1 = (\lambda + 1)^2 = 0$$

which yields the single eigenvalue $\lambda_1 = -1$ of multiplicity two. The eigenvector $\mathbf{u}_1$ can be obtained by applying the first equation of (A.1.5.19), which gives

$$\begin{bmatrix} -1 & 1 \\ -1 & 1 \end{bmatrix} \begin{bmatrix} u_{11} \\ u_{12} \end{bmatrix} = 0.$$

Hence

$$\mathbf{u}_1 = \begin{bmatrix} 1 \\ 1 \end{bmatrix}$$

with $u_{11} = 1$, a conveniently assigned value. This is the only eigenvector for λ_1.

---

* For proof of linear independence of the vectors defined in (A.1.5.19) refer to B. Friedman, *op. cit.*

Thus, applying the second equation of (A.1.5.20), we can obtain a generalized eigenvector $\mathbf{u}_2$ as follows:

$$\begin{bmatrix} -1 & 1 \\ -1 & 1 \end{bmatrix} \begin{bmatrix} u_{21} \\ u_{22} \end{bmatrix} = \begin{bmatrix} 1 \\ 1 \end{bmatrix}$$

which yields the equation

$$-u_{21} + u_{22} = 1.$$

For convenience let $u_{21} = 0$, and we find that

$$\mathbf{u}_2 = \begin{bmatrix} 0 \\ 1 \end{bmatrix}.$$

Therefore, $\mathbf{S}$ and its inverse $\mathbf{S}^{-1}$ are given by

$$\mathbf{S} = \begin{bmatrix} 1 & 0 \\ 1 & 1 \end{bmatrix}, \qquad \mathbf{S}^{-1} = \begin{bmatrix} 1 & 0 \\ -1 & 1 \end{bmatrix},$$

and finally we obtain

$$\mathbf{J} = \mathbf{S}^{-1}\mathbf{A}\mathbf{S} = \begin{bmatrix} 1 & 0 \\ -1 & 1 \end{bmatrix} \begin{bmatrix} -2 & 1 \\ -1 & 0 \end{bmatrix} \begin{bmatrix} 1 & 0 \\ 1 & 1 \end{bmatrix} = \begin{bmatrix} -1 & 1 \\ 0 & -1 \end{bmatrix}.$$

Quadratic Forms

Consider a linear transformation of the form

$$y_1 = a_{11}x_1 + a_{12}x_2 + \cdots + a_{1n}x_n$$
$$y_2 = a_{21}x_1 + a_{22}x_2 + \cdots + a_{2n}x_n$$
$$\vdots \qquad\qquad\qquad\qquad\qquad\qquad\qquad \text{(A.1.5.23)}$$
$$y_n = a_{n1}x_1 + a_{n2}x_2 + \cdots + a_{nn}x_n$$

which, in vector notation, can be expressed as

$$\mathbf{y} = \mathbf{A}\mathbf{x}, \qquad\qquad\qquad \text{(A.1.5.24)}$$

where

$$\mathbf{y} = \begin{bmatrix} y_1 \\ y_2 \\ \vdots \\ y_n \end{bmatrix}, \qquad \mathbf{A} = \begin{bmatrix} a_{11} & a_{12} & \cdots & a_{1n} \\ a_{21} & a_{22} & \cdots & a_{2n} \\ \vdots & & & \\ a_{n1} & a_{n2} & \cdots & a_{nn} \end{bmatrix}, \qquad \text{and} \qquad \mathbf{x} = \begin{bmatrix} x_1 \\ x_2 \\ \vdots \\ x_n \end{bmatrix}.$$

If the inner product of $\mathbf{x}$ and $\mathbf{y}$, denoted by F, is formed, we obtain

$$F = (\mathbf{x}, \mathbf{y}) = (\mathbf{x}, \mathbf{Ax}) = \mathbf{x}^T \mathbf{Ax}$$

$$= a_{11}x_1^2 + a_{12}x_2x_1 + \cdots + a_{1n}x_nx_1$$

$$+ a_{21}x_1x_2 + a_{22}x_2^2 + \cdots + a_{2n}x_nx_2 \qquad (A.1.5.25)$$

$$\vdots$$

$$+ a_{n1}x_1x_n + a_{n2}x_2x_n + \cdots + a_{nn}x_n^2.$$

The function of n variables in the right-hand side of (A.1.5.25) is called a *quadratic form* which can be abbreviated through the use of a double summation; thus

$$F = \sum_{i,j=1}^{n} a_{ij}x_ix_j. \qquad (A.1.5.26)$$

The matrix $\mathbf{A}$ of the transformation (A.1.5.23) is called the *matrix of the quadratic form*. Quadratic forms appear frequently in engineering applications of various types such as energy functions in electrical networks, Liapunov functions in problems of stability analysis, performance indices in problems of optimization, just to name a few. However, the common interest in quadratic forms usually lies in the testing of a given quadratic form to determine if it is positive definite or positive semidefinite. A quadratic form F is said to be *positive definite* (*negative definite*) if, for all nonzero vectors $\mathbf{x}$, the condition

$$F = \sum_{i,j=1}^{n} a_{ij}x_ix_j > 0 \qquad (<0) \qquad (A.1.5.27)$$

is satisfied. If $F \geq 0 \, (\leq 0)$, it is called *positive semidefinite* (*negative semidefinite*).

The quadratic form of a matrix $\mathbf{A}$ of second order can be expanded in the form

$$F = (\mathbf{x}, \mathbf{Ax}) = a_{11}x_1^2 + (a_{12} + a_{21})x_1x_2 + a_{22}x_2^2. \qquad (A.1.5.28)$$

Two important properties can be derived as follows. In (A.1.5.28), it is observed that both a_{12} and a_{21} are coefficients of x_1x_2. Thus, if we define a new symmetric matrix $\mathbf{B}$ such that its coefficients b_{ij} are given by

$$b_{ij} = \tfrac{1}{2}(a_{ij} + a_{ji}), \qquad i,j = 1, 2, \qquad (A.1.5.29)$$

we have

$$F = (\mathbf{x}, \mathbf{Bx}) = b_{11}x_1^2 + (b_{12} + b_{21})x_1x_2 + b_{22}x_2^2$$

$$= a_{11}x_1^2 + \tfrac{1}{2}(a_{12} + a_{21} + a_{21} + a_{12})x_1x_2 + a_{22}x_2^2$$

$$= a_{11}x_1^2 + (a_{12} + a_{21})x_1x_2 + a_{22}x_2^2$$

$$= (\mathbf{x}, \mathbf{Ax}). \qquad (A.1.5.30)$$

Of course, the same can be extended to an n-dimensional form by writing

$$b_{ij} = \tfrac{1}{2}(a_{ij} + a_{ji}), \qquad i, j = 1, 2, \ldots, n. \tag{A.1.5.31}$$

Thus, we conclude that the matrix associated with a quadratic form can always be converted into a symmetric matrix, and, without loss of generality, the matrix of the quadratic form can be assumed to be symmetric.

Another important property can be obtained by considering the *bilinear form*

$$(\mathbf{Ax}, \mathbf{y}) = y_1 \left(\sum_{j=1}^{n} a_{1j}x_j \right) + y_2 \left(\sum_{j=1}^{n} a_{2j}x_j \right) + \cdots + y_n \left(\sum_{j=1}^{n} a_{nj}x_j \right)$$

or, rearranging,

$$(\mathbf{Ax}, \mathbf{y}) = x_1 \left(\sum_{i=1}^{n} a_{i1}y_i \right) + x_2 \left(\sum_{i=1}^{n} a_{i2}y_i \right) + \cdots + x_n \left(\sum_{i=1}^{n} a_{in}y_i \right).$$

Hence, we have

$$(\mathbf{Ax}, \mathbf{y}) = (\mathbf{x}, \mathbf{A}^T\mathbf{y}), \tag{A.1.5.32}$$

where $\mathbf{A}^T$ is, of course, the transpose of $\mathbf{A}$. This important relation indicates that the effect of the transformation $\mathbf{A}$ on $\mathbf{x}$ is equivalent to that of $\mathbf{A}^T$ on $\mathbf{y}$ as far as the inner product is concerned.

Let us apply the property expressed in (A.1.5.32) to reduce a *real symmetric* matrix $\mathbf{A}$ with *distinct* eigenvalues to a diagonal form. Let the eigenvalues of $\mathbf{A}$ be denoted by $\lambda_1, \lambda_2, \ldots, \lambda_n$ and let $\mathbf{u}_1, \mathbf{u}_2, \ldots, \mathbf{u}_n$ be the corresponding set of eigenvectors which are *normalized* by the condition that

$$\|\mathbf{u}_i\| = 1, \qquad i = 1, \ldots, n. \tag{A.1.5.33}$$

According to Theorem A.1.5.3, the eigenvectors associated with distinct eigenvalues of a real symmetric matrix are orthogonal to each other. Hence, together with (A.1.5.33), the set of eigenvectors $\mathbf{u}_i$ forms an orthonormal set satisfying the following conditions

$$(\mathbf{u}_i, \mathbf{u}_j) = \delta_{ij}, \qquad i, j = 1, \ldots, n, \tag{A.1.5.34}$$

where δ_{ij}, *the Kronecker delta*, is defined by

$$\delta_{ij} = \begin{cases} 0 & (i \neq j) \\ 1 & (i = j). \end{cases} \tag{A.1.5.35}$$

In view of (A.1.5.34), if we define the matrix $\mathbf{S}$ by

$$\mathbf{S} = [\mathbf{u}_1 \quad \mathbf{u}_2 \quad \cdots \quad \mathbf{u}_n], \tag{A.1.5.36}$$

then the following relation is satisfied:

$$\mathbf{S}^T\mathbf{S} = \mathbf{S}\mathbf{S}^T = \mathbf{I}. \tag{A.1.5.37}$$

Expression (A.1.5.37) defines an *orthogonal matrix*. Thus, for an orthogonal matrix, the inverse is exactly equal to its transpose

$$\mathbf{S}^{-1} = \mathbf{S}^T. \tag{A.1.5.38}$$

Now, let us make a change of coordinates by the relations

$$x_1 = u_{11}y_1 + u_{21}y_2 + \cdots + u_{n1}y_n$$

$$x_2 = u_{12}y_1 + u_{22}y_2 + \cdots + u_{n2}y_n$$

$$\vdots \tag{A.1.5.39}$$

$$x_n = u_{1n}y_1 + u_{2n}y_2 + \cdots + u_{nn}y_n$$

or, in vector notation,

$$\mathbf{x} = \mathbf{S}\mathbf{y}. \tag{A.1.5.40}$$

In view of (A.1.5.32) and (A.1.5.40), the inner product $(\mathbf{x}, \mathbf{A}\mathbf{x})$ can be expanded into the following form:

$$(\mathbf{x}, \mathbf{A}\mathbf{x}) = (\mathbf{S}\mathbf{y}, \mathbf{A}\mathbf{S}\mathbf{y})$$

$$= (\mathbf{y}, \mathbf{S}^T\mathbf{A}\mathbf{S}\mathbf{y}). \tag{A.1.5.41}$$

To further simplify the above expression, we recall that $\mathbf{A}\mathbf{u}_i = \lambda_i\mathbf{u}_i$ and hence find that

$$\mathbf{A}\mathbf{S} = [\lambda_1\mathbf{u}_1 \quad \lambda_2\mathbf{u}_2 \quad \cdots \quad \lambda_n\mathbf{u}_n]$$

$$= \mathbf{S}\mathbf{\Lambda}$$

which, with the help of (A.1.5.38), yields

$$\mathbf{S}^{-1}\mathbf{A}\mathbf{S} = \mathbf{S}^T\mathbf{A}\mathbf{S} = \mathbf{\Lambda} = \mathrm{diag}(\lambda_1, \lambda_2, \ldots, \lambda_n). \tag{A.1.5.42}$$

Consequently, we obtain

$$(\mathbf{x}, \mathbf{A}\mathbf{x}) = (\mathbf{y}, \mathbf{\Lambda}\mathbf{y})$$

$$= \lambda_1 y_1^2 + \lambda_2 y_2^2 + \cdots + \lambda_n y_n^2 \tag{A.1.5.43}$$

which indicates that the quadratic form is positive definite if and only if all the eigenvalues of $\mathbf{A}$ are positive. Similarly, the quadratic form $(\mathbf{x}, \mathbf{A}\mathbf{x})$ is positive semidefinite if and only if all the eigenvalues of $\mathbf{A}$ are nonnegative and at least one of them has the value of zero. A *necessary and sufficient* condition (*Sylvester's criterion*) that the quadratic form $(\mathbf{x}, \mathbf{A}\mathbf{x})$, where $\mathbf{A}$ is a real symmetric matrix of order n, be positive definite is that the determinant of $\mathbf{A}$ as well as all the successive principal minors of $\mathbf{A}$ are positive; that is,

$$a_{11} > 0, \begin{vmatrix} a_{11} & a_{12} \\ a_{21} & a_{22} \end{vmatrix} > 0, \begin{vmatrix} a_{11} & a_{12} & a_{13} \\ a_{21} & a_{22} & a_{23} \\ a_{31} & a_{32} & a_{33} \end{vmatrix} > 0, \ldots, \det \mathbf{A} > 0 \tag{A.1.5.44}$$

Example A.1.5.9 The quadratic form

$$F = \begin{bmatrix} x_1 & x_2 \end{bmatrix} \begin{bmatrix} 1 & 1 \\ 1 & 2 \end{bmatrix} \begin{bmatrix} x_1 \\ x_2 \end{bmatrix} = x_1^2 + 2x_1x_2 + 2x_2^2$$

is positive definite since

$$a_{11} > 1 \quad \text{and} \quad \det \mathbf{A} = \begin{vmatrix} 1 & 1 \\ 1 & 2 \end{vmatrix} = 1 > 0.$$

The property (A.1.5.44) provides a simple procedure for testing whether a given quadratic form is positive definite without determining the eigenvalues of **A**—which can prove to be an extremely complicated task when the order of **A** is large.

Function of a Matrix

In this section we shall give a brief representation on functions of a matrix. In particular, we shall present a few theorems which find extensive applications in engineering and applied sciences without proof since the proofs can be easily obtained from most standard texts on matrix analysis.* Moreover, we are mainly interested in those matrix functions which can be expressed as a series of powers of a square matrix, for these are the type of matrix functions we frequently encounter in practice.

Consider the second-order matrix given by

$$\mathbf{A} = \begin{bmatrix} 1 & 0 \\ 3 & 1 \end{bmatrix}.$$

The corresponding characteristic equation is

$$g(\lambda) = \lambda^2 - 2\lambda + 1 = 0.$$

Substituting **A** for λ and the identity matrix **I** for unity in the above expression yields

$$g(\mathbf{A}) = \mathbf{A}^2 - 2\mathbf{A} + \mathbf{I}$$

$$= \begin{bmatrix} 1 & 0 \\ 6 & 1 \end{bmatrix} - 2 \begin{bmatrix} 1 & 0 \\ 3 & 1 \end{bmatrix} + \begin{bmatrix} 1 & 0 \\ 0 & 1 \end{bmatrix} = \begin{bmatrix} 0 & 0 \\ 0 & 0 \end{bmatrix}.$$

The last expression reveals the essence of the famous Cayley-Hamilton theorem which can be stated as follows:

Theorem A.1.5.5 *Every square matrix satisfies its own characteristic equation.*

Thus, for an $n \times n$-matrix **A**, the characteristic equation $g(\lambda)$ can be expressed as

$$g(\lambda) = (-1)^n\lambda^n + a_{n-1}\lambda^{n-1} + \cdots + a_1\lambda + a_0. \qquad (A.1.5.45)$$

---

* For example, refer to F. R. Gantmacher [GA 1], Vol. 1.

Since the matrix satisfies its own characteristic equation, we have

$$g(\mathbf{A}) = (-1)^n \mathbf{A}^n + a_{n-1}\mathbf{A}^{n-1} + \cdots + a_1\mathbf{A} + a_0\mathbf{I} = 0. \qquad \text{(A.1.5.46)}$$

Example A.1.5.10   Let us apply this theorem to evaluate A^{-1} of the matrix given by

$$\mathbf{A} = \begin{bmatrix} 3 & 6 \\ 0 & 1 \end{bmatrix}.$$

Here, the characteristic polynomial is

$$g(\lambda) = \lambda^2 - 4\lambda + 3.$$

Hence, by the Cayley-Hamilton theorem, we have

$$g(\mathbf{A}) = \mathbf{A}^2 - 4\mathbf{A} + 3\mathbf{I} = 0$$

which gives

$$3\mathbf{I} = -\mathbf{A}^2 + 4\mathbf{A}.$$

Premultiplying both sides of the equation by $\mathbf{A}^{-1}$ yields

$$3\mathbf{A}^{-1} = -\mathbf{A} + 4\mathbf{I}$$

or

$$\mathbf{A}^{-1} = -\tfrac{1}{3}\mathbf{A} + \tfrac{4}{3}\mathbf{I} = \begin{bmatrix} \frac{1}{3} & -2 \\ 0 & 1 \end{bmatrix}.$$

Next, we shall state another powerful theorem concerning the determination of the eigenvalues of a polynomial of a matrix.

Theorem A.1.5.6 (*Frobenius theorem*)   Let $\lambda_1, \lambda_2, \ldots, \lambda_n$ be the eigenvalues of an $n \times n$-matrix $\mathbf{A}$. Then the eigenvalues of the matrix polynomial

$$f(\mathbf{A}) = a_p\mathbf{A}^p + a_{p-1}\mathbf{A}^{p-1} + \cdots + a_1\mathbf{A} + a_0\mathbf{I}$$

are $f(\lambda_1), f(\lambda_2), \ldots, f(\lambda_n)$.

Example A.1.5.11   As an example, the eigenvalues of

$$\mathbf{A} = \begin{bmatrix} 1 & 1 \\ 0 & 2 \end{bmatrix}$$

are $\lambda_1 = 1$ and $\lambda_2 = 2$. According to the Frobenius theorem, the eigenvalues of

$$f(\mathbf{A}) = \mathbf{A}^3$$

are given by

$$f(\lambda_1) = \lambda_1^3 = 1 \qquad \text{and} \qquad f(\lambda_2) = \lambda_2^3 = 2^3 = 8.$$

Vector Differential Equations

Consider the scalar differential equation of the form

$$\frac{dx}{dt} = ax + bu(t), \qquad\qquad\qquad\qquad (A.1.5.47)$$

where a, b are arbitrary constants and $u(t)$ is assumed to be a continuous function of t for convenience. Multiplying both sides of the above equation by the integrating factor e^{-at}, we find that

$$e^{-at}\frac{dx}{dt} - ae^{-at}x = be^{-at}u(t)$$

or,

$$\frac{d}{dt}\{e^{-at}x\} = be^{-at}u(t)$$

which, when integrated from t_0 to t, gives

$$e^{-at}x\Big|_{t_0}^{t} = \int_{t_0}^{t} e^{-a\tau}bu(\tau)\, d\tau.$$

Solving the above equation for $x(t)$, we obtain

$$x(t) = e^{a(t - t_0)}x(t_0) + \int_{t_0}^{t} e^{a(t - \tau)}bu(\tau)\, d\tau, \qquad t \geq t_0. \qquad (A.1.5.48)$$

The scalar differential equation (A.1.5.47) can be regarded as a special case of the system of equations described by

$$\frac{dx_1}{dt} = a_{11}x_1 + a_{12}x_2 + \cdots + a_{1n}x_n + b_{11}u_1 + b_{12}u_2 + \cdots + b_{1r}u_r$$

$$\frac{dx_2}{dt} = a_{21}x_1 + a_{22}x_2 + \cdots + a_{2n}x_n + b_{21}u_1 + b_{22}u_2 + \cdots + b_{2r}u_r$$

$$\vdots \qquad\qquad\qquad\qquad\qquad\qquad\qquad\qquad (A.1.5.49)$$

$$\frac{dx_n}{dt} = a_{n1}x_1 + a_{n2}x_2 + \cdots + a_{nn}x_n + b_{n1}u_1 + b_{n2}u_2 + \cdots + b_{nr}u_r,$$

which, in vector form, can be expressed as

$$\dot{\mathbf{x}} \triangleq \frac{d\mathbf{x}}{dt} = \mathbf{A}\mathbf{x} + \mathbf{B}\mathbf{u}. \qquad\qquad\qquad (A.1.5.50)$$

In (A.1.5.50), $\mathbf{x}$ and $\mathbf{u}$ are vectors defined by

$$\mathbf{x} = \begin{bmatrix} x_1 \\ x_2 \\ \vdots \\ x_n \end{bmatrix}, \qquad \mathbf{u} = \begin{bmatrix} u_1 \\ u_2 \\ \vdots \\ u_r \end{bmatrix}$$

and $\mathbf{A}$ and $\mathbf{B}$ are *constant* matrices of orders $n \times n$ and $n \times r$, respectively.

By analogy with the scalar equation (A.1.5.47), we are tempted to expect the solution of (A.1.5.50) to be of the form

$$\mathbf{x}(t) = e^{\mathbf{A}(t - t_0)}\mathbf{x}(t_0) + \int_{t_0}^{t} e^{\mathbf{A}(t - \tau)}\mathbf{B}\mathbf{u}(\tau)\, d\tau, \qquad t \geq t_0. \qquad (A.1.5.51)$$

Indeed, (A.1.5.51) is actually the solution of (A.1.5.50). This can be demonstrated by first introducing the matrix series

$$e^{\mathbf{A}t} \triangleq \sum_{n=0}^{\infty} \frac{\mathbf{A}^n t^n}{n!}$$

$$= \mathbf{I} + \mathbf{A}t + \frac{\mathbf{A}^2 t^2}{2!} + \frac{\mathbf{A}^3 t^3}{3!} + \cdots, \qquad (A.1.5.52)$$

where $\mathbf{A}^0 \triangleq \mathbf{I}$, which provides an interpretation of the matrix function $e^{\mathbf{A}t}$. It can be shown that the series defined in (A.1.5.52) is uniformly convergent in any finite interval and that the sum is a continuous function of t for all finite t.* Differentiating both sides of (A.1.5.52), we have

$$\frac{d}{dt} e^{\mathbf{A}t} = \mathbf{A} + \mathbf{A}^2 t + \frac{\mathbf{A}^3 t^2}{2!} + \cdots$$

$$= \mathbf{A}\left(\sum_{n=0}^{\infty} \frac{\mathbf{A}^n t^n}{n!} \right)$$

$$= \mathbf{A}e^{\mathbf{A}t}. \qquad (A.1.5.53)$$

In exactly the same manner as we derive the scalar functional equation $e^{a(t + \tau)} = e^{at}e^{a\tau}$, we can easily show that the corresponding matrix functional equation is always valid; i.e.,

$$e^{\mathbf{A}(t + \tau)} = e^{\mathbf{A}t}e^{\mathbf{A}\tau}. \qquad (A.1.5.54)$$

Next, differentiating both sides of (A.1.5.51) with respect to t, we have

$$\frac{d\mathbf{x}}{dt} = \frac{d}{dt}\left[e^{\mathbf{A}(t - t_0)}\mathbf{x}(t_0) + \int_{t_0}^{t} e^{\mathbf{A}(t - \tau)}\mathbf{B}\mathbf{u}(\tau)\, d\tau \right]$$

$$= \frac{d}{dt}\left[e^{\mathbf{A}t}e^{-\mathbf{A}t_0}\mathbf{x}(t_0) + e^{\mathbf{A}t}\int_{t_0}^{t} e^{-\mathbf{A}\tau}\mathbf{B}\mathbf{u}(\tau)\, d\tau \right]$$

$$= \mathbf{A}e^{\mathbf{A}t}e^{-\mathbf{A}t_0}\mathbf{x}(t_0) + \mathbf{A}e^{\mathbf{A}t}\int_{t_0}^{t} e^{-\mathbf{A}\tau}\mathbf{B}\mathbf{u}(\tau)\, d\tau + e^{\mathbf{A}t}[e^{-\mathbf{A}\tau}\mathbf{B}\mathbf{u}(\tau)]_{\tau = t}$$

$$= \mathbf{A}\left[e^{\mathbf{A}(t - t_0)}\mathbf{x}(t_0) + \int_{t_0}^{t} e^{\mathbf{A}(t - \tau)}\mathbf{B}\mathbf{u}(\tau)\, d\tau \right] + \mathbf{B}\mathbf{u}(t)$$

or,

$$\dot{\mathbf{x}} = \mathbf{A}\mathbf{x} + \mathbf{B}\mathbf{u}$$

which verifies that (A.1.5.51) is the solution of (A.1.5.50).

---

* See, for example, R. Bellman [BE 1], Chapter 10.

If we define the function

$$\mathbf{\Phi}(t - t_0) = e^{\mathbf{A}(t - t_0)}, \tag{A.1.5.55}$$

then we can write the solution (A.1.5.51) of (A.1.5.50) as

$$\mathbf{x}(t) = \mathbf{\Phi}(t - t_0)\mathbf{x}(t_0) + \int_{t_0}^{t} \mathbf{\Phi}(t - \tau)\mathbf{B}\mathbf{u}(\tau)\, d\tau. \tag{A.1.5.56}$$

The function $\mathbf{\Phi}(t - t_0)$, which is the solution of the matrix differential equation

$$\frac{d\mathbf{\Phi}(t - t_0)}{dt} = \mathbf{A}\mathbf{\Phi}(t - t_0), \qquad \mathbf{\Phi}(0) = \mathbf{I}, \tag{A.1.5.57}$$

is referred to as the *state-transition matrix* of (A.1.5.50).

An extension of the system (A.1.5.50) leads to the following vector differential equation

$$\dot{\mathbf{x}} = \mathbf{A}(t)\mathbf{x} + \mathbf{B}(t)\mathbf{u} \tag{A.1.5.58}$$

in which the matrices $\mathbf{A}(t)$ and $\mathbf{B}(t)$ are assumed to be continuous matrix functions of t. Again, if we let $\mathbf{\Phi}(t, t_0)$ be the solution of the homogeneous equation

$$\dot{\mathbf{\Phi}}(t, t_0) = \mathbf{A}(t)\mathbf{\Phi}(t, t_0), \qquad \mathbf{\Phi}(t_0, t_0) = \mathbf{I}, \tag{A.1.5.59}$$

then $\mathbf{\Phi}(t, t_0)$ is called the *state-transition matrix* of the system (A.1.5.58) whose solution is given by

$$\mathbf{x}(t) = \mathbf{\Phi}(t, t_0)\mathbf{x}(t_0) + \int_{t_0}^{t} \mathbf{\Phi}(t, \tau)\mathbf{B}(\tau)\mathbf{u}(\tau)\, d\tau. \tag{A.1.5.60}$$

It should be pointed out that the state-transition matrix $\mathbf{\Phi}(t, t_0)$ satisfies the following important properties:

1. If the matrices $\int_{t_0}^{t} \mathbf{A}(\tau)\, d\tau$ and $\mathbf{A}(t)$ commute with respect to multiplication for all t, then

$$\mathbf{\Phi}(t, t_0) = e^{\int_{t_0}^{t} \mathbf{A}(\tau)\, d\tau}. \tag{A.1.5.61}$$

2. The property

$$\mathbf{\Phi}(t_1, t_2)\mathbf{\Phi}(t_2, t_3) = \mathbf{\Phi}(t_1, t_3) \tag{A.1.5.62}$$

for any t_1, t_2, and t_3 is always satisfied.

3. If $\mathbf{A}(t)$ is continuous for all finite intervals of t, then $\mathbf{\Phi}(t, t_0)$ is nonsingular for all finite t.

4. $\mathbf{\Phi}^{-1}(t_1, t_2) = \mathbf{\Phi}(t_2, t_1)$ \hfill (A.1.5.63)

It should be remarked that the system described by (A.1.5.58) is quite general and hence is applicable to many practical systems. However, we should point out that, except for a few simple cases, the only effective means for determining the solution of the so-called time-varying systems is the use of numerical techniques with the digital computer.

REFERENCES

1. F. E. Hohn, *Elementary Matrix Algebra*, Second Ed., Macmillan, 1964.

2. L. E. Fuller, *Basic Matrix Theory*, Prentice-Hall, 1962.

3. F. Ayres, Jr., *Theory and Problems of Matrices*, Schaum Publishing Co., 1962.

4. B. Friedman, *Principles and Techniques of Applied Mathematics,* Wiley, 1965, Chapters 1 and 2.

5. R. Bellman, *Introduction to Matrix Analysis*, McGraw-Hill, 1960, Chapters 2, 3, 4, 6, and 10.

PROBLEMS

A.1.1 Given:

$$A = \begin{bmatrix} 1 & -2 & 3 \\ 0 & 4 & 2 \end{bmatrix}, \qquad B = \begin{bmatrix} 1 & 4 & 5 \\ 2 & 3 & -2 \end{bmatrix}, \qquad C = \begin{bmatrix} 1 & 2 \\ 0 & -1 \\ -2 & 1 \end{bmatrix}.$$

Compute: a) $A + B$; b) $A - B$; c) AC; d) CA.

A.1.2 Compute AB and BA, if

$$A = \begin{bmatrix} 2 & 0 & -2 \\ -1 & 5 & 3 \\ 3 & 1 & 4 \end{bmatrix} \quad \text{and} \quad B = \begin{bmatrix} 1 & -3 & 0 \\ 0 & 4 & 2 \\ -2 & 2 & 3 \end{bmatrix}.$$

Is $AB = BA$?

A.1.3 Find the inverse, if it exists, of each of the following matrices:

$$\text{a) } A = \begin{bmatrix} 1 & 2 & -1 \\ 3 & 0 & 2 \\ 4 & 5 & 6 \end{bmatrix}, \quad \text{b) } B = \begin{bmatrix} 2 & -1 & 2 & 0 & 0 \\ 0 & 4 & 3 & 0 & 0 \\ 3 & 1 & 1 & 0 & 0 \\ 0 & 0 & 0 & 2 & 4 \\ 0 & 0 & 0 & 3 & 1 \end{bmatrix}, \quad \text{c) } C = \begin{bmatrix} -1 + 3j & 2 \\ 3 + 2j & -2j \end{bmatrix}.$$

A.1.4 Evaluate $\dfrac{d}{dt} A$ and $\int A \, dt$, if

$$A = \begin{bmatrix} e^{-3t} \cos 2t & te^{-4t} \\ e^{-5t} & \sin 4t \end{bmatrix}.$$

A.1.5 Given that

$$A = \begin{bmatrix} 2 & -1 & 3 \\ 1 & 0 & 1 \\ 4 & 5 & 1 \end{bmatrix} \quad \text{and} \quad B = \begin{bmatrix} 1 & 2 & 4 \\ 0 & 3 & 0 \\ 1 & 1 & 2 \end{bmatrix}.$$

Show that the following statements hold:

a) $(AB)^{-1} = B^{-1}A^{-1}$,   b) $(AB)' = B'A'$.

A.1.6   Determine which of the following sets of vectors are linearly dependent. If dependent, obtain a set of coefficients (not all zeros) c_1, c_2, and c_3 satisfying the relation

$$c_1 \mathbf{x}_1 + c_2 \mathbf{x}_2 + c_3 \mathbf{x}_3 = \mathbf{0}.$$

a) $\mathbf{x}_1 = \begin{bmatrix} 1 \\ 1 \\ 0 \end{bmatrix}$, $\mathbf{x}_2 = \begin{bmatrix} 1 \\ 2 \\ 1 \end{bmatrix}$, $\mathbf{x}_3 = \begin{bmatrix} 2 \\ 1 \\ 1 \end{bmatrix}$

b) $\mathbf{x}_1 = \begin{bmatrix} -1 \\ 1 \\ -1 \end{bmatrix}$, $\mathbf{x}_2 = \begin{bmatrix} 0 \\ 1 \\ 2 \end{bmatrix}$, $\mathbf{x}_3 = \begin{bmatrix} 3 \\ 1 \\ 1 \end{bmatrix}$

c) $\mathbf{x}_1 = \begin{bmatrix} 2 \\ 1 \\ 0 \end{bmatrix}$, $\mathbf{x}_2 = \begin{bmatrix} 1 \\ 1 \\ -1 \end{bmatrix}$, $\mathbf{x}_3 = \begin{bmatrix} -3 \\ -2 \\ 1 \end{bmatrix}$

A.1.7   Show that the vectors

$$\mathbf{x}_1 = \begin{bmatrix} 1 \\ 2 \\ -1 \end{bmatrix}, \quad \mathbf{x}_2 = \begin{bmatrix} 1 \\ 0 \\ 1 \end{bmatrix}, \quad \mathbf{x}_3 = \begin{bmatrix} 2 \\ 1 \\ 1 \end{bmatrix}, \quad \mathbf{x}_4 = \begin{bmatrix} 3 \\ 1 \\ 0 \end{bmatrix}$$

are linearly dependent by expressing the vector $\mathbf{x}_1$ as a linear combination of the other three vectors.

A.1.8   Show that the matrices $\mathbf{A}$ and $\mathbf{P}^{-1}\mathbf{AP}$ have the same characteristic equation.

A.1.9   Given the vectors

$$\mathbf{x}_1 = \begin{bmatrix} 1 \\ 0 \\ -1 \end{bmatrix} \quad \text{and} \quad \mathbf{x}_2 = \begin{bmatrix} 1 \\ 1 \\ 2 \end{bmatrix},$$

determine the vector $\mathbf{x}_3$ with a unity norm such that it is orthogonal to both $\mathbf{x}_1$ and $\mathbf{x}_2$.

A.1.10   Determine the diagonal matrices similar to:

a) $\begin{bmatrix} 1 & 1 \\ -1 & 2 \end{bmatrix}$,   b) $\begin{bmatrix} -1 & 0 & 0 \\ 4 & 3 & 4 \\ -2 & -2 & -3 \end{bmatrix}$,   c) $\begin{bmatrix} 1 & 0 & 0 \\ 1 & 0 & -2 \\ -1 & 1 & -3 \end{bmatrix}$.

A.1.11   Determine a Jordan canonical form similar to the matrix given by

$$\mathbf{A} = \begin{bmatrix} 0 & 1 & 2 \\ 5 & 0 & -4 \\ -3 & 1 & 4 \end{bmatrix}.$$

A.1.12 Given that

$$A = \begin{bmatrix} -1 & 0 & 0 \\ 4 & 3 & 4 \\ -2 & -2 & -3 \end{bmatrix}, \qquad g(A) = A^2 + 2A + I,$$

determine the eigenvalues of A and of $g(A)$.

A.1.13 Find a set of normalized eigenvectors (i.e., vectors normalized to unity norm) so that by means of the nonsingular transformation formed by these vectors the matrix

$$A = \begin{bmatrix} -1 & 0 & 0 \\ 4 & 3 & 4 \\ -2 & -2 & -3 \end{bmatrix}$$

can be reduced to a diagonal matrix.

A.1.14 Determine a set of generalized eigenvectors so that the matrix A of Problem A.1.5 can be reduced to a Jordan canonical form.

A.1.15 Repeat Problem A.1.14 for the matrix

$$A = \begin{bmatrix} -1 & 4 & -1 \\ 4 & 7 & -1 \\ -4 & -4 & 4 \end{bmatrix}.$$

A.1.16 Test for positive definiteness of the quadratic form given by

$$F = x_1^2 + 3x_3^2 - 6x_1x_2 + 4x_1x_3 + 8x_2x_3.$$

LAPLACE TRANSFORMATION AND FOURIER SERIES

A.2.1 INTRODUCTION

This appendix is intended to serve as a review of the properties of both the Laplace transformation and the Fourier series. Because of the limitation on the size of this chapter, rigorous mathematical proofs are again deleted since they can be found in most standard texts.*

A.2.2 DEFINITIONS

A *transformation* T of a function $f(t)$, written as $T\{f(t)\}$, is said to be *linear* if for any constants c_1 and c_2 and every pair of functions $f_1(t)$ and $f_2(t)$ the relation

$$T\{c_1 f_1(t) + c_2 f_2(t)\} = c_1 T\{f_1(t)\} + c_2 T\{f_2(t)\} \qquad (A.2.2.1)$$

is satisfied.

A function $f(t)$ is *sectionally continuous* on a finite interval if that interval can be subdivided into a finite number of parts (subintervals) such that in each of these subintervals the function $f(t)$ is continuous and has finite limits at the endpoints of the subinterval. As an example, the function $f(t)$ of Fig. A.2.2.1 is sectionally continuous on the interval $0 \leq t \leq 6$ since, in each subinterval, the function $f(t)$ is continuous and has finite limits at the endpoints.

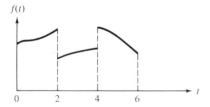

Fig. A.2.2.1 A sectionally continuous function on $0 \leq t \leq 6$.

A function $f(t)$ is of *exponential order* as t tends to infinity if the condition

$$|f(t)| \leq Me^{\alpha t} \qquad \text{for all} \quad t > T \qquad (A.2.2.2)$$

is satisfied for some real numbers M, α, and T. Thus, for instance, the function e^{t^2} does not satisfy (A.2.2.2) and hence is not a function of exponential order.

---

* For example, refer to R. V. Churchill [CH 6] for Laplace transformation and R. V. Churchill [CH 5] for Fourier series.

The *Laplace transform* of a function $f(t)$, written as $F(s) = \mathcal{L}\{f(t)\}$, is defined by the following integral transformation

$$F(s) \triangleq \int_0^\infty f(t)e^{-st}\,dt, \qquad (A.2.2.3)$$

where s, the complex variable, can be expressed as $s = \sigma + j\omega$ with σ and ω being the real and imaginary parts of s, respectively.

The existence of the Laplace transform of a function $f(t)$ is assured if the function $f(t)$ is both sectionally continuous in every finite interval for $t \geq 0$ and is of exponential order as $t \to \infty$. It should be noted that these existence conditions for the Laplace transform $F(s)$ of a function $f(t)$ are sufficient rather than necessary conditions.

Example A.2.2.1 (a) Consider the unit-step function $u(t)$ defined by the equations

$$u(t) = 1, \qquad t \geq 0;$$
$$= 0, \qquad t < 0.$$

Substituting these equations into (A.2.2.3), we have

$$U(s) = \mathcal{L}\{u(t)\}$$

$$= \int_0^\infty 1e^{-st}\,dt = -\frac{1}{s}e^{-st}\Big|_0^\infty$$

$$= -\frac{1}{s}\left[\lim_{t\to\infty} e^{-st} - 1\right].$$

In the last expression, we see that if σ, the real part of s, is positive,

$$\lim_{t\to\infty} e^{-st} = 0,$$

and hence the Laplace transform of $u(t)$, $U(s)$, exists and is given by

$$U(s) = \frac{1}{s}, \qquad \sigma = \mathrm{Re}\,s > 0$$

where $\mathrm{Re}\,s$ is read "the real part of s."

(b) Using the defining integral (A.2.2.3), it can be shown that the Laplace transform of e^{-at} for any real constant a is given by

$$\mathcal{L}\{e^{-at}\} = \frac{1}{s+a}, \qquad \mathrm{Re}\,s > -a.$$

Again, it should be emphasized that the Laplace transform of e^{-at} is defined only for $\mathrm{Re}\,s > -a$.

A.2.3 BASIC THEOREMS FOR THE LAPLACE TRANSFORMATION

Some of the more common properties of the Laplace transform are now given in the form of theorems. The corresponding proofs have been kept as brief and simple as possible. The reader interested in rigorous mathematical treatment should consult literature elsewhere.†

Consider any constants c_1 and c_2 and any *Laplace transformable functions*‡ $f_1(t)$ and $f_2(t)$. Then

$$\mathscr{L}\{c_1 f_1(t) + c_2 f_2(t)\} = \int_0^\infty [c_1 f_1(t) + c_2 f_2(t)]e^{-st}\, dt$$

$$= c_1 \int_0^\infty f_1(t)e^{-st}\, dt + c_2 \int_0^\infty f_2(t)e^{-st}\, dt$$

$$= c_1 \mathscr{L}\{f_1(t)\} + c_2 \mathscr{L}\{f_2(t)\}. \tag{A.2.3.1}$$

Comparing (A.2.3.1) and (A.2.2.1), we find that the Laplace transformation satisfies the property of linearity. Thus, writing this property in the form of a theorem, we have:

Theorem A.2.3.1 (*Linearity*) The Laplace transformation is a *linear* transformation.

Example A.2.3.1 Determine the Laplace transform of the function $f(t)$ defined by

$$f(t) = 3u(t) + 4e^{-5t}.$$

Since, in Example A.2.2.1, the Laplace transforms of $u(t)$ and e^{-5t} are found to be $1/s$ and $1/(s + 5)$, respectively, by applying Theorem A.2.3.1, we find that

$$F(s) = \mathscr{L}\{f(t)\}$$
$$= \mathscr{L}\{3u(t) + 4e^{-5t}\}$$
$$= 3\mathscr{L}\{u(t)\} + 4\mathscr{L}\{e^{-5t}\}$$
$$= \frac{3}{s} + \frac{4}{s + 5}.$$

The Laplace transform of the derivative of a function $f(t)$ can be easily obtained by applying the definition of the Laplace transform directly in the following manner:

$$\mathscr{L}\left\{\frac{df}{dt}\right\} = \int_0^\infty \frac{df}{dt} e^{-st}\, dt.$$

---

† For example, refer to R. V. Churchill [CH 6].
‡ A function $f(t)$ is said to be Laplace transformable if its Laplace transform exists.

Integrating by parts $\left(\int u\, dv = uv - \int v\, du \right)$, we have

$$\mathscr{L}\left\{\frac{df}{dt}\right\} = f(t)e^{-st}\Big|_0^\infty - \int_0^\infty f(t)e^{-st}(-s)\, dt$$

$$= -f(0) + s\int_0^\infty f(t)e^{-st}\, dt$$

$$= s\mathscr{L}\{f(t)\} - f(0)$$

which yields the theorem on differentiation.

Theorem A.2.3.2 (*Differentiation*) Let $F(s)$ be the Laplace transform of $f(t)$, then the Laplace transform of the derivative of $f(t)$ is given by

$$\mathscr{L}\left\{\frac{df}{dt}\right\} = sF(s) - f(0). \tag{A.2.3.2}$$

Example A.2.3.2 Let us apply Theorem A.2.3.2 to determine the solution of the differential equation

$$\frac{df}{dt} + 4f = 0, \qquad f(0) = 5.$$

Taking the Laplace transform of the differential equation and then substituting the initial condition into the resultant equation yields

$$[sF(s) - f(0)] + 4F(s) = 0$$

$$(s + 4)F(s) = 5$$

$$F(s) = \frac{5}{s + 4}.$$

Knowing the relation $\mathscr{L}\{e^{-4t}\} = 1/(s + 4)$, we arrive at the solution of the given differential equation:

$$f(t) = 5e^{-4t} \qquad \text{for all } t \geq 0.$$

Example A.2.3.3 As another example, the theorem on differentiation can be used to find the Laplace transform of the integral of a function $f(t)$. Let a function $g(t)$ be defined by

$$g(t) \triangleq \int_{-\infty}^t f(t)\, dt. \tag{A.2.3.3}$$

Then differentiating (A.2.3.3) with respect to t yields

$$\frac{dg}{dt} - f(t). \tag{A.2.3.4}$$

Applying Theorem A.2.3.2 to dg/dt, we obtain

$$\mathscr{L}\left\{\frac{dg}{dt}\right\} = \mathscr{L}\{f(t)\} = sG(s) - g(0). \qquad \text{(A.2.3.5)}$$

Since

$$G(s) = \mathscr{L}\{g(t)\} = \mathscr{L}\left\{\int_{-\infty}^{t} f(t)\,dt\right\}$$

and

$$g(0) = \int_{-\infty}^{0} f(t)\,dt,$$

we can write (A.2.3.5) in terms of $f(t)$ and its transform as

$$\mathscr{L}\{f(t)\} = s\mathscr{L}\left\{\int_{-\infty}^{t} f(t)\,dt\right\} - \int_{-\infty}^{0} f(t)\,dt \qquad \text{(A.2.3.6)}$$

or, solving for

$$\mathscr{L}\left\{\int_{-\infty}^{t} f(t)\,dt\right\},$$

we have

$$\mathscr{L}\left\{\int_{-\infty}^{t} f(t)\,dt\right\} = \frac{1}{s}\left[\mathscr{L}\{f(t)\} + \int_{-\infty}^{0} f(t)\,dt\right]. \qquad \text{(A.2.3.7)}$$

Thus, the theorem on integration is established.

Theorem A.2.3.3 (*Integration*) Let $F(s)$ be the Laplace transform of a function $f(t)$. Then the Laplace transform of the integral of $f(t)$ is given by

$$\mathscr{L}\left\{\int_{-\infty}^{t} f(t)\,dt\right\} = \frac{1}{s}\left[F(s) + \int_{-\infty}^{0} f(t)\,dt\right]. \qquad \text{(A.2.3.8)}$$

Note that if the function $f(t)$ is zero for $t < 0$, then

$$\int_{-\infty}^{0} f(t)\,dt = 0$$

and (A.2.3.8) reduces to

$$\mathscr{L}\left\{\int_{0}^{t} f(t)\,dt\right\} = \frac{1}{s}F(s). \qquad \text{(A.2.3.9)}$$

Example A.2.3.4 The Laplace transform of a unit-step function is $\mathscr{L}\{u(t)\} = 1/s$. The Laplace transform of the integral of the unit-step function, i.e., $tu(t)$, can be

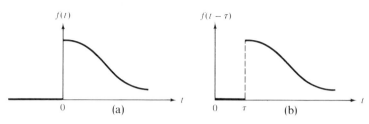

Fig. A.2.3.1 (a) A function $f(t)$ and (b) the same function delayed for τ units.

found by applying (A.2.3.9):

$$\mathscr{L}\left\{\int_0^t u(t)\,dt\right\} = \mathscr{L}\{tu(t)\} = \frac{1}{s}\left[\frac{1}{s}\right] = \frac{1}{s^2}.$$

In like manner, applying (A.2.3.9) again to $tu(t)$ yields

$$\mathscr{L}\left\{\int_0^t tu(t)\,dt\right\} = \mathscr{L}\left\{\frac{t^2}{2}u(t)\right\} = \frac{1}{s}\left[\frac{1}{s^2}\right]$$

or

$$\mathscr{L}\{t^2 u(t)\} = \frac{2!}{s^3}.$$

It is not difficult to show that, by repeated application of (A.2.3.9), the Laplace transform of t^n for any integer n is given by

$$\mathscr{L}\{t^n\} = \mathscr{L}\{t^n u(t)\} = \frac{n!}{s^{n+1}}. \qquad (A.2.3.10)$$

Next, we shall proceed to develop the theorem on real translation; i.e., the theorem for determining the Laplace transform of a function $f(t)$ after being delayed for τ units. To begin with, consider a function $f(t)$, *which is identically zero for* $t < 0$, as shown in Fig. A.2.3.1(a). Then the same function delayed by τ units is given by $f(t - \tau)$ as sketched in Fig. A.2.3.1(b). The Laplace transform of $f(t - \tau)$ is

$$\mathscr{L}\{f(t - \tau)\} = \int_0^\infty f(t - \tau)e^{-st}\,dt$$

$$= \int_\tau^\infty f(t - \tau)e^{-st}\,dt. \qquad (A.2.3.11)$$

Let the new variable t' be defined by the relation

$$t' = t - \tau. \qquad (A.2.3.12)$$

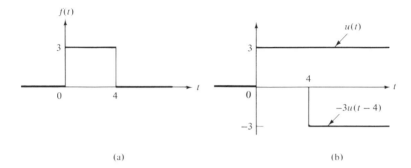

Fig. A.2.3.2 (a) A pulse function $f(t)$ and (b) its component functions.

Then (A.2.3.11) can be expressed in terms of t' as

$$\mathscr{L}\{f(t - \tau)\} = \int_0^\infty f(t')e^{-s(t' + \tau)}\, dt'$$

$$= e^{-s\tau}\int_0^\infty f(t')e^{-st'}\, dt'$$

$$= e^{-s\tau}\mathscr{L}\{f(t)\}$$

$$= e^{-s\tau}F(s). \tag{A.2.3.13}$$

Expression (A.2.3.13) establishes the theorem on real translation which can be expressed more concisely in terms of the unit-step function $u(t)$ as follows.

Theorem A.2.3.4 (*Real translation*) Let $F(s)$ be the Laplace transform of the function $f(t)u(t)$. Then the Laplace transform of the function $f(t - \tau)u(t - \tau)$ is given by

$$\mathscr{L}\{f(t - \tau)u(t - \tau)\} = e^{-s\tau}F(s). \tag{A.2.3.14}$$

Remark. The use of the unit-step function $u(t)$ in Theorem A.2.3.4 serves the purpose of requiring the function $f(t)$ to vanish for $t < 0$ which is necessary for the validity of the theorem. To further demonstrate this point, consider the function $f(t) \triangleq 1$ for all t. Application of (A.2.3.13) yields $\mathscr{L}\{f(t - \tau)\} = e^{-s\tau}/s$ which, we know, is false, since in this case, $f(t - \tau) = f(t) = 1$ and $\mathscr{L}\{f(t - \tau)\}$ must be the same as $\mathscr{L}\{f(t)\}$.

Example A.2.3.5 The pulse function $f(t)$ depicted in Fig. A.2.3.2(a) can be regarded as the sum of two component functions as shown in Fig. A.2.3.2.(b). Thus we have

$$f(t) = 3u(t) - 3u(t - 4).$$

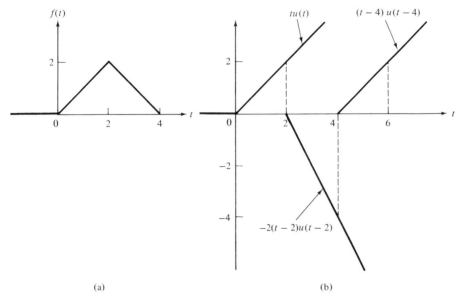

Fig. A.2.3.3 (a) A triangular pulse and (b) its component functions.

By means of (A.2.3.14), the Laplace transform of $f(t)$ can be readily shown to be

$$F(s) = \frac{3}{s}\left(1 - e^{-s4}\right)$$

Example A.2.3.6 The triangular pulse and its component functions are shown in (a) and (b) of Fig. A.2.3.3. In terms of component functions, $f(t)$ can be expressed as

$$f(t) = tu(t) - 2(t - 2)u(t - 2) + (t - 4)u(t - 4)$$

which, with the aid of Theorem A.2.3.4, has the Laplace transform

$$F(s) = \frac{1}{s^2} - \frac{2}{s^2}e^{-s2} + \frac{1}{s^2}e^{-s4}.$$

The next theorem to be derived is the theorem on complex translation which can be stated as follows:

Theorem A.2.3.5 (*Complex translation*) Let $F(s)$ be the Laplace transform of $f(t)$. Then the Laplace transform of the product $e^{-\alpha t}f(t)$ is given by

$$\mathcal{L}\{e^{-\alpha t}f(t)\} = F(s + \alpha), \qquad\qquad\qquad (A.2.3.15)$$

where α is a constant.

To verify the above theorem, let us apply the definition of Laplace transform directly to $e^{-\alpha t} f(t)$:

$$\mathcal{L}\{e^{-\alpha t} f(t)\} = \int_0^\infty f(t) e^{-\alpha t} e^{-st} \, dt$$

$$= \int_0^\infty f(t) e^{-(s+\alpha)t} \, dt,$$

which differs from $\int_0^\infty f(t) e^{-st} \, dt = F(s)$ in that the coefficient of $-t$ is $s + \alpha$ rather than s, and hence can be written as

$$\mathcal{L}\{e^{-\alpha t} f(t)\} = F(s + \alpha)$$

yielding the above theorem.

Example A.2.3.7   Let us use (A.2.3.15) to determine the Laplace transform of $e^{-\alpha t}$. Assigning $f(t) = u(t)$ with the corresponding Laplace transform $F(s) = 1/s$, then we find that

$$\mathcal{L}\{e^{-\alpha t}\} = \mathcal{L}\{e^{-\alpha t} u(t)\} = \frac{1}{s + \alpha}.$$

Example A.2.3.8   As another example, the Laplace transform of $te^{-\alpha t}$ is given by

$$\mathcal{L}\{te^{-\alpha t}\} = \frac{1}{(s + \alpha)^2}$$

since $\mathcal{L}\{t\} = \mathcal{L}\{tu(t)\} = 1/s^2$.

Another interesting property of the Laplace transformation can be obtained by differentiating both sides of (A.2.2.3) yielding

$$\frac{dF(s)}{ds} = \int_0^\infty f(t) e^{-st} (-t) \, dt$$

$$= - \int_0^\infty t f(t) e^{-st} \, dt$$

$$= - \mathcal{L}\{t f(t)\}$$

which can be summarized as stated in the next theorem.

Theorem A.2.3.6 (*Multiplication by t*)   Let $F(s)$ be the Laplace transform of $f(t)$. Then the Laplace transform of the product $tf(t)$ is given by

$$\mathcal{L}\{tf(t)\} = -\frac{dF(s)}{ds}. \qquad (A.2.3.16)$$

Example A.2.3.9 From Example A.2.3.6, we know that $\mathscr{L}\{e^{-\alpha t}\} = 1/(s + \alpha)$. The Laplace transform of $te^{-\alpha t}$ can be found by applying (A.2.3.16). Thus

$$\mathscr{L}\{te^{-\alpha t}\} = -\frac{d}{ds}\left[\frac{1}{s + \alpha}\right] = \frac{1}{(s + \alpha)^2}$$

which agrees with the result obtained in Example A.2.3.8.

The next two theorems to be discussed are very useful in determining the initial and the final values of a function $f(t)$ directly from its transform $F(s)$ without resorting to finding the expression for $f(t)$ itself. These two theorems are called the initial and the final value theorems.

Theorem A.2.3.7 (*Initial value*) If $F(s)$ is the Laplace transform of $f(t)$ and if

$$\lim_{s \to \infty} sF(s)$$

exists, then

$$\lim_{t \to 0} f(t) = \lim_{s \to \infty} sF(s). \qquad (A.2.3.17)$$

The proof of this theorem follows directly from Theorem A.2.3.2. That is, first we write

$$\mathscr{L}\left\{\frac{df}{dt}\right\} = \int_0^\infty \frac{df}{dt} e^{-st}\, dt = sF(s) - f(0).$$

Then, as s approaches ∞, we have

$$\lim_{s \to \infty} \int_0^\infty \frac{df}{dt} e^{-st}\, dt = \lim_{s \to \infty} [sF(s) - f(0)].$$

Consequently,

$$0 = \lim_{s \to \infty} [sF(s) - f(0)]. \qquad (A.2.3.18)$$

But

$$f(0) = \lim_{t \to 0} f(t)$$

which together with (A.2.3.18) yields (A.2.3.17). The final value theorem can be proved in a similar manner as demonstrated next.

Theorem A.2.3.8 (*Final value*) If $F(s)$ is the Laplace transform of $f(t)$ and if

$$\lim_{t \to \infty} f(t)$$

exists, then

$$\lim_{t \to \infty} f(t) = \lim_{s \to 0} [sF(s)]. \qquad (A.2.3.19)$$

Again, from Theorem A.2.3.2,

$$\lim_{s \to 0} \int_0^\infty \frac{df}{dt} e^{-st}\, dt = \lim_{s \to 0} [sF(s) - f(0)],$$

$$\int_0^\infty \frac{df}{dt}\, dt = \lim_{s \to 0} [sF(s) - f(0)]$$

which can be written as

$$\lim_{t \to \infty} f(t) - f(0) = \lim_{s \to 0} [sF(s) - f(0)].$$

After canceling $f(0)$, Theorem A.2.3.8 results.

Example A.2.3.10 For the function

$$f(t) = 5e^{-2t} + 4e^{-3t},$$

the initial value is 9 and the final value is zero. The same results can be determined from its Laplace transform which is given by

$$F(s) = \frac{5}{s + 2} + \frac{4}{s + 3} = \frac{9s + 23}{(s + 2)(s + 3)}$$

$$= \frac{9s + 23}{s^2 + 5s + 6}.$$

From Eq. (A.2.3.17),

$$f(0) = \lim_{s \to \infty} sF(s) = \lim_{s \to \infty} \frac{9s^2 + 23s}{s^2 + 5s + 6} = 9;$$

and from Eq. (A.2.3.19),

$$f(\infty) = \lim_{s \to 0} sF(s) = \lim_{s \to 0} \frac{9s^2 + 23s}{s^2 + 5s + 6} = 0.$$

Theorem A.2.3.9 (*Complex multiplication*) If $F_1(s)$ and $F_2(s)$ are the Laplace transforms of $f_1(t)$ and $f_2(t)$, respectively, then

$$F_1(s)F_2(s) = \mathcal{L}\left\{ \int_0^t f_1(t - \tau) f_2(\tau)\, d\tau \right\}. \tag{A.2.3.20}$$

The functions $f_1(t)$ and $f_2(t)$ in (A.2.3.20) are said to be *convolved* and the corresponding integral is referred to as the *convolution integral*. Denoting the product $F_1(s)F_2(s)$ by $F(s)$, then (A.2.3.20) can be written as

$$F(s) = F_1(s)F_2(s) = \mathcal{L}\left\{ \int_0^t f_1(t - \tau) f_2(\tau)\, d\tau \right\} \tag{A.2.3.21}$$

from which, the function $f(t)$, that is, the *inverse Laplace transform*† of $F(s)$, is given by

$$f(t) = \int_0^t f_1(t - \tau) f_2(\tau)\, d\tau$$

$$\triangleq f_1(t) * f_2(t), \tag{A.2.3.22}$$

where $f_1(t) * f_2(t)$ denotes the convolution of $f_1(t)$ and $f_2(t)$.

Equation (A.2.3.20) can be readily derived by taking the Laplace transform of (A.2.3.22)

$$\int_0^\infty \left[\int_0^t f_1(t - \tau) f_2(\tau)\, d\tau \right] e^{-st}\, dt = F(s) \tag{A.2.3.23}$$

which can be expressed as

$$\int_0^\infty \left[\int_0^\infty f_1(t - \tau) f_2(\tau) u(t - \tau)\, d\tau \right] e^{-st}\, dt = F(s)$$

since $u(t - \tau) = 0$ for $\tau > t$. Changing the order of integration gives

$$\int_0^\infty f_2(\tau) \left[\int_0^\infty f_1(t - \tau) u(t - \tau) e^{-st}\, dt \right] d\tau = F(s)$$

which, in turn, can be expressed in terms of the new variable $t' \triangleq t - \tau$ as

$$\int_0^\infty f_2(\tau) \left[\int_{-\tau}^\infty f_1(t') u(t') e^{-s(t' + \tau)}\, dt' \right] d\tau = F(s).$$

The unit-step function $u(t')$ makes the value of the inner integral zero for t' less than zero. Thus, the last expression reduces to

$$\int_0^\infty f_2(\tau) \left[\int_0^\infty f_1(t') e^{-s(t' + \tau)}\, dt' \right] d\tau = F(s)$$

which can be separated as a product of two integrals

$$\left[\int_0^\infty f_2(\tau) e^{-s\tau}\, d\tau \right]\left[\int_0^\infty f_1(t') e^{-st'}\, dt' \right] = F(s). \tag{A.2.3.24}$$

However, the left-hand side of (A.2.3.24) is simply equal to $F_2(s)F_1(s)$. Hence, with

---

† If $F(s)$ is the Laplace transform of $f(t)$, then $f(t)$ is referred to as the inverse Laplace transform of $F(s)$ and is given by the *inversion integral*

$$f(t) = \mathcal{L}^{-1}\{F(s)\} = \frac{1}{2\pi j} \int_{\sigma_1 - j\infty}^{\sigma_1 + j\infty} F(s) e^{st}\, ds$$

where σ_1 is a real constant to be chosen such that the integral exists. More on this subject will be discussed later.

the aid of (A.2.3.23), (A.2.3.24) reduces to

$$F_2(s)F_1(s) = \mathscr{L}\left\{ \int_0^t f_1(t - \tau)f_2(\tau)\,d\tau \right\}$$

which, of course, is what we wish to verify.

Example A.2.3.11 The inverse Laplace transform of

$$F(s) = \frac{1}{(s + \alpha)^2}$$

can be determined by Theorem A.2.3.9 as follows. Let $F_1(s) = F_2(s) = 1/(s + \alpha)$, which in the t-domain corresponds to $f_1(t) = f_2(t) = e^{-\alpha t}$. Thus, applying (A.2.3.22) yields

$$f(t) = \int_0^t f_1(t - \tau)f_2(\tau)\,d\tau = \int_0^t e^{-\alpha(t-\tau)}e^{-\alpha\tau}\,d\tau$$

$$= \int_0^t e^{-\alpha t}\,d\tau = e^{-\alpha t}\int_0^t d\tau = te^{-\alpha t}$$

which, of course, is the same time function used in Example A.2.3.7.

A.2.4 PARTIAL FRACTION EXPANSION

In general, it is easier to solve an algebraic equation than an ordinary differential equation of similar complexity. Through the use of the Laplace transform method, we can transform problems described by ordinary differential equations with certain time functions $f_i(t)$ to be determined into equivalent algebraic equations with transform functions $F_i(s)$. We then simplify the image functions $F_i(s)$ and carry out the inverse Laplace transformations so that we can obtain the solutions of the original problems.

As we pointed out in the preceding section, a function $f(t)$ and its transform $F(s)$ are related by the pair of equations

$$F(s) = \mathscr{L}\{f(t)\} = \int_0^\infty f(t)e^{-st}\,dt \tag{A.2.4.1}$$

and

$$f(t) = \mathscr{L}^{-1}\{F(s)\} = \frac{1}{2\pi j}\int_{\sigma_1 - j\infty}^{\sigma_1 + j\infty} F(s)e^{-st}\,ds, \tag{A.2.4.2}$$

where σ_1 is a real constant to be selected such that the inversion integral (A.2.4.2) exists. Theoretically, once the function $F(s)$ is known, the inverse transform $f(t)$ can be determined uniquely by the inversion integral. However, in practice,

Table A.2.4.1 Laplace Transform Pairs.

	$f(t), \quad t \geq 0$	$F(s)$
1	$\delta(t)$, unit impulse	1
2	$\delta^{(n)}(t)$, $n = $ positive integer	s^n
3	$u(t)$, unit step	$\dfrac{1}{s}$
4	$tu(t)$, unit ramp	$\dfrac{1}{s^2}$
5	$\dfrac{t^n}{n!}$, $n = $ positive integer	$\dfrac{1}{s^{n+1}}$
6	e^{-at}, $a = $ constant	$\dfrac{1}{s+a}$
7	$\dfrac{t^n}{n!}e^{-at}$, $a = $ constant, $n = $ positive integer	$\dfrac{1}{(s+a)^{n+1}}$
8	$\sin bt$, $b = $ constant	$\dfrac{b}{s^2+b^2}$
9	$\cos bt$	$\dfrac{s}{s^2+b^2}$
10	$e^{-at}\sin bt$	$\dfrac{b}{(s+a)^2+b^2}$
11	$e^{-at}\cos bt$	$\dfrac{s+a}{(s+a)^2+b^2}$

instead of using (A.2.4.2), it is far easier to determine $f(t)$ by decomposing the given function $F(s)$ into a set of simpler functions in s (by partial fraction expansion) whose inverse Laplace transforms can be readily recognized through the use of a table of Laplace transform pairs. One such table is shown in Table A.2.4.1. A detailed discussion of partial fraction expansion by the method of residues will be presented in the following paragraphs.

The basic assumption which the function $F(s) = N(s)/D(s)$ under consideration must satisfy is that $F(s)$ be a rational function; that is, a ratio of two polynomials. To simplify the discussion, rational functions are classified into different cases depending on the relative degrees of the numerator and denominator polynomials as well as the order of complexity of the poles of $F(s)$ [or, the roots of the equation $D(s) = 0$]; that is, whether the poles of $F(s)$ are simple (namely, of order 1) or of higher order.

For purposes of discussion, let the function under consideration be described by

$$F(s) = \frac{N(s)}{D(s)} = \frac{b_m s^m + b_{m-1} s^{m-1} + \cdots + b_1 s + b_0}{a_n s^n + a_{n-1} s^{n-1} + \cdots + a_1 s + a_0}, \tag{A.2.4.3}$$

where the constants $a_1, a_2, \ldots, a_n, b_1, b_2, \ldots, b_m$ are assumed to be real numbers. Expressing $D(s)$ in factored form, we find that

$$F(s) = \frac{N(s)}{D(s)} = \frac{N(s)}{a_n(s - s_1)(s - s_2) \cdots (s - s_n)} \tag{A.2.4.4}$$

where $s_1, s_2, \ldots, s_n$ are the n roots of the equation $D(s) = 0$. Let us consider the following three cases.

Case 1. If the roots $s_1, s_2, \ldots, s_n$ in (A.2.4.4) are *distinct* and if the degree of $N(s)$ is less than the degree of $D(s)$, then the partial fraction expansion of $F(s)$ is

$$F(s) = \frac{N(s)}{D(s)} = \frac{K_1}{s - s_1} + \frac{K_2}{s - s_2} + \cdots + \frac{K_n}{s - s_n}, \tag{A.2.4.5}$$

where the K's are constants called the *residues* of $F(s)$. To determine the values of these residues, say K_1, multiply both sides of (A.2.4.5) by the factor $(s - s_1)$ yielding

$$(s - s_1)F(s) = K_1 + \left(\frac{K_2}{s - s_2} + \cdots + \frac{K_n}{s - s_n} \right)(s - s_1),$$

from which we obtain K_1 by setting $s = s_1$:

$$K_1 = (s - s_1)F(s)\big|_{s = s_1} \tag{A.2.4.6}$$

It appears that K_1 is equal to zero in (A.2.4.6). However, it should be noted that the function $F(s)$ contains the factor $(s - s_1)$ in the denominator in (A.2.4.5) which will cancel with the one in the numerator before the variable s is set to s_1, thus making K_1 nonzero. If the subscript 1 is replaced by i with $i = 1, 2, \ldots, n$, we obtain the general formula for evaluating any one of the residues; that is,

$$K_i = (s - s_i)F(s)\big|_{s = s_i}, \qquad i = 1, 2, \ldots, n. \tag{A.2.4.7}$$

The following example will illustrate the process of determining the partial fraction expansion.

Example A.2.4.1 Consider the rational function

$$F(s) = \frac{2s + 1}{s^2 + 5s + 4} = \frac{2s + 1}{(s + 1)(s + 4)}$$

$$= \frac{K_1}{s + 1} + \frac{K_2}{s + 4}.$$

Residues K_1 and K_2 can be determined by applying (A.2.4.7) with $i = 1, 2$. Thus,

$$K_1 = (s + 1)F(s)\Big|_{s=-1} = \frac{2s + 1}{s + 4}\Big|_{s=-1} = -\frac{1}{3},$$

$$K_2 = (s + 4)F(s)\Big|_{s=-4} = \frac{2s + 1}{s + 1}\Big|_{s=-4} = \frac{7}{3},$$

and the partial fraction expansion is

$$F(s) = \frac{-\frac{1}{3}}{s + 1} + \frac{\frac{7}{3}}{s + 4}.$$

Case 2. If a root, say s_1, of $D(s) = 0$ is of order r, and the other roots are simple with the degree of $N(s)$ again less than that of $D(s)$, then the partial fraction expansion of $F(s)$ is

$$F(s) = \frac{N(s)}{D(s)} = \frac{N(s)}{a_n(s - s_1)^r(s - s_2)\cdots(s - s_m)}$$

$$= \frac{K_{11}}{s - s_1} + \frac{K_{12}}{(s - s_1)^2} + \cdots + \frac{K_{1(r-1)}}{(s - s_1)^{r-1}} + \frac{K_{1r}}{(s - s_1)^r}.$$

$$+ \frac{K_2}{s - s_2} + \cdots + \frac{K_m}{s - s_m}, \qquad\qquad (A.2.4.8)$$

where $K_2, K_3, \ldots, K_m$ can be determined by applying (A.2.4.7) as in Example A.2.4.1, and the formula for evaluating the constants $K_{11}, K_{12}, \ldots, K_{1r}$ can be derived as follows. First, multiplying both sides of (A.2.4.8) by $(s - s_1)^r$, we obtain

$$(s - s_1)^r F(s) = K_{11}(s - s_1)^{r-1} + K_{12}(s - s_1)^{r-2} + \cdots + K_{1(r-1)}(s - s_1)$$

$$+ K_{1r} + \left(\frac{K_2}{s - s_2} + \cdots + \frac{K_m}{s - s_m}\right)(s - s_1)^r. \qquad (A.2.4.9)$$

Next, differentiating both sides of (A.2.4.9) with respect to s yields

$$\frac{d}{ds}[(s - s_1)^r F(s)] = (r - 1)K_{11}(s - s_1)^{r-2} + (r - 2)K_{12}(s - s_1)^{r-3}$$

$$+ \cdots + K_{1(r-1)} + \frac{d}{ds}\left[\left(\frac{K_2}{s - s_2} + \cdots + \frac{K_m}{s - s_m}\right)(s - s_1)^r\right]. \qquad (A.2.4.10)$$

Now, putting $s = s_1$ in both (A.2.4.9) and (A.2.4.10) leads to the expressions for determining K_{1r} and $K_{1(r-1)}$, respectively, as

$$K_{1r} = (s - s_1)^r F(s)\big|_{s=s_1} \qquad\qquad (A.2.4.11)$$

and

$$K_{1(r-1)} = \frac{d}{ds}[(s - s_1)^r F(s)]_{s=s_1}. \tag{A.2.4.12}$$

Differentiating (A.2.4.10) again with respect to s and then setting $s = s_1$ will lead to the formula for determining $K_{1(r-2)}$ as

$$K_{1(r-2)} = \frac{1}{2!}\left\{\frac{d^2}{ds^2}[(s - s_1)^r F(s)]\right\}_{s=s_1}. \tag{A.2.4.13}$$

It can be readily shown that repeated application of the above process (that is, differentiating and setting $s = s_1$) will lead to the general expression

$$K_{1i} = \frac{1}{(r-i)!}\left\{\frac{d^{(r-i)}}{ds^{(r-i)}}[(s - s_1)^r F(s)]\right\}_{s=s_1}, \qquad i = 1,\ldots,r \tag{A.2.4.14}$$

from which all the constants $K_{11}, K_{12}, \ldots, K_{1(r-1)}, K_{1r}$ can be evaluated by letting $i = 1, 2, \ldots, r - 1, r$, respectively.

Example A.2.4.2 The partial fraction expansion of the function

$$F(s) = \frac{s^2 + 1}{(s + 1)^3(s + 2)}$$

is

$$F(s) = \frac{K_{11}}{s + 1} + \frac{K_{12}}{(s + 1)^2} + \frac{K_{13}}{(s + 1)^3} + \frac{K_2}{s + 2},$$

where, with the aid of (A.2.4.7) and (A.2.4.14), the constants are given by

$$K_2 = (s + 2)F(s)\Big|_{s=-2} = \frac{s^2 + 1}{(s + 1)^3}\Big|_{s=-2} = -5,$$

$$K_{13} = (s + 1)^3 F(s)\Big|_{s=-1} = \frac{s^2 + 1}{s + 2}\Big|_{s=-1} = 2,$$

$$K_{12} = \frac{1}{1!}\left\{\frac{d}{ds}[(s + 1)^3 F(s)]\right\}_{s=-1} = \left\{\frac{d}{ds}\left[\frac{s^2 + 1}{s + 2}\right]\right\}_{s=-1}$$

$$= \frac{2s(s + 2) - 1(s^2 + 1)}{(s + 2)^2}\Big|_{s=-1} = \frac{s^2 + 4s - 1}{(s + 2)^2}\Big|_{s=-1} = -4,$$

$$K_{11} = \frac{1}{2!}\left\{\frac{d^2}{ds^2}[(s + 1)^3 F(s)]\right\}_{s=-1} = \frac{1}{2!}\left\{\frac{d}{ds}\left[\frac{s^2 + 4s - 1}{(s + 2)^2}\right]\right\}_{s=-1}$$

$$= \frac{1}{2}\left\{\frac{(2s + 4)(s + 2)^2 - 2(s + 2)(s^2 + 4s - 1)}{(s + 2)^4}\right\}_{s=-1} = 5.$$

In the derivation of (A.2.4.14) only one multiple root was assumed for the denominator polynomial of $F(s)$. However, it should be noted that (A.2.4.14) is still applicable even when the denominator polynomial of $F(s)$ contains two or more multiple roots as the following example will demonstrate.

Example A.2.4.3 The function

$$F(s) = \frac{1}{(s + 1)^3(s + 2)^2}$$

has a pole of order 3 at $s = -1$ and a pole of order 2 at $s = -2$. The partial fraction expansion of $F(s)$ can be expressed as

$$F(s) = \frac{K_{11}}{s + 1} + \frac{K_{12}}{(s + 1)^2} + \frac{K_{13}}{(s + 1)^3} + \frac{K_{21}}{s + 2} + \frac{K_{22}}{(s + 2)^2}$$

which consists of a chain of *three* terms for the *third-order* pole $s = -1$ and a chain of *two* terms for the *second-order* pole at $s = -2$. The coefficients K's are given by

$$K_{13} = (s + 1)^3 F(s)\Big|_{s=-1} = \frac{1}{(s + 2)^2}\Big|_{s=-1} = 1,$$

$$K_{12} = \left\{\frac{d}{ds}[(s + 1)^3 F(s)]\right\}_{s=-1} = \left\{\frac{d}{ds}\frac{1}{(s + 2)^2}\right\}_{s=-1} = \frac{-2}{(s + 2)^3}\Big|_{s=-1} = -2,$$

$$K_{11} = \frac{1}{2!}\left\{\frac{d^2}{ds^2}[(s + 1)^3 F(s)]\right\}_{s=-1} = \frac{1}{2!}\left\{\frac{d}{ds}\left[\frac{-2}{(s + 2)^3}\right]\right\}_{s=-1}$$

$$= \frac{6}{2!}\left[\frac{1}{(s + 2)^4}\right]_{s=-1} = 3,$$

$$K_{22} = (s + 2)^2 F(s)\Big|_{s=-2} = \frac{1}{(s + 1)^3}\Big|_{s=-2} = -1,$$

$$K_{21} = \left\{\frac{d}{ds}[(s + 2)^2 F(s)]\right\}_{s=-2} = \left\{\frac{d}{ds}\frac{1}{(s + 1)^3}\right\}_{s=-2} = \frac{-3}{(s + 1)^4}\Big|_{s=-2} = -3.$$

Case 3. If the degree of $N(s)$ is equal to, or greater than, the degree of $D(s)$, then the function $F(s) = N(s)/D(s)$ is called an *improper fraction* and the partial fraction expansion of $F(s)$ can be obtained by first dividing $D(s)$ into $N(s)$ until the remainder is of degree less than that of $D(s)$. As a result of the division, $F(s)$ can be written as

$$F(s) = \sum_{i=0}^{m-n} c_i s^i + \frac{N_1(s)}{D(s)} = \sum_{i=0}^{m-n} c_i s^i + G(s), \tag{A.2.4.15}$$

where $G(s)$, called a *proper fraction*,* is a rational function with the degree of

---

* Thus rational functions belonging to Cases 1 or 2 are all defined as proper fractions, whereas all those that are classified under Case 3 are improper fractions by definition.

numerator polynomial $N_1(s)$ less than that of the denominator polynomial $D(s)$, and m and n are, respectively, the degrees of $N(s)$ and $D(s)$. Depending upon whether the roots of $D(s)$ are simple, the partial fraction expansion of $G(s)$ can be determined either as in Case 1 or 2. The next example will illustrate the steps involved.

Example A.2.4.4   The degree of $N(s)$ is higher than that of $D(s)$ for the function

$$F(s) = \frac{N(s)}{D(s)}$$

$$= \frac{s^3 + 6s^2 + 11s + 5}{s^2 + 5s + 4}$$

which can be written in the form of (A.2.4.15) by dividing $D(s)$ into $N(s)$, yielding

$$F(s) = s + 1 + \frac{2s + 1}{s^2 + 5s + 4}.$$

The rational function on the right-hand side of the above equation is identical to that in Example A.2.4.1. Hence the partial fraction expansion of $F(s)$ is

$$F(s) = s + 1 - \frac{\frac{1}{3}}{s + 1} + \frac{\frac{7}{3}}{s + 4}.$$

A.2.5 APPLICATIONS BY THE LAPLACE TRANSFORMATION METHOD

The procedure for solving a differential equation (or a set of differential equations) by the method of Laplace transformation can be briefly summarized as follows:

Step 1. Take the Laplace transform of the given differential equation with the unknown function $f(t)$, yielding an algebraic equation with the transformed function $F(s)$.

Step 2. Substitute the initial conditions into the algebraic equation found in Step 1.

Step 3. Express $F(s)$ as a rational function and expand it into partial fractions.

Step 4. Take the inverse Laplace transform of $F(s)$ to obtain $f(t)$—the solution of the differential equation. This step can be readily accomplished by referring to a table of Laplace transform pairs such as Table A.2.4.1.

The following examples will illustrate the ideas involved.

Example A.2.5.1   Consider the problem of determining the solution of the second-order differential equation

$$\frac{d^2 f}{dt^2} + 3 \frac{df}{dt} + 2f = 4u(t) \tag{A.2.5.1}$$

with the initial conditions $f(0) = 0$ and $(df/dt)(0) = 3$. Taking the Laplace transform of (A.2.5.1), we have

$$s^2F(s) - sf(0) - \frac{df}{dt}(0) + 3[sF(s) - f(0)] + 2F(s) = \frac{4}{s}$$

or

$$(s^2 + 3s + 2)F(s) = \frac{4}{s} + (s + 3)f(0) + \frac{df}{dt}(0) \qquad (A.2.5.2)$$

in which the expression $\mathcal{L}\{d^2f/dt^2\} = s^2F(s) - sf(0) - (df/dt)(0)$ has been applied
(see Problem A.2.1). Substituting the initial conditions into (A.2.5.2) and then
expressing $F(s)$ in the form of a rational function, we obtain

$$F(s) = \frac{3s + 4}{s(s^2 + 3s + 2)} = \frac{3s + 4}{s(s + 1)(s + 2)}$$

which belongs to Case 1. Therefore, with the aid of (A.2.4.7), the partial fraction
expansion of $F(s)$ is

$$F(s) = \frac{2}{s} - \frac{1}{s + 1} - \frac{1}{s + 2}.$$

Using the Laplace transform pairs in Table A.2.4.1, $f(t)$ can be readily shown
to be

$$f(t) = 2 - e^{-t} - e^{-2t}, \qquad t \geq 0$$

which is the solution of the differential equation (A.2.5.1).

Example A.2.5.2 As another example, consider the set of simultaneous differential
equations

$$\frac{dy}{dt} + y + 2z = 0$$

$$\frac{dz}{dt} + 4z + 2y = u(t) \qquad (A.2.5.3)$$

with the initial conditions $y(0) = 3$ and $z(0) = 1$. The problem is to find both $y(t)$
and $z(t)$ for $t \geq 0$. Again, taking the Laplace transform of both equations in
(A.2.5.4), we find

$$(s + 1)Y(s) + 2Z(s) = y(0)$$

$$2Y(s) + (s + 4)Z(s) = \frac{1}{s} + z(0) \qquad (A.2.5.4)$$

from which one can solve for $Y(s)$ and $Z(s)$:

$$Y(s) = \frac{\begin{vmatrix} y(0) & 2 \\ 1/s + z(0) & s + 4 \end{vmatrix}}{\begin{vmatrix} s + 1 & 2 \\ 2 & s + 4 \end{vmatrix}} = \frac{y(0)(s + 4) - 2[1/s + z(0)]}{s(s + 5)} \qquad \text{(A.2.5.5)}$$

$$Z(s) = \frac{\begin{vmatrix} s + 1 & y(0) \\ 2 & 1/s + z(0) \end{vmatrix}}{s(s + 5)} = \frac{(s + 1)[1/s + z(0)] - 2y(0)}{s(s + 5)} \qquad \text{(A.2.5.6)}$$

Substituting the initial conditions in (A.2.5.5) and (A.2.5.6), we can show that the partial fraction expansions for $Y(s)$ and $Z(s)$ are given by

$$Y(s) = \frac{3s^2 + 10s - 2}{s^2(s + 5)} = \frac{\frac{52}{25}}{s} - \frac{\frac{2}{5}}{s^2} + \frac{\frac{23}{25}}{s + 5}$$

and

$$Z(s) = \frac{s^2 - 4s + 1}{s^2(s + 5)} = \frac{-\frac{21}{25}}{s} + \frac{\frac{1}{5}}{s^2} + \frac{\frac{46}{25}}{s + 5}$$

whose inverse Laplace transforms are the solutions of the problem, viz.,

$$y(t) = \tfrac{52}{25} - \tfrac{2}{5}t + \tfrac{23}{25}e^{-5t}$$

and

$$z(t) = -\tfrac{21}{25} + \tfrac{1}{5}t + \tfrac{46}{25}e^{-5t}$$

valid for $t \geq 0$.

Example A.2.5.3 Consider the differential equation

$$\frac{d^2y}{dt^2} + 2\frac{dy}{dt} + 5y = u(t)$$

with $y(0) = (dy/dt)(0) = 1$. Taking the Laplace transform of the differential equation, we find that

$$s^2Y(s) - sy(0) - \frac{dy}{dt}(0) + 2[sY(s) - y(0)] + 5Y(s) = \frac{1}{s}$$

or

$$(s^2 + 2s + 5)Y(s) = \frac{1}{s} + sy(0) + \frac{dy}{dt}(0) + 2y(0).$$

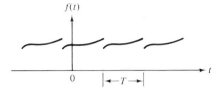

$f(t)$

0    $|\leftarrow T \rightarrow|$ t

Fig. A.2.6.1 A periodic function.

Substituting the initial conditions into the last equation above, we obtain the partial fraction expansion for $Y(s)$

$$Y(s) = \frac{s^2 + 3s + 1}{s(s^2 + 2s + 5)}$$

$$= \frac{\frac{1}{5}}{s} + \frac{\frac{8}{20} + j\frac{9}{20}}{s + 1 + j2} + \frac{\frac{8}{20} - j\frac{9}{20}}{s + 1 - j2}.$$

Using the result of Problem A.2.2, the inverse transform of $Y(s)$ is given by

$$y(t) = \tfrac{1}{5} + 2e^{-t}(\tfrac{8}{20}\cos 2t + \tfrac{9}{20}\sin 2t)$$

$$= \tfrac{1}{5} + \tfrac{1}{10}e^{-t}(8\cos 2t + 9\sin 2t).$$

A.2.6 ORTHOGONALITY CONDITIONS AND FOURIER SERIES

A function $f(t)$ is said to be a *periodic function* of period T if the condition

$$f(t) = f(t + T) \tag{A.2.6.1}$$

is satisfied for all values of t. A typical waveform fulfilling this requirement is shown in Fig. A.2.6.1 in which the pattern of the graph is repeated for every interval of length T. Since periodic functions that occur in practice are usually rather complicated, to simplify the solution of a given problem it is often desirable to represent a complicated periodic function in terms of a set of simpler functions such as

$$1, \qquad \sin\frac{n2\pi}{T}t, \qquad \text{and} \qquad \cos\frac{n2\pi}{T}t, \qquad n = 1, 2, \dots$$

which have the same *period T*. It can be shown* that if a function $f(t)$ satisfies condition (a) $f(t) = f(t + T)$ for all t, and condition (b) $f(t)$ is sectionally continuous in the interval $0 \le t \le T$, then $f(t)$ can be represented by the *Fourier series*

$$f(t) = a_0 + \sum_{n=1}^{\infty}(a_n \cos n\omega t + b_n \sin n\omega t), \tag{A.2.6.2}$$

---

* For example, refer to R. V. Churchill [CH 5], Chapter 4.

where $\omega \triangleq 2\pi/T$ and

$$a_0 = \frac{1}{T} \int_0^T f(t)\, dt, \tag{A.2.6.3}$$

$$a_n = \frac{2}{T} \int_0^T f(t) \cos n\omega t\, dt, \tag{A.2.6.4}$$

$$b_n = \frac{2}{T} \int_0^T f(t) \sin n\omega t\, dt \tag{A.2.6.5}$$

for all values of t. The only exception is that at the points of discontinuity where the function has a right- and left-hand derivative* the Fourier series converges to the value

$$\tfrac{1}{2}[f(t+0) + f(t-0)].$$

To derive the expressions for the Fourier coefficients a_0, a_n, and b_n as given in (A.2.6.3) through (A.2.6.5), we need a set of orthogonality conditions. Two functions $f_1(t)$ and $f_2(t)$ are said to be *orthogonal* to each other over the interval $a \leq t \leq b$ if the condition

$$\int_a^b f_1(t) f_2(t)\, dt = 0 \tag{A.2.6.6}$$

is satisfied. It can be readily shown that the following orthogonality conditions are valid: for $\omega = 2\pi/T$,

$$\int_0^T \sin n\omega t\, dt = 0 \qquad (n = 1, 2, \ldots), \tag{A.2.6.7}$$

$$\int_0^T \cos n\omega t\, dt = 0 \qquad (n = 1, 2, \ldots)$$
$$= T \qquad (n = 0), \tag{A.2.6.8}$$

---

* The right-hand derivative of $f(t)$ at t_0 is defined as

$$\frac{df}{dt}(t_0 + 0) \triangleq \lim_{\Delta t \to 0} \frac{f(t_0 + \Delta t) - f(t_0 + 0)}{\Delta t},$$

where $f(t_0 + 0) \triangleq \lim_{\Delta t \to 0} f(t_0 + \Delta t)$. Similarly, the left-hand derivative is

$$\frac{df}{dt}(t_0 - 0) \triangleq \lim_{\Delta t \to 0} \frac{f(t_0 - 0) - f(t_0 - \Delta t)}{\Delta t}$$

with $f(t_0 - 0) \triangleq \lim_{\Delta t \to 0} f(t_0 - \Delta t)$.

$$\int_0^T \cos n\omega t \cos m\omega t \, dt = 0 \qquad (m \neq n)$$

$$= \frac{T}{2} \qquad (m = n, n = 1, 2, \ldots), \qquad \text{(A.2.6.9)}$$

$$\int_0^T \sin n\omega t \sin m\omega t \, dt = 0 \qquad (m \neq n)$$

$$= \frac{T}{2} \qquad (m = n, n = 1, 2, \ldots), \qquad \text{(A.2.6.10)}$$

$$\int_0^T \cos n\omega t \sin m\omega t \, dt = 0 \qquad (m, n = 1, 2, \ldots). \qquad \text{(A.2.6.11)}$$

To derive (A.2.6.3), we integrate both sides of (A.2.6.2) with respect to t over the interval $0 \leq t \leq T$ yielding

$$\int_0^T f(t) \, dt = \int_0^T a_0 \, dt + \sum_{n=1}^{\infty} \left(a_n \int_0^T \cos n\omega t \, dt + b_n \int_0^T \sin n\omega t \, dt \right).$$

Applying (A.2.6.7) and (A.2.6.8) to the above equation results in

$$\int_0^T f(t) \, dt = a_0 T$$

which is (A.2.6.3). Next, multiplying (A.2.6.2) by $\cos m\omega t$ and integrating gives

$$\int_0^T f(t) \cos m\omega t \, dt = a_0 \int_0^T \cos m\omega t \, dt + \sum_{n=1}^{\infty} \left(a_n \int_0^T \cos n\omega t \cos m\omega t \, dt \right.$$

$$\left. + b_n \int_0^T \sin n\omega t \cos m\omega t \, dt \right)$$

which, by virtue of (A.2.6.8) through (A.2.6.10), reduces to

$$\int_0^T f(t) \cos m\omega t \, dt = \frac{a_m T}{2} \qquad (m = 1, 2, \ldots),$$

an expression identical to (A.2.6.4) after m is replaced by n.

Similarly, multiplying (A.2.6.2) by $\sin m\omega t$ and integrating gives

$$\int_0^T f(t) \sin m\omega t \, dt = \frac{b_m T}{2} \qquad (m = 1, 2, \ldots)$$

which, after replacing m by n, is identical to (A.2.6.5).

Example A.2.6.1 The coefficients of the Fourier series representing the triangular waveform shown in Fig. A.2.6.2 can be determined as follows. By inspection, the period T is 2π. Hence, $\omega = 2\pi/T = 1$ and, with the aid of (A.2.6.3) through

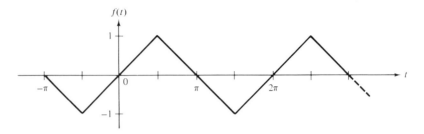

Fig. A.2.6.2 A triangular waveform.

(A.2.6.5), the coefficients in the Fourier series are

$$a_0 = \frac{1}{2\pi} \int_0^{2\pi} f(t)\, dt = \frac{1}{2\pi} \left[\int_0^{\pi/2} \left(\frac{2}{\pi} t\right) dt + \int_{\pi/2}^{3\pi/2} \left(-\frac{2}{\pi} t + 2\right) dt \right.$$

$$\left. + \int_{3\pi/2}^{2\pi} \left(\frac{2}{\pi} t - 4\right) dt \right] = 0,$$

$$a_n = \frac{1}{\pi} \int_0^{2\pi} f(t) \cos nt\, dt = \frac{1}{\pi} \left[\int_0^{\pi/2} \frac{2}{\pi} t \cos nt\, dt \right.$$

$$\left. + \int_{\pi/2}^{3\pi/2} \left(-\frac{2}{\pi} t + 2\right) \cos nt\, dt + \int_{3\pi/2}^{2\pi} \left(\frac{2}{\pi} t - 4\right) \cos nt\, dt \right] = 0,$$

$$b_n = \frac{1}{\pi} \int_0^{2\pi} f(t) \sin nt\, dt = \frac{1}{\pi} \left[\int_0^{\pi/2} \frac{2}{\pi} t \sin nt\, dt \right.$$

$$\left. + \int_{\pi/2}^{3\pi/2} \left(-\frac{2}{\pi} t + 2\right) \sin nt\, dt + \int_{3\pi/2}^{2\pi} \left(\frac{2}{\pi} t - 4\right) \sin nt\, dt \right]$$

$$= \frac{8}{n^2 \pi^2} \sin n \frac{\pi}{2}.$$

Substituting the Fourier coefficients into (A.2.6.2), we find the Fourier series representing the triangular waveform to be

$$f(t) = \sum_{n=1}^{\infty} b_n \sin nt = \sum_{n=1}^{\infty} \frac{8}{\pi^2} \left(\frac{\sin n \dfrac{\pi}{2}}{n^2} \right) \sin nt$$

$$= \frac{8}{\pi^2} \left(\sin t - \frac{1}{3^2} \sin 3t + \frac{1}{5^2} \sin 5t - \cdots \right) \qquad (A.2.6.12)$$

Based on a number of interesting properties of certain classes of periodic functions, the evaluation of the Fourier coefficients can sometimes be greatly

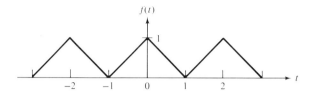

Fig. A.2.6.3 An example of an even function.

simplified depending upon the waveform under consideration. These interesting properties can be briefly summarized as follows.

Case 1. The average value of $f(t)$ is zero. The term a_0 in the Fourier series (A.2.6.2) represents the average value of the function $f(t)$. Thus, if the area enclosed by the positive part of $f(t)$ equals that enclosed by the negative part, the Fourier coefficient a_0 assumes the value of zero. An inspection of Fig. A.2.6.2 immediately leads to the conclusion that the term a_0 is zero for the triangular waveform.

Case 2. The function $f(t)$ is an even function; that is, $f(t)$ satisfies the condition

$$f(t) = f(-t) \qquad \text{for all } t. \tag{A.2.6.13}$$

A simple example of an even function is that shown in Fig. A.2.6.3.
     It can be readily shown that when $f(t)$ is an even function, the Fourier coefficients can be simplified to

$$a_0 = \frac{2}{T} \int_0^{T/2} f(t)\, dt, \tag{A.2.6.14}$$

$$a_n = \frac{4}{T} \int_0^{T/2} f(t) \cos n\omega t\, dt \qquad (n = 1, 2, \ldots), \tag{A.2.6.15}$$

$$b_n = 0 \qquad (n = 1, 2, \ldots) \tag{A.2.6.16}$$

and the series in (A.2.6.2) becomes the *Fourier cosine series*

$$f(t) = a_0 + \sum_{n=1}^{\infty} a_n \cos n\omega t. \tag{A.2.6.17}$$

Case 3. The function $f(t)$ is an odd function; that is, $f(t)$ satisfies the condition

$$f(t) = -f(-t) \qquad \text{for all } t. \tag{A.2.6.18}$$

Simple algebraic manipulations will lead to the following expressions for the Fourier coefficients

$$a_n = 0 \qquad (n = 0, 1, \ldots) \tag{A.2.6.19}$$

$$b_n = \frac{4}{T} \int_0^{T/2} f(t) \sin n\omega t\, dt \qquad (n = 1, 2, \ldots), \tag{A.2.6.20}$$

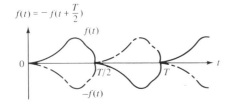

$f(t) = -f\left(t + \dfrac{T}{2}\right)$

Fig. A.2.6.4 An example of a half-wave symmetric function.

and the series in (A.2.6.2) reduces to the *Fourier sine series*

$$f(t) = \sum_{n=1}^{\infty} b_n \sin n\omega t. \qquad (A.2.6.21)$$

The triangular waveform of Fig. A.2.6.2 is an example of an odd function and hence the series representation of $f(t)$ is in the form of a Fourier sine series. An inspection of (A.2.6.12) will substantiate this fact.

Case 4. The function $f(t)$ is a half-wave symmetric function; that is, $f(t)$ satisfies the condition

$$f(t) = -f\left(t + \frac{T}{2}\right). \qquad (A.2.6.22)$$

A function $f(t)$ possesses this property when the negative portion of the waveform is the mirror image of the positive portion displaced horizontally a distance equal to half a period. The waveform shown in Fig. A.2.6.4 is an example of a half-wave symmetric function. The Fourier coefficients in this case can be shown to be

$$a_0 = 0 \qquad (A.2.6.23)$$

$$a_n = \begin{cases} 0 & (n = 2, 4, 6, \ldots) \\ \dfrac{4}{T} \displaystyle\int_0^{T/2} f(t) \cos n\omega t \, dt & (n = 1, 3, 5, \ldots) \end{cases} \qquad (A.2.6.24)$$

$$b_n = \begin{cases} 0 & (n = 2, 4, 6, \ldots) \\ \dfrac{4}{T} \displaystyle\int_0^{T/2} f(t) \sin n\omega t \, dt & (n = 1, 3, 5, \ldots) \end{cases} \qquad (A.2.6.25)$$

and the corresponding Fourier series reduces to

$$f(t) = \sum_{n=1,3,5} (a_n \cos n\omega t + b_n \sin n\omega t) \qquad (A.2.6.26)$$

A.2.7 DIFFERENT FORMS OF FOURIER SERIES

As we discussed in the previous section, the general Fourier series for a periodic function $f(t)$ with period T is expressed in the form

$$f(t) = a_0 + \sum_{n=1}^{\infty} (a_n \cos n\omega t + b_n \sin n\omega t), \qquad (A.2.7.1)$$

where

$$\omega \triangleq \frac{2\pi}{T}, \qquad (A.2.7.2)$$

$$a_0 = \frac{1}{T} \int_0^T f(t)\, dt, \qquad (A.2.7.3)$$

$$a_n = \frac{2}{T} \int_0^T f(t) \cos n\omega t\, dt \qquad (n = 1, 2, \ldots), \qquad (A.2.7.4)$$

$$b_n = \frac{2}{T} \int_0^T f(t) \sin n\omega t\, dt \qquad (n = 1, 2, \ldots). \qquad (A.2.7.5)$$

The Fourier series in (A.2.7.1) is usually referred to as the *trigonometric form*. It is sometimes convenient to combine the sine and cosine terms for each value of n and write the Fourier series in *phase-angle and magnitude form*; that is,

$$f(t) = C_0 + \sum_{n=1}^{\infty} C_n \cos(n\omega t + \phi_n). \qquad (A.2.7.6)$$

To determine the constants C_0, C_n, and ϕ_n, $(n = 1, 2, \ldots)$, we write

$$C_n \cos(n\omega t + \phi_n) = C_n (\cos n\omega t \cos \phi_n - \sin n\omega t \sin \phi_n)$$

which can be substituted into (A.2.7.6) and then compared with (A.2.7.1) term-by-term to obtain

$$C_0 = a_0,$$

$$C_n \cos \phi_n = a_n, \qquad (A.2.7.7)$$

$$-C_n \sin \phi_n = b_n.$$

The relations (A.2.7.7) can be solved to give

$$C_n = \sqrt{a_n^2 + b_n^2} \qquad (n = 1, 2, \ldots) \qquad (A.2.7.8)$$

and

$$\phi_n = \tan^{-1} \frac{-b_n}{a_n} \qquad (n = 1, 2, \ldots). \qquad (A.2.7.9)$$

The term a_0 in (A.2.7.1) or C_0 in (A.2.7.6) represents the average value of the periodic function $f(t)$. If the independent variable t is regarded as time measured in seconds, then $\omega = 2\pi/T$, a constant for a fixed T, is called the *fundamental frequency* measured in radians per second, and C_n and ϕ_n are called, respectively, the *amplitude* and the *phase* of the nth *harmonic* $C_n \cos(n\omega t + \phi_n)$. The form (A.2.7.6) as well as the Fourier exponential series to be discussed are quite useful in determining the rate of convergence of a Fourier series which will be studied in the next section.

Another alternative form for the Fourier series can be obtained by expressing the sine and cosine terms in (A.2.7.1) in terms of exponential functions. Thus

$$f(t) = a_0 + \sum_{n=1}^{\infty} \left(a_n \frac{e^{jn\omega t} + e^{-jn\omega t}}{2} + b_n \frac{e^{jn\omega t} - e^{-jn\omega t}}{2j} \right)$$

which becomes

$$f(t) = a_0 + \sum_{n=1}^{\infty} \left\{ \left(\frac{a_n - jb_n}{2} \right) e^{jn\omega t} + \left(\frac{a_n + jb_n}{2} \right) e^{-jn\omega t} \right\}. \qquad \text{(A.2.7.10)}$$

Let the complex constant D_n be defined by the relation

$$D_n \triangleq \tfrac{1}{2}(a_n - jb_n) \qquad (n = 1, 2, \ldots). \qquad \text{(A.2.7.11)}$$

Then, with the aid of (A.2.7.4) and (A.2.7.5), D_n takes the form

$$D_n = \frac{1}{T} \int_0^T f(t)(\cos n\omega t - j \sin n\omega t)\, dt$$

$$= \frac{1}{T} \int_0^T f(t) e^{-jn\omega t}\, dt \qquad \text{(A.2.7.12)}$$

from which we obtain, by replacing n by $-n$,

$$D_{-n} = \frac{1}{T} \int_0^T f(t) e^{jn\omega t}\, dt. \qquad \text{(A.2.7.13)}$$

It is evident that D_{-n} is equal to D_n^*, the complex conjugate of D_n, and can be written as

$$D_{-n} = D_n^*$$

$$= \tfrac{1}{2}(a_n + jb_n). \qquad \text{(A.2.7.14)}$$

Note also that if D_0 is defined as $D_0 \triangleq a_0$, then, by (A.2.7.3),

$$D_0 = \frac{1}{T} \int_0^T f(t)\, dt. \qquad \text{(A.2.7.15)}$$

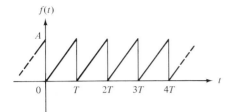

Fig. A.2.7.1 A sawtooth waveform.

Expressing (A.2.7.10) in terms of the complex constants D_0, D_n, and D_{-n}, we have

$$f(t) = D_0 + \sum_{n=1}^{\infty} (D_n e^{jn\omega t} + D_{-n} e^{-jn\omega t})$$

which we may write in the form

$$f(t) = D_n e^{jn\omega t}\big|_{n=0} + \sum_{n=1}^{\infty} D_n e^{jn\omega t} + \sum_{n=-1}^{-\infty} D_n e^{jn\omega t}$$

or, combining the terms on the right-hand side, we find

$$f(t) = D_0 + (D_1 e^{j\omega t} + D_2 e^{j2\omega t} + \cdots) + (D_{-1} e^{-j\omega t} + D_{-2} e^{-j2\omega t} + \cdots)$$

$$= \sum_{n=-\infty}^{\infty} D_n e^{jn\omega t}, \tag{A.2.7.16}$$

where D_n is given by

$$D_n = \frac{1}{T} \int_0^T f(t) e^{-jn\omega t}\, dt \qquad (n = 0, \pm 1, \pm 2, \ldots). \tag{A.2.7.17}$$

Equations (A.2.7.16) and (A.2.7.17) define the *Fourier series in (complex) exponential form.*

Example A.2.7.1 The waveform shown in Fig. A.2.7.1 is known as a *sawtooth wave.* The coefficients D_n for the Fourier exponential series can be determined by applying (A.2.7.17) as follows:

$$D_n = \frac{1}{T} \int_0^T \frac{A}{T} t\, e^{-jn\omega t}\, dt$$

$$= \frac{A}{T^2} \left\{ \frac{e^{-jn\omega t}}{(-jn\omega)^2} (-jn\omega t - 1) \right\}_0^T = \frac{jA}{2\pi n} \qquad (n \neq 0), \tag{A.2.7.18}$$

where the relation $\omega = 2\pi/T$ has been used. For $n = 0$,

$$D_0 = \frac{1}{T} \int_0^T \frac{A}{T} t\, dt = \frac{A}{T^2} \frac{t^2}{2}\bigg|_0^T = \frac{A}{2}. \tag{A.2.7.19}$$

Substituting (A.2.7.18) and (A.2.7.19) into (A.2.7.16), we find that

$$f(t) = \frac{A}{2} + \sum_{n=-\infty}^{\infty} j\frac{A}{2\pi n} e^{jn\omega t}$$

$$= \frac{A}{2} + \frac{A}{2\pi}j\left\{ \left(\frac{e^{j\omega t} - e^{-j\omega t}}{1}\right) + \left(\frac{e^{j2\omega t} - e^{-j2\omega t}}{2}\right) + \left(\frac{e^{j3\omega t} - e^{-j3\omega t}}{3}\right) + \cdots \right\}$$

(A.2.7.20)

from which the Fourier series in phase-angle and magnitude form can be easily derived as

$$f(t) = \frac{A}{2} - \frac{A}{\pi}\left\{ \left(\frac{e^{j\omega t} - e^{-j\omega t}}{2j}\right) + \frac{1}{2}\left(\frac{e^{j2\omega t} - e^{-j2\omega t}}{2j}\right) + \frac{1}{3}\left(\frac{e^{j3\omega t} - e^{-j3\omega t}}{2j}\right) + \cdots \right\}$$

$$= \frac{A}{2} - \frac{A}{\pi}\left\{ \sin \omega t + \frac{1}{2}\sin 2\omega t + \frac{1}{3}\sin 3\omega t + \cdots \right\}$$

$$= A\left[\frac{1}{2} + \sum_{n=1}^{\infty} \frac{1}{\pi n}\cos(n\omega t + \pi) \right].$$

(A.2.7.21)

Comparing the like terms between (A.2.7.6) and (A.2.7.21), we can see that the Fourier coefficient C's are given by

$$C_0 = \frac{A}{2},$$

(A.2.7.22)

$$C_n = \frac{A}{\pi n} \qquad (n = 1, 2, \ldots),$$

(A.2.7.23)

and

$$\phi_n = \pi \qquad (n = 1, 2, \ldots).$$

(A.2.7.24)

A.2.8 CONVERGENCE IN TRUNCATED SERIES

As discussed in previous sections, a Fourier series representing a given periodic function, in general, consists of an infinite number of terms. From a practical viewpoint in the analysis of engineering problems involving Fourier series, a given Fourier series is usually approximated by another series with a finite number of terms. Of course, the exact number of terms used in the approximating series depends upon the degree of accuracy desired for a given problem. Let us consider a periodic function $f(t)$ that has the Fourier series

$$f(t) = C_0 + \sum_{n=1}^{\infty} C_n \cos(n\omega t + \phi_n),$$

(A.2.8.1)

where the coefficients C_0 and C_n are determined by (A.2.7.7) and (A.2.7.8), respectively. Let $g_N(t)$ be the finite series

$$g_N(t) = A_0 + \sum_{n=1}^{N} A_n \cos{(n\omega t + \phi_n)} \tag{A.2.8.2}$$

used to approximate $f(t)$. The *mean-square error* for approximating $f(t)$ by $g_N(t)$ is defined as

$$J \triangleq \frac{1}{T} \int_0^T [f(t) - g_N(t)]^2 \, dt. \tag{A.2.8.3}$$

Let us consider the problem of determining the coefficients A_0 and A_n such that the mean-square error J is a minimum; that is, to find $g_N(t)$ such that it is the best approximation to $f(t)$ in the sense of least squares.

With the aid of (A.2.8.1) and (A.2.8.2), the integral J can be written as

$$J = \frac{1}{T} \left\{ \int_0^T [f(t)]^2 \, dt - \int_0^T 2f(t)g_N(t) \, dt + \int_0^T [g_N(t)]^2 \, dt \right\}$$

$$= \frac{1}{T} \left\{ \int_0^T [f(t)]^2 \, dt - \int_0^T 2\left[C_0 + \sum_{m=1}^{\infty} C_m \cos{(m\omega t + \phi_m)} \right] \right.$$

$$\times \left[A_0 + \sum_{n=1}^{N} A_n \cos{(n\omega t + \phi_n)} \right] dt$$

$$\left. + \int_0^T \left[A_0 + \sum_{m=1}^{N} A_m \cos{(m\omega t + \phi_m)} \right]\left[A_0 + \sum_{n=1}^{N} A_n \cos{(n\omega t + \phi_n)} \right] dt \right\}$$

$$= \frac{1}{T} \int_0^T [f(t)]^2 \, dt + A_0^2 + \sum_{n=1}^{N} \frac{A_n^2}{2} - 2C_0 A_0 - \sum_{n=1}^{N} C_n A_n. \tag{A.2.8.4}$$

Completing the squares in the right-hand side of (A.2.8.4) by adding and subtracting the term $C_0^2 + \sum_{n=1}^{N}(C_n^2/2)$, we have

$$J = \frac{1}{T} \int_0^T [f(t)]^2 \, dt - C_0^2 - \sum_{n=1}^{N} \frac{C_n^2}{2} + (A_0 - C_0)^2 + \sum_{n=1}^{N} \frac{(A_n - C_n)^2}{2}. \tag{A.2.8.5}$$

It is evident that the value of J has the least value when $A_0 = C_0$ and $A_n = C_n$ $(n = 1, 2, \ldots, N)$. Under these conditions, the minimum value of J is found to be

$$J = \frac{1}{T} \int_0^T [f(t)]^2 \, dt - \left[C_0^2 + \sum_{n=1}^{N} \frac{C_n^2}{2} \right]. \tag{A.2.8.6}$$

From the fact that $J \geq 0$, together with (A.2.8.6), we obtain the so-called *Bessel's inequality*, viz.,

$$\frac{1}{T} \int_0^T [f(t)]^2 \, dt \geqq C_0^2 + \sum_{n=1}^{N} \frac{C_n^2}{2}. \tag{A.2.8.7}$$

Of course, this inequality can also be derived by substituting the Fourier series (A.2.8.1) into the left-hand side of (A.2.8.7), yielding

$$\frac{1}{T} \int_0^T [f(t)]^2 \, dt = C_0^2 + \sum_{n=1}^{\infty} \frac{C_n^2}{2} \tag{A.2.8.8}$$

from which (A.2.8.7) follows immediately.

The relation (A.2.8.8) is known as *Parseval's theorem*. Since the mean-square value of the term $C_n \cos(n\omega t + \phi_n)$ is simply $C_n^2/2$ and that of C_0 is C_0^2, Parseval's theorem says that the *mean-square value of a periodic function $f(t)$ is equal to the sum of the mean-square values of all the components in the Fourier series representing the function.*

In the above discussion, it has been shown that the Nth-order approximation to a function $f(t)$ in the sense of least squares is obtained by using the Fourier coefficients; that is,

$$f_N(t) = C_0 + \sum_{n=1}^{N} C_n \cos(n\omega t + \phi_n). \tag{A.2.8.9}$$

The number of terms used in approximating a function depends not only upon the waveform of the function under consideration but also upon the degree of accuracy desired. In other words, the waveform of a periodic function is closely related to the rate of convergence of the Fourier series as the following example will illustrate.

Example A.2.8.1 Consider the triangular waveform in Example A.2.6.1 with the corresponding Fourier series repeated here for convenience:

$$f(t) = \frac{8}{\pi^2} \left(\sin t - \frac{1}{3^2} \sin 3t + \frac{1}{5^2} \sin 5t + \cdots \right). \tag{A.2.8.10}$$

The fifth-order approximation to $f(t)$ in the least-square sense is given by

$$f_5(t) = \frac{8}{\pi^2} \left(\sin t - \frac{1}{3^2} \sin 3t + \frac{1}{5^2} \sin 5t \right). \tag{A.2.8.11}$$

By means of (A.2.8.6), the mean-square error in the approximation is found to be

$$J = \frac{4}{2\pi} \int_0^{\pi/2} \left(\frac{2}{\pi} t \right)^2 dt - \frac{1}{2} \left(\frac{8}{\pi^2} \right)^2 \left[1 + \left(\frac{1}{3^2} \right)^2 + \left(\frac{1}{5^2} \right)^2 \right]$$

$$= 0.3333 - 0.3330$$

$$= 0.0003 \quad \text{or} \quad 0.03\%.$$

Similarly, the Fourier series for the sawtooth waveform in Example A.2.7.1 with $A = 1$ and its corresponding fifth-order approximation in the sense of least squares are, respectively, given by

$$f(t) = \frac{1}{2} - \frac{1}{\pi} \left\{ \sin \omega t + \frac{1}{2} \sin 2\omega t + \frac{1}{3} \sin 3\omega t + \cdots \right\} \qquad (\text{A.2.8.12})$$

and

$$f_5(t) = \frac{1}{2} - \frac{1}{\pi} \left\{ \sin \omega t + \frac{1}{2} \sin 2\omega t + \frac{1}{3} \sin 3\omega t + \frac{1}{4} \sin 4\omega t + \frac{1}{5} \sin 5\omega t \right\}. \qquad (\text{A.2.8.13})$$

The mean-square error in this case becomes

$$J = \frac{1}{T} \int_0^T \left(\frac{1}{T} t \right)^2 dt - \left[\left(\frac{1}{2} \right)^2 + \frac{1}{2\pi^2} \left(1 + \frac{1}{4} + \frac{1}{9} + \frac{1}{16} + \frac{1}{25} \right) \right]$$

$$= 0.3333 - 0.3241 = .0092 \qquad \text{or} \qquad 0.92\%.$$

The above example shows that, even though both functions have the same mean-square value, the fifth-order approximation for the triangular waveform is much more accurate than that for the sawtooth waveform. This fact can be readily appreciated by noting that the rate of convergence for the triangular waveform is much more rapid than that for the sawtooth (since the Fourier coefficients of the former are proportional to $1/n^2$, while those of the latter are proportional to $1/n$).

REFERENCES

1. R. V. Churchill, *Operational Mathematics*, Second Ed., McGraw-Hill, 1958, Chapters 1, 2, 5, and 6.

2. R. V. Churchill, *Fourier Series and Boundary Value Problems*, Second Ed., McGraw-Hill, 1963.

PROBLEMS

A.2.1   Show that the Laplace transform of the second derivative of a function $f(t)$ is given by

$$\mathscr{L}\left\{ \frac{d^2 f(t)}{dt^2} \right\} = s^2 F(s) - sf(0) - \frac{df}{dt}(0),$$

and then show that the Laplace transform of the nth derivative of a function $f(t)$ is given by

$$\mathscr{L}\left\{ \frac{d^n f(t)}{dt^n} \right\} = s^n F(s) - s^{n-1} f(0) - s^{n-2} \frac{df}{dt}(0) - \cdots - \frac{d^{n-1} f}{dt^{n-1}}(0).$$

A.2.2 Verify the Laplace transform pairs of the following table:

$f(t)$ for $t > 0$	$F(s)$
a) $\sin \omega t$	$\dfrac{\omega}{s^2 + \omega^2}$
b) $\cos \omega t$	$\dfrac{s}{s^2 + \omega^2}$
c) $e^{-\sigma t} \sin \omega t$	$\dfrac{\omega}{(s + \sigma)^2 + \omega^2}$
d) $e^{-\sigma t} \cos \omega t$	$\dfrac{s + \sigma}{(s + \sigma)^2 + \omega^2}$
e) $\sinh \sigma t$	$\dfrac{\sigma}{s^2 - \sigma^2}$
f) $\cosh \sigma t$	$\dfrac{s}{s^2 - \sigma^2}$

A.2.3 Determine the Laplace transform of the square wave shown in Fig. P.A.2.3. If possible, express your answer in closed form.

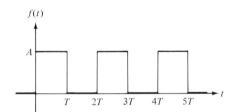

Figure P.A.2.3

A.2.4 Determine the Laplace transform of the function $f(t)$ defined by

$$f(t) = u(\cos \tfrac{1}{2}t),$$

where $u(t)$ is the unit-step function.

A.2.5 Determine the Laplace transform of each of the following functions:

a) $f_1(t) = A \cos(\omega t + \phi)$, A, ω, and ϕ are constants,
b) $f_2(t) = u(\tau - t)$, τ is a positive real number,
c) $f_3(t) = u(\tau_1 - t)u(t - \tau_2)$, $\tau_1 > \tau_2 > 0$.

A.2.6 Use the convolution integral to determine the inverse Laplace transform of

$$F(s) = \frac{2}{s(s^2 + 4)}.$$

A.2.7   Repeat Problem A.2.6 for the function

$$F(s) = \frac{1}{(s^2 + 4)^2}.$$

A.2.8   Show that the inverse Laplace transform of

$$F(s) = \frac{c + jd}{s + a + jb} + \frac{c - jd}{s + a - jb}$$

is given by

$$f(t) = 2e^{-at}[c \cos bt + d \sin bt],$$

where a, b, c, and d are real constants.

A.2.9   Let $F(s)$ be a rational function defined by

$$F(s) = \frac{N(s)}{D(s)} = \frac{N(s)}{(s - s_1)(s - s_2)\cdots(s - s_n)}$$

with $s_1, s_2, \ldots, s_n$ being distinct constants. Show that the residues $K_i(i = 1, 2, \ldots, n)$ are given by

$$K_i = \frac{N(s)}{(d/ds)D(s)}\bigg|_{s = s_i} \qquad (i = 1, \ldots, n).$$

A.2.10   Find the inverse transforms of the functions listed below, where a_1, a_2, and a_3 are distinct constants.

a) $F_1(s) = \dfrac{1}{(s + a_1)(s + a_2)(s + a_3)}$,   b) $F_2(s) = \dfrac{s}{(s + a_1)(s + a_2)}$

A.2.11   Repeat Problem A.2.10 for the following functions.

a) $F_1(s) = \dfrac{1 + e^{-3s}}{s(s + 1)}$

b) $F_2(s) = \dfrac{s^5 + 7s^4 + 19s^3 + 25s^2 + 16s + 6}{(s + 1)^3(s + 2)^2}$

c) $F_3(s) = \dfrac{s + 2}{s^2(s^2 + 4)}$

Solve the differential equations in A.2.12 through A.2.16 using the Laplace transformation method.

A.2.12 $\dfrac{d^2y}{dt^2} + 3\dfrac{dy}{dt} + 2y = 0$, $y(0) = 0$, $\dfrac{dy}{dt}(0) = 1$

A.2.13 $\dfrac{dy}{dt} + 4y = 5$, $\dfrac{dy}{dt}(0) = 2$

A.2.14 $\dfrac{d^2y}{dt^2} + 2\dfrac{dy}{dt} + y = 1$, $y(0) = 3$, $\dfrac{dy}{dt}(0) = 2$

A.2.15 $\dfrac{d^2y}{dt^2} + 3\dfrac{dy}{dt} + 2y = e^{-t}, \qquad y(0) = 0, \qquad \dfrac{dy}{dt}(0) = 1$

A.2.16 $\dfrac{dy}{dt} + 2y + 2z = 0$

$\qquad \dfrac{dz}{dt} + z + y = 2, \qquad y(0) = 1, \qquad z(0) = 2$

A.2.17   Let $f(t)$ be an arbitrary function defined for all t. Show that $f_1(t) \triangleq \tfrac{1}{2}[f(t) + f(-t)]$ is an even function and that $f_2(t) \triangleq \tfrac{1}{2}[f(t) - f(t)]$ is an odd function.

A.2.18   Show that any function defined for all t can be expressed as the sum of an even and an odd function of t.

A.2.19   Let a function $f(t)$ be represented by a Fourier series in trigonometric form. What can be said about the Fourier coefficients a_o, a_n, and b_n if $f(t)$ satisfies one of the following conditions:

a) $f(t) = f\left(t + \dfrac{T}{2}\right),$

b) $f(t) = -f\left(\dfrac{T}{2} - t\right)$ and $f(t) = f(-t).$

A.2.20   Show that the Fourier series representing the function $f(t)$ with the waveform shown in Fig. P.A.2.20 consists of the constant term a_o and the sine terms $b_n \sin n\omega t, n = 1, 2, \ldots,$ only.

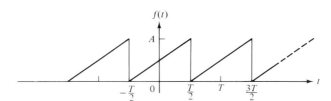

Figure P.A.2.20

A.2.21   Determine the Fourier series in trigonometric form for the waveform of Fig. P.A.2.20.

A.2.22   Determine the Fourier series in both exponential and phase-angle and magnitude forms for the waveform of Fig. P.A.2.20.

A.2.23   Repeat Problem A.2.21 for the waveform of Fig. P.A.2.23.

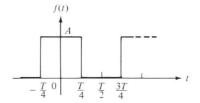

Figure P.A.2.23

A.2.24   Repeat Problem A.2.22 for the waveform of Fig. P.A.2.23.

A.2.25   Determine the Fourier series for the waveform of Fig. P.A.2.25 in

    a) trigonometric form,

    b) exponential form, and

    c) phase-angle and magnitude form.

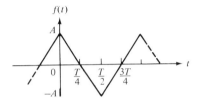

Figure P.A.2.25

A.2.26   Verify Eq. (A.2.6.26).

BASIC NUMERICAL METHODS

A.3.1 INTRODUCTION

Much of the material discussed in this Appendix is either covered in a required undergraduate numerical analysis course or incorporated into courses in computer applications or programming techniques as part of the undergraduate engineering curriculum in a great number of universities and colleges today. Therefore, our main purpose here is to provide a brief review of some of the basic numerical methods that we used, either directly or indirectly, throughout the two chapters on computer-aided network analysis. Examples will be used wherever applicable to illustrate the methods as they are presented in the subsequent sections. However, derivations and proofs will not be given due to the limited space available and to the fact that such developments will be beyond the scope of this text. The interested reader may find a more detailed treatment of the subject matter in a standard numerical analysis text.*

A.3.2 SIMPSON'S RULE

Numerical integration is essentially a numerical process or algorithm used to determine the value of a definite integral $\int_a^b f(x)\,dx$ whenever the function $f(x)$ cannot be integrated in finite terms or the evaluation of its integral is too cumbersome. Many reliable and practical algorithms are available and *Simpson's rule* is one of them that can be found in almost every numerical analysis text.

This particular rule is based on the simple formula

$$A = \frac{h}{3}(y_0 + 4y_1 + y_2) \qquad (A.3.2.1)$$

which represents the area under the curve of the parabola

$$y = C_1 x^2 + C_2 x + C_3 \qquad (A.3.2.2)$$

between $x = -h$ and $x = +h$ as shown in Fig. A.3.2.1.

---

* For example, refer to D. D. McCracken and W. S. Dorn [MC 2] and S. C. Conte [CO 1].

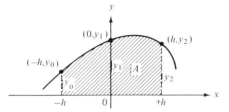

Figure A.3.2.1 Illustration of Simpson's rule.

Expression (A.3.2.1) may be obtained by first integrating (A.3.2.2) between $-h$ and $+h$, yielding

$$A = \int_{-h}^{+h} (C_1 x^2 + C_2 x + C_3)\, dx$$

$$= \frac{h}{3}(2C_1 h^2 + 6C_3).$$

(A.3.2.3)

Next, since the curve passes through the three points $(-h, y_0)$, $(0, y_1)$, and (h, y_2), we find that

$$y_0 = C_1 h^2 - C_2 h + C_3,$$

$$y_1 = C_3,$$

$$y_2 = C_1 h^2 + C_2 h + C_3,$$

from which we obtain

$$C_3 = y_1,$$

$$2C_1 h^2 = y_0 - 2y_1 + y_2.$$

(A.3.2.4)

Substituting (A.3.2.4) into (A.3.2.3), we discover that the area A (Fig. A.3.2.1) may now be expressed in terms of y_0, y_1, and y_2 yielding

$$A = \frac{h}{3}[(y_0 - 2y_1 + y_2) + 6y_1] = \frac{h}{3}(y_0 + 4y_1 + y_2)$$

which is the desired result, namely (A.3.2.1).

Simpson's rule is readily obtained by applying (A.3.2.1) to successive portions of the curve $y = f(x)$ between $x = a$ and $x = b$, with each portion covering an x-interval of width $2h$ approximated by a segment of a parabola through the three specified points, namely its ends and its midpoint. Since each sub-area under a parabolic segment is given by an expression similar to (A.3.2.1) with only changes in subscripts for the ordinates (that is, y_0, y_1, y_2 for the first sub-area, y_2, y_3, y_4 for second, etc.), the approximate value of $\int_a^b f(x)\, dx$ is obtained by adding all the sub-areas to give

$$\int_a^b f(x)\,dx = \frac{h}{3}(y_0 + 4y_1 + 2y_2 + 4y_3 + 2y_4 + \cdots + 2y_{n-2} + 4y_{n-1} + y_n),$$

$$(A.3.2.5)$$

where

$$y_i = f(x_i), \qquad i = 0, 1, 2, \ldots, n;$$

$$x_0 = a,$$

$$x_1 = a + h,$$

$$x_2 = a + 2h,$$

$$\vdots$$

$$x_n = a + nh = b;$$

and

$$h = \frac{(b-a)}{n}$$

with n being an *even* integer.

Expression (A.3.2.5) is known as *Simpson's rule*, a numerical method of approximating a definite integral.

Example A.3.2.1   Using Simpson's rule with $n = 4$, determine the approximate value of the definite integral

$$\int_0^1 (1 + x + x^2)\,dx.$$

Since $x_0 = a = 0$, $x_n = b = 1$, and $n = 4$, the step size h is

$$h = \frac{b-a}{n} = \frac{1-0}{4} = \frac{1}{4} = 0.25$$

and we need to evaluate a total of $n + 1 = 5$ points of the function $y = f(x) = 1 + x + x^2$ from $x = 0$ to $x = 1$ at $x = k(\frac{1}{4})(k = 0, 1, \ldots, 4)$. Hence

$$y_0 = f(0) = 1,$$

$$y_1 = f(\tfrac{1}{4}) = 1.3125,$$

$$y_2 = f(\tfrac{1}{2}) = 1.7500,$$

$$y_3 = f(\tfrac{3}{4}) = 2.3125,$$

$$y_4 = f(1) = 3.$$

Using (A.3.2.5), we have

$$\int_0^1 [1 + x + x^2]\, dx = \frac{1/4}{3}[1 + 4(1.3125) + 2(1.7500) + 4(2.3125) + 3]$$
$$\cong 1.8333.$$

(A.3.2.6)

Now, by actual integration, we obtain

$$\int_0^1 [1 + x + x^2]\, dx = \left[x + \frac{x^2}{2} + \frac{x^3}{3} \right]\Bigg|_0^1 \cong 1.83333$$

which is identical to (A.3.2.6).

A.3.3 RUNGE-KUTTA AND ADAMS-MOULTON ALGORITHMS

As in the case of numerical integration, many numerical methods for solving the differential equation

$$y' \equiv \frac{dy}{dx} = f(x, y), \qquad y(x_0) = y_0$$

may be found in the literature. Among the existing numerical techniques, one particular group, known as the *Runge-Kutta methods*, has found extensive use in digital computation. One of the reasons that this group of methods is more practical than others such as the Taylor series solution, for example, is the fact that they do not require the evaluation of any derivative of the function $f(x, y)$, but only of the function itself. Other distinguishing characteristics of the Runge-Kutta methods include the fact that they are "one-step" algorithms [that is, to determine y_{m+1}, one needs only to know the values of the preceding point (x_m, y_m)], and that formulas for the Runge-Kutta type of any order (depending upon the number of terms used in the series expansion of the function under consideration) can be derived using a method analogous to the one used to develop the formulas for the two lowest orders, namely, those of the first* and second order. The derivations of the Runge-Kutta formulas are beyond the scope of our intended coverage.† We shall therefore content ourselves with stating the fourth-order formula, which is the most popular and most commonly used of this type as follows.

Runge-Kutta Method of Order 4. The recursion formula for generating approximations y_n to $y(x_0 + nh)$ for a fixed step size h (and for $n = 0, 1, 2, \ldots$) of the differential equation $y' = f(x, y)$, $y(x_0) = y_0$ is given by

$$y_{n+1} = y_n + \tfrac{1}{6}(k_1 + 2k_2 + 2k_3 + k_4),$$

(A.3.3.1)

---

* The Runge-Kutta method of the first order reduces to the well-known *Euler's method*.
† The interested reader may refer to the paper of A. Ralston [RA 2], or a standard numerical analysis text such as McCracken and Dorn [MC 2], for a more detailed discussion.

where

$$k_1 = hf(x_n, y_n),$$
$$k_2 = hf(x_n + \tfrac{1}{2}h, y_n + \tfrac{1}{2}k_1),$$
$$k_3 = hf(x_n + \tfrac{1}{2}h, y_n + \tfrac{1}{2}k_2), \qquad \text{(A.3.3.2)}$$
$$k_4 = hf(x_n + h, y_n + k_3),$$

with $x_{n+1} = x_n + h$. The following example will serve to illustrate the application of this method.

Example A.3.3.1   By the application of the fourth-order Runge-Kutta method, solve the differential equation

$$f(x, y) = y' = 3 - 100y, \qquad \text{(A.3.3.3)}$$

with $y(0) = 0$ and step size $h = 0.0002$, for the first iteration (namely, the value of y_{n+1} with $y_n = y(0) = 0$). Using (A.3.3.2) with (A.3.3.3) and the initial value $y_0 = y(0) = 0$, we take the case for $n = 0$, and find that

$$f(x_n, y_n) = f(0, 0) = 3,$$
$$k_1 = 2 \times 10^{-4}(3) = 6 \times 10^{-4},$$
$$k_2 = 2 \times 10^{-4}(2.97) = 5.94 \times 10^{-4},$$
$$k_3 = 2 \times 10^{-4}(2.9703) = 5.9406 \times 10^{-4},$$
$$k_4 = 2 \times 10^{-4}(2.940594) = 5.881188 \times 10^{-4}.$$

Thus, for $n = 0$, the first iteration gives

$$y_{n+1} = y_1 = 0 + \tfrac{1}{6}[6 + 2(5.94) + 2(5.9406) + 5.881188] \times 10^{-4}$$

or

$$y_1 = 0.000594. \qquad \text{(A.3.3.4)}$$

It can be shown that the analytical solution to (A.3.3.3) is given by

$$y = \frac{3}{100}(1 - e^{-100x}). \qquad \text{(A.3.3.5)}$$

Thus, at $x = 0.0002$, (A.3.3.5) gives

$$y|_{x=0.0002} = \frac{3}{100}[1 - e^{-100(0.0002)}] = \frac{3}{100}[1 - e^{-0.02}]$$

$$\cong \frac{3}{100}[1 - 0.98020] = \frac{3}{100}[0.01980] = 0.000594$$

which is in close agreement with the result in (A.3.3.4).

In contrast to the Runge-Kutta methods, the so-called *predictor-corrector methods* are all multi-step methods. Each consists of a pair of formulas, one called the predictor and the other the corrector, to be used iteratively to first predict and then correct the values by comparison until the magnitude of the difference between successive correctors falls within a specified amount. The *Adams-Moulton methods* constitute one of the most widely used classes of such multi-step predictor-corrector algorithms for solving initial value problems. We shall state the specifications for the Adams-Moulton method as follows.

The Adams-Moulton Predictor-Corrector Method. Consider the first-order differential equation $y' = f(x, y)$. Let $y_i \triangleq y(x_i)$ and $f_i \triangleq f(x_i, y_i)$ for $i = 0, 1, 2, \ldots$. Then for a specified step size h together with the known pairs (y_0, f_0), (y_1, f_1), (y_2, f_2), (y_3, f_3), and (y_4, f_4), the steps for solving the differential equation $y' = f(x, y)$ can be stated as follows:

1. Compute $y_{n+1}^{(0)}$, using the *predictor formula*

$$y_{n+1}^{(0)} = y_n + \frac{h}{720}(1901f_n - 2984f_{n-1} + 2616f_{n-2} - 1274f_{n-3} + 251f_{n-4}),$$

$$n = 4, 5, \ldots, N - 1 \qquad \text{(A.3.3.6)}$$

where N is the number of solution points desired.

2. Compute $f_{n+1}^{(0)} = f(x_{n+1}, y_{n+1}^{(0)})$.
3. Compute $y_{n+1}^{(k)}$, using the *corrector formula*

$$y_{n+1}^{(k)} = y_n + \frac{h}{720}\{251f(x_{n+1}, y_{n+1}^{(k-1)}) + 646f_n - 264f_{n-1} + 106f_{n-2} - 19f_{n-3}\},$$

$$k = 1, 2, \ldots \qquad \text{(A.3.3.7)}$$

4. For a specified ε (the value of ε to be specified depends upon the degree of accuracy desired), iterate on k until the following inequality is satisfied:

$$\frac{|y_{n+1}^{(k)} - y_{n+1}^{(k-1)}|}{|y_{n+1}^{(k)}|} < \varepsilon \qquad \text{(A.3.3.8)}$$

5. Then set $y_{n+1} = y_{n+1}^{(k)}$, and increase the value of n by 1 so that the above steps can be repeated for evaluating the next value of y. The process is completed after the last solution point y_N is obtained.

Again, we shall not include the derivation of the Adams-Moulton method here. The interested reader, however, may refer to literature elsewhere.* The next example will illustrate the steps of this algorithm.

Example A.3.3.2 Let us consider the differential equation (A.3.3.3) used in Example A.3.3.1. We shall apply the Adams-Moulton method following the steps

---

* See S. D. Conte [CO 1] or T. R. McCalla [MC 1], for example.

outlined above with $h = 0.0002$, $x_{n+1} = 0.0010$, and the solution values

$$y_{n-4} = 0,$$
$$y_{n-3} = 0.000594,$$
$$y_{n-2} = 0.0011763, \qquad (A.3.3.9)$$
$$y_{n-1} = 0.0017472,$$
$$y_n = 0.0023064.$$

We first evaluate the derivatives at the five known solution values given in (A.3.3.9) by substituting each of these values into (A.3.3.7), yielding

$$f_{n-4} = y'_{n-4} = 3,$$
$$f_{n-3} = y'_{n-3} = 3 - 100(0.000594) = 2.9406,$$
$$f_{n-2} = y'_{n-2} = 3 - 100(0.0011763) = 2.88237, \qquad (A.3.3.10)$$
$$f_{n-1} = y'_{n-1} = 3 - 100(0.0017472) = 2.82528,$$
$$f_n = y'_n = 3 - 100(0.0023064) = 2.76936.$$

Next, substituting the values of (A.3.3.10) into the predictor formula (A.3.3.6), we find that

$$y_{n+1}^{(0)} = 0.0023064 + \frac{0.0002}{720}\{1901\,(2.76936) - 2984\,(2.82528)$$

$$+\ 2616\,(2.88237) - 1274\,(2.9406) + 251(3)\}$$

which reduces to

$$y_{n+1}^{(0)} = 0.0026900. \qquad (A.3.3.11)$$

With this *predicted* value of y_{n+1} we now calculate the derivative $y_{n+1}^{(0)'} = f_{n+1}^{(0)}$ as defined in Step 2 of the procedure, yielding

$$f_{n+1}^{(0)} = f(x_{n+1}, y_{n+1}^{(0)}) = 3 - 100\,(0.0026900) = 2.73100.$$

Next, we evaluate the *corrected* value $y_{n+1}^{(k)}$ of the solution point by substituting the required values of the derivatives in (A.3.3.10) and the predicted value in (A.3.3.11) into the corrector formula (A.3.3.7) with $k = 1$, as follows:

$$y_{n+1}^{(1)} = 0.0023064 + \frac{0.0002}{720}\{251\,(2.73100) + 646\,(2.76936)$$

$$-\ 264\,(2.82528) + 106\,(2.88237) - 19\,(2.9406)\}.$$

After simplifications, we have

$$y_{n+1}^{(1)} = 0.00285592. \qquad (A.3.3.12)$$

Table A.3.3.1 A comparison between Runge-Kutta and Adams-Moulton methods.

Runge-Kutta (one-step) methods	Adams-Moulton (multi-step) methods
Advantages	*Advantages*
1. They are self-starting, since they do not use information from previously calculated points.	1. They provide an automatic estimate of accuracy (truncation error) at each step since they do require information from previously calculated points.
2. They permit easy changes in step size.	2. They are fast as they require only two evaluations of the function $f(x, y)$ per step.
3. They are stable and provide good accuracy.	
4. They require relatively small amounts of core storage when adapted into a computer program.	
Disadvantages	*Disadvantages*
1. They require several evaluations of the function $f(x, y)$ per integration step, and hence are time-consuming.	1. They are not self-starting.
2. They provide no easily obtainable estimate of accuracy (truncation error).	2. Since special techniques are required for starting and for changing the step size, they occupy a large amount of core storage for computer implementation.

To illustrate Step 4 for $k = 1$, we use (A.3.3.8) and obtain

$$\frac{|y_{n+1}^{(1)} - y_{n+1}^{(0)}|}{y_{n+1}^{(1)}} = \frac{|0.00285592 - 0.0026900|}{0.00285592}$$

$$= 0.0580.$$

If the specified ε value is greater than or equal to 0.00580, then the value for $y_{n+1}^{(1)}$ in (A.3.3.12) is taken as the desired solution for the fixed n value, i.e., y_{n+1}. Otherwise, Step 3 must be repeated for $k = 2, 3, \ldots$, until the inequality in (A.3.3.8) is satisfied.

For each of the solution values corresponding to the subsequent values of n, the entire process, as described in Example A.3.3.2, will then be repeated.

While there are many methods available for numerical solution of differential equations in the literature, the fourth-order Runge-Kutta method and the Adams-Moulton predictor-corrector method presented in this section are among those most popularly used in the United States.

Although we must realize that no one method can be said to be always better than another for all problems, it is not difficult to point out certain distinguishing

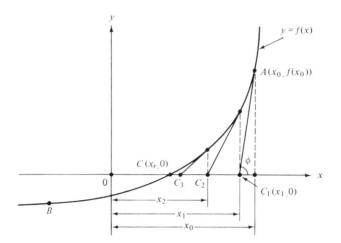

Fig. A.3.4.1 Illustration of the Newton-Raphson method for $f(x) = 0$.

properties of each of these types as advantages and disadvantages for general-purpose applications. Table A.3.3.1 serves to give a brief comparison between the one-step methods (as typified by the fourth-order Runge-Kutta algorithm) and the multi-step predictor-corrector methods (such as the Adams-Moulton methods).

Because of their complementary nature as outlined in Table A.3.3.1, the Runge-Kutta and the Adams-Moulton methods constitute a fairly ideal combination, when used together, for the numerical solution of ordinary differential equations.

A.3.4 THE NEWTON-RAPHSON METHOD

The *Newton-Raphson method* is a simple iterative procedure for numerically approximating the roots of an equation $f(x) = 0$. Consider a segment AB of the curve $y = f(x)$, which includes a root x_r of the equation $f(x) = 0$ as shown in Fig. A.3.4.1. Point $A(x_0, f(x_0))$ is a starting point which is assumed to be sufficiently close to the point $C(x_r, 0)$ where x_r is a root of $f(x) = 0$ so that $f(x_r) = 0$. Point $C_1(x_1, 0)$ is the intercept of the tangent $\overline{AC_1}$ to the curve $y = f(x)$ at A making an angle ϕ with the x-axis where x_1 is the x-intercept which is used as the first approximation to the root x_r. From Fig. A.3.4.1, we have

$$f'(x_0) = \tan \phi = \frac{f(x_0)}{x_0 - x_1}, \qquad (A.3.4.1)$$

where

$$f'(x_0) \equiv \frac{d}{dx} f(x) \bigg|_{x = x_0}$$

is the derivative of $f(x)$ with respect to x evaluated at $x = x_0$. Thus, solving (A.3.4.1) for x_1, we obtain

$$x_1 = x_0 - \frac{f(x_0)}{f'(x_0)} \qquad (A.3.4.2)$$

from which the following iterative formula can be derived:

$$x_{n+1} = x_n - \frac{f(x_n)}{f'(x_n)}, \qquad (A.3.4.3)$$

where $n = 0, 1, 2, \ldots$.

Expression (A.3.4.3) is the celebrated *Newton-Raphson formula*. The conditions for convergence of this iterative formula are: (a) the starting value x_0 is sufficiently close to a root of $f(x) = 0$; (b) the second derivative, $f''(x)$, of $f(x)$ with respect to x does not become exceedingly large; and (c) the first derivative, $f'(x)$, of $f(x)$ is not too close to zero (that is, no two roots are too close together).

It should be evident that if the function $f(x)$ is real-valued and if the initial guess x_0 is real, only real numbers will be involved in the computations using the Newton-Raphson formula (A.3.4.3). However, if x_0 is a complex number, then the values $x_i (i = 1, 2, \ldots)$ resulting from succeeding iterations may also be complex. Indeed, the same principle applies when the function $f(z)$ under consideration is a function of a complex variable z.

In an engineering problem, we often encounter functions which are polynomials of a complex variable of the form

$$P(z) = a_n z^n + a_{n-1} z^{n-1} + \cdots + a_1 z + a_0,$$

where $a_i (i = 0, 1, 2, \ldots, n)$ are real. We recall from algebra that if $\alpha + j\beta$ is a root of $P(z) = 0$ (where $j = \sqrt{-1}$), then $\alpha - j\beta$ is also a root. Hence $P(z)$ can be written in factored form as

$$P(z) = [z^2 - 2\alpha z + (\alpha^2 + \beta^2)] P_1(z),$$

where $P_1(z)$ is a polynomial of degree $n - 2$. Thus, once the roots $\alpha \pm j\beta$ have been determined, $P_1(z)$ may be found by noting

$$P_1(z) = \frac{P(z)}{z^2 - 2\alpha z + (\alpha^2 + \beta^2)}$$

and the same procedure may now be repeated with $P(z)$ replaced by $P_1(z)$, until all the roots are determined.

Example A.3.4.1 Illustrate the procedure of the Newton-Raphson method by performing only one iteration step for the polynomial

$$P(z) = z^2 - z - 2 \qquad (A.3.4.4)$$

with an estimated root $z_i = -0.9$. Writing $z = x + jy$, (A.3.4.4) becomes

$$P(z) = (x + jy)^2 - (x + jy) - 2$$
$$= (x^2 - y^2 - x - 2) + j(2xy - y)$$
$$= U + jV,$$

where

$$U = x^2 - y^2 - x - 2$$
$$V = 2xy - y$$
$$= y(2x - 1). \tag{A.3.4.5}$$

Taking the derivative of (A.3.4.4) with respect to z, we have

$$P'(z) = 2z - 1$$

which can be expressed as

$$P'(z) = 2(x + jy) - 1$$
$$= (2x - 1) + j2y$$
$$= Q + jR$$

with

$$Q = 2x - 1 \quad \text{and} \quad R = 2y. \tag{A.3.4.6}$$

With $z_i = x_i = -0.9$, we obtain from (A.3.4.5) and (A.3.4.6)

$$U_i = (.81 + .9 - 2) = -0.29, \qquad V_i = 0,$$
$$Q_i = 2(-.9) - 1 = -2.8, \qquad R_i = 0.$$

Substituting these values into the Newton-Raphson iterative formula (A.3.4.3), we get our next estimated root of $P(z) = 0$ as follows:

$$x_{i+1} = x_i - \frac{U_i Q_i + V_i R_i}{Q_i^2 + R_i^2}$$

$$= -0.9 - \frac{(-0.29)(-2.8)}{(-2.8)^2}$$

$$= -0.9 - \frac{0.29}{2.8}$$

$$= -0.9 - 0.10357 = -1.00357$$

$$y_{i+1} = 0 + \frac{U_i R_i - V_i Q_i}{Q_i^2 + R_i^2}$$

$$= 0 + 0 = 0.$$

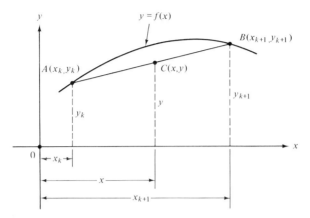

Fig. A.3.5.1 Illustration of the method of linear interpolation.

Thus

$$z_{i+1} = x_{i+1} + jy_{i+1} = -1.00357 \qquad \text{(A.3.4.7)}$$

which is the desired value of z for the first iteration step.

In the Newton-Raphson iterative process at a given step, the resulting evaluated value is used to compute the subsequent estimated value of the root until the difference $d_{i,i+1} = |z_{i+1} - z_i|$ between two consecutive estimated values z_{i+1} and z_i falls within a prescribed number. Then the last estimated value z_{i+1} is taken as the desired approximated root.

A.3.5 LINEAR INTERPOLATION

Consider a segment of the curve $f(x)$ between two points $A(x_k, y_k)$ and $B(x_{k+1}, y_{k+1})$ as illustrated in Fig. A.3.5.1. To compute the approximate value of $f(x)$ corresponding to a given x, we draw a straight line $\overline{AB}$ between points A and B and write the expression for the ordinate y of the point $C(x, y)$ on $\overline{AB}$ for a given x as follows. First, we write the two-point formula for the straight line $\overline{AB}$ through points A and B

$$\frac{y - y_k}{x - x_k} = \frac{y_{k+1} - y_k}{x_{k+1} - x_k}, \qquad \text{(A.3.5.1)}$$

and then solve (A.3.5.1) for y to obtain

$$y = y_k + \left(\frac{y_{k+1} - y_k}{x_{k+1} - x_k}\right)(x - x_k). \qquad \text{(A.3.5.2)}$$

Finally, rearranging (A.3.5.2), we have

$$y = \frac{y_k(x - x_{k+1}) - y_{k+1}(x - x_k)}{x_k - x_{k+1}} \tag{A.3.5.3}$$

which is one standard form of the formula for approximating $y = f(x)$ for a given $x(x_k \leq x \leq x_{k+1})$ by the method of *linear interpolation*.

A.3.6 THE GAUSS-SEIDEL ITERATIVE METHOD

Consider the system of n linear algebraic equations in n variables

$$a_{11}x_1 + a_{12}x_2 + a_{13}x_3 + \cdots + a_{1n}x_n = b_1$$
$$a_{21}x_1 + a_{22}x_2 + a_{23}x_3 + \cdots + a_{2n}x_n = b_2$$
$$\vdots \tag{A.3.6.1}$$
$$a_{n1}x_1 + a_{n2}x_2 + a_{n3}x_3 + \cdots + a_{nn}x_n = b_n$$

If the diagonal elements $a_{ii}(i = 1, 2, \ldots, n)$ are all nonzero, then (A.3.6.1) can be rewritten in the form

$$x_1 = \frac{1}{a_{11}}(b_1 - a_{12}x_2 - a_{13}x_3 - \cdots - a_{1n}x_n)$$

$$x_2 = \frac{1}{a_{22}}(b_2 - a_{21}x_1 - a_{23}x_3 - \cdots - a_{2n}x_n) \tag{A.3.6.2}$$

$$\vdots$$

$$x_n = \frac{1}{a_{nn}}(b_n - a_{n1}x_1 - a_{n2}x_2 - \cdots - a_{n,n-1}x_{n-1}).$$

An iterative procedure for solving the given system of equations (A.3.6.1) based on the form of (A.3.6.2) can be established as follows.

1. Choose a set of arbitrary values $x_1^0, x_2^0, x_3^0, \ldots, x_n^0$ as the starting point.
2. Substitute these arbitrary values for $x_1, x_2, x_3, \ldots, x_n$ on the right-hand sides of (A.3.6.2).
3. Compute the values of the expressions on the right-hand sides of (A.3.6.2) for $x_1, x_2, x_3, \ldots, x_n$ on the left and denote the computed values by $x_1^1, x_2^1, x_3^1, \ldots, x_n^1$.
4. Using these newly computed values x_i^1 ($i = 1, 2, \ldots, n$), repeat Steps 2 and 3 with (A.3.6.2) to compute the next set of values, namely x_i^2 ($i = 1, 2, \ldots, n$), and so on.

The procedure described above leads directly to the following set of recursion formulas:

$$x_1^{k+1} = \frac{1}{a_{11}}(b_1 - a_{12}x_2^k - a_{13}x_3^k - \cdots - a_{1n}x_n^k)$$

$$x_2^{k+1} = \frac{1}{a_{22}}(b_2 - a_{21}x_1^k - a_{23}x_3^k - \cdots - a_{2n}x_n^k)$$

$$\vdots$$ \hfill (A.3.6.3)

$$x_n^{k+1} = \frac{1}{a_{nn}}(b_n - a_{n1}x_1^k - a_{n2}x_2^k - \cdots - a_{n,n-1}x_{n-1}^k) \qquad (k = 0, 1, 2, \ldots)$$

which is known as *Jacobi's iterative method* for solving systems of linear algebraic equations.

If Jacobi's recursion formulas (A.3.6.3) are slightly modified so that as each x_i^{k+1} is determined it is immediately used in computing $x_{i+1}^{k+1}, x_{i+2}^{k+1}, \ldots, x_n^{k+1}$, then we have

$$x_1^{k+1} = \frac{1}{a_{11}}(b_1 - a_{12}x_2^k - a_{13}x_3^k - \cdots - a_{1n}x_n^k)$$

$$x_2^{k+1} = \frac{1}{a_{22}}(b_2 - a_{21}x_1^{k+1} - a_{23}x_3^k - \cdots - a_{2n}x_n^k)$$

$$x_3^{k+1} = \frac{1}{a_{33}}(b_3 - a_{31}x_1^{k+1} - a_{32}x_2^{k+1} - a_{34}x_4^k - \cdots - a_{3n}x_n^k)$$ \hfill (A.3.6.4)

$$\vdots$$

$$x_n^{k+1} = \frac{1}{a_{nn}}(b_n - a_{n1}x_1^{k+1} - a_{n2}x_2^{k+1} - \cdots - a_{n,n-1}x_{n-1}^{k+1})$$

which are known as the *Gauss-Seidel recursion formulas*.

A *sufficient* condition for the convergence of the Gauss-Seidel method (A.3.6.4) is given by

$$\max_i \left\{ \frac{1}{a_{ii}} \sum_{j \neq 1} |a_{ij}| \right\} < 1 \qquad (i = 1, 2, \ldots, n). \hfill (A.3.6.5)$$

It can be shown* that the rate of convergence for the Gauss-Seidel method is twice that of Jacobi's formulas.

Example A.3.6.1 For the system of linear equations

$$\begin{bmatrix} 3 & -1 & -1 \\ -1 & 4 & -1 \\ -1 & -1 & 3 \end{bmatrix} \begin{bmatrix} I_1 \\ I_2 \\ I_3 \end{bmatrix} = \begin{bmatrix} 2 \\ 3 \\ -3 \end{bmatrix}, \hfill (A.3.6.6)$$

---

* See J. Todd [TO 1], p. 404.

use the Gauss-Seidel iterative formulas to solve for I_1, I_2, and I_3 starting with the set of initial values $I_1^k = 1$, $I_2^k = 1$, and $I_3^k = -1$.

Applying (A.3.6.3) to (A.3.6.6), we obtain

$$I_1^{k+1} = \tfrac{1}{3}[2 + I_2^k + I_3^k],$$

$$I_2^{k+1} = \tfrac{1}{4}[3 + I_1^{k+1} + I_3^k], \qquad\qquad (A.3.6.7)$$

$$I_3^{k+1} = \tfrac{1}{3}[-3 + I_1^{k+1} + I_2^{k+1}],$$

which, with $I_1^k = 1$, $I_2^k = 1$, $I_3^k = -1$, become

$$I_1^{k+1} = \tfrac{1}{3}[2 + 1 - 1] = 0.66666,$$

$$I_2^{k+1} = \tfrac{1}{4}[3 + 1 - 1] = 0.75000, \qquad\qquad (A.3.6.8)$$

$$I_3^{k+1} = \tfrac{1}{3}[-3 + \tfrac{2}{3} + \tfrac{3}{4}] = -0.52777.$$

In like manner, the next iteration gives

$$I_1^{k+2} = \tfrac{1}{3}[2 + 0.75000 - 0.52777] = 0.73333,$$

$$I_2^{k+2} = \tfrac{1}{4}[3 + 0.73333 - 0.52777] = 0.801388, \qquad\qquad (A.3.6.9)$$

$$I_3^{k+2} = \tfrac{1}{3}[-3 + 0.73333 + 0.801388] = -0.48842.$$

For the purpose of checking the results from these two iterations, we can solve the matrix equation (A.3.6.6) for I_1, I_2, and I_3 directly to get

$$I_1 = \tfrac{19}{24} = 0.791666,$$

$$I_2 = \tfrac{20}{24} = 0.83333, \qquad\qquad (A.3.6.10)$$

$$I_3 = -\tfrac{11}{24} = -0.458333.$$

Comparing the results in (A.3.6.8) and (A.3.6.9) with the values in (A.3.6.10), it is evident that the iteration steps indeed approach the true solution very rapidly.

REFERENCES

1. S. D. Conte, *Elementary Numerical Analysis*, McGraw-Hill, 1965.

2. T. R. McCalla, *Introduction to Numerical Methods and FORTRAN Programming*, John Wiley & Sons, 1967.

3. D. D. McCracken and W. S. Dorn, *Numerical Methods and FORTRAN Programming*, John Wiley & Sons, 1964.

4. N. Macon, *Numerical Analysis*, John Wiley & Sons, 1963.

5. S. S. Kuo, *Numerical Methods and Computers*, Addison-Wesley, 1965.

6. F. Scheid, *Theory and Problems of Numerical Analysis*, Schaum's Outline Series, McGraw-Hill, 1968.

7. J. Todd, *Survey of Numerical Analysis*, McGraw-Hill, New York, 1962.

PROBLEMS

A.3.1 Using Simpson's rule with $n = 4$, determine the approximate value of the definite integral

$$\int_0^\pi \sin x \, dx.$$

Check the result by actual integration.

A.3.2 Repeat Problem A.3.1 with $n = 8$ and compare the results.

A.3.3 Repeat Problem A.3.1 for the integral

$$\int_0^1 (1 + x^2 + x^3 + x^4) \, dx.$$

Discuss the problem of accuracy and the choice of the value of n.

A.3.4 By applying the fourth-order Runge-Kutta method, determine the solution of the differential equation

$$f(x, y) = y' = 3 - 100y$$

at $x = 0.0004$, given that $h = 0.0002$, and $y = 0.000595$ at $x = 0.0002$.

A.3.5 Suppose that in Example A.3.3.2 the specified allowable error ε is smaller than the calculated value of 0.0580. Continue to repeat Step 3 in the solution for $k = 2$.

A.3.6 Using the result in Example A.3.4.1, improve the estimated value of the root in the example, by performing an additional iteration step of the Newton-Raphson method.

A.3.7 Illustrate the Newton-Raphson method of root finding by performing only one iteration step for the polynomial

$$P(s) = s^2 + 2s + 2$$

with an estimated root $s_i = -0.9 + j1.0$.

A.3.8 Using the Gauss-Seidel iteration method, continue the solution of the equation (A.3.6.6) in Example A.3.6.1 for an additional iteration step.

A.3.9 Using the Gauss-Seidel iteration method, determine the approximate solution of the system

$$\begin{bmatrix} 3 & -1 \\ -1 & 2.5 \end{bmatrix} \begin{bmatrix} I_1 \\ I_2 \end{bmatrix} = \begin{bmatrix} 1 \\ 1 \end{bmatrix}.$$

A.3.10 Repeat Problem A.3.7 for the polynomial

$$P(s) = s^4 - 1$$

with the estimated root $s_i = 0.9$.

BIBLIOGRAPHY

[AT 1] Atabekov, G. I., *Linear Network Theory*, translated by J. Yeoman, Pergamon Press, New York, 1965.

[AT 2] Athens, M. and P. Falb, *Optimal Control: An Introduction to the Theory and Its Applications*, McGraw-Hill, New York, 1966.

[AY 1] Ayres, F., Jr., *Theory and Problems of Matrices*, Schaum Publishing Co., New York, 1962.

[BE 1] Bellman, R., *Introduction to Matrix Analysis*, McGraw-Hill, New York, 1960.

[BE 2] Berge, C., *The Theory of Graphs and Its Applications*, John Wiley & Sons, New York, 1962.

[BL 1] Blackwell, W. A., *Mathematical Modeling of Physical Networks*, Macmillan, New York, 1968.

[BO 1] Bode, H., *Network Analysis and Feedback Amplifier Design*, Van Nostrand, New York, 1950.

[BR 1] Brenner E. and M. Javid, *Analysis of Electric Circuits*, Second Ed., McGraw-Hill, New York, 1967.

[CA 1] Calahan, D. A., "Linear Network Analysis and Realization Digital Computer Programs: An Instruction Manual," *University of Illinois Bulletin 472*, Vol. 62, No. 58, February 1965.

[CA 2] Carlin, H. J. and A. B. Giordano, *Network Theory—An Introduction to Reciprocal and Nonreciprocal Circuits*, Prentice-Hall, Englewood Cliffs, N.J., 1964.

[CH 1] Chan, S. P., *Introductory Topological Analysis of Electrical Networks*, Holt, Rinehart and Winston, New York, 1969.

[CH 2] Chang, S. S. L., *Synthesis of Optimum Control Systems*, McGraw-Hill, New York, 1961.

[CH 3] Chestnut, H. and R. W. Mayer, *Servomechanisms and Regulating System Design*, Second Ed., John Wiley & Sons, New York, 1963.

[CH 4] Chirlian, P. M., *Integrated and Active Network Analysis and Synthesis*, Prentice-Hall, Englewood Cliffs, N.J., 1967.

[CH 5] Churchill, R. V., *Fourier Series and Boundary Value Problems*, Second Ed., McGraw-Hill, New York, 1963.

[CH 6] Churchill, R. V., *Operational Mathematics*, Second Ed., McGraw-Hill, New York, 1958.

[CO 1] Conte, S. D., *Elementary Numerical Analysis*, McGraw-Hill, New York, 1965.

[CR 1] Craig, E. J., *Laplace and Fourier Transforms for Electrical Engineers*, Holt, Rinehart and Winston, New York, 1964.

[CR 2] Cruz, J. B., Jr. and M. E. VanValkenburg, *Introductory Signals and Circuits*, Blaisdell, Waltham, Mass., 1967.

[DE 1] DeRusso, P. M., R. J. Roy, and C. M. Close, *State Variables for Engineers*, John Wiley & Sons, New York, 1967.

[DE 2] Desoer, C. A. and E. S. Kuh, *Basic Circuit Theory*, McGraw-Hill, New York, 1969.

[FR 1] Freeman, H., *Discrete-Time Systems*, John Wiley & Sons, New York, 1965.

[FR 2] Friedman, B., *Principles and Techniques of Applied Mathematics*, John Wiley & Sons, New York, 1965.

[FU 1] Fuller, L. E., *Basic Matrix Theory*, Prentice-Hall, Englewood Cliffs, N.J., 1962.

[GA 1] Gantmacher, F. R., *Applications of the Theory of Matrices*, John Wiley & Sons, New York, 1959.

[GI 1] Gilbert, E. G., "Controllability and Observability in Multivariable Control Systems," *SIAM J. Control*, Vol. 1, No. 2, pp. 128–151, 1963.

[GU 1] Guillemin, E. A., *Introductory Circuit Theory*, John Wiley & Sons, New York, 1953.

[GU 2] Guillemin, E. A., *The Mathematics of Circuit Analysis*, John Wiley & Sons, New York, 1949.

[HO 1] Hohn, F. E., *Elementary Matrix Algebra*, Second Ed., Macmillan, New York, 1964.

[HU 1] Huelsman, L. P., *Active Filters*, McGraw-Hill, New York, 1970.

[HU 2] Huelsman, L. P., *Circuits, Matrices and Linear Vector Spaces*, McGraw-Hill, New York, 1963.

[HU 3] Huelsman, L. P., *Digital Computations in Basic Circuit Theory*, McGraw-Hill, New York, 1968.

[HU 4] Huelsman, L. P., *Theory and Design of Active RC Circuits*, McGraw-Hill, New York, 1968.

[IB 1] IBM, *1620 Electronic Circuit Analysis Program (ECAP) (1620-EE-02X) User's Manual*, IBM Corporation, 1965.

[JE 1] Jensen, R. W. and M. D. Lieberman, *IBM Electronic Circuit Analysis Program*, Prentice-Hall, Englewood Cliffs, N.J., 1968.

[KA 1] Karni, S., *Network Theory: Analysis and Synthesis*, Allyn and Bacon, Boston, Mass., 1966.

[KO 1] Koenig, H. E., Y. Tokad, and H. K. Hesavan, *Analysis of Discrete Physical Systems*, McGraw-Hill, New York, 1967.

[KU 1] Kuh, E. S. and R. A. Rohrer, "The State-Variable Approach to Network Analysis," *Proceedings of the IEEE*, Vol. 53, No. 7, pp. 672–686, July 1965.

[KU 2] Kuo, B. C., *Analysis and Synthesis of Sampled-Data Control Systems*, Prentice-Hall, Englewood Cliffs, N.J., 1963.

[KU 3] Kuo, B. C., *Linear Networks and Systems*, McGraw-Hill, New York, 1967.

[KU 4] Kuo, F. F., *Network Analysis and Synthesis*, Second Ed., John Wiley & Sons, New York, 1966.

[KU 5] Kuo, F. F. and J. F. Kaiser, *System Analysis by Digital Computer*, John Wiley & Sons, New York, 1966.

[KU 6]   Kuo, F. F. and W. G. Magnuson, *Computer Oriented Circuit Design*, Prentice-Hall, Englewood Cliffs, N.J., 1969.

[KU 7]   Kuo, S. S., *Numerical Methods and Computers*, Addison-Wesley, Reading, Mass., 1965.

[KU 8]   Kuratowski, C., "Sur le Problème des Courbes Gauches en Topologie," *Fundamentae Mathematicae*, Vol. 15, pp. 271–283, 1930.

[KU 9]   Kurokawa, K., "Power Waves and the Scattering Matrix," *IEEE Transactions on Microwave Theory and Techniques*, Vol. MTT-13, p. 194, March 1965.

[LE 1]   Leon, B. J. and P. A. Wintz, *Basic Linear Networks for Electrical and Electronics Engineers*, Holt, Rinehart and Winston, New York, 1970.

[MA 1]   Macon, N., *Numerical Analysis*, John Wiley & Sons, New York, 1963.

[MA 2]   Mason, S. J., "Feedback Theory—Some Properties of Signal Flowgraphs," *Proc. IRE*, Vol. 41, pp. 1144–1156, September 1953.

[MC 1]   McCalla, T. R., *Introduction to Numerical Methods and FORTRAN Programming*, John Wiley & Sons, New York, 1967.

[MC 2]   McCracken, D. D. and W. S. Dorn, *Numerical Methods and FORTRAN Programming*, John Wiley & Sons, New York, 1964.

[MC 3]   McNamee, L. P. and H. Potash, "A User's Guide and Programmer's Manual for NASAP," *Report No. 68-38*, University of California, Los Angeles, 1968.

[ME 1]   Merriam, C. W. III, *Analysis of Lumped Electrical Systems*, John Wiley & Sons, New York, 1969.

[MI 1]   Mitra, S. K., *Analysis and Synthesis of Linear Active Networks*, John Wiley & Sons, New York, 1969.

[MO 1]   Morse, A. S. and L. P. Huelsman, "A Gyrator Realization Using Operational Amplifiers," *Proc. IRE*, Vol. CT-11, No. 2, pp. 277–278, June 1964.

[NE 1]   Newcomb, R. W., *Active Integrated Circuit Synthesis*, Prentice-Hall, Englewood Cliffs, N.J., 1968.

[OG 1]   Ogata, K., *State Space Analysis of Control Systems*, Prentice-Hall, Englewood Cliffs, N.J., 1967.

[PE 1]   Percival, W. S., "Solution of Passive Electrical Networks by Means of Mathematical Trees," *Proceedings of the IRE* (London), Vol. 100, Pt. III, pp. 143–150, May 1953.

[PO 1]   Pontryagin, L. S., V. G. Boltyanskii, R. V. Gamkrelidze, and E. F. Mishchenko, *The Mathematical Theory of Optimal Processes*, Interscience Publishers, New York, 1962.

[PO 2]   Pottle, C., "A 'Textbook' Computerized State-Space Network Analysis Algorithm," *IEEE Transactions on Circuit Theory*, Vol. CT-16, pp. 566–568, Nov. 1969.

[RA 1]   Ragazzini, J. R. and G. F. Franklin, *Sampled-data Control Systems*, McGraw-Hill, New York, 1958.

[RA 2]   Ralston, A., "Runge-Kutta Methods with Minimum Error Bounds," *Mathematics of Computation*, Vol. 16, pp. 431–437, 1962.

[RO 1]   Roe, P. H., *Networks and Systems*, Addison-Wesley, Reading, Mass., 1966.

[RU 1]   Rudenberg, R., *Transient Performance of Electric Power Systems*, McGraw-Hill, New York, 1950.

[RU 2]   Ruston, H. and J. Bordogna, *Electric Networks: Functions, Filters, Analysis*, McGraw-Hill, New York, 1966.

[SC 1]   Scheid, F., *Theory and Problems of Numerical Analysis*, Schaum's Outline Series, McGraw-Hill, New York, 1968.

[SC 2]   Scott, R. E., *Elements of Linear Circuits*, Addison-Wesley, Reading, Mass., 1965.

[SE 1]   Seshu, S. and M. B. Reed, *Linear Graphs and Electrical Networks*, Addison-Wesley, Reading, Mass., 1961.

[SU 1]   Su, K. L., *Active Network Synthesis*, McGraw-Hill, New York, 1965.

[TE 1]   Tellegen, B. D. H., "A General Network Theorem, with Applications," *Philips Research Reports*, Vol. 7, pp. 259–269, 1952.

[TE 2]   Tellegen, B. D. H., "The Gyrator: a New Network Element," *Philips Research Reports*, Vol. 3, pp. 81–101, April 1948.

[TO 1]   Todd, J., *Survey of Numerical Analysis*, McGraw-Hill, New York, 1962.

[VA 1]   Van Valkenburg, M. E., *Network Analysis*, Second Ed., Prentice-Hall, Englewood Cliffs, N.J., 1964.

[YA 1]   Yanagisawa, T. and Y. Kawashima, "Active Gyrator," *Electronics Letters*, Vol. 3, pp. 105–107, March 1967.

[YO 1]   Youla, D. C., "On Scattering Matrices Normalized to Complex Port Numbers," *Proc. IRE*, Vol. 49, No. 7, p. 1221, July 1961.

[ZA 1]   Zadeh, L. A. and C. A. Desoer, *Linear System Theory*, McGraw-Hill, New York, 1963.

INDEX

INDEX

A BCDE FGH798765432

$$V_1 = \frac{1}{C_1} \int i_1 \, dt$$

$$V_2 = \frac{1}{C_2} \int i_2 \, dt$$

$$\dot{V}_1 = \frac{1}{C} i_1$$

$$\dot{V}_2 = \frac{1}{C_2} i_2 = \frac{1}{C_2}(i_1 - i_3) = \frac{1}{C_2} i_1 - \frac{V_2}{R}$$

$$\cdot V = V_1 + V_2$$

$$i_1 = i_2 + i_3$$